Supplementary Cementitious Materials in Concrete

Supplementary Cementitious Materials in Concrete

Editors

Alessandro P. Fantilli
Daria Jóźwiak-Niedźwiedzka

MDPI • Basel • Beijing • Wuhan • Barcelona • Belgrade • Manchester • Tokyo • Cluj • Tianjin

Editors
Alessandro P. Fantilli
Politecnico di Torino,
Department of Structural
Italy

Daria Jóźwiak-Niedźwiedzka
Institute of Fundamental
Technological Research of the
Polish Academy of Sciences
Poland

Editorial Office
MDPI
St. Alban-Anlage 66
4052 Basel, Switzerland

This is a reprint of articles from the Special Issue published online in the open access journal *Materials* (ISSN 1996-1944) (available at: https://www.mdpi.com/journal/materials/special_issues/supplementary_cementitious).

For citation purposes, cite each article independently as indicated on the article page online and as indicated below:

LastName, A.A.; LastName, B.B.; LastName, C.C. Article Title. *Journal Name* **Year**, *Volume Number*, Page Range.

ISBN 978-3-0365-1481-9 (Hbk)
ISBN 978-3-0365-1482-6 (PDF)

© 2021 by the authors. Articles in this book are Open Access and distributed under the Creative Commons Attribution (CC BY) license, which allows users to download, copy and build upon published articles, as long as the author and publisher are properly credited, which ensures maximum dissemination and a wider impact of our publications.

The book as a whole is distributed by MDPI under the terms and conditions of the Creative Commons license CC BY-NC-ND.

Contents

About the Editors ix

Preface to "Supplementary Cementitious Materials in Concrete" xi

Alessandro P. Fantilli and Daria Jóźwiak-Niedźwiedzka
Special Issue: Supplementary Cementitious Materials in Concrete, Part I
Reprinted from: *Materials* **2021**, *14*, 2291, doi:10.3390/ma14092291 1

Alessandro P. Fantilli, Francesco Tondolo, Bernardino Chiaia and Guillaume Habert
Designing Reinforced Concrete Beams Containing Supplementary Cementitious Materials
Reprinted from: *Materials* **2019**, *12*, 1248, doi:10.3390/ma12081248 7

Giyeol Lee and Okpin Na
Assessment of Mechanical, Thermal and Durability Properties of High-Volume GGBS Blended Concrete Exposed to Cryogenic Conditions
Reprinted from: *Materials* **2021**, *14*, 2129, doi:10.3390/ma14092129 19

Carmen Andrade, Ana Martínez-Serrano, Miguel Ángel Sanjuán and José Antonio Tenorio Ríos
Reduced Carbonation, Sulfate and Chloride Ingress Due to the Substitution of Cement by 10% Non-Precalcined Bentonite
Reprinted from: *Materials* **2021**, *14*, 1300, doi:10.3390/ma14051300 41

Ghafur H. Ahmed, Hawreen Ahmed, Babar Ali and Rayed Alyousef
Assessment of High Performance Self-Consolidating Concrete through an Experimental and Analytical Multi-Parameter Approach
Reprinted from: *Materials* **2021**, *14*, 985, doi:10.3390/ma14040985 61

Babar Ali, Rawaz Kurda, Bengin Herki, Rayed Alyousef, Rasheed Mustafa, Ahmed Mohammed, Ali Raza, Hawreen Ahmed and Muhammad Fayyaz Ul-Haq
Effect of Varying Steel Fiber Content on Strength and Permeability Characteristics of High Strength Concrete with Micro Silica
Reprinted from: *Materials* **2020**, *13*, 5739, doi:10.3390/ma13245739 83

Viet-Anh Vu, Alain Cloutier, Benoît Bissonnette, Pierre Blanchet and Christian Dagenais
Steatite Powder Additives in Wood-Cement Drywall Particleboards
Reprinted from: *Materials* **2020**, *13*, 4813, doi:10.3390/ma13214813 101

Roman Jaskulski, Daria Jóźwiak-Niedźwiedzka and Yaroslav Yakymechko
Calcined Clay as Supplementary Cementitious Material
Reprinted from: *Materials* **2020**, *13*, 4734, doi:10.3390/ma13214734 115

Jung-Geun Han, Jin-Woo Cho, Sung-Wook Kim, Yun-Suk Park and Jong-Young Lee
Characteristics of CO_2 and Energy-Saving Concrete with Porous Feldspar
Reprinted from: *Materials* **2020**, *13*, 4204, doi:10.3390/ma13184204 151

Adrian Ionut Nicoara, Alexandra Elena Stoica, Mirijam Vrabec, Nastja Šmuc Rogan, Saso Sturm, Cleva Ow-Yang, Mehmet Ali Gulgun, Zeynep Basaran Bundur, Ion Ciuca and Bogdan Stefan Vasile
End-of-Life Materials Used as Supplementary Cementitious Materials in the Concrete Industry
Reprinted from: *Materials* **2020**, *13*, 1954, doi:10.3390/ma13081954 169

Hyun-Min Yang, Seung-Jun Kwon, Nosang Vincent Myung, Jitendra Kumar Singh, Han-Seung Lee and Soumen Mandal
Evaluation of Strength Development in Concrete with Ground Granulated Blast Furnace Slag Using Apparent Activation Energy
Reprinted from: *Materials* **2020**, *13*, 442, doi:10.3390/ma13020442 189

Katarzyna Kalinowska-Wichrowska, Marta Kosior-Kazberuk and Edyta Pawluczuk
The Properties of Composites with Recycled Cement Mortar Used as a Supplementary Cementitious Material
Reprinted from: *Materials* **2020**, *13*, 64, doi:10.3390/ma13010064 203

Alessandro P. Fantilli, Lucia Paternesi Meloni, Tomoya Nishiwaki and Go Igarashi
Tailoring Confining Jacket for Concrete Column Using Ultra High Performance-Fiber Reinforced Cementitious Composites (UHP-FRCC) with High Volume Fly Ash (HVFA)
Reprinted from: *Materials* **2019**, *12*, 4010, doi:10.3390/ma12234010 223

Hyuk Lee, Vanissorn Vimonsatit, Priyan Mendis and Ayman Nassif
Study of Strain-Hardening Behaviour of Fibre-Reinforced Alkali-Activated Fly Ash Cement
Reprinted from: *Materials* **2019**, *12*, 4015, doi:10.3390/ma12234015 235

Shafiq Ishak, Han-Seung Lee, Jitendra Kumar Singh, Mohd Azreen Mohd Ariffin, Nor Hasanah Abdul Shukor Lim and Hyun-Min Yang
Performance of Fly Ash Geopolymer Concrete Incorporating Bamboo Ash at Elevated Temperature
Reprinted from: *Materials* **2019**, *12*, 3404, doi:10.3390/ma12203404 249

Linda Monfardini, Luca Facconi and Fausto Minelli
Experimental Tests on Fiber-Reinforced Alkali-Activated Concrete Beams Under Flexure: Some Considerations on the Behavior at Ultimate and Serviceability Conditions
Reprinted from: *Materials* **2019**, *12*, 3356, doi:10.3390/ma12203356 267

Mirjana Vukićević, Miloš Marjanović, Veljko Pujević and Sanja Jocković
The Alternatives to Traditional Materials for Subsoil Stabilization and Embankments
Reprinted from: *Materials* **2019**, *12*, 3018, doi:10.3390/ma12183018 285

Izabela Hager, Tomasz Tracz, Marta Choińska and Katarzyna Mróz
Effect of Cement Type on the Mechanical Behavior and Permeability of Concrete Subjected to High Temperatures
Reprinted from: *Materials* **2019**, *12*, 3021, doi:10.3390/ma12183021 307

Michał A. Glinicki, Daria Jóźwiak-Niedźwiedzka and Mariusz Dabrowski
The Influence of Fluidized Bed Combustion Fly Ash on the Phase Composition and Microstructure of Cement Paste
Reprinted from: *Materials* **2019**, *12*, 2838, doi:10.3390/ma12172838 321

Viet-Anh Vu, Alain Cloutier, Benoit Bissonnette, Pierre Blanchet and Josée Duchesne
The Effect of Wood Ash as a Partial Cement Replacement Material for Making Wood-Cement Panels
Reprinted from: *Materials* **2019**, *12*, 2766, doi:10.3390/ma12172766 335

Piotr Woyciechowski, Paweł Woliński and Grzegorz Adamczewski
Prediction of Carbonation Progress in Concrete Containing Calcareous Fly Ash Co-Binder
Reprinted from: *Materials* **2019**, *12*, 2665, doi:10.3390/ma12172665 347

Victoria Eugenia García-Vera, Antonio José Tenza-Abril, José Miguel Saval and Marcos Lanzón
Influence of Crystalline Admixtures on the Short-Term Behaviour of Mortars Exposed to Sulphuric Acid
Reprinted from: *Materials* **2019**, *12*, 82, doi:10.3390/ma12010082 . . . . **365**

About the Editors

Alessandro P. Fantilli is an Associate Professor in the Department of Structural, Building, and Geotechnical Engineering of Politecnico di Torino, Turin, Italy. He received his MS and PhD from Politecnico di Torino. His research interests include nonlinear analysis of reinforced concrete structures, as well as structural application and sustainability of high-performance fiber-reinforced cementitious concrete.

Daria Jóźwiak-Niedźwiedzka is an Associate Professor at the Institute of Fundamental Technological Research, Polish Academy of Sciences in Warsaw, Poland. She received her MS, PhD and DSc in civil engineering in 1999, 2005 and 2016, respectively. Her research interests include durability of cement-based materials (AAR as well as ionizing radiation exposure) and microstructure analysis.

Preface to "Supplementary Cementitious Materials in Concrete"

The substitution strategy, consisting of the partial replacement of Portland cement with supplementary cementitious materials (SCMs), or the more common application of blended cements, is an effective way to improve the sustainability of the construction industries. In fact, they foster the introduction of new and more environmentally friendly construction materials, with improved physical and mechanical properties, to be used in building engineering. Therefore, various research issues regarding SCMs are herein considered and described in detail through both research papers and state-of-the art reviews.

Alessandro P. Fantilli, Daria Jóźwiak-Niedźwiedzka
Editors

Editorial

Special Issue: Supplementary Cementitious Materials in Concrete, Part I

Alessandro P. Fantilli [1] and Daria Jóźwiak-Niedźwiedzka [2,*]

[1] Department of Structural Geotechnical and Building Engineering, Politecnico di Torino-DISEG, Corso Duca degli Abruzzi 24, 10129 Torino, Italy; alessandro.fantilli@polito.it

[2] Department of Experimental Mechanics, Institute of Fundamental Technological Research, Polish Academy of Sciences, Pawińskiego 5B, 02-106 Warsaw, Poland

* Correspondence: djozwiak@ippt.pan.pl; Tel.: +48-22-826-1281

Citation: Fantilli, A.P.; Jóźwiak-Niedźwiedzka, D. Special Issue: Supplementary Cementitious Materials in Concrete, Part I. *Materials* **2021**, *14*, 2291. https://doi.org/10.3390/ma14092291

Received: 6 April 2021
Accepted: 27 April 2021
Published: 28 April 2021

Publisher's Note: MDPI stays neutral with regard to jurisdictional claims in published maps and institutional affiliations.

Copyright: © 2021 by the authors. Licensee MDPI, Basel, Switzerland. This article is an open access article distributed under the terms and conditions of the Creative Commons Attribution (CC BY) license (https://creativecommons.org/licenses/by/4.0/).

1. Introduction

The environmental impact of the Portland cement production and the large use of cement-based building materials is a growing problem. The substitution strategy, consisting of the partial replacement of Portland cement with supplementary cementitious materials (SCMs), or the more common application of blended cements, is an effective way to improve the sustainability of the construction industries.

To date, the most common SCM is siliceous fly ash [1], a by-product of coal burning in power plants. However, fly ash, which is used in the production of both cement and concrete, is slowly losing primacy due to the progressive decommissioning of coal-fired power plants. Granulated blast furnace slag [2], generally used to reduce the clinker content, cannot replace fly ash due to its properties and limited availability. On the other hand, fly ash from biomass combustion [3] or natural (pumice, volcanic tuffs) [4] or artificial pozzolans (metakaolin) [5] are increasingly being considered within the building materials industry.

The issues related to the possibility of extending the range of SCMs, including, for instance, calcareous fly ash [6], wood ash [7], or activated copper tailings [8], are related to their physical and chemical properties, which in turn can enhance some concrete properties (performance strategy). Indeed, due to the use of SCMs, differences in the microstructure are observed in cement and concrete, consisting in the reduction of the total volume of open pores in the hardened cement paste and in the contact zone between paste and aggregate grains. This improves the performances of cement-based composites, especially in terms of durability, or of resistance to an aggressive environment (due to carbonation [9], presence of chloride ions [10], sulphates [11], etc.), by increasing, for instance, water tightness [12].

Accordingly, in this Special Issue of Materials, aimed at recognizing the current state of knowledge and development in the use of SCMs within the substitution and performance strategies, the following aspects are investigated:

- Measuring the chemical, physical, and mineralogical properties of SCMs, before and after hydration.
- Defining the amounts and the types of SCMs in accordance with the desired effects on fresh and hardened concrete performances.
- Designing structural elements made with normal and high-performance concretes containing SCMs.
- Assessing the durability and environmental impact of cement-based composites containing SCMs.

Hence, various research issues regarding SCMs are herein considered and described in detail through both research papers and state-of-the art reviews. The articles featured in this Special Issue cover the aspects of design, testing, and application of various types of supplementary cementitious materials in concrete. The results of the research, conducted by over 45 international universities and scientific centers, prove the great interest in the SCM topic. In fact, they foster the introduction of new and more environmentally friendly construction materials, with improved physical and mechanical properties, to be used in building engineering.

2. Short Description of the Articles Presented in This Special Issue

The issues of the original research papers and stat-of-the art reviews published in this Special Issue of Materials, and coming from over 45 international universities and scientific centres, can be divided as follows:

- permeability and diffusivity, which are directly related to the quality of concrete and to the durability in various aggressive environments [13–17];
- properties of modified cement matrix composites reinforced with fibers [14,18–20];
- application of new types of supplementary cementitious materials [13,16,21–28];
- characterization of various types of modified cement based materials [18,21,23,25–31].

In the study presented by Andrade et al. [13] a common clay–bentonite was used as concrete additive. The Authors not only analysed the influence of such clay on the mechanical properties of concrete, but also its chemical resistance to sulphates, carbonation and chlorides. They revealed that in the bentonite-bearing material, the resistance to carbonation can be lower than in reference plain Portland cement.

Ahmed et al. [32] showed detailed characteristic of the high-performance self-con solidating concrete incorporating waste mineral materials (i.e., micro-silica and fly ash). They analysed fresh and hardened properties of concrete, and introduced a multi-parameter analytical approach to identify the optimum concrete mixture in terms of cost, workability, strength, and durability.

In the paper presented by Ali et al. [14] the individual and combined incorporation of steel fiber and micro-silica in high-strength concrete were investigated. The tailoring of concrete were performed according to the results of mechanical and permeability tests. By varying the fiber dosage, a mixed effects on the permeability of concrete (water absorption and chloride ion penetration) occurs. However, the presence of micro-silica minimalized the negative effects of high fiber dosage on the properties of concrete.

Research performed by Vu et al. [21] concerned the introduction of new drywall wood-based particleboard as an alternative to gypsum board. More precisely, the use of wood particles in combination with steatite powder and Portland cement was investigated. Both screw withdrawal resistance and bending properties were improved with respect to gypsum board having a similar density. Authors also revealed that wood-cement-steatite powder particleboard could be classified as a quasi-non-combustible material.

Han et al. [31] investigated the application of porous feldspar to reduce the use of cement and sand in the heat storage concrete layer. The mechanical and chemical activation methods were used to compensate the reduction of strength, due to the lower content of cement. With respect to a reference cement mortar, the compressive strength was approximately twice when chemical activation was performed after reducing the cement content by 5% and replacing the sand with porous feldspar. In a large-scale model experiment, the heat storage layer containing the porous feldspar exhibited better thermal properties than those of heat storage layers made with ordinary cement mortar.

The study presented by Yang et al. [29] was aimed at introducing a precise strength evaluation technique. The apparent activation energy of ground granulated blast furnace slag (GGBFS) was calculated through several experiments and used to set up a prediction model. The latter, based on the thermodynamic reactivity of GGBFS within a concrete system cured at different temperature, was able to estimate the compressive strength of GGBFS concrete in accordance to the experimental results.

Kalinowska-Wichrowska et al. [22] presented the results of using recycled cement mortar, obtained from old concrete, as a supplementary cementitious material. Authors showed that by means of a thermal treatment of concrete rubble, a high-quality fine fraction can be obtained. In particular the fine material has pozzolanic properties and can be used as a partial cement replacement in new mortar and concrete.

The parameters affecting the fibre pullout capacity and strain-hardening behavior of fibre-reinforced alkali-activated cement-based composites were investigated by Lee et al. [18]. They used fly ash as a common aluminosilicate source in alkali-activated cementitious composite, whose compressive and flexural strengths were analyzed in addition to the strain-hardening behavior. In particular, the composite critical energy release rate was determined with a nanoindentation approach.

Fantilli et al. [19] investigated the use of ultra-high performance fibre-reinforced cementitious composites (UHP-FRCC), made with various replacement ratio of cement with fly ash, as a reinforcement material of existing concrete columns. Relationships between the size of the UHP-FRCC jacket, the percentage of cement replaced by fly ash, and the strength of the columns were measured and analyzed by means of the eco-mechanical approach. They found that replacement of approximately half of cement with fly ash, and a suitable thickness of the ultra-high performance fibre-reinforced cementitious composites jacket, could ensure the lowest environmental impact without lowering the mechanical properties.

In the study presented by Ishak et al. [23], the influence of the high temperature (from 200 °C to 800 °C) on the fly ash geopolymer concrete incorporating bamboo ash was investigated. When 5% bamboo ash is added to fly ash, geopolymers exhibited more than 50% improvement in residual strength. Moreover, bamboo origin fly ash could be one of the alternatives to fly ash when geopolymer concrete is exposed to high temperature.

Research performed by Monfardini et al. [20] was focused on the flexural behavior of structural elements made with alkali-activated concrete containing class F fly ash, and reinforced with conventional steel rebars in combination with fibers randomly distributed within the concrete matrix. The benefical effects of the hybrid reinforcement was measured through experimental analyses performed on full-scale strutcure at ultimate and serviceability limit states.

Hager et al. [15] analyzed the influence of the binder type (Portland cement and slag cement) on the mechanical and transport properties of heated concrete. The compressive and tensile strength, as well as the static modulus of elasticity and permeability, were measured after the exposure to elevated temperatures (from 200 °C to 1000 °C). The damage of concrete and crack growth due to high temperatures were quantified in accordance with the variation of the static modulus of elasticity. Test results clearly showed the existence of an exponential increment of permeability with damage for both the types of cement.

Vukićević et al. [24] proposed the use of alternative waste materials and hydraulic binders for the soft soil stabilization. High plasticity clay stabilization using fly ash, as well as engineering properties of ash and ash-slag mixtures, were investigated. Test results showed the positive effects of clay stabilization using fly ash, in terms of increasing strength and stiffness and reducing expansivity. As the mechanical properties of fly ash and ash-slag mixtures were comparable with those of sands, they can be used as sustainable fill materials for embankments.

Glinicki et al. [25] investigated the influence of fluidized bed combustion fly ash on the phase composition and microstructure of cement paste. They observed a significant changes in portlandite content and only moderate changes in the content of ettringite, especially when a quantitative evaluation of the phase composition, as a function of fluidized bed fly ash content, was performed.

The aim of the research presented by Vu et al. [26] concerned the use of biomass wood ash as a partial replacement of cement material in wood-cement particleboards. Test results indicated that water demand increased with the increasing of the ash content, and the mechanical properties decreased slightly with an increase of the ash content. The heat capacity increased with the wood ash content as well. The replacement of cement to an extent of approximately 30% by weight was found to give the optimum result.

Woyciechowski et al. [16] presented a new self-terminatin model of carbonation, where the content of calcareous fly ash was taken into consideration as binder component. Also, the idea of developing models for various concrete compositions, as a tool for designing concrete cover thickness of reinforced elements, was proposed.

Fantilli et al. [30] investigated the structural behavior of reinforced concrete elements made with fly ash substitution. A new procedure was introduced with the aim of fulfilling a new limit state of sustainability, in accordance with the serviceability and ultimate limit states currently required by building codes. The proposed approach showed that the CO_2 emission of a reinforced concrete beam was not a monotonic function of the substitution rate of cement with fly ash. On the contrary, there were favorable values of such substitution rates.

García-Vera et al. [17] presented a research on the effect produced by an aggressive environment (containing sulphuric acid solution) on mortars containing different percentages of a crystalline admixture. After a sulphuric acid exposure, mortars made with crystalline admixtures showed higher compressive strength than the reference mortars, besides exhibiting lower mass loss. However, the crystalline admixture did not produce any significant effect on the capillary water absorption coefficient. Whereas, in the short term analysis made in a nonaggressive environment, the crystalline admixture did not have a significant effect neither on the compressive strength and on the capillary water absorption coefficient, nor on the ultrasonic pulse velocity.

The state-of-the-art review presented by Jaskulski et al. [27] concerned various aspects of calcined clays application as a supplementary cementitious material. In more than 200 recent research papers, the authors discussed in detail the idea of replacing Portland cement with large amounts of calcined clay.

Finally, Nicoara et al. [28] reviewed a series of papers regarding the use of end-of-life materials as SCMs in the concrete industry. Ordinary Portland Cement can be effectively substituted by several industrial end-of-life products that contain calcareous, siliceous and aluminous materials, as well as by natural pozzolanic materials like sugarcane bagasse ash, palm oil fuel ash, rice husk ash, mine tailings, marble dust, and construction and demolition debris. Authors revealed that the application of the above-mentioned waste materials as SCMs would decrease the amount of cement used in the production of concrete, and reduce the carbon emissions associated with cement production.

3. Conclusions

The application of supplementary cementitious materials in cement-based composites was an appropriate Special Issue choice, as evidenced by the wide number of research papers published on this subject. They primarily concerned the characterization of new types of SCMs and their possible use in cement-based composites. The properties of matrixes containing SCMs could be further improved by the presence of a fiber reinforcement. Moreover, a better durability in various aggressive environments was also observed. Thus, new building materials containing SCMs can be effectively tailored with the aim of substituting traditional virgin raw materials and to increase their performances, espcially in terms of reducing CO_2 and NO_x emissions.

Author Contributions: Conceptualization, writing—original draft preparation, and writing—review and editing: D.J.N. and A.P.F. All authors have read and agreed to the published version of the manuscript.

Funding: This research received no external funding.

Institutional Review Board Statement: Not applicable.

Informed Consent Statement: Not applicable.

Conflicts of Interest: The authors declare no conflict of interest.

References

1. Mahmud, S.; Manzur, T.; Samrose, S.; Torsha, T. Significance of Properly Proportioned Fly Ash Based Blended Cement for Sustainable Concrete Structures of Tannery Industry. *Structures* **2021**, *29*, 1898–1910. [CrossRef]

2. Wang, Y.; Suraneni, P. Experimental Methods to Determine the Feasibility of Steel Slags as Supplementary Cementitious Materials. *Constr. Build. Mater.* **2019**, *204*, 458–467. [CrossRef]
3. Fořt, J.; Šál, J.; Ševčík, R.; Doleželová, M.; Keppert, M.; Jerman, M.; Záleská, M.; Stehel, V.; Černý, R. Biomass Fly Ash as an Alternative to Coal Fly Ash in Blended Cements: Functional Aspects. *Constr. Build. Mater.* **2021**, *271*, 121544. [CrossRef]
4. Lemougna, P.N.; Wang, K.T.; Tang, Q.; Nzeukou, A.N.; Billong, N.; Melo, U.C.; Cui, X.M. Review on the Use of Volcanic Ashes for Engineering Applications. *Resour. Conserv. Recycl.* **2018**, *137*, 177–190. [CrossRef]
5. Elavarasan, S.; Priya, A.K.; Ajai, N.; Akash, S.; Annie, T.J.; Bhuvana, G. Experimental Study on Partial Replacement of Cement by Metakaolin and GGBS. *Mater. Today Proc.* **2021**, *37*, 3527–3530. [CrossRef]
6. Jóźwiak-Niedźwiedzka, D. Microscopic Observations of Self-Healing Products in Calcareous Fly Ash Mortars. *Microsc. Res. Tech.* **2015**, *78*, 22–29. [CrossRef]
7. Acordi, J.; Luza, A.; Fabris, D.C.N.; Raupp-Pereira, F.; De Noni, A., Jr.; Montedo, O.R.K. New Waste-Based Supplementary Cementitious Materials: Mortars And Concrete Formulations. *Constr. Build. Mater.* **2020**, *240*, 117877. [CrossRef]
8. Vargas, F.; Lopez, M. Development of a New Supplementary Cementitious Material from the Activation of Copper Tailings: Mechanical Performance and Analysis of Factors. *J. Clean. Prod.* **2018**, *182*, 427–436. [CrossRef]
9. Jóźwiak-Niedźwiedzka, D.; Sobczak, M.; Gibas, K. Carbonation of Concretes Containing Calcareous Fly Ashes. *Roads Bridges Drogi i Mosty* **2013**, *12*, 223–236. [CrossRef]
10. Han, X.; Feng, J.; Shao, Y.; Hong, R. Influence of a Steel Slag Powder-Ground Fly Ash Composite Supplementary Cementitious Material on the Chloride and Sulphate Resistance of Mass Concrete. *Powder Technol.* **2020**, *370*, 176–183. [CrossRef]
11. Elahi, M.M.A.; Shearer, C.R.; Reza, A.N.R.; Saha, A.K.; Khan, M.N.N.; Hossain, M.M.; Sarker, P.K. Improving the Sulfate Attack Resistance of Concrete by Using Supplementary Cementitious Materials (SCMs): A Review. *Constr. Build. Mater.* **2021**, *281*, 122628. [CrossRef]
12. Jóźwiak-Niedźwiedzka, D.; Gibas, K.; Glinicki, M.A.; Nowowiejski, G. Influence of High Calcium Fly Ash on Permeability of Concrete in Respect to Aggressive Media. *Roads Bridges Drogi i Mosty* **2011**, *10*, 39–61.
13. Andrade, C.; Martínez-Serrano, A.; Sanjuán, M.Á.; Tenorio Ríos, J.A. Reduced Carbonation, Sulfate and Chloride Ingress Due to the Substitution of Cement by 10% Non-Precalcined Bentonite. *Materials* **2021**, *14*, 1300. [CrossRef] [PubMed]
14. Ali, B.; Kurda, R.; Herki, B.; Alyousef, R.; Mustafa, R.; Mohammed, A.; Raza, A.; Ahmed, H.; Fayyaz Ul-Haq, M. Effect of Varying Steel Fiber Content on Strength and Permeability Characteristics of High Strength Concrete with Micro Silica. *Materials* **2020**, *13*, 5739. [CrossRef] [PubMed]
15. Hager, I.; Tracz, T.; Choińska, M.; Mróz, K. Effect of Cement Type on the Mechanical Behavior and Permeability of Concrete Subjected to High Temperatures. *Materials* **2019**, *12*, 3021. [CrossRef] [PubMed]
16. Woyciechowski, P.; Woliński, P.; Adamczewski, G. Prediction of Carbonation Progress in Concrete Containing Calcareous Fly Ash Co-Binder. *Materials* **2019**, *12*, 2665. [CrossRef] [PubMed]
17. García-Vera, V.E.; Tenza-Abril, A.J.; Saval, J.M.; Lanzón, M. Influence of Crystalline Admixtures on the Short-Term Behaviour of Mortars Exposed to Sulphuric Acid. *Materials* **2019**, *12*, 82. [CrossRef] [PubMed]
18. Lee, H.; Vimonsatit, V.; Mendis, P.; Nassif, A. Study of Strain-Hardening Behaviour of Fibre-Reinforced Alkali-Activated Fly Ash Cement. *Materials* **2019**, *12*, 4015. [CrossRef]
19. Fantilli, A.P.; Paternesi Meloni, L.; Nishiwaki, T.; Igarashi, G. Tailoring Confining Jacket for Concrete Column Using Ultra High Performance-Fiber Reinforced Cementitious Composites (UHP-FRCC) with High Volume Fly Ash (HVFA). *Materials* **2019**, *12*, 4010. [CrossRef]
20. Monfardini, L.; Facconi, L.; Minelli, F. Experimental Tests on Fiber-Reinforced Alkali-Activated Concrete Beams Under Flexure: Some Considerations on the Behavior at Ultimate and Serviceability Conditions. *Materials* **2019**, *12*, 3356. [CrossRef]
21. Vu, V.-A.; Cloutier, A.; Bissonnette, B.; Blanchet, P.; Dagenais, C. Steatite Powder Additives in Wood-Cement Drywall Particleboards. *Materials* **2020**, *13*, 4813. [CrossRef] [PubMed]
22. Kalinowska-Wichrowska, K.; Kosior-Kazberuk, M.; Pawluczuk, E. The Properties of Composites with Recycled Cement Mortar Used as a Supplementary Cementitious Material. *Materials* **2020**, *13*, 64. [CrossRef]
23. Ishak, S.; Lee, H.-S.; Singh, J.K.; Ariffin, M.A.M.; Lim, N.H.A.S.; Yang, H.-M. Performance of Fly Ash Geopolymer Concrete Incorporating Bamboo Ash at Elevated Temperature. *Materials* **2019**, *12*, 3404. [CrossRef]
24. Vukićević, M.; Marjanović, M.; Pujević, V.; Jocković, S. The Alternatives to Traditional Materials for Subsoil Stabilization and Embankments. *Materials* **2019**, *12*, 3018. [CrossRef]
25. Glinicki, M.A.; Jóźwiak-Niedźwiedzka, D.; Dąbrowski, M. The Influence of Fluidized Bed Combustion Fly Ash on the Phase Composition and Microstructure of Cement Paste. *Materials* **2019**, *12*, 2838. [CrossRef] [PubMed]
26. Vu, V.-A.; Cloutier, A.; Bissonnette, B.; Blanchet, P.; Duchesne, J. The Effect of Wood Ash as a Partial Cement Replacement Material for Making Wood-Cement Panels. *Materials* **2019**, *12*, 2766. [CrossRef] [PubMed]
27. Jaskulski, R.; Jóźwiak-Niedźwiedzka, D.; Yakymechko, Y. Calcined Clay as Supplementary Cementitious Material. *Materials* **2020**, *13*, 4734. [CrossRef] [PubMed]
28. Nicoara, A.I.; Stoica, A.E.; Vrabec, M.; Šmuc Rogan, N.; Sturm, S.; Ow-Yang, C.; Gulgun, M.A.; Bundur, Z.B.; Ciuca, I.; Vasile, B.S. End-of-Life Materials Used as Supplementary Cementitious Materials in the Concrete Industry. *Materials* **2020**, *13*, 1954. [CrossRef] [PubMed]

29. Yang, H.-M.; Kwon, S.-J.; Myung, N.V.; Singh, J.K.; Lee, H.-S.; Mandal, S. Evaluation of Strength Development in Concrete with Ground Granulated Blast Furnace Slag Using Apparent Activation Energy. *Materials* **2020**, *13*, 442. [CrossRef] [PubMed]
30. Fantilli, A.P.; Tondolo, F.; Chiaia, B.; Habert, G. Designing Reinforced Concrete Beams Containing Supplementary Cementitious Materials. *Materials* **2019**, *12*, 1248. [CrossRef]
31. Han, J.-G.; Cho, J.-W.; Kim, S.-W.; Park, Y.-S.; Lee, J.-Y. Characteristics of CO_2 and Energy-Saving Concrete with Porous Feldspar. *Materials* **2020**, *13*, 4204. [CrossRef] [PubMed]
32. Ahmed, G.H.; Ahmed, H.; Ali, B.; Alyousef, R. Assessment of High Performance Self-Consolidating Concrete through an Experimental and Analytical Multi-Parameter Approach. *Materials* **2021**, *14*, 985. [CrossRef] [PubMed]

Article

Designing Reinforced Concrete Beams Containing Supplementary Cementitious Materials

Alessandro P. Fantilli [1],*, Francesco Tondolo [1], Bernardino Chiaia [1] and Guillaume Habert [2]

1   Politecnico di Torino, 10129 Torino, Italy; francesco.tondolo@polito.it (F.T.); bernardino.chiaia@polito.it (B.C.)
2   ETH Zurich, 8093 Zurich, Switzerland; habert@ibi.baug.ethz.ch
*   Correspondence: alessandro.fantilli@polito.it; Tel.: +39-011-090-4900

Received: 4 April 2019; Accepted: 12 April 2019; Published: 16 April 2019

Abstract: If supplementary cementitious materials (SCMs) are used as binders, the environmental impact produced by cement-based composites can be reduced. Following the substitution strategy to increase sustainability, several studies have been carried out with the aim of measuring the mechanical properties of different concrete systems, in which a portion of Portland cement was substituted with SCMs, such as fly ashes. On the other hand, studies on the structural behavior of reinforced concrete (RC) elements made with SCMs are very scarce. For this reason, in this paper, a new procedure is introduced with the aim of fulfil a new limit state of sustainability, in accordance with the serviceability and ultimate limit states required by building codes. Although the environmental impact of concrete decreases with the reduction of cement content, the proposed approach shows that the carbon dioxide emission of an RC beam is not a monotonic function of the substitution rate of cement with SCMs. On the contrary, there are favorable values of such substitution rates, which fall within a well-defined range.

Keywords: fly ash; substitution strategy; structural concrete; steel reinforcement; limit states; RC beams in bending; carbon footprint

---

1. Introduction

Reinforced concrete (RC) structures are currently designed to satisfy ultimate and serviceability limit states [1]. Nevertheless, as stated by Model Code 2010 [2], the design of structures is a process of developing a suitable solution in which not only must safety and functionality be guaranteed during service life, but also sustainability must be assured. Although green concrete structures are achieved via different approaches [3], two possible strategies can be applied to better fulfill environmental requirements [4]:

- Material performance strategy, aimed at the reduction of clinker and thus of the volume of structures, by increasing the mechanical performance of concrete.
- Material substitution strategy, which consists of substituting clinker with cementitious and/or pozzolanic mineral admixtures (e.g., fly ashes, silica fumes, etc.).

In several cases, these two strategies are contemporarily used, such as in the substitution of cement with supplementary cementitious materials (SCMs), which can be byproducts of the industrial process. For instance, coal fly ashes, deriving from the combustion of coal in power plants and which can be used to partially substitute Portland cement, can also enhance the strength and the durability of traditional concrete [5].

From a practical point of view, the abovementioned strategies are not well integrated into the current limit state design approach. In other words, there is not a single procedure capable of assuring structural safety while also minimizing the environmental impact of concrete elements. In almost all cases, after designing the mechanical performance of RC structures, the environmental impact is assessed through broad-based green building rating schemes [2]. As the most common rating systems grant a posteriori (i.e., after building the structure) sustainability certificate, the sustainability and the mechanical performances of different concretes cannot be compared [6–8]. Hence, the European Union (EU) target to reduce the greenhouse gasses GHG emissions by 20% [9] cannot be fulfilled by the cement and concrete industry if the current mechanical and environmental approaches used to design RC structures are not integrated.

In the opinion of the authors, to design more sustainable reinforced concrete structures, a new limit state has to be introduced and used in combination with the traditional limit states. In this way, a code-specific language addressing sustainability practices, which is one of the key objectives of the American Concrete Institute ACI Concrete Sustainability Forum [10], can be developed. Thus, here, a simply supported beam is designed not only to satisfy the bearing capacity and deflection limits, but also to reduce, as much as possible, the environmental impact and fulfill the EU target [9]. Specifically, an integrated ecological and mechanical procedure is herein proposed to select the best concrete with the optimal replacement rate of cement with fly ash.

2. The Sustainability of Materials

In the material performance strategy, the CO_2 emitted per cubic meter of concrete increases with the concrete strength. According to Habert's and Roussel's [4] model (see Figure 1a), a quadratic function can define this relationship:

$$\beta = \delta \sqrt{f_c} \tag{1}$$

where β = mass of CO_2 emitted by the production of a cubic meter of concrete (whose binder is only cement); f_c = average compressive strength of concrete (whose binder is only cement); and δ = coefficient of proportionality.

Conversely, the application of the substitution strategy, e.g., replacing part of the cement with fly ash, produces a decrement of the initial values of CO_2 emission, β_A, and concrete strength, f_{cA}, in a specific concrete system (Figure 1b).

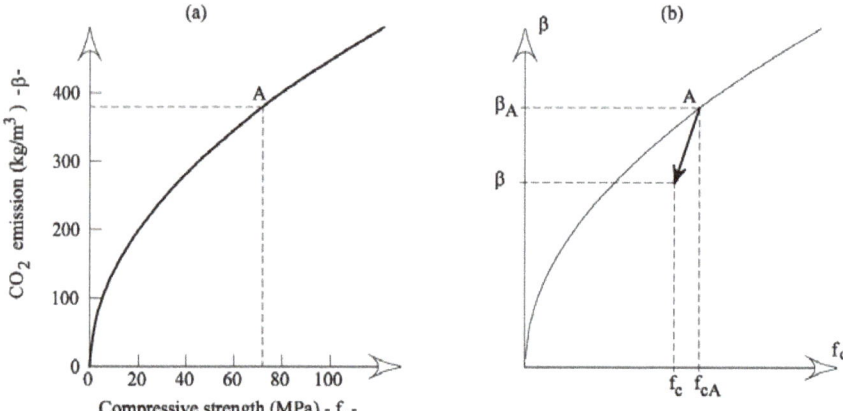

Figure 1. The impact of concrete: (a) the quadratic function proposed by Habert and Roussel [4]; (b) the decrement of β and f_c due to the substitution of cement with fly ash in a specific concrete system, whose initial values of CO_2 emission and average compressive strength are β_A and f_{cA}, respectively.

The new values of f_c and β of concrete in which part of the cement is substituted by fly ash, depend on the initial values β_A and f_{cA} (of a concrete made by only cement) and on the rate of substitution S.

Thus, for given values of β_A, f_{cA}, and S, by means of the following functions, both f_c and β can be evaluated:

$$f_c = (1 + \alpha \cdot S) f_{cA} \qquad (2)$$

$$\beta = (1 + \gamma \cdot S) \beta_A \qquad (3)$$

where S = is the substitution rate of cement with fly ash that modifies f_{cA} and β_A into f_c and β, respectively; α = strength coefficient; and γ = sustainability coefficient. Obviously, for a specific concrete system, the three coefficients α, δ, and γ have to be evaluated through the regression analyses of the available experimental data.

The Tests of Lam et al. [11]

Lam et al. [11] investigated the effects of replacing cement by fly ash on the compressive strength of concrete. The investigation included 15 concretes, having 3 sets of water/cement ratios and containing low and high volumes of fly ash. The mixtures taken into consideration are reported in Table 1. The same Table also shows the results of compressive strength measured on the cylindrical specimens at 28 days. To evaluate the impact of the concrete components, in terms of CO_2 released into the atmosphere, the data reported in Table 2 are assumed herein [8].

Table 1. The concretes tailored and tested by Lam et al. [11].

Mix	w/c	Cement (kg/m³)	Fly Ash (kg/m³)	Aggregate (kg/m³)	Superplasticizer (kg/m³)	f_c (MPa)
S1-0	0.3	500	0	1810	7.5	82.5
S1-15	0.3	425	75	1810	7.5	77.9
S1-25	0.3	375	125	1810	7.5	79.1
S1-45	0.3	275	225	1810	7.5	64
S1-55	0.3	225	275	1810	7.5	57.1
S2-0	0.4	400	0	1810	7.5	55.8
S2-15	0.4	340	60	1810	7.5	44.8
S2-25	0.4	300	100	1810	7.5	44.1
S2-45	0.4	220	180	1810	7.5	32.7
S2-55	0.4	180	220	1810	7.5	32.4
S3-0	0.5	410	0	1810	7.5	42.6
S3-15	0.5	348.5	61.5	1810	7.5	38.1
S3-25	0.5	307.5	102.5	1810	7.5	35.2
S3-45	0.5	225.5	184.5	1810	7.5	30.4
S3-55	0.5	184.5	225.5	1810	7.5	25.9

Table 2. The environmental impact of the components of reinforced concrete (RC) structures [8].

Materials	Unit	Global Warming Potential (GWP) CO_2 (kg)
Cement Type I 52.5	kg	0.832
Ground limestone	kg	0.0191
Fly ash	kg	-
Silica fume	kg	-
Aggregates	kg	0.00246
Steel	kg	1.50
Water	kg	0.000318
Superplasticizer	kg	0.720
Air entraining	kg	0.0860
Retarder	kg	0.0760

Accordingly, the following values can be obtained through least squares approximation of the experimental data reported in Tables 1 and 2:

- δ = 48.088 kg CO_2/(m³ MPa$^{0.5}$);
- α = −0.006732;
- γ = −0.009731.

Such parameters, to be used in Equations (1)–(3), seem to be independent of the water/cement ratio and are included in the procedure illustrated in Figure 2, herein used to evaluate the curves f_c-S and β-S of a specific concrete system. For instance, the diagram depicted in Figure 3 shows the results of the proposed procedure applied to the three series of specimens tested by Lam et al. [11].

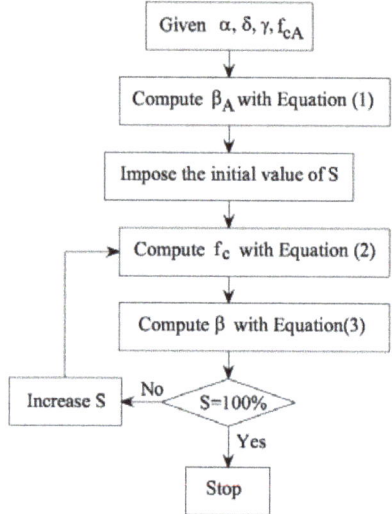

Figure 2. The procedure used to obtain the functions f_c-S and β-S.

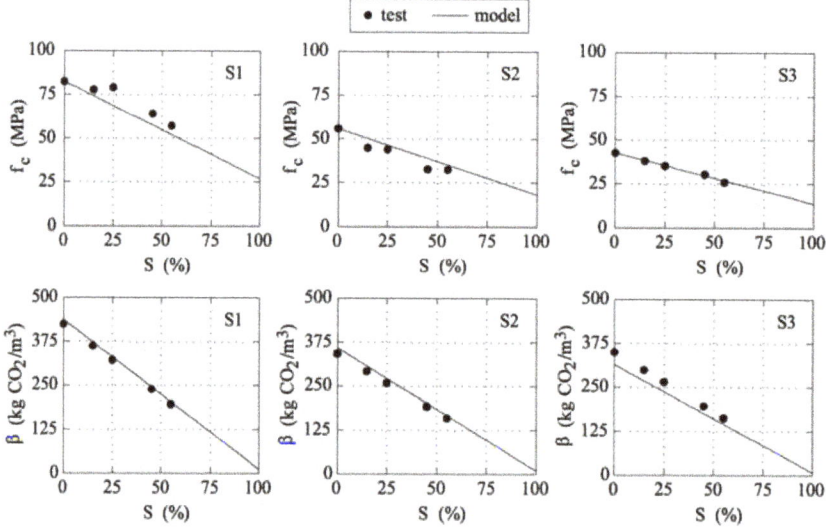

Figure 3. The proposed functions f_c-S and β-S compared with test data measured by Lam et al. [11].

3. The Limit States of an RC Beam in Bending

According to Eurocode 2 (EC2) [1] the ultimate limit states of RC beams in bending (Figure 4a,b) depend on the constitutive relationships of materials. For normal-weight concrete of a class lower than 50 MPa, the parabola–rectangle relationship illustrated in Figure 4c can be used. The bilinear elastic–perfectly plastic relationship is assumed for steel in tension (Figure 4d). In the latter, after yielding (i.e., $\varepsilon_s > \varepsilon_{yd} = f_{yd}/E_s$, where E_s = 200 GPa = elastic modulus of steel), the stress is constant and equal to the yielding strength, regardless of the strain.

The design strengths of both materials are computed in accordance with the partial safety factors given by Eurocode 2 [1]:

$$\sigma_{cd} = 0.85 \frac{f_{ck}}{\gamma_c} \qquad (4)$$

$$f_{yd} = \frac{f_{yk}}{\gamma_s} \qquad (5)$$

where f_{ck} = characteristic compressive cylinder strength of concrete at 28 days; f_{yk} = characteristic yield strength of reinforcement; γ_c = 1.5 = partial safety factor of concrete; and γ_s = 1.15 = partial safety factor of steel.

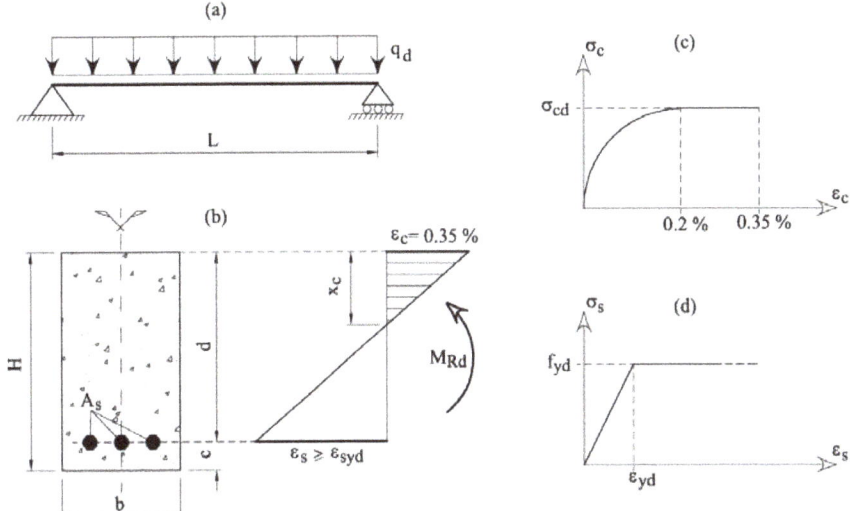

Figure 4. The ultimate limit state in reinforced concrete beams in bending: (**a**) a simply supported beam subjected to distributed loads; (**b**) the limit state profile in a cross-section; (**c**) the parabola–rectangle relationship for concrete; and (**d**) the elastic–perfectly plastic relationship for steel.

With the constitutive laws illustrated in Figure 4c,d, an RC cross-section can be designed in order to satisfy the following condition:

$$M_{Rd} \geq M_{Ed} \qquad (6)$$

where M_{Ed} = design bending moment applied to the cross-section and produced by the external actions and M_{Rd} = design bending moment capacity of the cross-section.

The value of M_{Rd} can be analytically computed assuming the limit strain conditions illustrated in Figure 4b. Specifically, the maximum strain of concrete is reached in the compressed edge of the beam, whereas the strain of steel in tension should be larger than or equal to that at yielding (i.e., $\varepsilon_s \geq \varepsilon_{yd}$).

Under these assumptions, the equilibrium and compatibility equations provide [12]:

$$\omega = 0.81\, \xi \qquad (7a)$$

$$\mu_{Rd} = 0.81\, \xi (1 - 0.42\xi) \qquad (7b)$$

where, according to the symbols reported in Figure 4b, the following non-dimensional geometrical and mechanical properties are taken into consideration

$$\xi = \frac{x_c}{d} \tag{8}$$

$$\omega = \frac{A_s\, f_{yd}}{b\, d\, \sigma_{cd}} \tag{9}$$

$$\mu_{Rd} = \frac{M_{Rd}}{b\, d^2\, \sigma_{cd}} \tag{10}$$

If the value of ξ is fixed, the optimal values of ω and μ_{Rd} can be calculated through Equation (7).

Generally, code rules fix the minimum and the maximum value of the reinforcement area [1,2] as follows:

$$k_1 \frac{b\, d}{f_{yk}} \leq A_s \leq k_2 \frac{b\, d}{f_{yk}} \tag{11}$$

where $k_1 = 1.4$ and $k_2 = 3.5$ are the values used in Italy.

To reduce the volume of the cross-section, it is better to design the area A_s close to the upper bound of Equation (11), thus:

$$\omega = \frac{k_2}{\sigma_{cd}\, \gamma_s} \tag{12}$$

If Equation (12) is substituted into Equation (7a), the optimal value of ξ can be obtained:

$$\xi = \frac{k_2}{0.81\, \sigma_{cd}\, \gamma_s} \tag{13}$$

It must be noted that in the case of concrete C25 (which is the most used in Italy), the value of $\xi = 0.25$ is obtained when $k_2 = 3.5$ and $\gamma_s = 1.15$. As stated by EC2 [1], the plastic analysis of beams, frames, and slabs can be performed without the explicit verification of the required ductility when $\xi \leq 0.25$ for concrete strength classes lower than C50.

Finally, by substituting Equation (9) into Equation (12) and Equation (10) and Equation (13) into Equation (7b), the following formulae can be obtained:

$$A_s = \frac{b\, d\, k_2}{f_{yd}\, \gamma_s} \tag{14a}$$

$$M_{Rd} = b\, d^2 \frac{k_2}{\gamma_s}\left(1 - 0.42 \frac{k_2}{0.81\, \sigma_{cd}\, \gamma_s}\right) \tag{14b}$$

As the direct computation of deflection is not always necessary [1], the span/depth ratio is herein limited for avoiding deflection problems in RC beams. In other words:

$$H \geq \frac{L}{\psi} \tag{15}$$

where L = span length of the beam (Figure 4a); H = height of the beam (Figure 4a); and ψ = coefficient.

The depth of the concrete cover c is related to durability requirements. Thus, it depends on the environmental conditions (i.e., the class of exposition), and it can be assumed as a fraction of the height H:

$$c \geq \frac{H}{\rho} \tag{16}$$

where ρ = coefficient.

4. A New Design Procedure for RC Beams in Bending

When a concrete system is introduced (and, therefore, δ, α and γ are known), it is possible to select a specific value of strength f_c (herein assumed as the average value of strength) and the corresponding coefficient β. For the beam depicted in Figure 4a, the length of the span L, the density of concrete

De, and the applied load q_d are the input data. The values of the depth H and concrete cover c can be obtained from the coefficients ψ and ρ, regarding the serviceability (control of deflection) and the durability requirements, respectively.

Under these conditions, to obtain the geometry of the beam, only the width b and the area of the reinforcement A_s have to be calculated. Such values mainly depend on the maximum bending moment acting on the beam:

$$M_{Rd} = M_{Ed} = (1.3\, b\, H\, De + 1.5\, q_d)\frac{L^2}{8} \tag{17}$$

where 1.3 and 1.5 are the partial safety factors of the structural weight and service load.

If Equation (17) is substituted into Equation (14b), and assuming $d = H - c$, then the width b can be obtained:

$$b = \frac{1.5\, q_d\, L^2}{\left[8\,(H-c)^2 \frac{k_2}{\gamma_s}\left(1 - 0.42\frac{k_2}{0.81\,\sigma_{cd}\,\gamma_s}\right) - 1.3\, H\, De\, L^2\right]} \tag{18}$$

The area of reinforcement in tension is then computed with Equation (14a), and the global impact of the beam BI, in terms of CO_2 released into the atmosphere, is:

$$BI = \beta(b\, H - A_s) + \phi\, A_s \tag{19}$$

where ϕ is the environmental impact of steel as obtained from Table 2.

The procedure illustrated in Figure 2 and used to calculate the f_c-S and β-S functions can now be extended to calculate the relationships b-S, A_s-S, and BI-S of the RC beam illustrated in Figure 4a. The flow chart of the new procedure is drawn in Figure 5, whereas Figure 6 shows the curves computed in the case of $f_{ckA} = 25$ MPa (f_{ckA} = the characteristic value of strength in the absence of cement substitution = f_{cA} − 8 MPa [1,2]) and:

- $\delta = 48.088$ kg CO_2/(m^3 MPa$^{0.5}$);
- $\alpha = -0.006732$;
- $\delta = -0.009731$;
- $\psi = 0.1$;
- $\rho = 0.07$;
- $L = 5000$ mm;
- $De = 25$ kN/m^3;
- $q_d = 46.5$ kN/m;
- $k_2 = 3.5$;
- $\phi = 1174.525$ kg CO_2 /m^3.

As shown in Figure 3, β (and thus f_c) linearly decreases with S (see also Figure 6a). Consequently, the geometrical dimensions of the beam increase as the substitution rate of cement with fly ash increases. As a matter of fact, the width of the beam b becomes larger as S grows. Nevertheless, the b-S function (Figure 6b) is not linear as is β-S (Figure 6a). In particular, when $S > 75\%$ the width of the beam drastically increases for small increments of S, and Figure 6b shows a vertical asymptote when $S \rightarrow 100\%$.

The above observations are also valid for the area of the steel used to reinforce the tensile zone of the RC beam. Namely, Figure 6c reveals a monotonic increment of A_s with S, but the A_s-S function shows two different slopes before and after $S \cong 75\%$ (Figure 6c). As a result, the global impact of an RC beam decreases when $S < 75\%$, whereas BI grows when $S > 75\%$ (Figure 6d). In other words, although the unitary impact of concrete always decreases with S (see Figure 6a), the global impact of a beam BI is not a monotonic function of S (see Figure 6d). For the given initial strength and impact (i.e., f_{cA} and BI_0), the values of BI have a minimum, BI_{min}, in correspondence to the substitution rate S_F (where $0 < S_F < 100\%$).

It must be noted that the shape of the functions *BI-S* strongly depends on f_{ckA}. As shown in Figure 7, where five *BI-S* functions, corresponding to five different values of f_{ckA}, are reported, BI_{min} tends to decrease and S_F tends to increase if the initial strength of the concrete increases. However, BI_0 becomes larger as f_{ckA} increases, and, when $S < S_F$, although the beam can be cast with a low amount of concrete (and steel, as well), the impact is higher due to the high content of cement. On the contrary, when $S > S_F$, the impact increases despite the low amount of cement (and low concrete strength), because large amounts of concrete and steel are needed. Finally, the proposed model reveals that for high values of f_{ckA}, the best substitution rate of cement with fly ash can be 100% (i.e., $S_F = 100\%$).

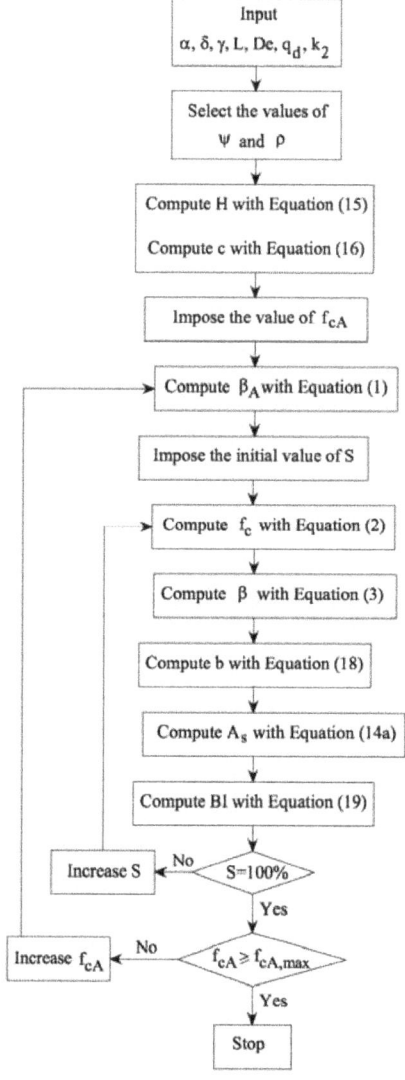

Figure 5. The procedure to compute the functions β-*S*, *b*-*S*, A_s-*S*, and *BI*-*S* in concrete systems with an average compressive strength in the absence of cement substitution lower than $f_{cA,max}$.

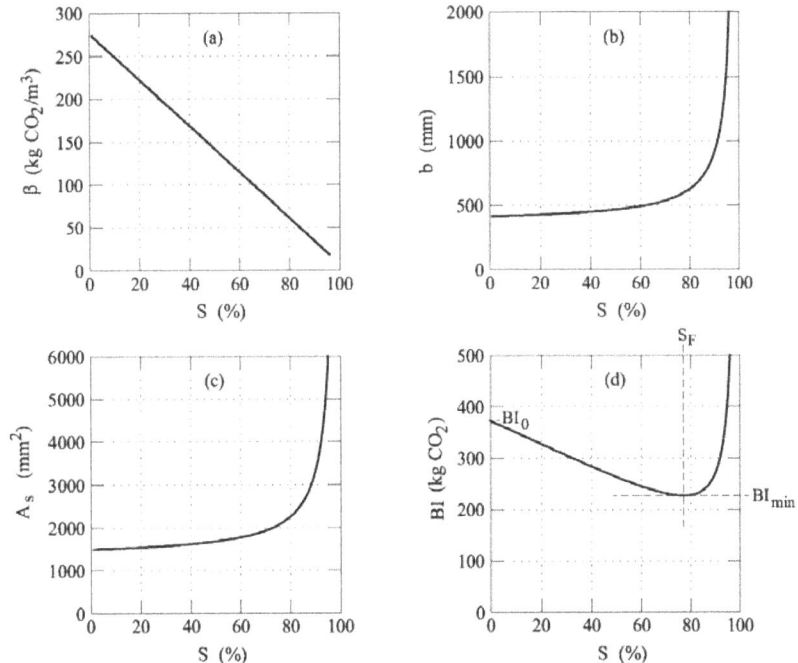

Figure 6. The result of the procedure herein proposed to design RC beams in bending with f_{ckA} = 25 MPa: (a) β-S function; (b) b-S function, (c) A_s-S function; and (d) BI-S function.

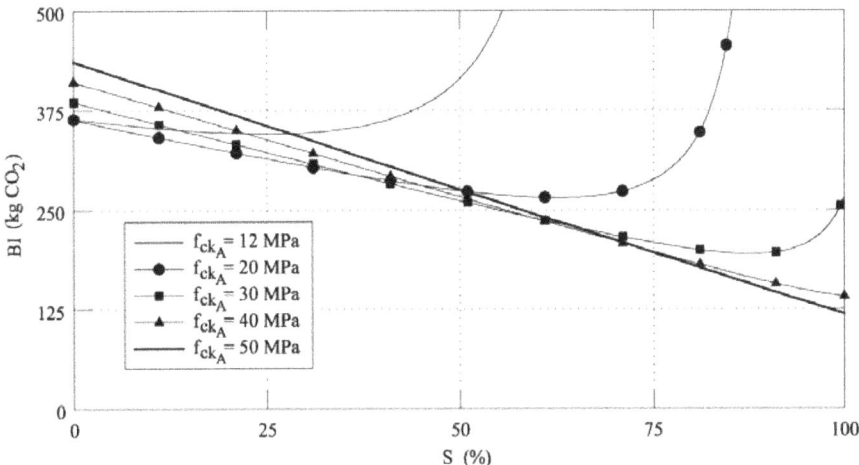

Figure 7. BI-S functions obtained by substituting cement with fly ashes in concrete systems with different f_{ckA}.

From a practical point of view, the substitution rate cannot be too high, because some problems occur in the concrete system, whose early strength decreases with S [13]. Thus, to reduce the emission of CO_2, a new limit state of sustainability, corresponding to the maximum environmental impact of a structure, is herein introduced. For instance, code rules or tenders can require a concrete in which the substitution of cement with fly ashes leads to a reduction of the carbon dioxide emission of larger than 20% (as suggested in [9]), with respect to the emission produced by the same concrete system when $S = 0$. Referring to Figure 8, where the concrete strength f_{ckA} is 25 MPa, a new limit $BI_{max} = 80\% \ BI_0$ must be introduced. It defines a range of the admissible S, where the optimal substitution rate of cement with fly ash (or others SCMs) can be selected. The best S does not necessarily coincide with S_F, because, for large substitutions, the RC beams and the area of rebar are too large to be used in practice. Moreover, higher rates of substitution would provide a decrease in the early strength of concrete. Thus, some building codes impose lower limits on the usage rates of fly ash than the feasibility rates measured by laboratory tests.

Finally, it must be noted that though the proposed approach herein applies to fly ashes only, it can be easily generalized to other SCMs. Indeed, the procedure illustrated in Figure 5 can be used in all cases, if the parameters of Equations (2) and (3) are experimentally measured for the supplementary cementitious material taken into consideration.

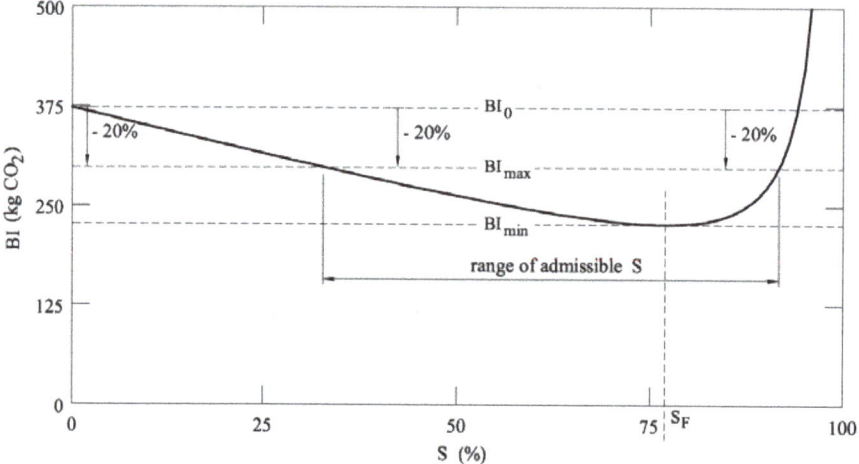

Figure 8. Application of a possible sustainability limit state and definition of the range of admissible S (f_{ckA} = 25 MPa).

5. Conclusions

According to the results obtained by applying the design procedure previously described, the following conclusions are drawn:

- The use of SCMs as cement replacement can be directly integrated within the current design procedure of RC structures, as long as specific experimental analysis on concrete systems provides the function f_c-S and β-S (Figure 1).
- In the new approach, the design of an RC beam in bending (Figure 4), performed in accordance with the traditional ultimate and serviceability limit states, also includes the evaluation of the environmental impact BI, herein computed as a function of the substitution rate of cement with SCMs.
- In absence of cement substitution (i.e., $S = 0$), BI increases with the initial strength f_{cA}. Nevertheless, the relative minimum of the curve BI-S moves towards higher S. As BI_{min} decreases when f_{cA} increases, it seems more convenient to use high strength concrete systems (i.e., with the highest f_{cA}) but with the maximum substitution rate of cement with fly ash.
- If a new limit state of sustainability (i.e., BI_{max}) is introduced, the reduction of the carbon dioxide emission can be achieved also in the case of low values of S.

Finally, future works will be devoted to calculating *BI-S* functions in more complex structures, such as frames and slabs, as well as considering the effects of other actions (e.g., shrinkage, seismic loads, etc.).

Author Contributions: The authors equally contributed to develop research and write this article.

Funding: This research received no external funding

Acknowledgments: The results of a joint research developed by Politecnico di Torino and ETH Zurich are reported in this paper. The authors wish to thank the Italian Laboratories University Network of Seismic Engineering (ReLUIS) for supporting this research.

Conflicts of Interest: The authors declare no conflict of interest.

References

1. EN 1992-1-1:2004. *Eurocode 2: Design of Concrete Structures. General Rules and Rules for Buildings*; European Committee for Standardization: Brussels, Belgium, 2004.
2. Fib. *Model Code for Concrete Structures 2010*, 1st ed.; Ernst & Sohn: Berlin, Germany, 2013.
3. Fib. *Bulletin No. 67. Guidelines for Green Concrete Structures*; Fédération Internationale du Béton: Lausanne, Switzerland, 2012.
4. Habert, G.; Roussel, N. Study of two concrete mix-design strategies to reach carbon mitigation objectives. *Cem. Concr. Compos.* **2009**, *31*, 397–402. [CrossRef]
5. ACI Committee 130. *ACI 130R-19: Report on the Role of Materials in Sustainable Concrete Construction*; American Concrete Institute: Farmington Hills, MI, USA, 2019.
6. Müller, H.S.; Breiner, R.; Moffatt, J.S.; Haist, M. Design and properties of sustainable concrete. *Procedia Eng.* **2014**, *95*, 290–304. [CrossRef]
7. Fantilli, A.P.; Chiaia, B. Eco-mechanical performances of cement-based materials: An application to self-consolidating concrete. *Constr. Build. Mater.* **2013**, *40*, 189–196. [CrossRef]
8. Chiaia, B.; Fantilli, A.P.; Guerini, A.; Volpatti, G.; Zampini, D. Eco-mechanical index for structural concrete. *Constr. Build. Mater.* **2014**, *67*, 386–392. [CrossRef]
9. The Roadmap to a Resource Efficient Europe. Available online: http://ec.europa.eu/environment/resource_efficiency/about/roadmap/index_en.htm (accessed on 10 March 2019).
10. Sakai, K.; Buffenbarger, J.K. ACI Concrete Sustainability Forum XI: Concrete sustainability is entering a new stage! *ACI Concr. Int.* **2019**, *41*, 47–51.
11. Lam, L.; Wong, Y.L.; Poon, C.S. Effect of fly ash and silica fume on compressive and fracture behaviors of concrete. *Cem. Concr. Res.* **1998**, *28*, 271–283. [CrossRef]
12. Nilson, A.; Darwin, D.; Dolan, C. *Design of Concrete Structures*, 12th ed.; McGraw-Hill: New York, NY, USA, 2009; pp. 80–94.
13. Mehta, P.K.; Monteiro, P.J.M. *Concrete: Microstructure, Properties, and Materials*, 3rd ed.; McGraw-Hill: New York, NY, USA, 2006; pp. 485–491.

© 2019 by the authors. Licensee MDPI, Basel, Switzerland. This article is an open access article distributed under the terms and conditions of the Creative Commons Attribution (CC BY) license (http://creativecommons.org/licenses/by/4.0/).

Article

Assessment of Mechanical, Thermal and Durability Properties of High-Volume GGBS Blended Concrete Exposed to Cryogenic Conditions

Giyeol Lee [1] and Okpin Na [2,*]

[1] Department of Landscape Architecture, College of Agriculture and Life Science, Chonnam National University, Gwangju 61186, Korea; gylee@jnu.ac.kr
[2] R&D Division, Hyundai E&C, Yongin-si 16891, Korea
* Correspondence: okpin.na@hdec.co.kr

Citation: Lee, G.; Na, O. Assessment of Mechanical, Thermal and Durability Properties of High-Volume GGBS Blended Concrete Exposed to Cryogenic Conditions. *Materials* 2021, 14, 2129. https://doi.org/10.3390/ma14092129

Academic Editor: Alessandro P. Fantilli

Received: 1 April 2021
Accepted: 20 April 2021
Published: 22 April 2021

Publisher's Note: MDPI stays neutral with regard to jurisdictional claims in published maps and institutional affiliations.

Copyright: © 2021 by the authors. Licensee MDPI, Basel, Switzerland. This article is an open access article distributed under the terms and conditions of the Creative Commons Attribution (CC BY) license (https://creativecommons.org/licenses/by/4.0/).

Abstract: The purpose of this study is to suggest the optimum mix design with a high volume of GGBS (Ground Granulated Blast-furnace Slag) replacement and the procedure of the cryogenic test to consider mechanical and thermal properties, and durability performance. To decide the optimum mix design, four mix designs with high-volume of GGBS replacement were suggested, in terms of the slump and retention time. Based on the test results, with respect to the workability and compressive strength, the mixtures with 65% of GGBS (C40-2 and C40-4) were better than the mixtures with 50% and 60% of GGBS (C40-1 and C40-3). After selecting two mixtures, two types of cryogenic test methods were conducted under one-cycle cryogenic condition (Test A) and 50-cycles cryogenic condition (Test B). As a result, in Test A, the compressive strength and elastic modulus of the C40-2 and C40-4 mixtures tended to be decreased over time, because of the volume expansion of ice crystals contained in the capillary pores. In Test B, the mechanical properties of the C40-4 mixture were better than those of the C40-2 mixture, in terms of the reduction rate of compressive strength and elastic modulus. In the view of the heat of hydration, the semi-adiabatic test was conducted. In the results, the C40-4 mixture was better to control the thermal cracks. Thus, the C40-4 mixture would be more suitable for cryogenic concrete and this procedure could be helpful to decide the mixture of cryogenic concrete. In the future, the long-term performance of cryogenic concrete needs to be investigated.

Keywords: cryogenic condition; GGBS; compressive strength; thermal conductivity; semi-adiabatic test

1. Introduction

LNG (Liquefied Natural Gas) has been regarded as the most realistic alternative to reduce global warming from petroleum energy because it emits very little sulfurous acid gas (SO_2) recognized as a major cause of environmental problems. LNG demand is expected to be determined by the climate change response activities of each country around the world. Most of all, Asia is the world's largest importing region of LNG and accounts for two-thirds of global consumption [1–4].

Natural gas is cooled to about −163 °C to convert the gaseous gas to a liquid state for storage. Therefore, many kinds of research have been conducted to minimize heat loss and ensure safety against gas leakage [5–7].

LNG storage tanks can be largely divided into above-ground, in-ground and underground depending on the installation location. According to the type of inner tank, it can be roughly classified into a 9%-nickel steel tank and membrane tank. Based on the definition of NFPA 59 A (2001), BS EN 1473 (1996) and BS 7777 (1993), a 9%-Ni steel tank is classified into single, double and fully containment LNG storage tanks. The full containment LNG storage tank with relatively high safety is a double tank structure in which the inner tank and the outer tank can independently store LNG at cryogenic temperatures [8].

The inner tank stores cryogenic LNG under normal operating conditions and the outer tank is located between 1 m and 2 m from the inner tank and functions as a dike as well as has a function to support the outer roof. The tank has a prestressed concrete outer container with a flexible inner container and insulation supported by an outer tank wall. Prestressed concrete (PC) structures are suitable for storing LNG and their design incorporates special loads and a special performance at cryogenic temperatures [4,8–10].

Nevertheless, the inner and outer storage tanks of LNG involve several potential risks. Sudden failure of LNG storage tank is not acceptable, since the escaped liquid would vaporize, mix with air and form an explosive cloud. The explosions or fires resulting from such an event could lead to an unacceptable loss of life and damage to the plant and environment. To reduce the potential risks and safety issues of a concrete storage tank, design guidance is given in BS 7777, NFPA 95A and BS EN 1473 [10,11]. Moreover, Jeon et al. (2003) studied the liquid tightness design associated with the cryogenic temperature under the emergency condition of LNG leakage [6]. After then, Jean et al. (2004) focused on the major factors deciding the shape of the large LNG tank [12]. Hoyle (2013) in Chevron presented the modular design of the precast concrete outer wall, instead of the in-situ concrete wall for the full containment storage tank. It concluded that this modular concept could replace the 9%-nickel steel with concrete and reduced the material cost and construction time [13]. Based on the composite concrete cryogenic tank (C3T) of Chevron, Jeon et al. (2014) and Jo et al. (2015) also studied the precast concrete module with outer liners to shorten the construction period [14,15]. Even though the concrete module could make construction time to be saved, the connection between concrete panels and countermeasures of emergency leakage should be supported with sufficient researches.

Generally, impermeable insulations such as PUF (Poly Urethane Foam) have been located between the inner tank and outer concrete wall to prevent direct contact as demonstrated in Figure 1. Recently, to remove the impermeable insulation, cryogenic rebars as reinforcement at the inner surface of the concrete wall were partially used. Yoon (2012) suggested the application of cryogenic rebar to LNG storage outer tank [16]. Cryogenic steel rebar is used to prevent the brittle failure of concrete outer wall when the leakage of LNG occurs. Cryogenic steel rebar is a specially-designed concrete reinforcing steel for cryogenic applications and is suitable for use in storage tanks with temperatures down to −170 °C in accordance with EN 14620-3:2006. In fact, cryogenic rebar has been manufactured by Commercial Metals Company (CMC) in USA and ArcelorMittal in Luxembourg City.

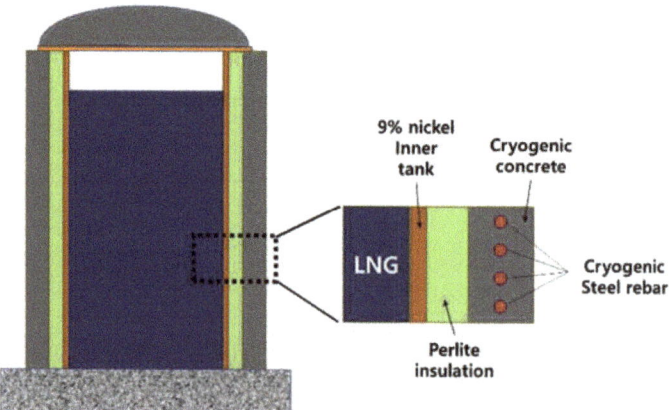

Figure 1. Full containment LNG storage tank.

Concrete used as the outer shell can be directly exposed to the LNG leakage and the inner surface of outer concrete can be cooled lower than −165 °C of cryogenic temperature. The concrete used to contain the liquefied natural gas must withstand sub-arctic temperatures as low as −165 °C and is called "cryogenic concrete". However, concrete behavior at cryogenic conditions has been not elucidated. Kogbara et al. (2013) reviewed the concrete properties under cryogenic temperatures such as permeability, coefficient of thermal expansion (CTE), tensile strength, bonding strength, compressive strength and so on [17]. Kogbara et al. (2014) investigated the damaged microstructure of concrete due to cryogenic temperature including the effect of aggregate type and introduced the design method of a damage-resistance cryogenic concrete. To demonstrate the damage effects before and after freezing, acoustic emission (AE) and X-ray computed tomography (XRCT) methods were employed. The results indicated that microcracking resistance of concrete after the cryogenic condition was very related to the type of coarse aggregate [18]. Dahmari et al. (2007) mentioned the basic cause of concrete failure under cyclic freezing resulted from the transition to ice from free water in the pores and lead to the reduction in strength and structural damage [19]. Especially, Kwak et al. (2008) studied to measure the change in mechanical properties of concrete exposed to a specific temperature range from −20 to −60 °C [7]. To reveal the fracture properties at temperatures ranging from 20 to −170 °C, Rocco et al. (2001) conducted three-point bending tests on notched beams and determined the fracture parameters with the cohesive crack model, in terms of tensile strength, fracture energy, softening curve (stress vs. crack opening), characteristic length and modulus of elasticity [20]. Recently, for developing outer concrete, Kim et al. (2018) investigated the flexural and cracking behavior of ultra-high-performance fiber-reinforced concrete (UHPFRC) before and after exposure to cryogenic temperatures through four-point bending tests. The test results indicated that UHPFRC had higher resistance to microcrack formation and better flexural performance rather than normal concrete [21]. Moreover, Mazur et al. (2019) also carried out laboratory tests to reveal the negative effect under low temperature and suggest the improved ways of mix design with respect to decease in w/c ratio, type of cement and aggregate and use of aeration admixture [22].

As shown in Figure 1, concrete outer wall in LNG tank is generally designed as mass concrete with 1 m thick and more and most of LNG tanks are located in coastal area. Thus, this concrete outer wall can be easily exposed to chloride-rich environment. To enhance the concrete durability in corrosive environment, high-volume of GGBS (Ground Granulated Blast-furnace Slag) should be added. Based on ACI 233 and some references, basically, more than 50% of GGBS replacement in concrete mixture has an influence on the improvement of the durability and the reduction of the heat of hydration in mass concrete [23–27]. Rashad et al. (2017) and Rachel et al. (2019) investigated the mechanical and durability properties of the high-volume GGBS mixture with metakaolin and flyash. Even though the replacement of GGBS increased, test results of RCPT (Rapid Chloride Permeability Test), sorptivity and water permeability were lower than those of conventional concrete mixture [25,26]. Recently, Lee et al. (2020) evaluated the optimal CaO content range to secure the durability performance. As a result, the optimal CaO content was within range of about 55% and the replacement ratio of GGBS was about 50% [27]. Therefore, high-volume of GGBS replacement would be necessary to improve the durability performance of concrete outer tank installed on the coastal area.

Despite the progress of many types of research about the properties of cryogenic concrete, there are some limitations of the researches using the composition of high-volume GGBS binder in mix design. In addition, sufficient research results about the practical procedure of cryogenic tests have not been provided significantly with focus on the concrete properties. Therefore, the purpose of this study is to suggest the optimum mix design with a high volume of GGBS replacement and the procedure of the cryogenic test to consider mechanical and thermal properties, and durability performance based on the review of ACI 376 [28].

Above all, ACI 376 was reviewed to define the investigation items about mechanical and durability properties under cryogenic environment. Then, all raw materials used in mix design were tested to compare the test results with requirements in ASTM and BS codes. Particularly in this study, high-volume of GGBS and air entrainer admixture were used to the control of heat of hydration and durability for the increase of freeze-thawing resistance in accordance with emergency condition of LNG leakage. Two types of cryogenic conditions were employed, and various specimens were tested to measure the mechanical, thermal and durability properties. Finally, with mock-up specimens, productivity and semi-adiabatic tests were carried out.

2. Experimental Plan

2.1. Materials and Mix Design

All materials used in concrete followed the related standards in Table 1. Cement was combined with Ground Granulated Blast-furnace Slag (GGBS) in conformity to ASTM C 595 [29]. Table 2 shows the requirements for cement from ASTM C 150 and all test results satisfied with the requirements are provided by the supplier [30].

Table 1. Concrete Materials and Standards [30–35].

Materials	Related Standard
Cements	ASTM C 150
Mineral Admixtures (GGBS)	ASTM C 989
Water	ASTM C 94
Fine Aggregate	ASTM C 33
Coarse Aggregate	ASTM C 33
Chemical Admixture (HWRA)	ASTM C 494
Air Entraining Admixture	ASTM C 260

Table 2. Properties, requirements and test results for cement according to ASTM C 150 [30].

Cement—Type I Properties	Requirements	Test Result
Magnesium oxide (MgO)	Max. 6.0%	1.22%
Sulfur trioxide (SO_3)	Max. 3.0%	1.67%
Loss on ignition	Max. 3.0%	1.09%
Insoluble residue	Max. 0.75%	0.19%
Equivalent alkalies ($Na_2O + 0.658K_2O$)	Max. 0.6%	0.52%
Fineness (Air Permeability)	Min. 260 m^2/kg	315.7 m^2/kg
3 days Compressive strength	Min. 12 MPa	23.2 MPa
7 days Compressive strength	Min. 19 MPa	36.7 MPa
Time of setting—initial	Not less than 45 min	135 min
Time of setting—final	Not more than 375 min	170 min
Compressive strength—28 days	Min. 28 MPa	46.4 MPa

In order to achieve the strength and durability of the concrete, the replacement of GGBS as a mineral admixture is necessary. GGBS decreases a permeability of concrete and improves chemical resistance such as chlorides and sulfates. It also reduces the heat of hydration related to Delayed Ettringite Formation (DEF). The slag constituent shall not exceed 70% of the mass of total cementitious material in the concrete mix. Table 3 shows the requirements for GGBS from ASTM C 989 and all test results satisfy the requirements [31].

Table 3. Properties, requirements, and test results for GGBS according to ASTM C 989.

Properties	Requirements	Test Results
Amount retained when wet screened on a 45-µm (No. 325) sieve	Max. 20%	8.7%
Fineness by air permeability	No limit	546.7 m^2/kg
Air content of slag mortar	Max. 12%	6.5%
Activity Index—Grade 100—7 day	Min. 75%	93.2%
Activity Index—Grade 100—28 day	Min. 95%	109.1%
Sulfide sulfur (S)	Max. 2.5%	0.024%
Sulfate (SO$_3$)	N/A	0.97
Magnesium Oxide (MgO)	N/A	7.51
Acid soluble chloride ion content	N/A	0.01

Unwashed original sand contains many fine particles less than the sieve number 200 (0.075 mm) which induce more water and admixture consumption and then weaken the strength and durability of concrete. Sand as fine aggregate was washed at the washing plant before supplied for the test. The grading and the requirement of fine aggregates are shown in Figure 2 and Table 4, respectively. Coarse aggregates were washed gravels or crushed stones in accordance with Table 5. Coarse aggregates were combined with two types of single size, 20 mm and 10 mm and supplied from local providers. In Table 5, all test results for coarse aggregate were compared with the requirements.

In order to improve workability, strength and setting time, a high range and retarding super-plasticizer as a chemical admixture was used, complying with ASTM C 494. Air entrainer admixture shall comply with ASTM C 260. Micro-air 100 as an air entrainer admixture was used for improving a freeze-thaw resistance under a cryogenic environment. Mixing water was used without oil, acid, alkaline and organic matters or deleterious substances, complying with ASTM C 94 as shown in Table 6.

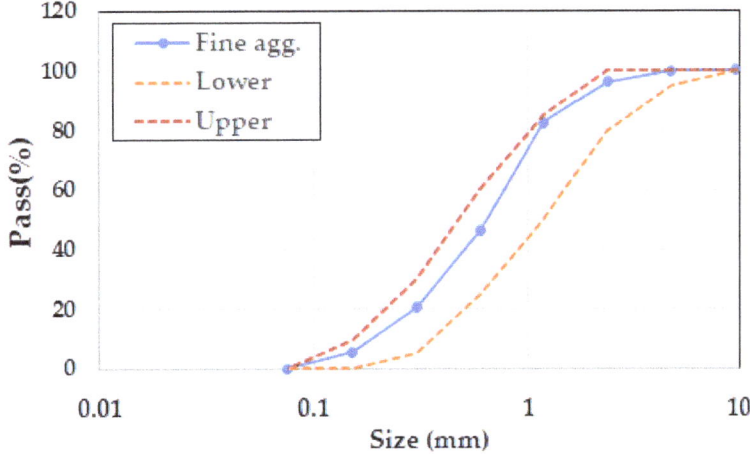

Figure 2. Grading of fine aggregate.

Table 4. Properties, requirements and test results for fine aggregates [36–42].

Properties	Requirements	Test Results	Related Standard
Clay lumps and friable particles	Max. 1.0%	0.5%	ASTM C 142
Coal and lignite	Max. 0.25%	0.2%	ASTM C 123
Material finer than 75-um	Max. 3.0%	2.7%	AASHTO T 11
Organic Impurities	Lighter than that of reference standard Color Solution		ASTM C 40
Specific gravity on saturated surface-dry basis	Min. 2.6 g/cm^3	2.613	ASTM C 128
Water soluble chloride ion content	Max. 0.01%	0.0096%	BS 812:Part 117
Acid soluble sulphate content as SO_3	Max. 0.3%	0.0553%	BS 812:Part 118

Table 5. Properties, requirements and test results for coarse aggregates [36,43–48].

Properties	Requirements	Test Results 20 mm	Test Results 10 mm	Related Standard
Clay lumps and friable particles	Max. 1.0%	Nil	Nil	ASTM C 142
Specific gravity on saturated surface-dry basis—calcareous	Min 2.65 g/cm^3	2.688	2.680	ASTM C 127
Water absorption	Max. 1.0%	0.5%	0.6%	AASHTO T 85
Los Angeles loss	Max. 30%	21.2%	21.8%	AASHTO T 96
Acid soluble chloride ion content	Max. 0.01%	0.0085%	0.0081%	AASHTO T 260
Acid soluble sulphate content as SO_3	Max. 0.4%	0.0115%	0.0168%	BS 812:Part 118
Soundness using Sodium sulphate	Max. 12%	2.4%	2.5%	ASTM C 88
Alkali reactivity	Max. 0.04%	Innocuous	Innocuous	ASTM C 1293

Table 6. Properties, requirements and test results for mixing water.

Properties	Requirements	Test Results
PH Value	6.0–8.0	7.0
Residue Content	6.0–7.6%	6.8%

2.2. Mix Design for Cryogenic Concrete

The specified compressive strength of cryogenic concrete was 40 MPa and target compressive strength was 50 MPa which was determined by adding 10 MPa to the specified compressive strength due to variations in materials, operations and testing. The maximum size of aggregate was 20 mm. The target slump and slump flow were 220 ± 25 mm and 620 ± 75 mm, respectively. The water/binder ratio was selected with 0.28, and total cementitious content was varied between 475 kg/m^3 to 495 kg/m^3. Water content was started from 128 to 133 kg/m^3 for cryogenic concrete. For water reduction, two types of high-range water-reducing chemical admixtures were applied: Daracem 208 (GCP applied technologies, Cambridge, MA, USA) as naphthalene type and Baxel PC 650 (Baxel, Sharq, Kuwait) as polycarboxylate type. Air content of 4 ± 1.5% was also achieved using a proper air-entraining admixture. Mix proportions are listed as shown in Table 7. Concrete materials were mixed by following ASTM C 192.

Table 7. Trial mix proportions of cryogenic concrete.

No.	W/B (%)	Unit Weight (kg/m³)							Admixture (Liter)		
		Water	Binder			Coarse Agg.		Fine Agg.	Type I	Type II	AE
			Total	OPC	GGBS	20 mm	10 mm	Sand			
C40-1	28	131	490	245 (50%)	245 (50%)	690	480	580	9.0–11.0	-	0.25–1.5
C40-2	28	133	495	175 (35%)	320 (65%)	680	470	580	9.0–11.0	-	0.25–1.5
C40-3	28	128	475	190 (40%)	285 (60%)	690	480	610	-	6.0–7.0	0.25–1.5
C40-4	28	128	475	166 (35%)	309 (65%)	690	480	610	-	6.0–7.0	0.25–1.5

Type I: Naphthalene type (Daracem 208), Type II: Polycarboxylate type (Baxel PC 650).

2.3. Preparation of SPECIMENS and Test Methods

For fresh concrete, air content and retention time of concrete slump were measured. For hardened concrete, mechanical properties such as compressive strength and elastic modulus were carried out. A total of 15 specimens were prepared in each mixture including reserved samples as shown in Table 8.

Table 8. Test method and concrete specimens.

No.	Fresh Concrete		Hardened Concrete				Reserved Samples	Total No. Samples
	Air Content (%)	Slump (mm)	Compressive Strength (MPa)					
			1 d	3 d	7 d	28 d		
C40-1	4 ± 1.5	Initial	3	3	3	3	3	15
C40-2		30 min.	3	3	3	3	3	15
C40-3		60 min. 90 min.	3	3	3	3	3	15
C40-5		120 min	3	3	3	3	3	15

For one-time cryogenic test (Test A method), a total of 38 specimens were prepared as referred to Table 9. Test A method consisted of five specified tests: compressive and tensile strength, elastic modulus, thermal expansion coefficient and thermal conductivity. Each test was conducted under four different temperature conditions. For freeze-thaw cyclic test (Test B method), a total of 12 specimens were cast as referred to Table 10. Test B method was conducted to investigate the compressive strength and elastic modulus after 50-times freeze-thaw cycles.

2.4. Cryogenic Test Methods

According to ACI 376, cryogenic concrete should be assessed the material properties as follows: compressive strength, elastic modulus, poisson's ratio, thermal conductivity and durability such as resistance to cycles of freezing and thawing [28]. With taking all properties into consideration, cryogenic tests were performed according to the procedure and criteria of one cycle of cryogenic temperature condition up to $-196\ °C$ (Test A method) and cyclic temperature condition between $5–-20\ °C$ (Test B method).

Table 9. Test method and concrete specimens under cryogenic condition.

Test A Method—One-Time Cryogenic Cycle	Curing (Day)	Temperature Conditions			
		Ambient (0 min)	−50 °C (15 min)	−120 °C (30 min)	−196 °C (60 min)
Compressive strength	28	3	3	3	3
Tensile strength	28	3	-	-	3
Elastic modulus	28	3	3	3	3
Thermal expansion coefficient	28	2	-	-	2
Thermal conductivity	28	2	-	-	2
Total	-	13	6	6	13

Table 10. Test method and concrete specimens after 50 freeze-thaw cycles.

Test B Method—50-Times Freeze/Thaw Cycles	Temperature Conditions		
	Curing (Day)	Ambient	5−−20 °C
Compressive strength after cycling	28	3	3
Elastic modulus after cycling	28	3	3
Total		6	6

2.4.1. Test A Method: One Cycle of Cryogenic Temperature Condition Up to −196 °C

The test specimens were subject to a single cycle at very low temperature and subsequently tested. The test results were compared with the concrete characteristic strength (40 MPa). The description and specimens of the test were specified as shown in Table 9. Test specimens were directly immersed into liquid nitrogen for 15, 30 and 60 min. at −196 °C and put out from the insulated storage. The cooled specimens were stored at the curing room (23 °C and RH 95%) where the temperature and moisture could be constantly controlled for 48 h. Reference specimens were kept in the curing room at the same time. After 2 days, the mechanical properties of cryogenic concrete were measured in terms of compressive strength, elastic modulus, poisson ratio, splitting tensile strength, length change for thermal expansion coefficient and thermal conductivity. For measuring the temperature automatically, two thermocouples were installed inside a spare specimen and one sensor was set up outside the spare specimen. The installed locations of thermocouples from the surface of a concrete sample were 25 mm and 75 mm, respectively. The temperature was recorded with a data logger and the temperatures of 15 min, 30 min and 60 min were corresponded to test specimens exposed to the cryogenic condition. The temperature recording was finished before immersing all test specimens into liquid nitrogen. As shown in Figure 3, the cryogenic tests were carried out as follows: compressive strength, tensile strength, elastic modulus, moisture content, length change and thermal conductivity.

Figure 3. Test A method under cryogenic condition.

2.4.2. Test B Method: Cyclic Temperature Condition between 5–−20 °C

In Figure 4, test specimens were subject to 50 cycles between 5–−20 °C and subsequently tested and test results were compared with the concrete characteristic strength (40 MPa). After finishing freezing and thawing, all samples were maintained in the curing chamber and crushed at the same time as the cooled specimen. The temperature change rate might not be greater than 10 °C per hour. The freeze-thaw cycling test was carried out according to the ASTM C 666. The cooling and heating procedure of the cycle met conditions defined in ASTM C 666. The temperature of specimens was monitored throughout the test used by the thermocouples. The cycled specimens were taken out from the chamber and cycled and un-cycled specimens were crushed at the same time. After finishing the compressive strength tests, the moisture content was determined with the crushed debris dried in an oven for 24 h.

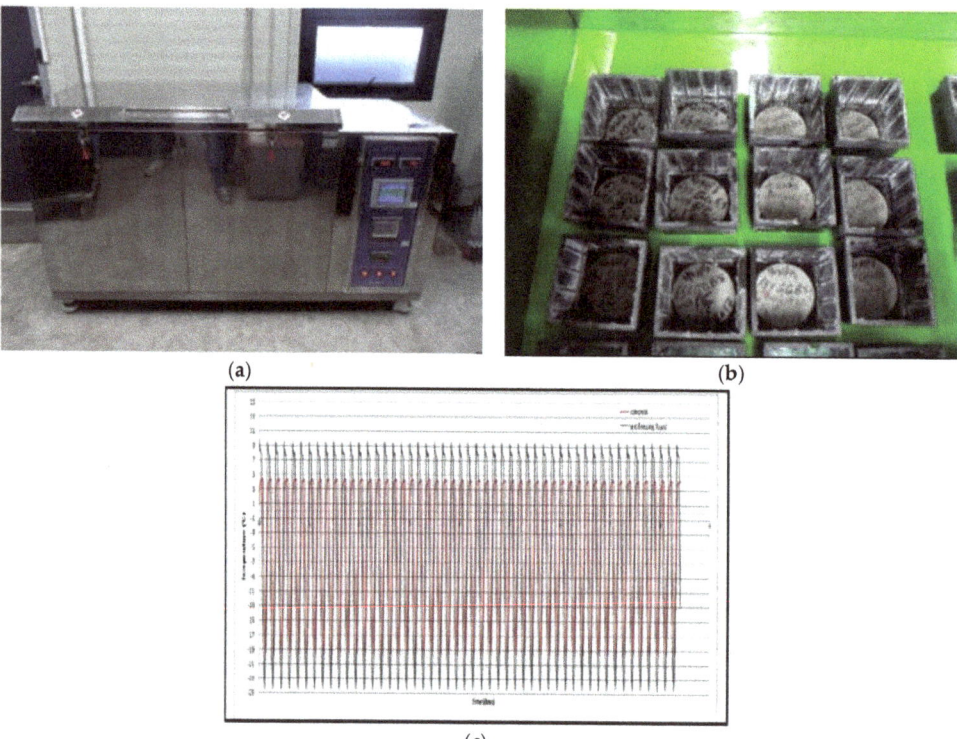

Figure 4. Test B method under cryogenic condition: (**a**) freezing-thawing equipment, (**b**) test set-up with specimens and (**c**) freeze-thaw cycles.

3. Test Results and Discussions

3.1. Selection of Optimum Mix Design with Slump and Compressive Strength

During 120 min after casting, slump tests with admixture Daracem 208 were performed every 30 min, and the testing value of cryogenic concrete was described in Figure 5. The initial slump of the C40-1 mixture exceeded the allowable tolerance, but the other slumps were in the range of it. On the other hand, the slump value of the C40-2 mixture was satisfied during the retention time of 120 min.

The chemical admixture of C40-3 and C40-4 was used with a polycarboxylate type (Baxel PC 650). The values of slump flow for 120 min were demonstrated in Figure 6. Based on the results, the initial slump flow of the C40-4 mixture with 65% GGBS was satisfied with the tolerance, but the C40-3 mixture with 60% GGBS was not. After 120 min, both of them were in the range of slump flow, 595 mm and 570 mm, respectively. These values were contained within the target range of 620 ± 75 mm. The tolerance of slump flow was within ±75 mm and the C40-4 mixture was more suitable rather than C40-3 one, in terms of retention time and workability.

All test samples were cast in accordance with ASTM C 172 and several times of compressive strength tests were conducted for 28 days [49]. The target compressive strength was determined to be more than 50 MPa, including a 10 MPa margin. As shown in Figure 7, the target strengths of all mix designs were sufficiently developed at the age of 7 days. All specimens at the age of 28 days already satisfied the compressive strength of more than 60 MPa. For describing in detail, the binder content of C40-1 and C40-2 was greater than that of C40-3 and C40-4 by 15 kg/m³, but the compressive strength of C40-1 and C40-2 was lower than that of C40-3 and C40-4. Therefore, with respect to the workability of fresh concrete, C40-2 and C40-4 were better than C40-1 and C40-3. In view of the development of compressive strength, C40-2 and C40-4 were superior to C40-1 and C40-3. In the next step for cryogenic tests, C40-2 and C40-4 mixture were chosen.

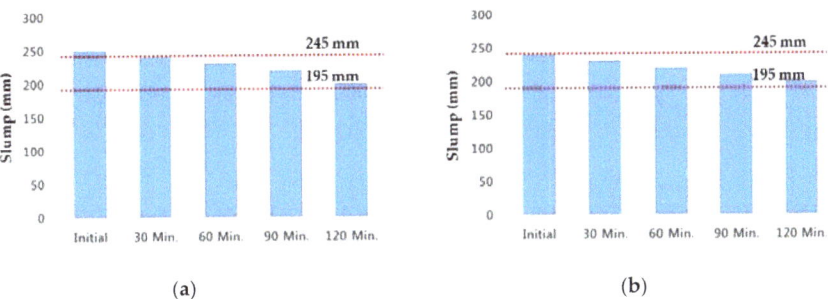

Figure 5. Slump test (**a**) C40-1 (GGBS 50%, Daracem 208) (**b**) C40-2 (GGBS 65%, Daracem 208).

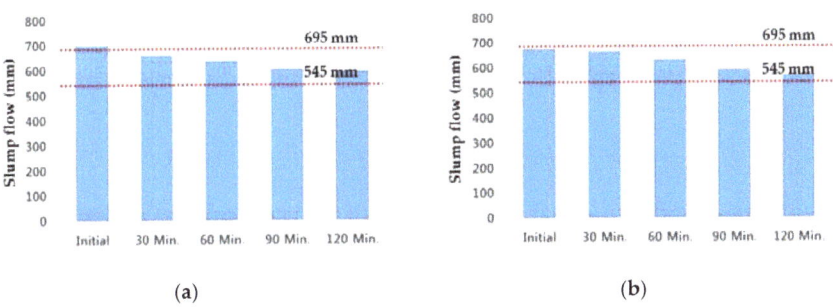

Figure 6. Slump flow test (**a**) C40-3 (GGBS 60%, Baxel PC 650) (**b**) C40-4 (GGBS 65%, Baxel PC 650).

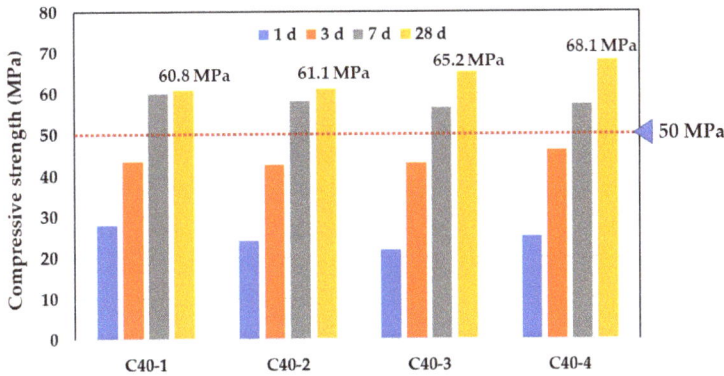

Figure 7. Compressive Strength of concrete mix design over time.

3.2. Mechanical, Thermal and Durability Properties under Cryogenic Condition

3.2.1. Mechanical and Thermal Properties after Exposed to a Cryogenic Condition

To figure out the characteristics of the concrete exposed to the cryogenic condition, firstly, two thermocouples in the specimen were installed as shown in Figure 8a. Then, the concrete specimen was put into liquid nitrogen to measure the temperature variation over time. As the result, the initial temperature was started from 22 ± 2 °C and after 45 min, the temperature was dropped down up to −190 ± 2 °C as shown in Figure 8b. The measured temperature variation was regarded as the temperature of the other specimens without thermo-couples. That is, after 15 min, the temperature of the concrete surface went down up to −100 °C. If the specimen was immersed in the liquid nitrogen storage for 15 min, it would be equally considered to be exposed to −100 °C.

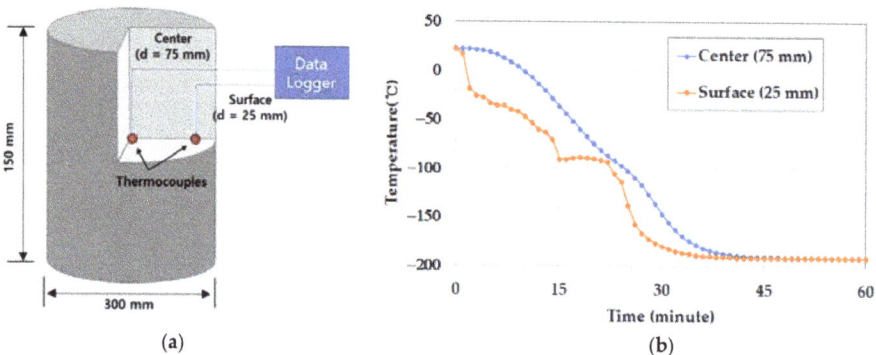

Figure 8. Thermocouple installation and temperature variation: (**a**) sample size and sensor location and (**b**) temperature variation over time.

In Table 11, the compressive strength and elastic modulus of concrete specimens under cryogenic temperature tended to be decreased over time. When the specimen was frozen under very low temperature, the expansion of ice crystal made some cracks in the capillary pores. Because the compression was loaded to concrete specimens after melting the ice crystal under ambient temperature, compressive strength of the specimens decreased. However, this trend was not the same as mentioned in ACI 376. That is, many studies had shown that compressive strength and elastic modulus increased as the temperature went down [28,50–53]. As the temperature decreased, the moisture contained in the capillary pores was changed into ice, and the internal structure of the concrete become tight and dense, resulting in an effect of increasing strength. The compressive strength of concrete exposure to cryogenic temperature rose up to about three times compared with ambient temperature. That is, the increment rate in compressive strength of concrete increased as the moisture content increased and the temperature decreased. In particular, below −120 °C, the deviation of compressive strength increased and the increment rate decreased. This was because the volume of ice crystals rapidly reduced around −120 °C or below [52].

Table 11. Test results of compressive strength, elastic modulus, poisson ratio and absorption.

Test Item	Immersed Time (min.)	Compressive Strength (MPa)	Elastic Modulus (GPa)	Poisson Ratio	Absorption (%)
C40-2	0 min. (ambient)	61.8	36.1	0.1483	3.57
	15 min.	52.8	35.1	0.1440	3.50
	30 min.	53.9	34.2	0.1437	2.28
	60 min.	53.9	33.9	0.1440	2.40
C40-4	0 min. (ambient)	62.17	39.8	0.1517	2.93
	15 min.	60.0	35.8	0.1437	2.17
	30 min.	57.9	35.7	0.1430	2.23
	60 min.	57.5	35.4	0.1437	2.21

In Table 11, the reason for the decline of mechanical properties was the expansion of volume. Freezing moisture expanded its volume by about 10%. The expanded volume of ice in the capillary pores caused pressure increase and the excess of the tensile strength of the pore walls impacted on the occurrence of cracks of concrete microstructure. Additionally, the degree of water saturation had a significant effect on the frost resistance of the concrete mix. As shown in Figure 9, the compressive strength and elastic modulus declined over time. That is, in the case of compressive strength, the strength reduction of the C40-2 mixture was about 15% in Figure 9a, but in the case of elastic modulus, 10% of reduction of the C40-4 mixture was observed in Figure 9b. This is because freezing moisture in concrete pores induced the cracks in the pore walls and this cracking resulted in the reduction of mechanical properties [50].

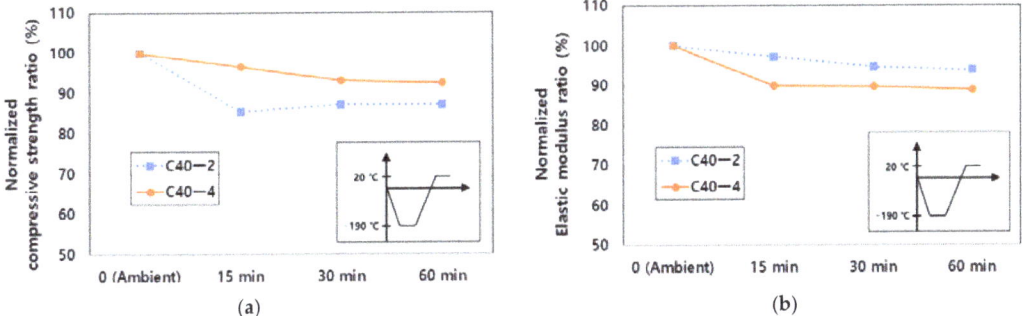

Figure 9. Normalized mechanical properties exposed to cryogenic temperature: (**a**) normalized compressive strength over temperature and (**b**) normalized elastic modulus over temperature.

For calculating the coefficient of temperature expansion (CTE), the length change of concrete specimens was measured as shown in Table 12. The CTE of cryogenic concrete was derived as shown in Equation (1) and the CTE of C40-2 and C40-4 was -1.503 and $-1.605 \times 10^{-6}/°C$, respectively.

$$C = \frac{(R_h - R_l)}{G \cdot \Delta T} \quad (1)$$

where, C = coefficient of linear thermal expansion of the concrete ($10^{-6}/°C$), R_h = length reading at higher temperature (mm), R_l = length reading at lower temperature (mm), G = gage length between inserts (mm) and ΔT = difference in temperature of specimen between the two length readings (°C).

Table 12. Test results of length change and splitting tensile strength.

Test Item	Immersed Time (min.)	Length Change (mm)	Splitting Tensile Strength (MPa)
C40-2	0 min. (Ambient)	0.0015	4.30
	60 min.	−0.063	3.44
C40-4	0 min. (Ambient)	0.0002	4.43
	60 min.	−0.0690	3.21

For the measure of the thermal conductivity, GHP (Guarded Hot Plate) 456 Titan manufactured by NETZSCH in Selb, Germany was used in Figure 10 and was employed with standardized guarded hot plate technique according to ASTM C 177 [54]. The temperature range of the equipment was −160–(+250) °C and the range of thermal conductivity was 0.003 to 2 W/(m·K). The two samples of each mix design were prepared and the size of it was a square with 300 mm sides and with 90 mm thick due to the measurement limit of equipment. GHP (Guarded Hot Plate) was based on the absolute measurement method without calibration and correction. The thermal conductivity value resulted in the stationary state and was derived from Equation (2) as follows:

$$\lambda = \frac{\dot{Q} \times d}{2A \times \Delta T} \quad (2)$$

 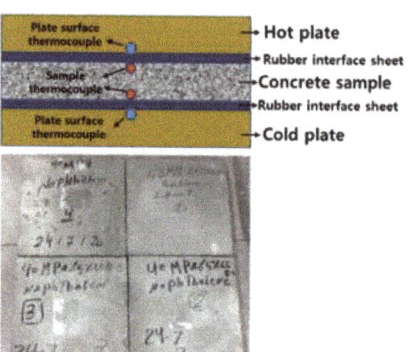

Figure 10. GHP (Guarded Hot Plate) 456 Titan and set-up of the sample, thermal sensors and plate.

In Equation (1), $\dot{Q}$ is precisely measured total power input into the hot plate, d is average sample thickness, A is measurement area and ΔT is mean temperature difference along the sample.

In ACI 376, the moisture content in concrete have an effect on the thermal conductivity. As temperature goes down, the thermal conductivity rises up linearly. In detail, the thermal conductivity of partially saturated normal-weight concrete increases from approximately 3.2 W/(m·K) at 25 °C to 4.71 W/(m·K) at −155 °C [28]. Table 13 demonstrated the thermal conductivity of concrete exposed to ambient and cryogenic temperature.

Table 13. Test results of thermal conductivity.

Test Item	Temperature (°C)	Thermal Conductivity (W/m·K)
		300 mm × 300 mm × 90 mm
C40-2	20 °C	1.512
	−160 °C	0.643
C40-4	20 °C	1.485
	−160 °C	0.723

As a result, the thermal conductivity of C40-2 and C40-4 at ambient temperature (20 °C) was about 1.5 W/(m·K) and that of C40-2 and C40-4 at very low temperature (−160 °C) was 0.643 and 0.723, respectively. This result was opposite to what ACI 376 mentioned. The factors affecting the thermal conductivity were the ratio of aggregate volume, water-cement ratio, moisture content and curing period. That is, as the volume fraction of aggregate and moisture content increased as well as water-cement ratio and curing period deceased, the thermal conductivity tended to be increased [55]. On the contrary, the test error could be decreased with the thicker specimen. For verifying this tendency, additional thermal conductivity tests were carried out with thinner sample as shown in Figure 11. As the sample thickness was decreased up to 50 mm, the thermal conductivity was down up to 50% or more.

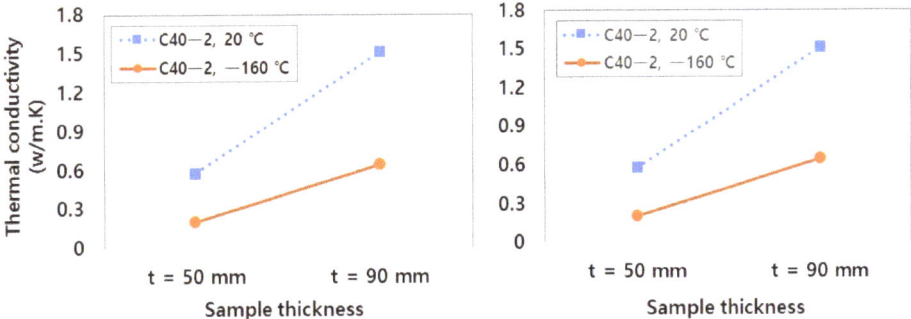

Figure 11. Effect of sample thickness on thermal conductivity.

3.2.2. Mechanical Properties after Exposed to Cyclic Low Temperature

The test B indicated that the cyclic temperature was repeated for 50 cycles in the range of 5 °C to −20 °C, in accordance with ASTM C 666. After exposed to the freeze-thaw conditions, the compressive strength and elastic modulus tests were carried out with two types of mix designs, C40-2 and C40-4. In Table 14, as the freeze-thaw cycles increased, the compressive strength and elastic modulus decreased because the volume expansion of ice crystal in the pores induced the microcracks under a low temperature [7,20]. Thus, the mechanical properties of the C40-4 mixture were better than those of the C40-2 mixture, in terms of the reduction rate of compressive strength and elastic modulus. The reduction rate of normalized mechanical properties was shown in Figure 12. In detail, the strength reduction was about 10%, and in the case of elastic modulus, the reduction was about 5% less.

Table 14. Test results of compressive strength, elastic modulus, poisson ratio and absorption.

Test Item	Freezing Thawing (Cycle)	Compressive Strength (MPa)	Elastic Modulus (GPa)	Poisson Ratio	Absorption (%)
C40-2	0	59.2	38.1	0.147	2.94
	50	53.3	36.5	0.142	3.42
C40-4	0	66.0	42.2	0.156	2.69
	50	60.7	41.5	0.150	3.01

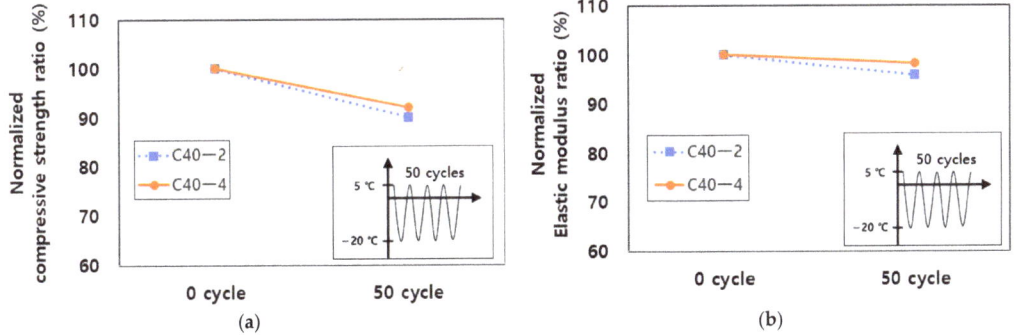

Figure 12. Normalized mechanical properties exposed to cyclic low temperature: (**a**) normalized compressive strength over F-T cycles and (**b**) normalized elastic modulus over F-T cycles.

3.3. Mock-Up Test for Semi-Adiabatic Temperature Monitoring

3.3.1. Preparation of Mock-Up Specimen

A mock-up test was performed to verify the heat of hydration of mass concrete. The two numbers of mock-up specimens were casted with the best optimum mixes for cryogenic concrete such as C40-2 and C40-4. Detailed sizes of Specimens were shown in Figure 13. The size of mock-up specimen was 2.0 m by 2.0 m by 1.5 m and the side surface of it was surrounded with insulation board of 200 mm thick. Concrete placing work and casting specimen had been taken for 30 min after produced. The vibrating works were carefully applied for good consolidation of poured concrete during concrete placing as shown in Figure 14. Temperature of fresh concrete was measured in accordance with AASHTO T 309 [56]. The initial and 30 min temperatures of concrete were controlled less than 32 °C because it was produced in summer season. Table 15 indicated that the slump (flow) and air content of both C40-2 and C40-4 mixtures were satisfied on the target requirement. A wooden form and polystyrene insulation were removed in 21 days after concrete placement. The top surface of concrete had cured with the moisture curing method such as wet blankets and plastic films.

3.3.2. Semi-Adiabatic Temperature Monitoring

All temperature sensors were installed and mock-up test was pretested before concrete pouring. A total of five temperature sensors (if required, two more spare sensors at the surface and center of specimens) were installed and positioned in concrete specimen at various depths and locations as shown in Figure 15. Temperatures of concrete specimen were measured every 30 min for the first 48 h and then every 1–2 h for 21 days. Ambient temperature was also recorded.

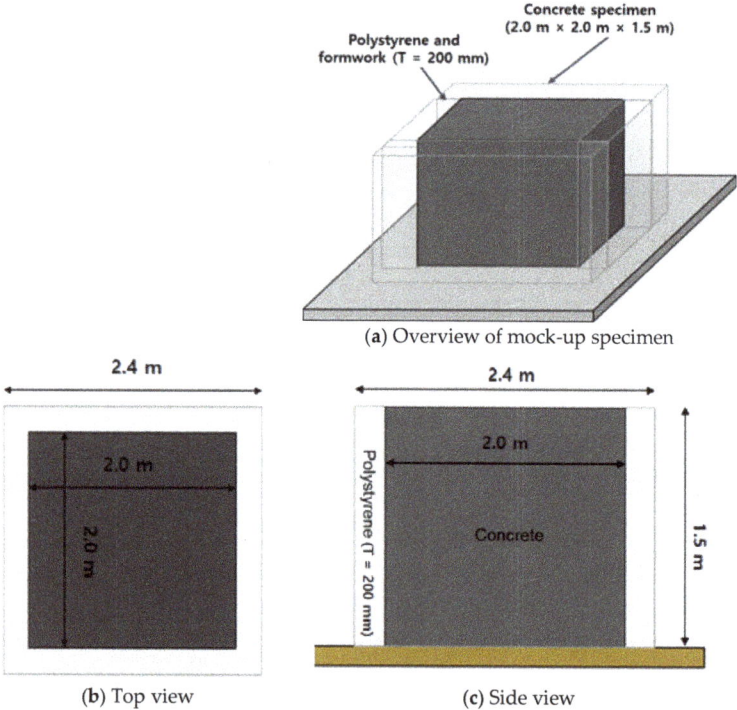

Figure 13. Dimension of Mock-up specimen.

(a) Placing concrete

(b) Vibrating concrete

Figure 14. Placing and vibrating of concrete.

Table 15. Test result of fresh concrete produced from batch plant.

Test Item	Temperature (°C)		Slump (Flow) (mm)		Air Content (%)		Density (kg/m^3)	
	Initial	30 Min.	Initial	30 Min.	Initial	30 Min.	Initial	30 Min.
C40-2	26.0	27.1	240	220	5.5	5.2	2303	2369
C40-4	28.9	29.2	650	620	4.5	4.0	2386	2393

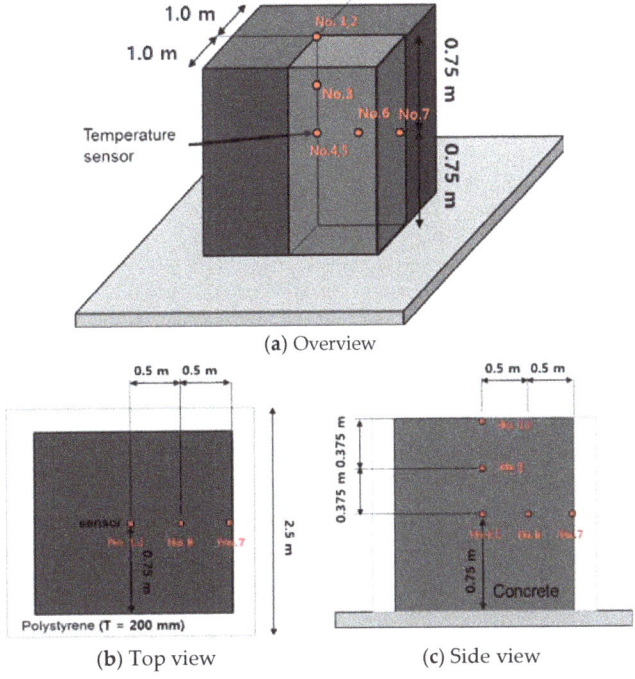

(a) Overview

(b) Top view

(c) Side view

Figure 15. Locations of temperature sensors.

The maximum temperature was captured in the center of the mock-up specimen and the temperature difference was measured in the center and surface of mock-up specimen. The maximum temperature of the hardened mass concrete usually occurred between 1 to 3 days after placement and then gradually decreased. According to ACI 308-16, ACI 207.1R and CS 163, in the case of blended cement, the maximum temperature should be controlled in the range of 70 °C and 85 °C, and the temperature difference should not exceed 19 °C [57–59]. Table 16 and Figures 16 and 17 show the measured temperature data. For C40-2 with Daracem 208 (naphthalene), the maximum temperature of the center location was 70.85 °C and the temperature difference between center and side surface was 21.85 °C. For C40-4 with Baxel PC 650, the maximum temperature of the center location was 70.8 °C and the temperature difference between center and side surface was 16.95 °C. The maximum temperatures of C40-2 and C40-4 were controlled by less than 75 °C, but the temperature difference of C40-2 did not satisfy the requirement with the exceed of 19 °C. Moreover, the binder amount of the C40-2 and C40-4 mixtures was applied with 495 kg/m^3 and 475 kg/m^3, respectively. That meant that the amount of binder for the C40-2 mixture was 20 kg/m^3 more than that of C40-4. With regard to the heat of hydration, C40-4 mixture was better to control the thermal cracks. Thus, the mix design of C40-4 (GGBS 65% with Baxel PC650) was more suitable for cryogenic concrete, in terms of workability, mechanical and thermal properties under cryogenic conditions and heat of hydration.

Table 16. Results of measured temperature data (replacement of GGBS 65%).

No.	Admixture	Temp. Max. at Center (°C)		Temp. Difference (°C)
		Surface	Center	
C40-2	Daracem 208	49.00	70.85	21.85
C40-4	Baxel PC 650	53.85	70.80	16.95

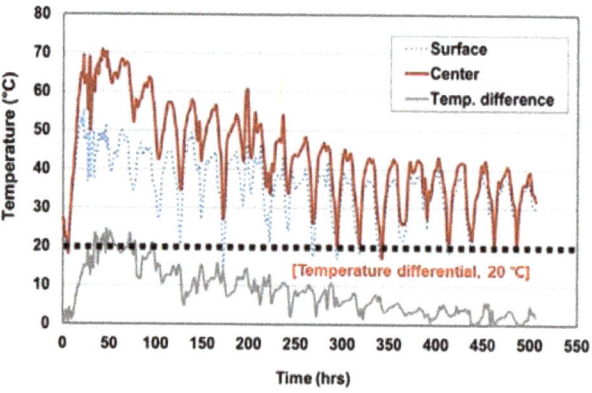

Figure 16. Cryogenic concrete with 65% GGBS and naphthalene type admixture.

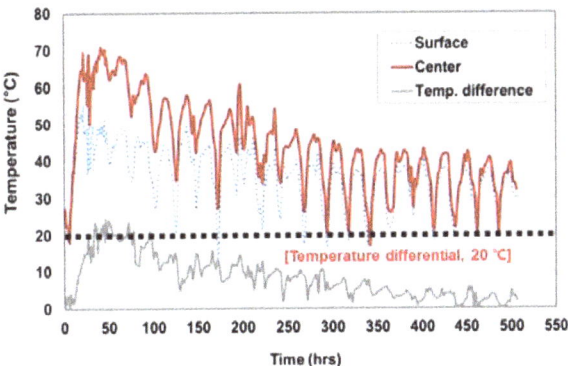

Figure 17. Cryogenic concrete with 65% GGBS and Polycarboxylate type admixture.

4. Conclusions

The purpose of this study is to suggest the optimum mix design with a high volume of GGBS replacement and the procedure of the cryogenic test to consider mechanical and thermal properties, and durability performance.

Above all, many research efforts including ACI 376 were reviewed to define the investigation items about mechanical and durability properties under cryogenic environment. Following this, all raw materials were tested to compare the test results with requirements. Particularly in this study, to the control of heat of hydration, the high-volume of GGBS replacement was adopted. For the improvement of freeze-thaw resistance, air entrainer admixture was used. With respect to the emergency condition such as LNG leakage, two types of cryogenic test methods were employed under one-cycle cryogenic condition (Test A) and 50-cycles cryogenic condition (Test B). Next, a mock-up test was conducted to find out the productivity and semi-adiabatic properties. The test results were summarized as below:

(1) With raw materials satisfied the requirements, four mix designs were suggested. To decide the optimum mix design, the slump and retention time of fresh concrete were investigated and the compressive strength of hardened specimens was measured. In this process, with respect to the workability of fresh concrete, C40-2 (GGBS 65% with Daracem 208) and C40-4 (GGBS 65% with Baxel PC650) were better than C40-1 (GGBS 50% with Daracem 208) and C40-3 (GGBS 60% with Baxel PC650). In view of the development of compressive strength, C40-2 and C40-4 were superior to C40-1 and C40-3. In the next step for cryogenic tests, the C40-2 and C40-4 mixture were selected.

(2) After one cycle of cryogenic temperature, the compressive strength and elastic modulus of the C40-2 and C40-4 mixtures tended to be decreased over time, because of the volume expansion of ice crystals contained in the capillary pores. In addition, the degree of water saturation had a significant effect on the frost resistance of the concrete mix.

(3) After exposed to the 50-times freeze-thaw cycles, the compressive strength and elastic modulus tests were carried out the mechanical properties of the C40-4 mixture (GGBS 65% with Baxel PC650) were better than those of the C40-2 mixture (GGBS 65% with Daracem 208), in terms of the reduction rate of compressive strength and elastic modulus. In detail, the strength reduction rate was about 10%, and in the case of elastic modulus, the reduction was about 5% less.

(4) The maximum temperatures of C40-2 and C40-4 were controlled by less than 75 °C, but the temperature difference of C40-2 did not satisfy the requirement with the exceed of 19 °C. Moreover, the binder amount of the C40-2 and C40-4 mixtures was applied with 495 kg/m^3 and 475 kg/m^3, respectively. That meant that the amount of

binder for the C40-2 mixture was 20 kg/m^3 more than that of C40-4. With regard to the heat of hydration, the C40-4 mixture was better to control the thermal cracks.

Thus, the mix design of C40-4 (GGBS 65% with Baxel PC650) was more suitable for cryogenic concrete, in terms of workability, mechanical and thermal properties under cryogenic conditions and heat of hydration. This test procedure would be helpful to select the better cryogenic mix design and to define the trend of mechanical, thermal and durability properties and test methods. In the future, the long-term performance of cryogenic concrete needs to be investigated.

Author Contributions: Conceptualization, G.L. and O.N.; Data curation, O.N.; Formal analysis, O.N.; Funding acquisition, G.L.; Investigation, G.L. and O.N.; Methodology, G.L. and O.N.; Supervision, G.L. and O.N.; Validation, G.L. and O.N.; Visualization, O.N.; Writing—original draft, G.L.; Writing—review & editing, O.N. All authors have read and agreed to the published version of the manuscript.

Funding: This research received no external funding.

Institutional Review Board Statement: Not applicable.

Informed Consent Statement: Not applicable.

Data Availability Statement: Data sharing is not applicable to this article.

Conflicts of Interest: The authors declare no conflict of interest.

References

1. Fourlis, A. *Energy Outlook*; Energy Outlook 2021 Report; Brunel: Amsterdam, The Netherlands, 2021.
2. Kim, D.H. A Study on the Economic Analysis of the LNG Terminal Vaporizer System. Master's Thesis, Hanyang University, Seoul, Korea, 2019.
3. Thierçault, J. Cryogenic Above Ground Storage Tanks: Full Containment and Membrane Comparison of Technologies. 2018. Available online: https://www.gti.energy/wp-content/uploads/2018/12/Storage-2-Jerome_Thiercault-LNG17-Poster.pdf (accessed on 22 April 2021).
4. Han, S.M.; Song, Y.C.; Lee, D.H.; Park, K.S.; Miura, T. Property and Applied Technology of Concrete at Very Low Temperature. *Mag. KCI* **2005**, *17*, 10–17.
5. Han, S.M.; Cho, M.S.; Song, Y.C. The Influence of Storage at Very Temperatures on the Deterioration of Concrete. *Doboku Gakkai Ronbunshu* **2002**, *704*, 13–25. [CrossRef]
6. Jeon, S.J.; Kim, Y.J.; Chung, C.H.; Jin, B.M.; Kim, S.W. A Study on the Liquid Tightness Design of LNG Tank Incorporating Cryogenic Temperature-induced Stresses. In Proceedings of the KCI Spring Conference, Chungju, Korea, 3 May 2003; pp. 128–131.
7. Kwak, J.H.; Nam, J.H.; Lee, S.S.; Lee, Y.; Yoon, S.J. An Experimental Study on the Compression Behavior of Concrete Experienced Extremely Low Cyclic Temperature. In Proceedings of the KSCE Conference, Daejeon, Korea, 30–31 October 2008; pp. 429–432.
8. Kwon, B.G.; Lee, S.L. History of Standard Development for LNG Storage Tank and Recent Trend. *Korea Gas Saf. Coop.* **2002**, *7*, 26–32.
9. Kwon, B.G.; Lee, S.L. Standard Development of Full Containment LNG Storage Tank. *Korea Gas Saf. Coop.* **2002**, *8*, 26–34.
10. Collins, C.; Patel, D.A.; Tarlowski, J. Developments in LNG Storage: Overview of LNG Storage Tanks. In Proceedings of the LNG 2000 Conference, London, UK, 16–17 February 2000.
11. Lee, S.L. Risk Comparison of Risk Evaluation for LNG Storage Tank. *Gas Saf. J.* **2011**, *2*, 14–19.
12. Jeon, S.J.; Jin, B.M.; Kim, Y.J. Design Basis for Large Above-ground LNG Tank. *Daewoo Constr. Technol. Rep.* **2004**, *26*, 19–29.
13. Hoyle, K.; Oliver, S.; Tsai, N. Composite Concrete Cryogenic Tank (C3T): A Precast Concrete Alternative for LNG Storage. In Proceedings of the 17th International Conference & Exhibition on Liquefied Natural Gas (LNG 17), Houston, TX, USA, 16–19 April 2013.
14. Jeon, S.J.; Jin, B.M.; Kim, Y.J. A Study on the Method for Shortening the Construction Period of LNG Storage Tanks. In Proceedings of the KCI Spring Conference, Jeju, Korea, 14–16 May 2014; pp. 931–932.
15. Cheol, J.H.; Hwi, K.J.; Won, L.K.; Hyun, O.S.; Gwan, H.H.; Mook, L.Y. Feasibility Study for Fast Constructing LNG Storage Tank Using Precast Concrete Method. In Proceedings of the KSCE Conference, Kunsan, Korea, 28–30 October 2015.
16. Yoon, S.I. Application of Cryogenic Re-bar for LNG Storage Outer Tank. In Proceedings of the KCI Spring Conference, Gyeongju, Korea, 2–4 May 2012; pp. 627–628.
17. Kogbara, R.B.; Iyengar, S.R.; Grasley, Z.C.; Rahman, S.; Masad, E.A.; Zollinger, D.G. Relating damage evolution of concrete cooled to cryogenic temperatures to permeability. *Cryogenics* **2014**, *64*, 21–28. [CrossRef]
18. Kogbara, R.B.; Iyengar, S.R.; Grasley, Z.C.; Masad, E.A.; Zollinger, D.G. A review of concrete properties at cryogenic temperatures: Towards direct LNG containment. *Constr. Build. Mater.* **2013**, *47*, 760–770. [CrossRef]

19. Dahmani, L.; Khenane, A.; Kaci, S. Behavior of the reinforced concrete at cryogenic temperatures. *Cryogenics* **2007**, *47*, 517–525. [CrossRef]
20. Rocco, C.; Planas, J.; Guinea, G.V.; Elices, M. Fracture Properties of Concrete in Cryogenic Conditions. In Proceedings of the Fracture Mechanics of Concrete Structures 4, Cachan, France, 28 May–1 June 2001; pp. 411–416.
21. Kim, S.; Kim, M.-J.; Yoon, H.; Yoo, D.-Y. Effect of cryogenic temperature on the flexural and cracking behaviors of ultra-high-performance fiber-reinforced concrete. *Cryogenics* **2018**, *93*, 75–85. [CrossRef]
22. Mazur, B.; Kotwa, A. Influence of Low Temperature on Concrete Properties. *IOP Conf. Ser. Mater. Sci. Eng.* **2019**, *471*, 032026. [CrossRef]
23. Divsholi, B.S.; Lim, T.Y.D.; Teng, S. Durability Properties and Microstructure of Ground Granulated Blast Furnace Slag Cement Concrete. *Int. J. Concr. Struct. Mater.* **2014**, *8*, 157–164. [CrossRef]
24. Karri, S.K.; Rao, G.V.R.; Raju, P.M. Strength and Durability Studies on GGBS Concrete. *Int. J. Civ. Eng.* **2015**, *2*, 34–41. [CrossRef]
25. Rashad, A.M.; Sadek, D.M. An investigation on Portland cement replaced by high-volume GGBS pastes modified with micro-sized metakaolin subjected to elevated temperatures. *Int. J. Sustain. Built Environ.* **2017**, *6*, 91–101. [CrossRef]
26. Rachel, P.P. Experimental Investigation on Strength and Durability of Concrete using High Volume Flyash, GGBS and M-Sand. *Int. J. Res. Appl. Sci. Eng. Technol.* **2019**, *7*, 396–403. [CrossRef]
27. Lee, J.; Lee, T. Durability and Engineering Performance Evaluation of CaO Content and Ratio of Binary Blended Concrete Containing Ground Granulated Blast-Furnace Slag. *Appl. Sci.* **2020**, *10*, 2504. [CrossRef]
28. ACI 376 Committee. *Code Requirements for Design and Construction of Concrete Structures for the Containment of Refrigerated Liquefied Gases and Commentary*; American Concrete Institute (ACI): Farmington Hills, MI, USA, 2011.
29. ASTM C 595. *Standard Specification for Blended Hydraulic Cements*; ASTM International: West Conshohocken, PA, USA, 2020.
30. ASTM C 150. *Standard Specification for Portland Cement*; ASTM International: West Conshohocken, PA, USA, 2020.
31. ASTM C 989. *Standard Specification for Slag Cement for Use in Concrete and Mortars*; ASTM International: West Conshohocken, PA, USA, 2018.
32. ASTM C 94. *Standard Specification for Ready-Mixed Concrete*; ASTM International: West Conshohocken, PA, USA, 2021.
33. ASTM C 33. *Standard Specification for Concrete Aggregates*; ASTM International: West Conshohocken, PA, USA, 2018.
34. ASTM C 494. *Standard Specification for Chemical Admixtures for Concrete*; ASTM International: West Conshohocken, PA, USA, 2019.
35. ASTM C 260. *Standard Specification for Air-Entraining Admixtures for Concrete*; ASTM International: West Conshohocken, PA, USA, 2016.
36. ASTM C 142. *Standard Test Method for Clay Lumps and Friable Particles in Aggregates*; ASTM International: West Conshohocken, PA, USA, 2017.
37. ASTM C 123. *Standard Test Method for Lightweight Particles in Aggregate*; ASTM International: West Conshohocken, PA, USA, 2014.
38. AASHTO T 11. *Standard Method of Test for Materials Finer Than 75-μm (No. 200) Sieve in Mineral Aggregates by Washing*; Association of State Highway and Transportation Officials: Washington, DC, USA, 2020.
39. ASTM C 40. *Standard Test Method for Organic Impurities in Fine Aggregates for Concrete, ASTM International*; ASTM International: West Conshohocken, PA, USA, 2020.
40. ASTM C 128. *Standard Test Method for Relative Density (Specific Gravity) and Absorption of Fine Aggregate*; ASTM International: West Conshohocken, PA, USA, 2015.
41. BS 812:Part 117. *Testing Aggregates. Method for Determination of Water-Soluble Chloride Salts*; British Standards Institution: London, UK, 1988.
42. BS 812:Part 118. *Testing Aggregates. Methods for Determination of Sulphate Content*; British Standards Institution: London, UK, 1988.
43. ASTM C 127. *Standard Test Method for Relative Density (Specific Gravity) and Absorption of Coarse Aggregate*; ASTM International: West Conshohocken, PA, USA, 2015.
44. AASHTO T 85. *Standard Method of Test for Specific Gravity and Absorption of Coarse Aggregate*; Association of State Highway and Transportation Officials: Washington, DC, USA, 2014.
45. AASHTO T 96. *Standard Method of Test for Resistance to Degradation of Small-Size Coarse Aggregate by Abrasion and Impact in the Los Angeles Machine*; Association of State Highway and Transportation Officials: Washington, DC, USA, 2002.
46. AASHTO T260. *Standard Method of Test for Sampling and Testing for Chloride Ion in Concrete and Concrete Raw Materials*; Association of State Highway and Transportation Officials: Washington, DC, USA, 2005.
47. ASTM C 88. *Standard Test Method for Soundness of Aggregates by Use of Sodium Sulfate or Magnesium Sulfate*; ASTM International: West Conshohocken, PA, USA, 2018.
48. ASTM C 1293. *Standard Test Method for Determination of Length Change of Concrete Due to Alkali-Silica Reaction*; ASTM International: West Conshohocken, PA, USA, 2020.
49. ASTM C 172. *Standard Practice for Sampling Freshly Mixed Concrete*; ASTM International: West Conshohocken, PA, USA, 2017.
50. Van de Veen, V. *Properties of Concrete at Very Low Temperatures*; Report; Delft University of Technology: Delft, The Netherlands, 1987.
51. Bamforth, P.B. The Structural Permeability of Concrete at Cryogenic Temperatures. Ph.D. Thesis, University of Aston, Birmingham, UK, 1987.
52. Kim, S.B.; Kim, D.H. Behavior of Concrete at Very Low Temperature. *Mag. KCI* **1997**, *9*, 31–41.
53. Han, S.M.; Cho, M.S.; Song, Y.C. The Influence of Storage at Very Low Temperatures on the Deterioration of Concrete. In Proceedings of the KCI Fall Conference, Seoul, Korea, 6–8 November 2002; pp. 931–932.

54. ASTM C 177. *Standard Test Method for Steady-State Heat Flux Measurements and Thermal Transmission Properties by Means of the Guarded-Hot-Plate Apparatus*; ASTM International: West Conshohocken, PA, USA, 2019.
55. Kim, K.-H.; Jeon, S.-E.; Kim, J.-K.; Yang, S. An experimental study on thermal conductivity of concrete. *Cem. Concr. Res.* **2003**, *33*, 363–371. [CrossRef]
56. AASHTO T 309. *Standard Method of Test for Temperature of Freshly Mixed Portland Cement Concrete*; Association of State Highway and Transportation Officials: Washington, DC, USA, 2020.
57. ACI Committee 308. *Guide to External Curing of Concrete*; ACI 308: Farmington Hills, MI, USA, 2016.
58. ACI Committee 207. *Guide to Mass Concrete*; ACI 207.1R: Farmington Hills, MI, USA, 2005.
59. The Concrete Society. *Guide to the Design of Concrete Structures in the Arabian Peninsula*; CS 163: Camberley, UK, 2008.

Article

Reduced Carbonation, Sulfate and Chloride Ingress Due to the Substitution of Cement by 10% Non-Precalcined Bentonite

Carmen Andrade [1], Ana Martínez-Serrano [2], Miguel Ángel Sanjuán [3] and José Antonio Tenorio Ríos [2,*]

1 International Center of Numerical Methods in Engineering (CIMNE)-UPC, 28010 Madrid, Spain; candrade@cimne.upc.edu
2 Institute of Construction Sciencies "Eduardo Torroja"-CSIC, 28033 Madrid, Spain; ana.martinez@ietcc.csic.es
3 Institute of Cement and Its Applications (IECA), 28003 Madrid, Spain; masanjuan@ieca.es
* Correspondence: tenorio@ietcc.csic.es

Citation: Andrade, C.; Martínez-Serrano, A.; Sanjuán, M.Á.; Tenorio Ríos, J.A. Reduced Carbonation, Sulfate and Chloride Ingress Due to the Substitution of Cement by 10% Non-Precalcined Bentonite. *Materials* **2021**, *14*, 1300. https://doi.org/10.3390/ma14051300

Academic Editor: Alessandro P. Fantilli

Received: 13 February 2021
Accepted: 3 March 2021
Published: 8 March 2021

Publisher's Note: MDPI stays neutral with regard to jurisdictional claims in published maps and institutional affiliations.

Copyright: © 2021 by the authors. Licensee MDPI, Basel, Switzerland. This article is an open access article distributed under the terms and conditions of the Creative Commons Attribution (CC BY) license (https://creativecommons.org/licenses/by/4.0/).

Abstract: The Portland cement industry is presently deemed to account for around 7.4% of the carbon dioxide emitted annually worldwide. Clinker production is being reduced worldwide in response to the need to drastically lower greenhouse gas emissions. The trend began in the nineteen seventies with the advent of mineral additions to replace clinker. Blast furnace slag and fly ash, industrial by-products that were being stockpiled in waste heaps at the time, have not commonly been included in cements. Supply of these additions is no longer guaranteed, however, due to restrained activity in the source industries for the same reasons as in clinker production. The search is consequently on for other additions that may lower pollutant gas emissions without altering cement performance. In this study, bentonite, a very common clay, was used as such an addition directly, with no need for precalcination, a still novel approach that has been scantly explored to date for reinforced structural concrete with structural applications. The results of the mechanical strength and chemical resistance (to sulfates, carbonation and chlorides) tests conducted are promising. The carbonation findings proved to be of particular interest, for that is the area where cement with mineral additions tends to be least effective. In the bentonite-bearing material analysed here, however, carbonation resistance was found to be as low as or lower than that observed in plain Portland cement.

Keywords: cement; bentonite; durability; clays

1. Introduction

The inordinate rise in the presence of greenhouse gases in the atmosphere is creating a pressing need to lower CO_2 emissions by, among other methods, reducing the proportion of clinker in cement [1–3]. The Portland cement industry is presently deemed to account for around 7.4% of the close to 2.9 Gt of carbon dioxide emitted annually worldwide (value for 2016) [4]. In light of such facts, the industry is assessing the measures that could be taken to ensure zero emissions by 2050. Under review are the processes involved in clinker, cement and concrete production, construction procedures and cement-based material carbonation during and after service life [5]. As the mitigation technologies presently in place are believed to be insufficient to hit the net zero carbon target by 2050, innovative measures are called for, including carbon dioxide capture, utilization and storage (CCUS) [6] and flameless mineral calcination systems. One new proposal for the latter, the tube-in-tube helical method, features use in concentrated solar power plants [7].

In the past, clinker content has been replaced with industrial waste such as fly ash, slag or silica fume with no adverse effect on concrete mechanical performance or durability [8]. With the abatement of the likewise carbon-intensive source industries, however, the availability of those substitute mineral additions is beginning to wane. Although the contribution of coal power plant-fueled energy to total consumption declined in Spain from 20.2% in 1990 to 9.8% in 2017 thanks to the growing use of alternative energies [9], coal continued to supply 38.5% of world demand in 2018. Concerns about greenhouse gas emissions cloud the future of coal, however, for it is pivotal to the debate on energy and climate policy. A number of countries, committed to net-zero greenhouse gas (GHG) emission targets by 2050, have established an end date for coal power generation. In others, however, coal plays a key role in the supply of affordable energy [10]. That notwithstanding, coal's share in the global power mix is expected to decline by 10% by 2050, with a concomitant downturn in the stock of fly ash. Hence, it is imperative to seek replacements for clinker in nature to broaden the spectrum of alternative materials. The study described hereunder explored the use of non-precalcined bentonite, a widely available clay, as one such alternative.

As an anionic clay present in nature, bentonite can be obtained at low cost. It is also highly water absorbent and thixotropic (gel-like when vibrated). It was first discovered in 1988 in the United States and more specifically at Fort Benton, Wyoming [11], after which it is named. Its composition consists primarily of magnesium silicate, montmorillonite and aluminium hydrate, the third in the form of colloid-sized crystallites. Each individual montmorillonite crystal, in turn, comprises an octahedral layer of aluminium sandwiched between two tetrahedral layers of silicon. It carries a negative charge associated with isomorphic substitutions, such as Al^{3+} for Mg^{2+}, in the crystallite network, which is offset by exchangeable alkaline metal cations [11].

Clay use as a mineral addition in cement is nothing new, although in most cases subject to precalcination [12]. A number of studies [13–15] have recently reported promising results around its application to replace more conventionally used additions such as fly ash or slag. Few studies on its non-precalcined use have been found in the literature, however, for that approach has consistently posed rheological problems [16–26]. That would explain why many building codes limit the presence of clay materials in aggregates and the much more common use of these materials in foundations and soil stabilisation than for structural applications [27–29]. Nonetheless, today's admixtures afford fresh concrete properties impossible to attain in the past, and modern laboratory techniques now in place can substantially shorten the time needed to design an optimal mix [30–33]. Non-precalcined bentonite has seldom been used to date in lieu of more conventional mineral additions [16–26]. This study was therefore designed to study the hydration mechanisms involved [19,24,26], fresh concrete properties such as flowability and bleeding [18,25], bentonite reactivity [19] and concrete water permeability [17,20,22] and compressive strength [16,25]. Durability was studied in terms of the replacement's effect on concrete resistance to freeze–thaw cycles and acid or sulfate attack [23]. Bentonite was reported to have a beneficial impact on preventing reinforcement corrosion [20] and carbonation resistance [21], although only one paper on each subject was located in the literature.

A review of the literature on the long-term performance of bentonite showed that it has exhibited excellent durability in underground works, where it has been used profusely as permanent formwork in concrete foundations often built long ago [27–29]. Additionally, whilst bentonite plays a non-calculated load-bearing role in such cases, none of the studies published report any long-term incompatibility between the two materials. It is likewise used in conjunction with concrete to generate impermeable slurry walls in highly radioactive waste storage facilities (designed to last for thousands of years) [34–36], where the caverns holding the radioactive waste are shotcreted and the encapsulated waste itself lies on a bed of bentonite. The use of such systems has given rise to research on how the alkaline nature of concrete may affect the long-term stability of bentonite clays. Such studies have verified the interaction between cement alkalinity and bentonite phases [34–37] or, equivalently, phase reactivity with clinker hydrated phases. From the standpoint of the role of clay as a clinker replacement, the findings have been initially promising thanks to the slow reactivity afforded by the high alkalinity of the pore solution, which dissolves the silicoaluminates in the bentonite. In terms of radioactive waste, such a result would be detrimental, however, if it affected the stability of the shotcrete/bentonite interface, given the many thousands of years they are intended to be in contact.

Standardised active additions such as fly ash, natural pozzolans, slag and silica fume, in turn, are known to effectively inhibit chloride and sulfate ingress [38,39], enhancing durability, although their presence in concrete impacts carbonation resistance adversely [13,16]. That is significant, for any decline in carbonate resistance is a primary long-term concern because it favours reinforcement corrosion [40–42] and the associated economic loss. The importance of seeking additions that either favour or at least are not detrimental to concrete durability cannot, therefore, be overstated.

Eluding the extra cost and additional handling involved in precalcining clays at temperatures of up to 1000 °C would carry obvious advantages. The present study consequently aims to explore the physical-mechanical properties and durability of concrete prepared with a non-precalcined bentonite as a substitute for clinker at different (wt/wt) replacement ratios.

This study sought to determine how replacing up to 30% clinker with non-precalcined bentonite may affect mortar mechanical properties and how carbonation depth and chloride and sulfate diffusion may be impacted by the presence of 10% of the clay with different types of binders bearing mineral additions. The findings are highly promising, particularly as regards carbonation, the weak point observed in other mineral additions. In the tests conducted, carbonation resistance either remained essentially unchanged or improved in the mortar prepared with the blended cement relative to the reference material. Plain Portland cement or cement bearing standardised additions was used throughout [43].

2. Materials and Methods

2.1. Materials

All the cements listed in Table 1 were used in the carbonation tests, whereas cement CEM I 52.5 R-SR 3 served as the basis for the mechanical strength and sulfate and chloride diffusion trials.

Table 1. Chemical composition of the cements used in the tests.

Cement	SiO_2	Al_2O_3	Fe_2O_3	CaO	MgO	SO_3	Na_2O	K_2O	LOI	IR	Cl^-
CEM I 42.5 R	20.24	3.99	2.92	62.88	1.41	3.47	0.08	0.86	2.78	0.10	0.02
CEM I 52.5R–SR 3	21.73	3.67	4.31	66.12	1.32	3.00	0.49	0.57	1.12	0.19	0.01
CEM II/A-P 42.5 R	29.09	5.29	2.98	51.60	1.78	2.82	0.39	0.53	2.94	-	0.06
CEM II/A-P 42.5 R	28.17	6.20	3.13	52.41	1.32	3.11	0.38	0.67	2.34	–	0.05
CEM II/A-S 42.5 N	23.24	5.74	2.46	61.82	2.29	2.83	0.46	0.59	-	-	0.05
CEM II/A- V 42.5 R	23.00	6.30	3.50	58.00	1.42	3.22	0.49	0.80	2.30	2.10	0.06
CEM III/A 42.5 N	24.55	6.42	2.14	57.14	3.00	2.80	0.40	0.50	0.91	0.21	0.05
CEM IV/A (V) 42.5 R-SR	27.14	5.25	3.20	53.10	1.58	2.82	0.37	0.49	2.70	-	0.06
IV/A(P) 42.5 R/MR	28.36	4.72	3.17	52.81	2.16	2.59	0.33	0.51	2.43	-	0.04
BL II/B-LL 42.5 R	18.70	3.83	2.64	61.70	1.29	2.97	0.06	0.81	10.86	0.33	0.02

A commercial bentonite (Mapeproof Seal), distributed by Mapei for purposes other than those studied here, was used to ensure consistent composition and particle size distribution throughout. According to the vendor's specifications sheet, the material contained over 95 wt % montmorillonite.

The particle size distribution and volume density curves were found by analysing bentonite powder on a Mastersizer 3000 laser diffractometer diffractor (Malvern Panalytical, Madrid, Spain) (Figure 1). Ninety per cent of the particles were <86 µm, whilst most lay within the 10 to 15 µm range (see the volume density curve). Such greater fineness than observed for the cement was initially deemed suitable, although optimisable. Addition fineness plays a significant role in the strength and rheology of composite cements, for the distribution curves for those materials complement the curve for the cement itself. This parameter must consequently be analyzed in depth in future research [43,44].

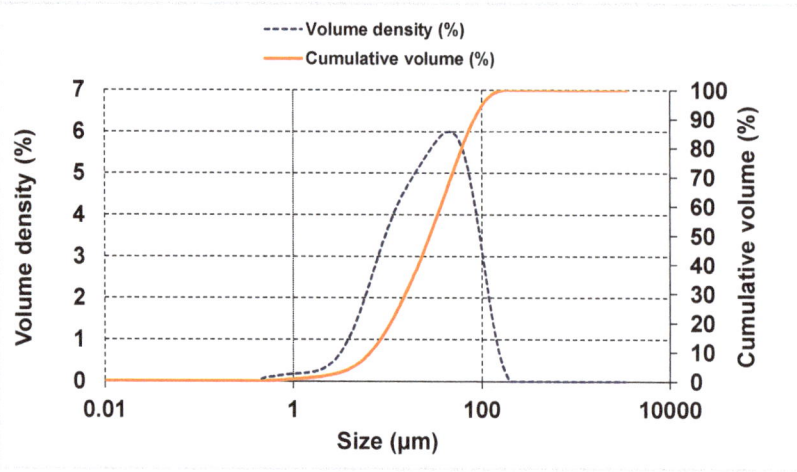

Figure 1. Particle size distribution and volume density plots for bentonite.

Another factor of particular interest in bentonite materials, which lies outside the scope of the present study, is water demand, affected not only by fineness but also by its sodium content [8].

In all the pastes and mortars prepared, bentonite replaced the corresponding binder content, and the water cement ratio was calculated as cement plus bentonite: w/cm.

2.2. Specimen Types

Different types of specimens were prepared, depending on the test.

- For mechanical strength and carbonation resistance testing, 10 × 10 × 60 mm cement paste specimens bearing 10%, 20% or 30% bentonite additions were prepared at a water/cement ratio of 0.5. They were cured in a climatic chamber at 90% relative humidity, first in the moulds for 24 h and after removal for 28 d prior to testing.
- For chloride diffusions, the 70 cubic mm cement mortar specimens used were prepared with a water/cement ratio of 0.5. They were cured in a climatic chamber at 90% relative humidity, first in the moulds for 24 h and after removal for 28 d prior to application of an electric current to test for chloride diffusion.

2.3. Test Methods

2.3.1. X-ray Diffraction

The mineralogical composition of the cement paste ground and sieved to 45 μm [40] was determined on a Bruker AXS DB Advance X-ray diffractor (Bruker, Madrid, Spain) configured without a monochromator, fitted with a 3 kW (Cu Kα1.2) copper anode X-ray source and a wolfram cathode. A 30 mA current was applied to the X-ray tube at a voltage of 40 kV. A 0.5 mm fixed divergence slit was used. The instrument was also fitted with a 2.5 rad primary Soller slit and a Lynx-eye X-ray super-speed detector diffractor (Hamamatsu, Hamamatsu, Spain) with a 3 mm anti-scatter slit, a 2.50 rad secondary Soller slit and a 0.5% Ni-K beta filter. The specific reflection peak used was $2\theta = 35°$.

2.3.2. Twenty-Eight Day Flexural and Compressive Strength

Testing for flexural strength [42] consisted in bending the prismatic specimens by applying a force perpendicular to their longitudinal axis, on a Netsch test frame specifically designed for small specimens.

The test was deemed valid only when the specimen failed across the middle.

The two halves of the specimens resulting from the flexural test were subsequently used for compression testing.

Compressive strength was found by exposing the specimens to two axial forces with equal modulus and orientation but coursing in opposite and convergent directions, on an Ibertest Autotest 200/10-SW test frame [45].

2.3.3. Carbonation in Natural Environments

The cements tested and their chemical compositions are given in Table 2. The 10 × 10 × 60 mm specimens were exposed to natural carbonation at the atmospheric CO_2 pressure prevailing in the city of Madrid, in an indoor laboratory environment and two outdoor environments, one sheltered and the other unsheltered from rainfall, i.e., environments with varying relative humidity and temperatures (Figure 2).

Table 2. Composition (%) of the cements used in the tests.

Cement	K	V	L/LL	S	P	Addition
CEM I 42.5 R	95					5
CEM I 52.5R–SR 3	95					5
CEM II/A-P 42.5 R	83				11	6
CEM II/A-P 42.5 R	80				16	4
CEM II/A-S 42.5 N	83			12		5
CEM II/A-V 42.5 R	80		15			5
CEM III/A 42.5 N	59			39		2
CEM IV/A (V) 42.5 R-SR	74	23				3
IV/A(P) 42.5 R/MR	87				13	0
BL II/B-LL 42.5 R	74		26			0

(a) (b) (c)

Figure 2. Exposure to natural carbonation: environments: (**a**) indoor (laboratory) environment, (**b**) outdoor unsheltered environment and (**c**) outdoor sheltered environment.

Carbonation depth as found with phenolphthalein, an acid-base indicator, was recorded for the 3-month and 6-month specimens, depicted in the three environments in Figure 2.

2.3.4. Sulfate Resistance: The Koch–Steinegger Method

Cement paste resistance to sulfate ions was tested on 10 × 10 × 60 mm specimens further to the Koch–Steinegger method, based on comparing flexural strength in such specimens soaked for 56 d in an aggressive solution (here, sodium sulfate at a concentration of 4.4 g/L) to the strength of analogous specimens soaked in water, likewise for 56 d (Figure 3). All the specimens had been cured in a humidity chamber for 28 d prior to testing.

Figure 3. Koch–Steinegger exposure to sulfate attack.

2.3.5. Accelerated Chloride Ingress

The accelerated chloride diffusion test described in Spanish standard UNE 83992-2 EX [36] was conducted on 70 cubic mm mortar specimens, each bearing an embedded steel bar. Performance by the samples with 10% bentonite was compared to the results observed for reference CEM I 42.5SR specimens of the same dimensions.

The test consisted in connecting specimens made with different types of mortar to an electrical current that accelerated chloride ion diffusion (migration) across the matrix toward the bar (see setup in Figure 4). The steel was assumed to begin to corrode when surface contact with the chlorides was electrochemically detected. That initial corrosion time and the amount of chloride on the bar surface were the parameters used to calculate the diffusion coefficient.

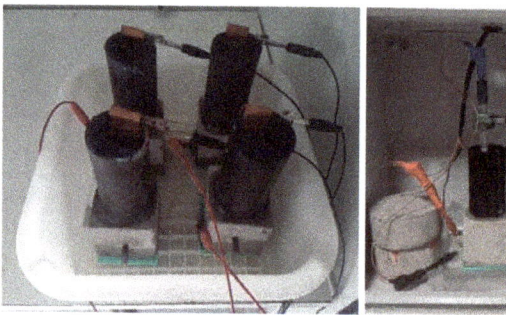

Figure 4. Accelerated corrosion test setup.

3. Results

3.1. Flexural and Comprenssive Strength

Additions should not alter, except to improve, mix mechanical performance. As Figure 5a shows, replacing 10% or 20% of cement CEM I 52.5R–SR 3 with bentonite raised 28 d flexural strength relative to the reference cement except at a replacement ratio of 30%. Adding 30% bentonite yielded lower compressive strength than in the reference and in the materials with 10% or 20% replacement.

While unaffected by bentonite at a replacement ratio of 10% (Figure 5b), compressive strength declined at ratios of 20% or 30%. Those findings informed the decision to use only the 10% bentonite in all the subsequent tests as the most conservative option.

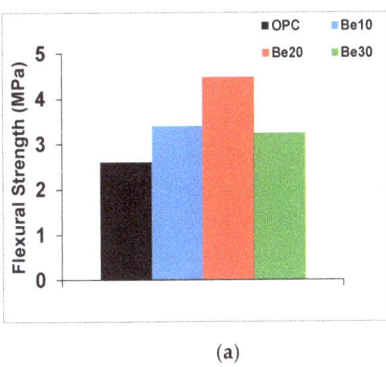
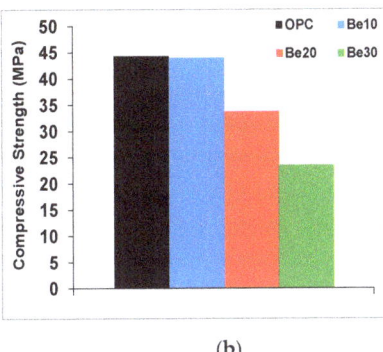

Figure 5. (a) Flexural and (b) compressive strength in cement pastes prepared with the reference and blended cements at replacement ratios of 10%, 20% or 30%.

3.2. X-ray Diffraction-Based Characterisation

The possible reactivity and stability of bentonite-bearing cement paste were also explored. The diffractograms for 28 d pastes bearing 10%, 20% and 30% bentonite are reproduced in Figure 6, whilst the relative content (counts, in per cent) of the various phases is graphed in Figure 7.

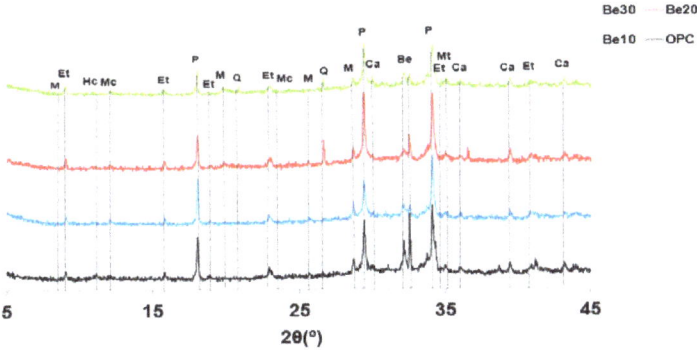

Figure 6. XRD patterns for unadditioned CEM I 52.5R–SR 3 and the same cement with 10%, 20% or 30% bentonite.

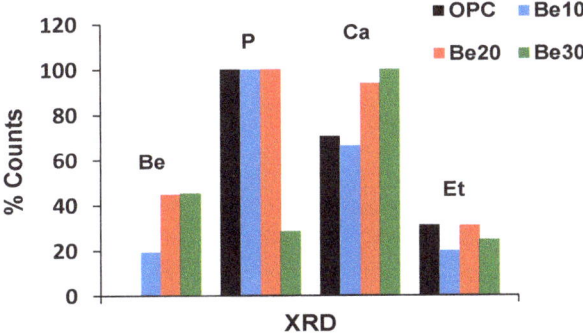

Figure 7. Crystalline phases identified on the XRD patterns for the reference and bentonite-bearing mixes: counts (%).

The reflections attributable to bentonite (montmorillonite and quartz) rose in intensity between 10% and 20% replacement, although no such rise was visible between 20% and 30% [44].

Portlandite content was similar to the reference in the former two mixes, but declined significantly at 30% replacement. According to [34], that decline might be explained by the formation of a calcium zeolite in the interaction between bentonite and the pore solution in the hydrated cement; confirmation of such a premise lies outside the scope of the present study.

Inasmuch as the carbonate phases would have been generated by carbonation occurring during the test, for the time being the rise in intensity with bentonite content need not be attributed to that higher proportion of the clay.

As ettringite content, in turn, followed neither an upward nor a downward pattern, its presence would be due to the cement and therefore may be deemed unaffected by the addition.

3.3. Sulfate Attack

The results for this test are deemed acceptable when the strength of the blended cement is greater than 70% of the value recorded for the control soaked in distilled water. Further to the flexural and compressive strengths of the reference specimen soaked in water and the specimens bearing 10% bentonite graphed in Figure 8, strength was higher in both the reference and in the specimen bearing the bentonite presence when soaked in the sulfate solution than when soaked in water. That rise in strength was attributed to the higher degree of hydration resulting from the difference in specimen ages: 56 d rather than 28 d. The values observed for the reference paste and the paste prepared with the blended cement were very similar. On the grounds of those data, the presence of bentonite may be deemed to have had no effect on cement resistance to sulfate attack.

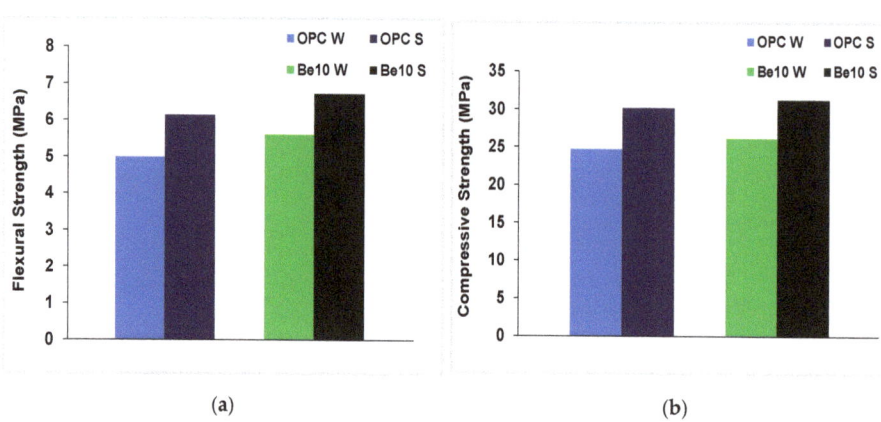

Figure 8. (a) Flexural and (b) compressive strength in reference specimens and specimens bearing 10% bentonite soaked in distilled water (W) or in sulfate (S).

3.4. Chloride Resistance Test

This test aimed to determine the effects of the bentonite addition on chloride transport in the mortar matrix and the chloride ion threshold at which the reinforcing steel began to corrode. Figure 9 shows corrosion potential and corrosion rate from time 0 until an abrupt change in tendency in the respective curves denoted the onset of reinforcement depassivation.

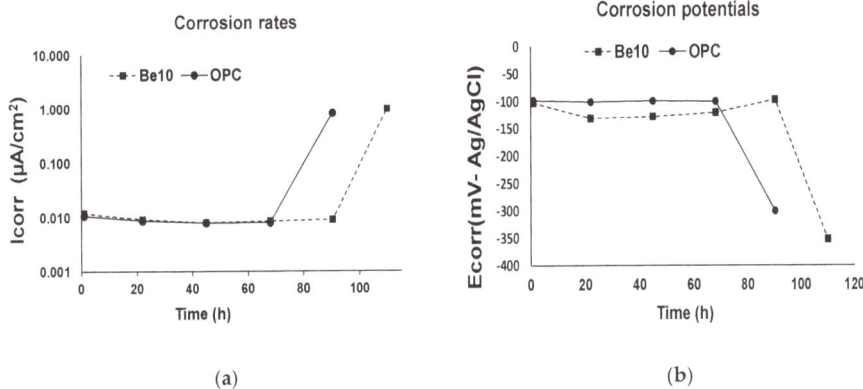

Figure 9. (a) Corrosion potential (0 to 100 h) and (b) corrosion rate (0 to 120 h), illustrating the abrupt change in tendency that denotes the onset of reinforcement depassivation.

The chloride diffusion coefficients calculated from the penetration depth of the colorimetric front depicted in Figure 10 are listed in Table 3. Much smaller values were observed for the mortar bearing 10% bentonite. Rather than penetration depth per se (the red line in Figure 10), the decline reflects differences in test times, for depassivation occurred in the reference earlier than in the bentonite-bearing material. In other words, it took much longer to reach the penetration shown in the figures in the bentonite-bearing than in the reference specimens, denoting higher electrical resistivity in the former.

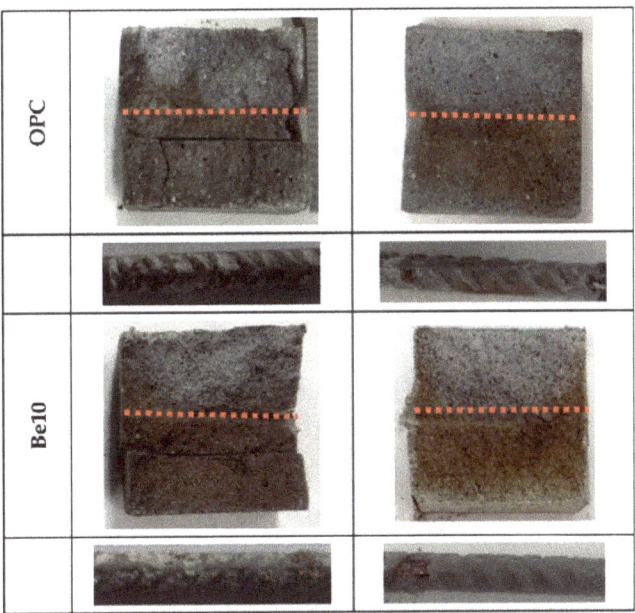

Figure 10. Chloride penetration front upon finalization of the experiment with the detection of the onset of corrosion.

Table 3. Chloride non-steady-state non steady-state diffusion coefficient (D_{ns}) in reference (OPC) specimens and samples bearing 10% bentonite (Be10).

Sample	Test Time (h)	Maximum Penetration (mm)	Mean Penetration (mm)	D_{ns} Diffusion Coefficient (cm^2/s)
OPC-1	110	34.04–33.03	33.535	17×10^{-12}
OPC-2	110	33.94–33.14	33.54	15×10^{-12}
Be10-1	90	34.12–34.06	21.015	4.75×10^{-12}
Be10-2	90	33.92–33.89	33.91	4.6×10^{-12}

Further to the chloride content data given in Table 4, surface concentration was higher, whilst the chloride threshold was, inversely, lower, in the presence of bentonite. This allows one to confirm that it is the transport phase expressed in the diffusion coefficient which controls the better behaviour of the mortar with bentonite.

Table 4. Concentration of chlorides in the surface of the specimen at the end of the experiment and in the surface of the steel bar.

	Bar in Reference	Bar in Be10
Chloride Surface concentration	1.10	1.64
Chloride threshold (% mass mortar)	0.35	0.13

3.5. Natural Carbonation

The cumulative rainfall recorded in Madrid in the 6 months the specimens were exposed to outdoor conditions was 160 L/m^2, whilst the mean temperature was on the order of 30 °C. The photographs in Table 5 depict the phenolphthalein staining in the specimens from which carbonation depth was deduced. Generally speaking, the shallowest depths were observed in the laboratory, intermediate penetration under outdoor sheltered conditions and the deepest in the specimens exposed to rainfall. That order of environmental aggressiveness is diametrically opposed to earlier reports. According to those data, penetration was deepest in specimens exposed to indoor environments or sheltered outdoor conditions, whilst carbonation was least intense in those exposed to rainfall, due to their higher or nearly optimal moisture content. The present findings were deemed accurate, however, for they were qualitatively identical in the 3-month and 6-month specimens. In this study, the specimens exposed for 6 months were also tested during the summer under high-temperature, low relative humidity conditions. In other seasons, with higher RH and more rain, the order may have differed. That is scantly relevant, however, for inasmuch as carbonation was intense in all the samples, the findings sufficed for the aim pursued, namely to compare the behaviour in the various cements.

Figure 11 plots the 3-month carbonation depths in the specimens bearing 10% bentonite against the respective references, and Figure 12 plots the same parameters in the 6-month samples. After 3 months, carbonation was less intense in a larger number of 10% bentonite than in reference specimens. The gap was smaller after 6 months, although in some cases carbonation was more intense in the blended cement than in the reference specimens.

Table 5. Phenolphthalein staining in the cements studied to determine carbonation depth.

CEMENT	3 MONTHS								6 MONTHS							
	INDOORS		OUTDOORS UNSHELTERD		OUTDOORS SHELTERED				INDOORS		OUTDOORS UNSHELTERD		OUTDOORS SHELTERED			
	OPC	10% Be	OPC	10% Be	OPC	10% Be			OPC	10% Be	OPC	10% Be	OPC	10% Be		
CEM I 42.5R																
CEM I 52.5R-SR3																
CEMII/AS42.5N																
CEMII/A-P(16) 42.5R																
CEMII/A-P(13) 42.5R																
CEMII/AV42.5R																
CEMIII/A42.5N																
CEMIV/A(V) 42.5R SR																
IV/A(P) 42.5R/MR																
BLII/B-LL42.5R																

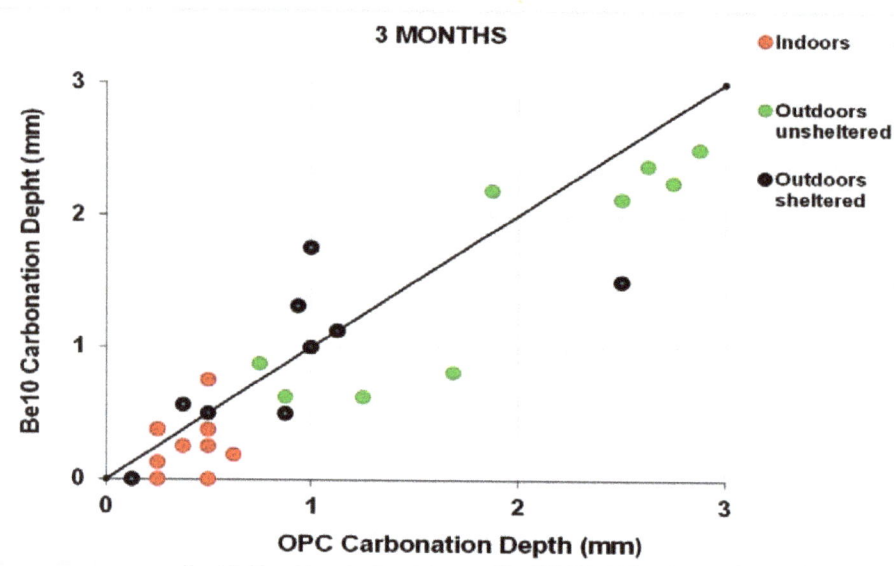

Figure 11. Three-month carbonation depth in reference OPC vs. 10% bentonite-bearing Be10.

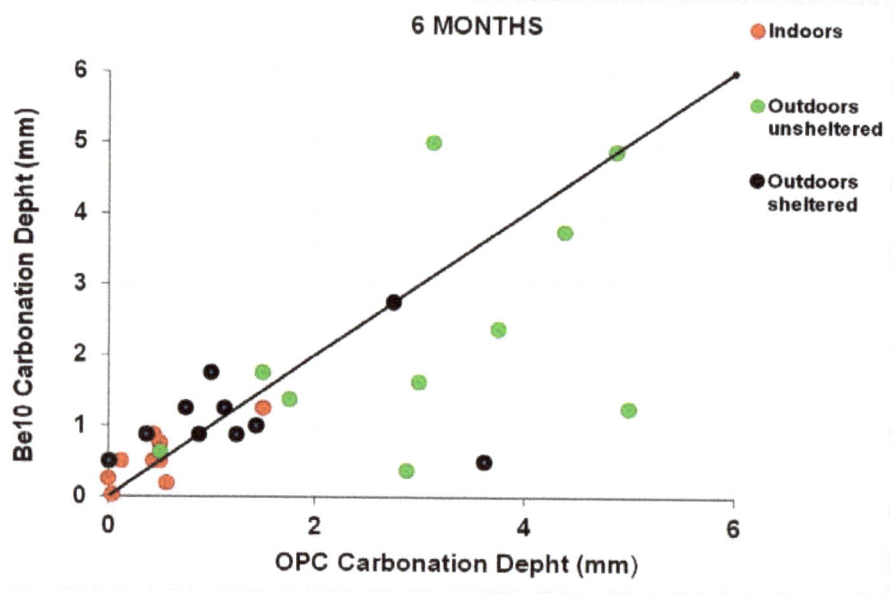

Figure 12. Six-month carbonation depth in reference OPC vs. 10% bentonite-bearing Be10.

As a rule, the flexural strengths (Figures 13–16) were fairly similar in the reference and blended samples. The compressive strength values were even closer in the two types of mortars. Inasmuch as the experiment was designed for purposes of comparison, the inference drawn from these findings is that replacing 10% bentonite in the cement mix had no material effect on mortar behaviour.

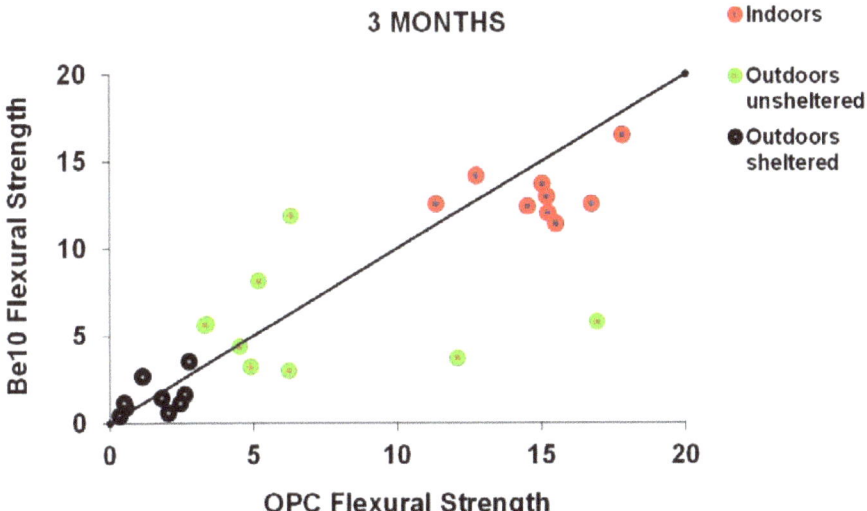

Figure 13. Three-month flexural strength: OPC vs. Be10.

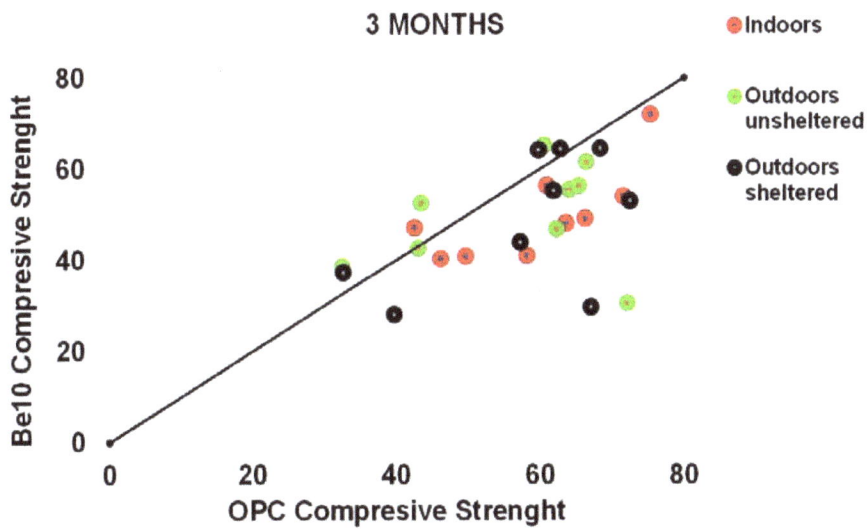

Figure 14. Three-month compressive strength: OPC vs. Be10.

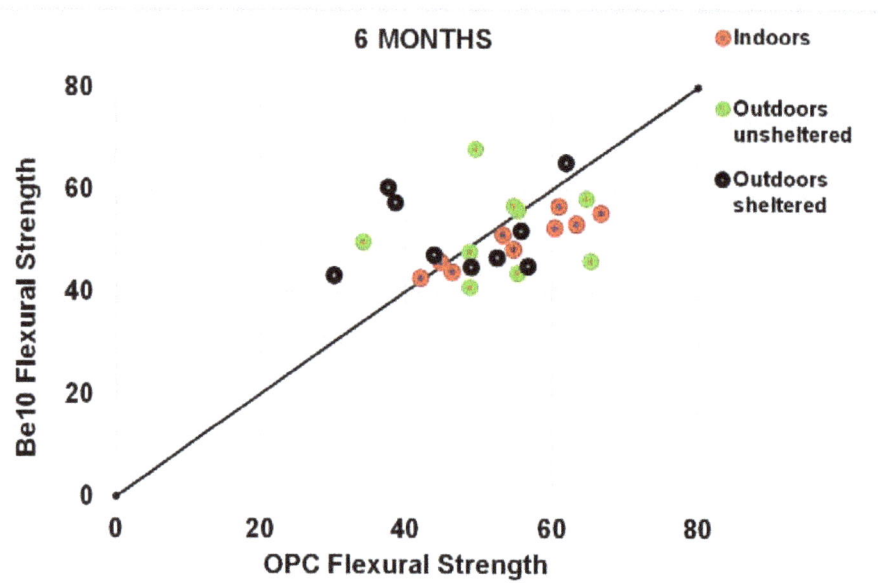

Figure 15. Six-month flexural strength: OPC vs. Be10.

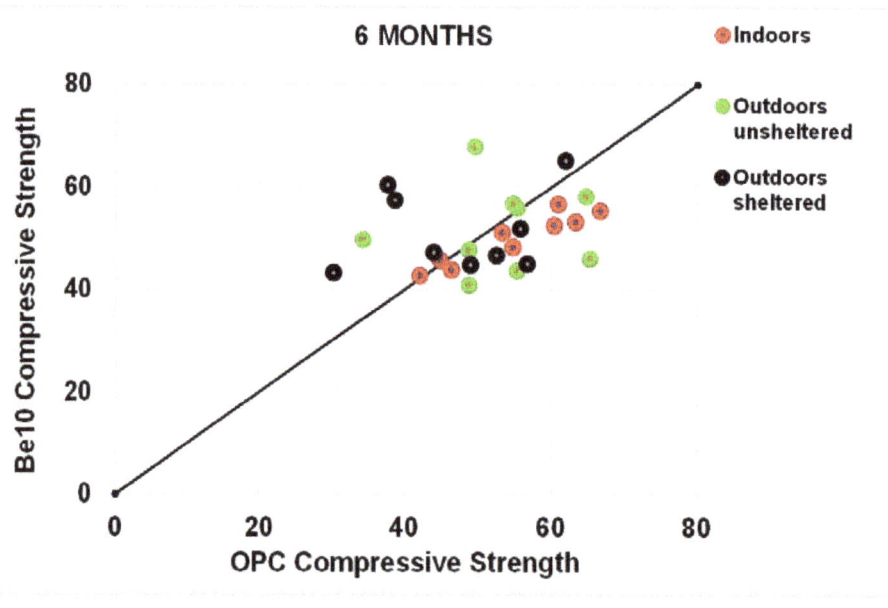

Figure 16. Six-month compressive strength: OPC vs. Be10.

4. Discussion

Rising to the challenge posed by the need to reduce the cement industry's carbon footprint may involve either large-scale technological change based on research into new manufacturing methods or adopting a more direct and technologically simple approach consisting in lowering the proportion of clinker in cements without altering their essential properties [1–3]. Replacing clinker with non-CO_2-emitting materials is the most immediate alternative open to the industry.

By-products such as blast furnace slag, that are in themselves cementitious, have long been deployed to reduce clinker content [1], as have acid materials (pozzolans and more recently fly ash and silica fume) that react with the calcium hydroxide released in cement hydration. The use of natural pozzolans began to decline in the wake of their depletion in some natural reserves or because of the adverse impact of quarrying on the environment. In contrast, as an industrial by-product, fly ash was much less costly, for it did not have to be mined and the enormous stockpiles of coal industry waste had to be pared down. At around the same time, the nineteen seventies oil crisis raised the price of the fuel used to manufacture clinker. That combination of factors led to a general trend to add minerals to the clinker, which, depending on the commercial and legislative conditions prevailing in any given country, were either milled directly with the clinker or added at concrete plants.

As noted in the introduction, the evidence of climate change and the gradual reduction of the stock of such by-products have driven a return to former paradigms, including the use of additions other than slag or fly ash, such as precalcined clays [12]. Even with the investment involved in precalcination, clay has become competitive due to the rising cost of emissions [14]. Indisputably, however, the initial cost would be even more competitive if cement performance could be ensured with no need for precalcination.

Such precalcination entails, among other consequences, the loss of the bound water in the constituent minerals present in clay [8–12], which is recovered during hydration. Such thermal dehydration affects clay reactivity, i.e., cement hydration kinetics, but should not in principle impact component stability, for the hydrated compounds at issue are the same as they would be if the clay were used without precalcination. That is one of the many matters in connection with the use of non-pre-dehydrated natural clays, bentonite among them, in need of more thorough research.

In light of its vast diversity and fairly widespread geographic availability, clay is just one more local raw material [17,20] deployed in cement manufacture. The very same clays used in clinker kilns might, in certain proportions, constitute compatible additions to lower milling-related CO_2 emissions (since milling clay is less energy-intensive than grinding clinker). That reasoning informed the initiative to undertake exploratory research along those lines, part of the results of which are described hereunder.

Surprisingly, very few studies [16–26] on the subject were found in the literature other than reports of the widespread joint use of cement and bentonite in underground works [28,29], such as the nuclear waste storage [34–38], soil stabilisation or impermeable slurry wall construction [27].

To be compatible with cement and usable in concrete, additions must meet a series of short- and long-term requisites, summarised below.

- They must be inert or at least not induce expansion or degenerative reactions.
- They must improve or at least not alter concrete volume stability (in terms of shrinkage and creep especially).
- They must improve or at least not alter mechanical performance.
- They must lengthen or at least not shorten concrete or steel durability.

This study addresses some but not all of those factors. The findings are deemed sufficiently promising to be made public, acknowledging, however, that the use of non-precalcined clays will call for considerable research, in light of their enormous variety.

The following paragraphs discuss the more or less basic features of the use of non-precalcined clays analysed here, i.e., the effect on mechanical strength, the nature of the hydration products forming and the impact of bentonites on resistance to sulfates, chloride ingress and carbonation.

Be it said from the outset in connection with flexural and compressive strength that bentonite thixotropy necessitates adjusting admixtures and a constant w/cm ratio to ensure suitable mix workability [31,32]. Such thixotropy, which has not been studied in depth, may be either a drawback or an advantage in terms of workability, depending on the intended application of the concrete (such as precasting or additive manufacturing, also known as 3D printing) at issue [43]. In this study, carboxylate admixtures [31,32] were the simplest choice to avoid mixing problems.

One of the most prominent findings of this research was the rise in flexural strength with the proportion of bentonite. In the absence of supplementary testing, no reasons for such a rise can be ventured at this time. Although compressive strength was observed to decline as the replacement ratio rose, that development was readily attributable to the concomitantly lower clinker content.

The XRD findings for the hydrated pastes revealed that at the ages studied the cement barely reacted with bentonite. Further to reports on underground structures, for the nuclear industry in particular, high cement alkalinity induces the formation of a certain proportion of calcium silicate hydrates and calcium zeolites [35,36]. At ambient temperatures, however, that reaction is apparently slow enough to deem bentonite a nearly inert substance.

Although the results of the Koch–Steinegger sulfate resistance tests might be dismissed for their failure to represent actual conditions, they are nonetheless indicative of relatively short-term anomalous and expansive reactions. Longer-term tests using different solutions would be required to confirm the present initially promising results in this regard.

The lower diffusion coefficient measured for chloride ingress, in turn, was attributable to the timing differences between the tests conducted with the reference and with bentonite [46]. As the clay retards chloride penetration significantly [20], its use as an addition would be beneficial, although further research is called for to determine the reasons for this behaviour. One possibility might be the reduction of porosity (parameter not measured here), whereas any reaction between bentonite and chlorides would be all but ruled out in light of the negative charge in the clay's interlayers, which would accommodate cations but not anions.

The effect of bentonite on carbonation depth must be assessed in the realisation that its action supplemented the action of other additions present in the cement. In other words, at least two mineral additions were in place in the tests conducted here, accounting in some cases for a substantial fraction of the total. In the two blended CEM I cements used, carbonation was the same or even lower than when no bentonite was present. Of the other cements, the ones bearing natural pozzolans appeared to perform better than those carrying fly ash or slag. In neither case did the use of bentonite induce clearly poorer performance than already observed in those cements. However, such behaviour cannot be attributed to a reaction between the clay and carbon dioxide, for as noted above, bentonite cannot accommodate anions in its interlayers [47–49].

This feature, the reduction or at least non-alteration of carbonation depth, is deemed to be the most relevant finding of this study. For the opposite, lower carbonate resistance is one of the shortcomings identified in mineral additions in general. Bentonite could consequently be used to advantage instead of the 5% of inert matter or the up to 10% of limestone routinely added to clinker. It may improve one or several properties of the end product. Confirmation of the foregoing will nonetheless call for much more testing, in particular to detect possible adverse effects on shrinkage or creep.

In the context of the pursuit of a circular economy and climate change mitigation, the cement industry is undertaking new strategies to reach a net zero emissions target by 2050. One such strategy, the production and use of blended cements with a high pozzolanic material content, makes the need to find new additions the more pressing [5]. Bentonite is a well-known clay consisting mostly of montmorillonite, an aluminium phyllosilicate mineral whose microscopic (~1 μm in diameter) plate-shaped particles afford the clay a large surface area. Precalcined clay has been standardised (European standard EN 197-1:2011 [50]) under the category 'natural calcined pozzolana (Q)', defined as thermally treated clays, shales, sedimentary rocks or materials of volcanic origin. Inasmuch as bentonite is a clay that requires no thermal activation, it might well be classified under the designatory letter 'Q'.

5. Conclusions

This article describes exploratory experimentation on some of the properties of pastes and mortars made with different proportions of bentonite. The conclusions that may be drawn from the findings include the following.

1. Replacing up to 20% cement with bentonite enhances flexural, and up to 10%, compressive strength.
2. At 10%, the addition induces no change in cement crystalline hydration products or in the components of bentonite itself (>95% montmorillonite further to supplier specifications).
3. Replacing 10% of the cement with bentonite:
    a. In a sulfate solution for 56 d raises cement paste mechanical strength relative to the same materials stored in distilled water, attributed to greater age in the absence of expansive reactions;
    b. Lowers the chloride diffusion coefficient significantly;
    c. Reduces or maintains the carbonation depth observed in the reference material, deemed to be a very promising development.

New and innovative measures must be taken by the cement industry worldwide to minimise its impact on climate change. One effective approach to reaching carbon neutrality consists in using new constituents to manufacture Portland cement. Insofar as bentonite, while a clay material, calls for no thermal activation, the present authors suggest that it be classified under the European standard EN 197-1:2011 [50] heading 'natural calcined pozzolana (Q)'.

Author Contributions: Conceptualization, C.A. and J.A.T.R.; methodology, C.A.; experimental procedures, A.M.-S.; formal analysis, A.M.-S. and C.A.; research, C.A., A.M.-S., M.Á.S. and J.A.T.R.; write-up: original draft, C.A. and A.M.-S.; review and editing: C.A., A.M.-S., M.Á.S. and J.A.T.R.; supervision, C.A. and J.A.T.R. All authors have read and agreed to the published version of the manuscript.

Funding: This research was funded by the Spanish Ministry of Science, Innovation and Universities under project ADIMULT [ref. BIA-2015-70350-R], and Spain's (Ministry of the Economy) training programme [ref. BES-2016-077157].

Institutional Review Board Statement: Not applicable.

Informed Consent Statement: Not applicable.

Data Availability Statement: Data sharing is not applicable to this article.

Acknowledgments: Use of Eduardo Torroja Institute for Construction Science (a National Research Council body) facilities to conduct the study is gratefully acknowledged. The authors wish to personally thank L. Caneda-Martínez. M.I. Sánchez de Rojas, M. Frías and M.T. Blanco for their assistance with testing.

Conflicts of Interest: The authors declare no conflict of interest.

References

1. Gartner, E.; Hirao, H. A review of alternative approaches to the reduction of CO_2 emissions associated with the manufacture of the binder phase in concrete. *Cem. Concr. Res.* **2015**, *78*, 126–142. [CrossRef]
2. Gartner, E. Industrially interesting approaches to "low-CO_2" cements. *Cem. Concr. Res.* **2004**, *34*, 1489–1498. [CrossRef]
3. Taylor, H.F.W. *Cement Chemistry*, 2nd ed.; Thomas Telford: London, UK, 1997.
4. Sanjuán, M.Á.; Andrade, C.; Mora, P.; Zaragoza, A. Carbon Dioxide Uptake by Cement-Based Materials: A Spanish Case Study. *Appl. Sci.* **2020**, *10*, 339. [CrossRef]
5. Sanjuán, M.A.; Argiz, C.; Mora, P.; Zaragoza, A. Carbon Dioxide Uptake in the Roadmap 2050 of the Spanish Cement Industry. *Energies* **2020**, *13*, 3452.
6. Plaza, M.G.; Martínez, S.; Rubiera, F. CO_2 Capture, Use, and Storage in the Cement Industry: State of the Art and Expectations. *Energies* **2020**, *13*, 5692. [CrossRef]
7. Haneklaus, N.; Zheng, Y.; Allelein, H.-J. Stop Smoking—Tube-In-Tube Helical System for Flameless Calcination of Minerals. *Processes* **2017**, *5*, 67. [CrossRef]
8. Schneider, M.; Romer, M.; Tschudin, M.; Bolio, H. Sustainable cement production: Present and future. *Cem. Concr. Res.* **2011**, *41*, 642–650. [CrossRef]

9. Secretary of State for Energy of the Ministry for the Ecological Transition (2019) Energy in Spain 2017. National State Administration Publications, Madrid, Spain. (In Spanish). Available online: https://energia.gob.es/balances/Balances/LibrosEnergia/Libro-Energia-2017.pdf (accessed on 2 February 2021).
10. International Energy Agency. IEA (2018) Coal 2018. IEA Publications: Paris, France. Available online: https://www.iea.org/reports/coal-2018 (accessed on 8 February 2021).
11. Van Olphen, H. *An Introduction to Clay Colloid Chemistry*; Interscience Publishers: New York, NY, USA, 1963.
12. Fernandez, R.; Martirena, F.; Scrivener, K.L.B. The origin of the pozzolanic activity of calcined clay minerals: A comparison between kaolinite, illite and montmorillonite. *Cem. Concr. Res.* **2011**, *41*, 113–122. [CrossRef]
13. Courard, L.; Darimont, A.; Schouterden, M.; Ferauche, F.; Willem, X.; Degeimbre, R. Durability of mortars modified with metakaolin. *Cem. Concr. Res.* **2003**, *33*, 1473–1479. [CrossRef]
14. Lima Souza, P.S.; Dal Molin, D.C.C. Viability of using calcined clays from industrial by-products. as pozzolans of high reactivity. *Cem. Concr. Res.* **2005**, *35*, 1993–1998. [CrossRef]
15. Pillai, R.G.; Gettu, R.; Santhanam, M.; Rengaraju, S.; Dhandapani, Y.; Rathnarajan, S.; Basavaraj, A.S. Service life and life cycle assessment of reinforced concrete systems with limestone calcined clay cement (LC3). *Cem. Concr. Res.* **2019**, *118*, 111–119. [CrossRef]
16. Memon, S.A.; Arsalan, R.; Khan, S.; Lo, T.Y. Utilization of Pakistani bentonite as partial replacement of cement in concrete. *Constr. Build. Mater.* **2012**, *30*, 237–242. [CrossRef]
17. Liu, M.; Hu, Y.; Lai, Z.; Yan, T.; He, X.; Wu, J.; Lu, Z.; Lv, S. Influence of various bentonites on the mechanical properties and impermeability of cement mortars. *Cem. Concr. Res.* **2020**, *241*, 118015. [CrossRef]
18. Huang, W.-H. Properties of cement-fly ash grout admixed with bentonite. Silica fume and organic fiber. *Cem. Concr. Res.* **1997**, *27*, 395–406. [CrossRef]
19. Wei, J.; Gencturk, B. Hydration of ternary Portland cement blends containing metakaolin and sodium bentonite. *Cem. Concr. Res.* **2019**, *123*, 105772. [CrossRef]
20. Masood, B.; Elahi, A.; Barbhuiya, S.; Ali, B. Mechanical and durability performance of recycled aggregate concrete incorporating low calcium bentonite. *Constr. Build. Mater.* **2020**, *237*, 117760. [CrossRef]
21. Rehman, S.-U.; Kiani, U.A.; Yaqub, M.; Ali, T. Controlling natural resources depletion through Montmorillonite replacement for cement-low cost construction. *Constr. Build. Mater.* **2020**, *232*, 117188. [CrossRef]
22. Yang, H.; Long, D.; Lai, Z.; He, Y.; Yan, T.; He, X.; Wu, J.; Lu, Z. Effects of bentonite on pore structure and permeability of cement mortar. *Constr. Build. Mater.* **2019**, *224*, 276–283. [CrossRef]
23. Zhao, N.; Wang, S.; Wang, C.; Quan, X.; Yan, Q.; Li, B. Study on the durability of engineered cementitious composites (ECCs) containing high-volume fly ash and bentonite against the combined attack of sulfate and freezing-thawing (F-T). *Constr. Build. Mater.* **2020**, *233*, 117313. [CrossRef]
24. Terzic, A.; Pezo, L.; Mijatovic, N.; Stojanovic, J.; Kragovic, M.; Milicic, L.; Andric, L. The effect of alternations in mineral additives (zeolite. bentonite. fly ash) on physico-chemical behavior of Portland cement based binders. *Constr. Build. Mater.* **2018**, *180*, 199–210. [CrossRef]
25. Sha, F.; Li, S.; Liu, R.; Li, Z.; Zhang, Q. Experimental study on performance of cement-based grouts admixed with fly ash, bentonite, superplasticizer and water glass. *Constr. Build. Mater.* **2018**, *161*, 282–291. [CrossRef]
26. Kalpokaité-Dickuvienė, R.; Lukosiuté, I.; Cesniené, J.; Brinkiené, K.; Baltusnikas, A. Cement substitution by organoclay. The role of organoclay type. *Cem. Concr. Compos.* **2015**, *62*, 90–96. [CrossRef]
27. Ata, A.A.; Salem, T.N.; Elkhawas, N.M. Properties of soil bentonite-cement bypass mixture for cutoff walls. *Constr. Build. Mater.* **2015**, *93*, 950–956. [CrossRef]
28. Garvinô, S.L.; Hayles, C.S. The chemical compatibility of cement-bentonite cut-off wall material. *Constr. Build. Mater.* **1999**, *13*, 329–341. [CrossRef]
29. Trivedi, D.P.; Holmes, R.G.G.; Brown, D. Monitoring the in-situ performance of a cement/bentonite cut-off wall at a low level waste disposal site. *Cem. Concr. Res.* **1992**, *22*, 339–349. [CrossRef]
30. Lei, L.; Plank, J. A study on the impact of different clay minerals on the dispersing force of conventional and modified vinyl ether based polycarboxylate superplasticizers. *Cem. Concr. Res.* **2014**, *60*, 1–10. [CrossRef]
31. Lei, L.; Plank, J. A concept for a polycarboxylate superplasticizer possessing enhanced clay tolerance. *Cem. Concr. Res.* **2012**, *42*, 1299–1306. [CrossRef]
32. Sahmaran, M.; Ozkan, N.; Keskin, S.B.; Uzal, B.; Yaman, I.O.; Erdem, T.K. Evaluation of natural zeolite as a viscosity-modifying agent for cement-based grouts. *Cem. Concr. Res.* **2008**, *38*, 930–937. [CrossRef]
33. Kaci, A.; Chaouche, M.; Andreani, P.-A. Influence of bentonite clay on the rheological behaviour of fresh mortars. *Cem. Concr. Res.* **2011**, *41*, 373–379. [CrossRef]
34. Mohammed, M.H.; Pusch, R.; Knutsson, S.; Warr, L.N. Hydrothermal alteration of clay and low pH concrete applicable to deep borehole disposal of high-level radioactive waste—A pilot study. *Constr. Build. Mater.* **2016**, *104*, 1–8. [CrossRef]
35. Dauzeres, A.; Le Bescop, P.; Sardini, P.; Cau Dit Coumes, C. Physico-chemical investigation of clayey/cement-based materials interaction in the context of geological waste disposal: Experimental approach and results. *Cem. Concr. Res.* **2010**, *40*, 1327–1340. [CrossRef]

36. Fernandez, R.; Cuevas, J.; Mader, U.K. Modeling experimental results of diffusion of alkaline solutions through a compacted bentonite barrier. *Cem. Concr. Res.* **2010**, *40*, 1255–1264. [CrossRef]
37. Hidalgo, A.; Llorente, I.; Alonso, C.; Andrade, C. *Study of Concrete/Bentonite Interaction Using Accelerated and Natural Leaching Tests*; RILEM Publications SARL: Paris, France, 2014.
38. Page, C.L.; Short, N.R.; El Tarras, A. Diffusion of chloride ions in hardened cement pastes. *Cem. Concr. Res.* **1981**, *11*, 395–406. [CrossRef]
39. Andrade, C.; d'Andrea, R.; Rebolledo, N. Chloride ion penetration in concrete: The reaction factor in the electrical resistivity model. *Cem. Concr. Compos.* **2014**, *47*, 41–46. [CrossRef]
40. Sanjuán, M.Á.; Estévez, E.; Argiz, C. Carbon Dioxide Absorption by Blast-Furnace Slag Mortars in Function of the Curing Intensity. *Energies* **2019**, *12*, 2346. [CrossRef]
41. González, J.A.; Algaba, S.; Andrade, C. Corrosion of reinforcing bars in carbonated concrete. *Br. Corros. J.* **1980**, *3*, 135–139. [CrossRef]
42. Alonso, C.; Andrade, C. Corrosion behavior of steel during accelerated carbonation of solutions which simulate the pore concrete solution. *Mater. Construcción* **1987**, *37*, 5–16.
43. Sanjuán, M.A.; Andrade, C.; Cheyrezy, M. Concrete carbonation tests in natural and accelerated conditions. Advances in. *Cem. Res.* **2003**, *15*, 171–180. [CrossRef]
44. Muggler, C.C.; Pape, T.H.; Buurman, P. Laser grain-size determination in soil genetic studies 2. Clay content, clay formation and aggregation in some brazilian oxisols. *Soil Sci.* **1997**, *162*, 219–228. [CrossRef]
45. Naswir, M.; Arita, S.; Salni, M. Characterization of bentonite by XRD and SEM-EDS and use to increase pH and color removal. Fe and organic substances in peat water. *J. Clean Energy Technol.* **2013**, *1*, 313–317. [CrossRef]
46. CEN-CENELEC. *European Standard EN 196-1. Methods of Resting Cement—Part 1: Determination of Strenght*; Cement and Building Limes: Brussels, Belgium, 2005.
47. *Tests on Concrete Durability. Test of Chloride Penetration and Reinforcement Corrosion: Accelerated Integral Test*; PrUNE 83992-2; Materials: Madrid, Spain, 2019.
48. Reales, O.A.M.; Duda, P.; Silva, E.C.; Paiva, M.D.; Toledo Filho, R.D. Nanosilica particles as structural buildup agents for 3D printing with Portland cement pastes. *Constr. Build. Mater.* **2019**, *219*, 91–100. [CrossRef]
49. Afzal, S.; Shahzada, K.; Fahad, M.; Saeed, S.; Ashraf, M. Assessment of early-age autogenous shrinkage strains in concrete using bentonite clay as internal curing technique. *Constr. Build. Mater.* **2014**, *66*, 403–409. [CrossRef]
50. *EN 197-1:2011. Cement—Part 1: Composition, Specifications and Conformity Criteria for Common Cement*; European Committee for Standardization (CEN): Brussels, Belgium, 2011.

Article

Assessment of High Performance Self-Consolidating Concrete through an Experimental and Analytical Multi-Parameter Approach

Ghafur H. Ahmed [1], Hawreen Ahmed [1,2,3,*], Babar Ali [4] and Rayed Alyousef [5]

1. Department of Highway and Bridge Engineering, Technical Engineering College, Erbil Polytechnic University, Erbil 44001, Iraq; ghafur.ahmed@epu.edu.iq
2. Scientific Research and Development Center, Nawroz University, Duhok 42001, Iraq
3. CERIS, Civil Engineering, Architecture and Georresources Department, Instituto Superior Técnico, Technical University of Lisbon, Av. Rovisco Pais, 1049-001 Lisbon, Portugal
4. Department of Civil Engineering, COMSATS University Islamabad, Sahiwal Campus, Sahiwal 57000, Pakistan; babar.ali@scetwah.edu.pk
5. Department of Civil Engineering, College of Engineering, Prince Sattam Bin Abdulaziz University, Alkharj 16273, Saudi Arabia; r.alyousef@psau.edu.sa
* Correspondence: hawreen.a@gmail.com

Citation: Ahmed, G.H.; Ahmed, H.; Ali, B.; Alyousef, R. Assessment of High Performance Self-Consolidating Concrete through an Experimental and Analytical Multi-Parameter Approach. *Materials* 2021, 14, 985. https://doi.org/10.3390/ma14040985

Academic Editor: Alessandro P. Fantilli

Received: 10 December 2020
Accepted: 29 January 2021
Published: 19 February 2021

Publisher's Note: MDPI stays neutral with regard to jurisdictional claims in published maps and institutional affiliations.

Copyright: © 2021 by the authors. Licensee MDPI, Basel, Switzerland. This article is an open access article distributed under the terms and conditions of the Creative Commons Attribution (CC BY) license (https://creativecommons.org/licenses/by/4.0/).

Abstract: High-performance self-consolidating concrete is one of the most promising developments in the construction industry. Nowadays, concrete designers and ready-mix companies are seeking optimum concrete in terms of environmental impact, cost, mechanical performance, as well as fresh-state properties. This can be achieved by considering the mentioned parameters simultaneously; typically, by integrating conventional concrete systems with different types of high-performance waste mineral admixtures (i.e., micro-silica and fly ash) and ultra-high range plasticizers. In this study, fresh-state properties (slump, flow, restricted flow), hardened-state properties (density, water absorption by immersion, compressive strength, splitting tensile strength, flexural strength, stress-strain relationship, modulus of elasticity, oven heating test, fire-resistance, and freeze-thaw cycles), and cost of high-performance self-consolidating concrete (HPSCC) prepared with waste mineral admixtures, were examined and compared with three different reference mixes, including normal strength-vibrated concrete (NSVC), high-strength self-compacted concrete (HSSCC), and high-performance highly-viscous concrete (HPVC). Then, a multi parameter analytical approach was considered to identify the optimum concrete mix in terms of cost, workability, strength, and durability.

Keywords: high performance concrete (HPC); self-consolidating concrete (SCC); flowability; durability; freeze-thaw cycle; fire resistance

1. Introduction

Self-consolidating concrete (SCC), also referred to as self-compacted concrete, is an innovative construction material with favorable rheological behavior that does not require vibration for placing and compaction. It can flow under its weight, filling in formworks, and achieving full compaction, even in the presence of complex-shaped concrete members with highly congested reinforcement [1–4]. Based on these properties, SCC may contribute to a significant improvement of the quality of concrete structures and opens up new fields for the application of concrete. The designation "self-compacting" is based on the fresh concrete properties of this material, which covers the mixture's degree of homogeneity, deformability, and viscosity. The yield point defines the force required to make the concrete flow. The speed of flow of SCC is associated with its plastic viscosity which describes the resistance of SCC to flow under external stresses [5–7]. SCC has become a preferred option for many projects that should satisfy strict fresh stage properties and quality assurance. To ensure stable and robust fresh stage properties, typically, a significant amount of fine materials has been incorporated into the mixture. In relative to traditional concrete, different durability characteristics can be expected for SCC because it can be produced with various mix compositions and the absence of vibration [8–10]. Due to the relatively short history of SCC in practical applications, there is a significant lack of information about long term performance in real structures. Such a concrete should have a relatively low yield value to ensure high flowability, a moderate viscosity to avoid segregation and bleeding, and must maintain its homogeneity during transportation, placing, and curing [11–13].

High-performance concrete (HPC) is engineered to meet specific needs of a project, including mechanical, durability, or constructability properties. The demand for HPC has been continuously increasing due to its superior mechanical and durability properties [14,15]. When considering the cost of concrete production, HPC is even better than ultra-high performance concrete (UHPC), since heat-curing restricts the applications of the latter and makes it mainly suitable for precast elements, not for ready-mix concrete [16,17]. The development of HPC started in the 1980s, and thereafter the global demand for its consumption has significantly increased over the recent years. HPC can be designed to have high workability and mechanical properties as well as improved durability [18,19]. It has been primarily used in bridges and tall buildings. In general, durability is the most important parameter to increase the service life of any concrete structure [20–22]. Most commonly durability of concrete is affected by sulfate or chloride attack, carbonation, high temperature, and freezing and thawing damage [23,24]. Scanning electron microscopic studies [25,26] show that the pore structure in powder type SCC, including the total pore volume, pore size distribution, and critical pore diameter, is very similar to HPC. Over the past decades, advancements in concrete technology has led to the development of a new generation of concrete (e.g., HPSCC) with significantly better properties in terms of strength, durability features, and rheology of fresh concrete mixtures. In comparison to ordinary concretes, the designing process of HPSCC mix is determined by the increased cement content, superplasticizers, and an additive of reactive materials, i.e., silica fume. HPSCC is thus characterized by its ability to fill a form with congested steel rebars and self-leveling without mechanical compaction and it yields exceptionally high strength and durability [27,28].

Abundant research can be found in the literature on the properties of SCC. Most of the previous works have tested the fresh SCC mixes for common workability tests in order to prove self-consolidation of the concrete. The investigated properties were flowability, deformability and passing-ability, through slump-cone flow, J-Ring, V-funnel, and L-box tests [11,28,29]. The rheological properties of SCC such as yield stress and plastic viscosity [30,31] have also been investigated. Some researchers focused on the mix design and mix proportions [1,8,11]. The influence of mineral admixtures (i.e., silica fume, fly ash, metakaolin, ground granulated blast furnace slag, ladle slag) [32–35] and chemical admixtures (i.e., superplasticizers and viscosity modifying admixtures) [3,9,12] have also been studied on the performance of SCC. Some studies investigated the hydration rate and microstructure of SCC [1,13,17,36]. Researchers have also studied the properties of SCC an HPC with the addition of glass fibers, steel fibers and carbon nanotubes [6,18,37–40]. The stability tests results of SCC, i.e., shrinkage, cracking resistance, and creep are also available in literature [7,15,41,42].

A study reported that the elastic modulus, creep and shrinkage of SCC did not differ significantly from the corresponding properties of normal strength concrete (NSC) [43]. Some of the durability tests, including chloride penetration, water permeability and absorption, gas permeability, carbonation, electrical resistivity, sulfate attack, acid attack, frost resistance, and scaling, have been investigated [17,19,23,44] and more especially the fire resistance, cooling methods, weight loss, and residual mechanical properties of SCC [5,45–47]. Only few studies were found in the literature that investigated HPSCC [2,9,48,49] and its optimization [50–52]; these studies had focused on mechanical properties with either porosity, workability, water penetration, rheological properties, exposure to elevated temperature, or one durability test; but frost or scaling resistance of SCC have rarely been investigated in the literature.

Regarding the novelty of this work, it can be clearly seen in the literature that the HPSCC has been investigated in the past two decades, but its practical application is still limited. This is due to the fact that its consolidated technical performance of HPSCC (e.g., mechanical strength, durability, and cost) is not fully understood, and there are often insufficient statements concerning its exact overall behavior. Existing research provides information only about the improvements in the properties of HPSCC mixes through the variation in the composition or addition of materials, but it does not inform what will happen to other parameters such as its consolidated economic and engineering performance. In the view of this understanding, this research was designed to present a comprehensive study regarding HPSCC's overall properties and comparing with three common reference concrete types, i.e., normal strength-vibrated concrete (NSVC), high-strength self-compacted concrete (HSSCC), and high-performance highly-viscous concrete (HPVC). The individual comparisons are based on the strength, workability and durability through 14 different types of tests. A new analytical approach has been proposed for a multi-parameter comparison between different types of concrete.

2. Material and Methods

2.1. Material Properties

The materials used for the concrete mixes were cement, silica fume, fly ash, fine and coarse aggregates, water, and superplasticizer. The cement was ordinary Portland cement type CEM-I 42.5R (, the micro-silica was MS90, which consisted of very fine SiO_2 particles (up to 93.1%). The fly ash was type F, primarily consisting of silica, alumina, iron, and calcium oxides. The chemical and the physical properties of binders are shown in Table 1. The fine aggregate was normal fluvial sand, comprising the average passing percentages shown in Table 2. Fluvial gravel with a nominal maximum particle size of 12.5 mm was used in concrete mixes, and the average grading of 3 samples is shown in Table 2. High-performance superplasticizer concrete admixture Sika Viscocrete–5930 was used for obtaining workable or flowable mix made with a low water to cement ratio. The product was a third-generation superplasticizer with a density of 1.095 kg/L. Regarding the manufactures, cement, aggregates, microsilica, fly ash and superplasticizer were provided by Mass-Kurdistan company (Erbil, Iraq), Kalak quarry Hawler company (Erbil, Iraq), Jordan DCP company (Amman, Jordan), Jordan DCP company (Amman, Jordan), and Sika company (Istanbul, Turkey), respectively.

Table 1. Chemical compositions and physical properties of cement, micro-silica, and fly ash.

Characteristics and Main Oxides	Cement	ASTM C150	Micro-Silica	ASTM C1240	Fly Ash	ASTMC 618
CaO (%)	63.12		0.34		1.43	
SiO_2 (%)	23.84		93.11	≥ 85.0	57.32	[Σ (SiO_2 + Al_2O_3 +
Al_2O_3 (%)	4.32		0.62	-	19.88	Fe_2O_3) = 88.9 > 70]
Fe_2O_3 (%)	3.36		1.28	-	11.67	
MgO (%)	1.38	≤ 6.0	1.04	-	1.36	-
SO_3 (%)	1.89	≤ 3.0	0.34	-	0.79	≤ 5.0
Na_2O (%)	-		0.28	-	-	-
H_2O (%)	-		1.08	≤ 3.0	0.24	≤ 3.0
Insoluble residue (%)	0.74	≤ 1.5	-		-	
LOI (%)	1.63	≤ 3.0	0.83	≤ 6.0	2.28	≤ 6.0
Initial setting time (min)	140	≥ 45	-	-	-	-
Final setting time (min)	245	≤ 375	-	-	-	-
Compressive strength in 3 days (MPa)	34.1	≥ 12.0	-	-	-	-
Compressive strength in 7 days (MPa)	42.7	≥ 19.0	-	-	-	-
Specific gravity	3.15		2.64		2.32	-
Fineness (m^2/kg)	316.2	≥ 160	21,700	$\geq 15,000$	-	-

Table 2. Grading of coarse and fine aggregates.

Material/Sieve Size (mm)	Coarse Aggregate (%)	ASTM C33-G7 Limits (%)	Fine Aggregate (%)	ASTM C33 Limits (%)
19	100	100		
12.5	94	90–100		
9.5	58	40–70	100	100
4.75	1	0–15	98	95–100
2.36	0	0–5	84	80–100
1.18			64	50–85
0.6			38	25–60
0.3			16	5–30
0.15			4	0–10
0.075			0	0–3
Fineness Modulus			3	2.3–3.1

2.2. Mix Types and Mix Proportions

Design and selection of the concrete components is the most important step, which subsequently indicates the class and properties of the concrete. The intended concrete class was HPSCC, while three additional reference mixes were selected from 16 trial mixes. The reference mixes were HSSCC, HPVC, and NSVC. The considered four main optimization principles for better concrete production and mix design were workability, strength, cost, and durability. Table 3 can explain that 3 mixes were of the same proportions between cement, sand, and gravel, while NSVC is a conventional normal strength mix. The parameter that changed the HPSCC to self-consolidating concrete was the increased ratio of water, when compared to HPVC, since the binder-to-aggregate ratio was 0.24 for both mixes. Furthermore, the only difference that made HPSCC as high-performance concrete is the admixture type, when compared to HSSCC, as both mixes had the water to binder ratio w/b of 0.35.

Table 3. Mix proportions and compositions for the concrete mixes.

Mix	Cement	Concrete Composition					Concrete Granular Structure				Variation in the Comparison Parameters			
	kg/m³	C	S	G	MS	FA	SP	S/G	B/A	W/B	Ad/C	Workability	Strength	Cost	Durability
NSVC (Reference)	316	316	848	1137	-	-	3.16	0.782	0.16	0.60	0.00	VC	Low	Low	Low
HPSCC	433	433	909	1039	35	-	4.33	0.875	0.24	0.35	0.08	SCC	High	Normal	High
HSSCC	396	396	831	950	-	158	3.96	0.875	0.31	0.35	0.40	SCC	High	Normal	Normal
HPVC	457	457	960	1096	55	-	4.57	0.875	0.24	0.23	0.12	VC	Extra-high	High	High

Note: C: Cement; S: Sand; G: Gravel; MS: Micro-silica; FA: Fly ash; SP: Superplasticizer; B/A: Binder-to-aggregate ratio; W/B: Water-to-binder ratio; Ad/C: Admixture-to-cement ratio.

2.3. Testing Fresh Concrete Properties

SCC is characterized by special fresh concrete properties. Many new tests have been developed to measure the SCC's flowability, viscosity, filling ability, passing ability, resistance to segregation, self-leveling, and stability of the mixture. In this project, the conventional slump test, slump flow test, and J-Ring test were performed. The slump test is acceptable to determine the workability of non-flowable concretes having a slump of 15–230 mm when the cone is raised. When concrete is non-plastic or it is not adequately cohesive, the slump test is no more reasonable. The slump test was performed according to ASTM C143 for NSVC and HPVC mixes (see Figure 1).

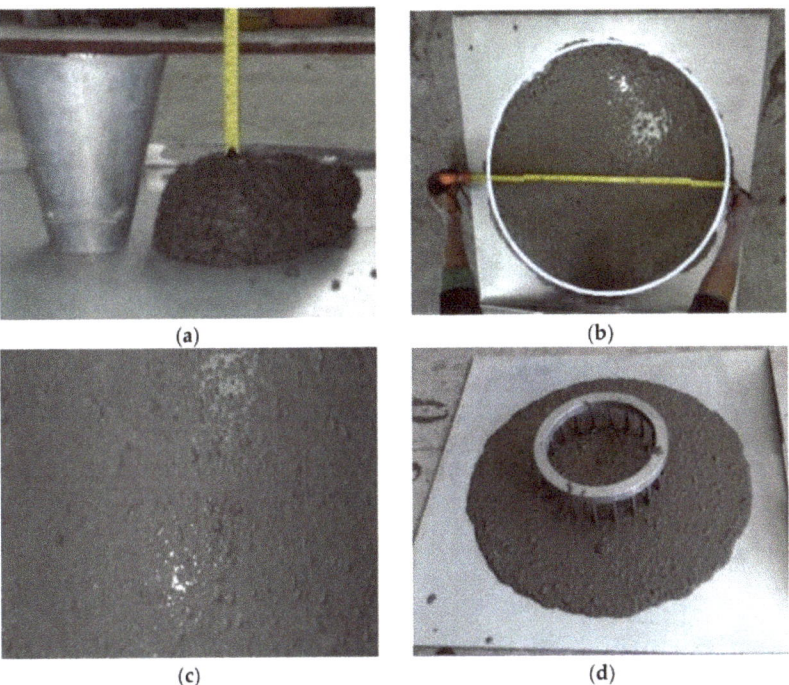

Figure 1. Fresh properties tests (**a**) slump test for vibrated concrete mixes, (**b**) slump flow test for self-consolidating concrete (SCC) mixes, (**c**) a SCC without segregation, and (**d**) restricted slump flow test.

HPVC had low water to binder ratio and more superplasticizer amount, therefore, the mix was very sticky, and needed additional effort for mixing, pouring, and casting. The slump flow test was performed for HPSCC and HSSCC, according to ASTM C1611 [53], to assess the flow rate in the absence of obstructions. During testing the accurate T_{500} (the time required for the slump flow patty to reach a 500 mm diameter) was recorded and when the concrete flow is stopped, the diameter of the spread at right angles is then measured and the mean is the slump flow (Figure 1). The restricted flow test was also performed according to ASTM C1621 for SCC classes. The J-Ring test represents the reinforcement inside the molds that restricts the flow of the concrete.

2.4. Testing Physical Properties of Hardened Concrete

Hardened density and absorption tests were performed for the four concrete mixes. The density of concrete was measured for different shapes and sizes and at different ages, in which the dimensions were measured to the accuracy of 0.01 mm, and the weights to 1 g. In the water absorption test, the concrete cubes were oven-dried at 60 °C for 48 h and the weights were recorded as oven-dry weights. After the cubes were submerged in water for 48 h, the surfaces were dried to represent saturated surface dry concrete.

2.5. Testing of Mechanical Properties

Strength tests are the most common for evaluation of different concrete classes; most of them were related to compressive strength by international standards. It is necessary to test as many as possible mechanical properties for special concrete classes, like HPC and SCC. To study the influence of shape and size of the specimens on compressive strength of different strength classes, 100 mm cubes and Ø100 mm cylinders were tested (Figure 2a).

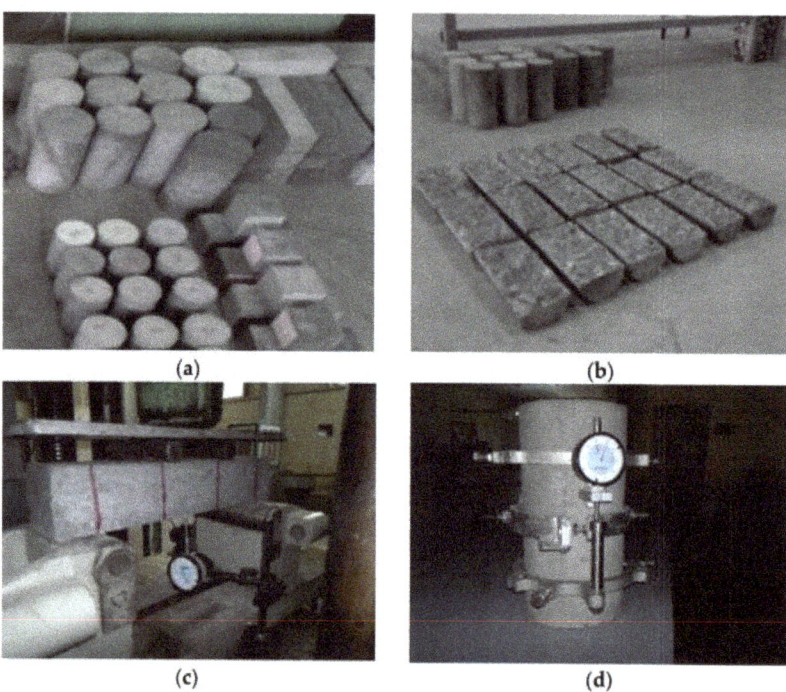

Figure 2. Mechanical properties tests (**a**) various size and shape specimens for compressive strength, (**b**) cylinders in splitting tensile strength test, (**c**) flexural strength test of concrete specimens, and (**d**) testing modulus of elasticity for Ø150 mm cylinders.

The age of concrete was also considered, and the tests were performed at 1, 3, 7, 28, 56, 90, and 180 days. Splitting tensile strength was carried out on Ø100 mm cylinders, in which three cylinders were tested for each mix (Figure 2b). Another most common test for evaluating concrete's tensile strength is the modulus of rupture. For this test, three prisms of 75 mm × 75 mm × 350 mm were prepared for each of the mixes and tested with 300 mm clear-span and third-point loading (Figure 2c). The compressive stress-strain relationship of concrete is the most basic constitutive relationship and is necessary for the understanding of structural response of concrete. The compressive stress-strain relationship was tested using Ø150 mm cylinders, that two cylinders for each of the mixes were tested (Figure 2d). Modulus of elasticity was calculated from the compressive stress-strain relationships.

2.6. Durability Tests of the Concrete Mixes

Heat resistance, direct exposure to the fire, freezing and thawing resistance, and scaling resistance were the tests carried out to assess the durability of the concrete mixes in extreme environments. The resistance of concrete to high temperature is one of the main characteristics of HPC mixes. The age of the concrete cubes of each mix at the time of testing was 36 days, and the maximum temperature of the oven shown in Figure 3 was 1200 °C. During exposure to high temperatures, the degree of strength-loss is dependent on the maximum temperature reached, heating/cooling rate, and the exposure duration. The heating rate was 200 °C/h up to 600 °C, 50 °C/h until 700 °C and whereas, the cooling rate was 25 °C/h. The specimens remained for 7.6 h at a temperature of +600 °C, and 2 h in +700 °C.

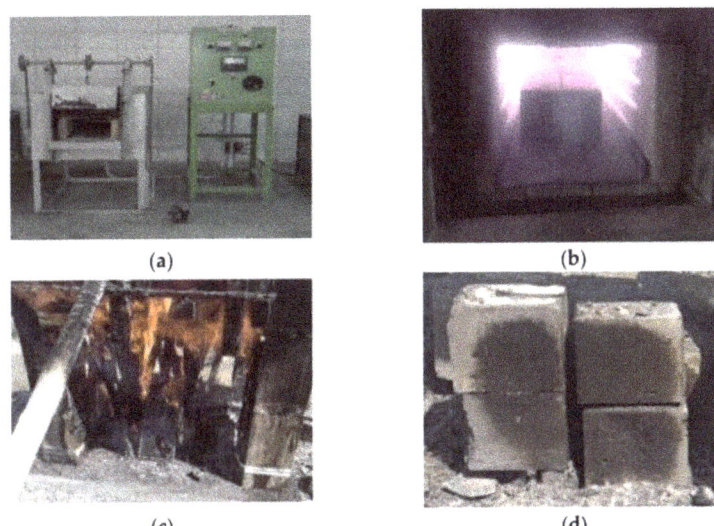

Figure 3. Heating of specimens. (**a**,**b**) Heat resistance test for concrete 100 mm cubes. (**c**,**d**) Exposure to direct fire flame test for concrete cubes.

A fire-attack is mostly considered as an accidental action, instead of a degradation process. For understanding the differences between direct fire resistance and oven heating, additional sets of cubes were subjected to direct fire (Figure 3). The test was performed for NSVC and HPSCC, and the average heating rate was 500 °C/0.5 h while the cooling rate was 95 °C/h. The specimens were exposed to direct fire for 1.70 h at a temperature of +400 °C and 0.75 h in +500 °C, with the maximum temperature reached, was 520 °C. The fire-temperature was regularly measured by a laser thermometer.

In this study, the freeze-thaw test was performed following the same procedure and temperature limitations in ASTM C666 [54], but only for 50 cycles, using 100 mm cubes, as shown in Figure 4a,b. The cubes were submerged in NaCl solution with a concentration of 40 g/L and then tested for loss in weight and strength at 225 days' age so that the possibility of interference of chemical reactions in the microstructure of concrete can be eliminated. The scaling test is used to determine the scaling-resistance of a horizontal concrete-surface exposed to 50 freeze-thaw cycles in the presence of de-icing chemicals. It is intended to evaluate the concrete's surface resistance qualitatively by visual examination as per ASTM C672 [55]. The prepared specimens for the tests were shown in Figure 4c,d of which, an aluminum frame was fixed to concrete specimens by a highly adhesive epoxy. Pans had an inside square dimension of 220 mm, and 25 mm was provided as a dike for the 6 mm depth of the solution.

2.7. Economic Assessment of the Concrete Mixes

Apart from the technical performance parameters, the cost is also an important factor to optimize the concrete mixes. In this study, the cost of concrete mixes was calculated without VAT (taxes). The data for economic assessment considerably vary between regions. This is because local conditions highly affect the cost of labor, and the market costs for recovered materials, as well as the transportation scenarios. In this study, the most probable case scenario for the city center (Erbil, capital of Kurdistan region in Iraq) was considered to estimate the cost of the concrete mixes. The distance between the concrete plant and the raw materials, namely cement and aggregates was 184 km and 110 km, respectively. Besides, the other raw materials are imported from Turkey.

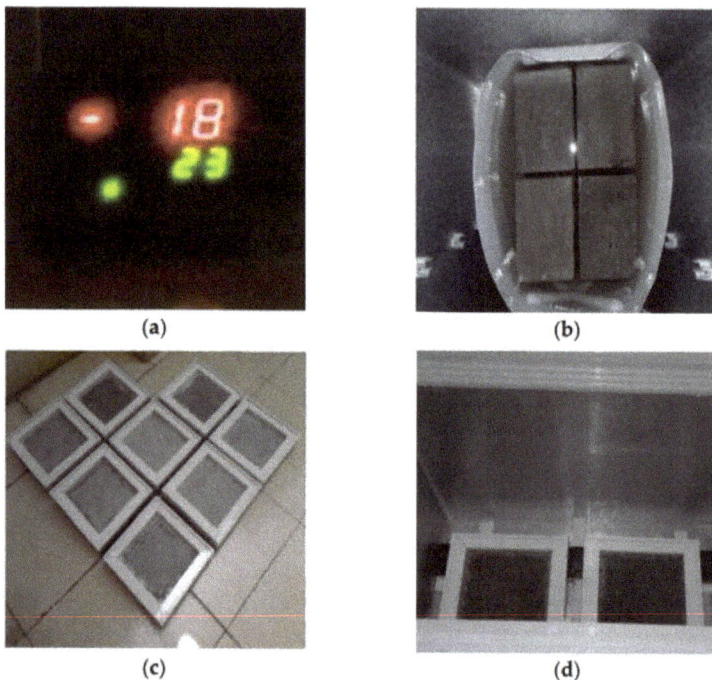

Figure 4. Freeze-thaw testing setup. (**a**,**b**) Freeze-thaw test for concrete cubes. (**c**,**d**) Scaling test for concrete specimens with aluminum frame.

3. Results and Discussions

3.1. Slump Test

The settlement time of NSVC was different from that of HPVC since the latter had contained the superplasticizer; it outspread at a slower rate until 8 s after the lifting of the cone, while NSVC was stable within 3 s. The slump test results were shown in Table 4. Both results were considered acceptable for good workability during casting of concrete, while here the weak point of the slump test can appear when the HPVC was behaving acceptable for the slump test, but the mixture was very stiff that could not perfectly fill the mold without extra vibration.

Table 4. Slump test result of normal strength-vibrated concrete (NSVC) and high-performance highly-viscous concrete (HPVC) mixes, a flow test result of high-performance self-consolidating concrete (HPSCC) and high-strength self-compacted concrete (HSSCC), and restricted flow test result for HPSCC and HSSCC.

	Mix		NSVC	HPVC
Slump test results	Slump	(mm)	190	155
	Stabilization time	(s)	3	8
	Average base diameter	(mm)	405	345
	Mix		HPSCC	HSSCC
Slump flow test results	Segregation Index	(SI)	0	0
	Average flow diameter	(mm)	810	750
	T500	(s)	2.6	3.8
	Mix		HPSCC	HSSCC
Restricted flow test results	Flow diameter	(mm)	760	740
	T500	(s)	3.6	4.1
	ΔH (inside & outside)	Avg. (mm)	3.7	4.5
		Accepted limit (mm)	0–10	0–10

3.2. Flow Test

The slump flow test results were shown in Table 4. The flowing of both concrete types was such that neither bleeding nor segregation had occurred. The flow diameter of micro silica concrete (HPSCC) was 810 mm with T500 of 2.62 s, while the flow diameter of fly ash concrete (HSSCC) was 750 mm, which was lower by 8%, but the T500 was 3.8 s and was higher by 44%. The results also proved that the micro-silica helped in achieving a better flowability (in addition to the higher strength) because the fly ash particles were relatively larger compared to microparticles of the silica. Research showed that the additional grinding of fly ash did not cause workability loss of SCC and the plastic viscosity has increased [31,56].

The flow of 740–900 mm is used as a conformity limit for highly congested structures, and the conformity criteria of 2–5 s is used for the T500 when the improvement of segregation-resistance is necessary. T500 of less than 2 s is applied for very congested structures, better surface finishing, and risk of bleeding or segregation. From a practical point of view, increasing the initial flow head can also increase the flow energy necessary to transport coarse aggregate [1,56]. The ability of SCC mixtures to resist segregation was determined based on the assigned segregation index (SI). If there is no obvious accumulation of coarse aggregate particles and no free water flowing around the concrete's perimeter, the mixture is assumed to have full segregation resistance (SI = 0). If the mixture exhibited an apparent accumulation of coarse aggregate or a small amount of water flowing, the mixture is unlikely to segregate (SI = 1). In case of obvious accumulation of coarse aggregate or free water, the SCC is likely to segregate (SI = 2). Finally, a large amount of accumulated coarse aggregate or a large amount of free water flowing indicates that the concrete will segregate, and the mixture must be rejected [25].

3.3. Restricted Flow Test

Results of the J-Ring test were shown in Table 4, noting that the restricted flow can decrease the flow diameter and T500, while the rate of restriction in the micro-silica concrete (HPSCC) was more than that of the fly ash containing concrete (HSSCC) compared to unrestricted flow. The difference in height inside and outside the ring was clearly showing a better flowability of the HPSCC, since the thickness of the concrete along the diameter of flow was almost homogeneous, neither bleeding nor segregation were observed. Restriction of concrete flow was causing the reduction in flow diameter by 6.2% and 1.3%, respectively for HPSCC and HSSCC, but the HSSCC mix had exhibited a little variation inside and outside the ring, not reaching the limit of segregation. The difference between the T500 values measured using the J-Ring test and the slump flow test should not be more than 2–4 s according to ASTM C1621 [57].

3.4. Density of the Hardened Concrete

Concrete density is an important property that is used in the design of concrete-structures through calculating the self-weight of the members. Table 5 shows the concrete density for the four concrete mixes considering the age of concrete. It can be observed that concrete densities were decreased with time. This phenomenon can be justified by continuous chemical reactions inside the concrete structure. The ratio of weight loss was between 0.5 to 2.5% when comparing 28-day densities with that of 6 months. It can also be noted that the two SCC mixes had less standard deviation (SD) than that of vibrated concrete mixes. When considering the shape and thickness of the concrete, neither systematic relation nor clear differences could be found between tile shaped specimens and the cubes, when comparing the 28-day densities in Table 5.

Table 5. Density of 100 mm cubes and tile shaped specimens, and absorption test results. Abbreviations: High-performance self-consolidating concrete (HPSCC), high-strength self-compacted concrete (HSSCC), normal strength-vibrated concrete (NSVC), and high-performance highly-viscous concrete (HPVC).

Mix	Density of Cubic Specimens (kg/m^3)				Wt. Loss (%)	Density of Tile Shaped Specimens at 28 Days (kg/m^3)		Tile Specimens Difference with the Cubes (%)	Absorption (%), at 28 Days
	28 Days		180 Days						
	Avg.	SD	Avg.	SD		Avg.	SD		
HPSCC	2475	13.5	2437	10.2	1.52	2507	8.4	+1.29	0.97
HSSCC	2393	11.9	2381	10.6	0.50	2417	8.0	+1.00	2.09
NSVC	2417	19.3	2357	17.3	2.48	2459	15.6	+1.74	2.32
HPVC	2534	16.9	2483	15.8	2.00	2526	12.5	−0.32	0.57

3.5. Water Absorption by Immersion

Results of the water absorption test are shown in Table 5, and each value in the table was representing the average of 4 tests. The results show that the water absorption of the NSVC was the largest due to the high void ratio, larger particle sizes, and less binder content compared to other mixes. Very low water absorption was observed in the HPVC mix due to the high binder content and improved packing of the particles as a result. Water absorption of the NSVC was 11% higher than that of the HSSCC; it was 2.4 and 4.1 times higher than that of HPSCC and HPVC, respectively. Similar results of about 2% water absorption were obtained in the previous study [28], while for higher fly ash replacements of 70 and 90%, the absorption was increased to 3.5 and 4.7%, respectively. Research showed that for an HPC with a water–binder ratio of 0.40 at 28 days, the water permeability was about 9 times higher than an HPC with a water–binder ratio of 0.23 [17].

3.6. Compressive Strength

The compressive strength results shown in Table 6 are average values of three 100 mm cubes. The rate of gaining strength is different between concrete types; the ratio of gaining strength at earlier ages (1 day) was 8% for NSVC, but it was 21% and 25% respectively for HPSCC and HPVC. At 28 days NSVC gained two-thirds of its 90 days' strength while the ratio was 72%, 79%, and 85% respectively for HSSCC, HPSCC, and HPVC type mixes.

Table 6. Compressive strength of 100 mm cubes at 7 ages, and Ø100 mm cylinders.

Mix	Compressive Strength (MPa) of 100 mm Cubic Specimens (fcu) in (t) Days (Gained Strength in Percent Relative to 90 Days' Strength)							Compressive Strength (MPa) of Ø100 mm Cylinders (fcy) at 90 Days	fcy/fcu at 90 Days
	1 Day	3 Days	7 Days	28 Days	56 Days	90 Days	180 Days		
HPSCC	20.9 (21)	38.6 (39)	58.1 (58)	79.4 (79)	94.4 (94)	100.2 (100)	104.4 (104)	90.3	0.901
HSSCC	8.4 (10)	31.3 (38)	46.0 (55)	59.9 (72)	74.8 (90)	82.9 (100)	86.2 (104)	71.3	0.860
NSVC	4.5 (08)	11.8 (22)	17.5 (33)	34.6 (65)	46.2 (86)	53.6 (100)	55.3 (103)	42.6	0.794
HPVC	30.2 (25)	48.9 (40)	67.8 (56)	103.5 (85)	116.2 (95)	121.9 (100)	126.7 (104)	117.5	0.964

No considerable differences were observed at 180-days. The water/binder ratio has a great influence on the compressive strength of SCC and VC, whereas, the subject is still controversial and the authors got different conclusions. Some studies on the mechanical behavior of SCC showed that for the same w/b ratio, SCC has generally lower mechanical strengths than traditional vibrated concrete [2]. However, other studies stated: Compared with the majority of the published test results the tendency becomes obvious that at the same w/c ratio, higher compressive strengths were reached for SCC [12]. Three cylinders with Ø100 mm had also been tested for each of the mixes, to study the influence of the shape of the specimen on compressive strength of concrete mixes. Results showed that the higher compressive strength mixes were less affected by specimen shape since the cube to cylinder factor for NSVC mix was 0.79, but it was 0.90 and 0.96 for HPSCC and HPVC mixes, respectively. The shape of specimens and loading direction during tests, were the two factors that controlling (fcy/fcu) ratio. Other authors have reported that the compressive strength ratio of cylinders to cubes is 0.80–0.85 for VC, but it is 0.90–1.00 for SCC [7]. Silica fume is the most commonly used admixture for the production of HPSCC. It has been reported that adding 10% silica fume to the mixtures can increase the compressive strength by 30–100%, 6–57%, or 5–24%, by different authors [15,29,43]. The ratio of the fly ash used for HSSCC was negatively affecting the compressive strength of the mix, since, it is determined that the optimum fly ash content is 25–35%. In essence, fly ashes with 10% do have positive influence on overall quality of SCC, which increases the workability, frost durability and an acceptable level of strength. Further increase in FA% led to reducing of the CaO content, which led to a lower level of hydration [10,25,33].

3.7. Splitting Tensile Strength

The average results of three 100 mm × 200 mm cylinders that tested for splitting tensile strength are shown in Table 7. The tensile strength of HPSCC was 1.53 times that for NSVC, while HSSCC had a tensile strength of only 14% higher than NSVC, remembering that its compressive strength was 67% higher. When comparing the ratio of tensile strength divided by the square root of compressive strength, NSVC had a value of 0.62, but a higher value of 0.65 was recorded for HPSCC mix, 0.66 for HPVC, and again HSSCC was lower and it was only 0.55. Piekarczyk recorded a ratio of 0.61 for an NSC and 0.65 for an SCC [1]. Others stated that the relationship between tensile and compressive strength of SCC is similar to that of VC [2]. Research showed that the average direct tensile strength of the SCC was found to be 3.5 MPa, whilst the average splitting tensile strength was found to be 3.8 MPa, which is only 8.6% higher [58].

Table 7. Splitting tensile strength of Ø100 mm specimens, Flexural strength of 75 mm × 75 mm × 350 mm concrete prisms, and Modulus of Elasticity of Ø150 mm specimens.

Mix	Compressive Strength f_{cy} (MPa)	Splitting Tensile Strength			Flexural Strength				Modulus of Elasticity			
		f_t (MPa)	f_{ct}/f_{cy} (%)	$f_t/\sqrt{f_{cy}}$	f_r (MPa)	f_t (MPa)	f_r/f_t	$f_r/\sqrt{f_{cy}}$	SP1 (MPa)	SP2 (MPa)	E (MPa)	$E/\sqrt{f'_c}$
HPSCC	90.3	6.21	6.88	0.654	7.15	6.21	1.151	0.752	43.24	42.10	42.7	4.49
HSSCC	71.3	4.60	6.45	0.545	4.82	4.60	1.047	0.571	32.91	32.65	32.8	3.88
NSVC	42.6	4.05	9.51	0.621	4.45	4.05	1.098	0.682	27.94	27.78	27.9	4.27
HPVC	117.5	7.19	6.12	0.663	8.41	7.19	1.169	0.776	45.31	46.00	45.7	4.22

3.8. Flexural Strength

The average test results of three 75 mm × 75 mm × 350 mm prisms at age of 90 days were presented in Table 7. HPSCC developed a flexural strength 15% lower than that of HPVC, while it was about 50% higher than both HSSCC and NSVC. The ratio was 0.75 in the case of HPSCC; it was 0.68 for an NSVC and 0.78 for an HPVC. When comparing the results of splitting tensile strength and the flexural strength, it can be noted that the concrete prisms with the higher strength behaved better against tensile stresses than the splitting cylinders.

3.9. Stress-Strain Relationships

The response of the concrete against stresses is different for the four mixes; when the strength of the concrete is higher, the strains were smaller for the same load level. Stress-strain relationships for the concrete mixes were shown in Figure 5. Two cylinders were tested up to 70–90% of the failure load for each of the mixes. The results are showing that the higher strength concrete mixes, especially for HPSCC and HPVC mixes were going in a straight path up to 80% of the load, whereas the case is not similar for NSVC, as it was starting deviation from linearity in the earlier stage of about 40% of the applied load. In 30 MPa stress level, and when comparing other mixes with HPSCC, the strain was higher by 26% and 100% for HSSCC and NSVC respectively, while the strain was less in HPVC by 12%.

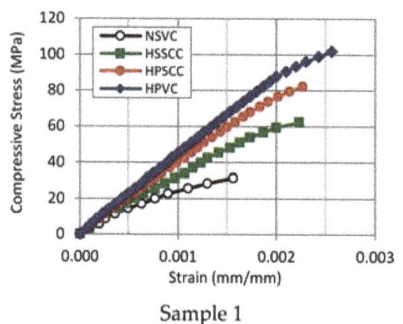
Sample 1

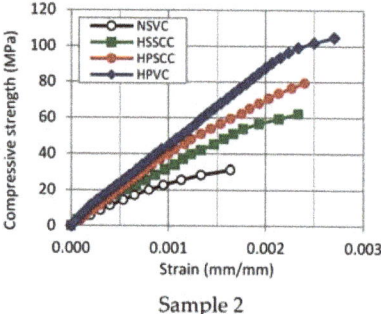
Sample 2

Figure 5. Stress-strain relationships for the four concrete mixes.

3.10. Modulus of Elasticity

Determination of modulus of elasticity was based on the 40% of ultimate load and 0.000050 strain level. The obtained results were arranged in Table 7. When comparing the values in the table, it can be noted that the elastic modulus is not only the function of the compressive strength, since the composition of the mixtures plays a great role. The elastic moduli of the HPVC and HPSCC were close in value despite their different compressive strengths. The ratio of $E/\sqrt{f'c}$ was 4.49 and 4.27 for HPSCC and NSVC mixes respectively. The recorded ratio by [6] was 4.20 and 4.33 for an SCC and NSC, respectively. As it is known, the modulus of elasticity depends on the proportion of Young's modulus of the individual components and their ratio by volume; thus, the modulus of elasticity of concrete increases with a high content of aggregates of high rigidity, whereas it decreases with increasing hardened paste content, and increasing porosity [7]. On the other side, packing of the particles and optimization of the mix composition leads to a higher elastic modulus, even with fewer or no coarse particles as in the case of UHPC [16]. Research showed that the modulus of elasticity of SCC seems to be very similar to that of VC, with an important but similar scatter present on the results for both types of concrete [59]; other authors concluded that the reduction in the elastic modulus of SCC compared to VC is 5% for SCC with high compressive strength (100 MPa) and up to 40% for those with the lowest strength (20 MPa) [2]. Meanwhile, the modulus of elasticity of SCC specimens with SF is increased with SF content increase [25], while it sensitively decreases with an increase in the FA replacement ratio [33].

3.11. Oven Heating Test

Concrete is a composite material that derives properties from its multiphase and multi-scale ingredients. These ingredients are thermally inconsistent and during fire conditions, start to dissociate, leading to degradation in its strength and durability. Although, the behavior of HPSCC subjected to fire has not been extensively studied and thus remains largely unknown [60]. The heating of concrete may be advantageous or causing a reduction in strength. Research showed that concrete specimens exposed to 300 °C might have an increased compressive strength by 18–22%, especially in earlier ages [2,16]. This increase refers mainly to the acceleration of the hydration process at an increased temperature. The limit at which transfers the heating of specimens changes from a useful to destructive factor depends on many parameters; however, in general, temperatures higher than 400 °C are regarded as destructive. In this study, the temperature of the oven was 700 °C, to prevent the explosion of the specimens. Results the Table 8 show that the loss of compressive strength was 79%, 63%, 52%, and 49%, respectively for NSVC, HSSCC, HPSCC, and HPVC. Thus, when a NSC exposed to +700 °C for 2 h it can resist only one fifth of its designed load, but HPC can still resist half of the load. The results of residual compressive strength were within the range of the database presented graphically in [47] which includes limits of codes and researchers' data. After initial heating to 400 °C in [5], the compressive strength decreased by 41–48% for an HSC containing 12.5% of silica fume. At 600 °C and 800 °C, the loss in strength was up to 44% and 79%, respectively. The strength loss at 400 °C was up to 18% whereas, at 600 °C and 800 °C, the strength loss was around 44% and 76%, respectively [19].

Table 8. Resistance of specimens exposed to high temperature, and specimens exposed to direct fire.

Mix	The Resistance of Specimens Exposed to High Temperature							The Resistance of Specimens Exposed to Direct Fire			
	Weight (g)			Compressive Strength (MPa)				Compressive Strength (MPa)			
	Dry in 60 °C	Heated to 700 °C	Weight Loss (%)	Control Cubes	Heated Cubes	Residual Strength (%)	Strength loss (%) by Heating (700 °C)	Control Cubes	Cubes Exposed to Fire (500 °C)	Residual Strength (%)	Strength Loss (%)
HPSCC	2462	2317	5.89	93.2	44.9	48.2	51.8	93.2	73.2	78.5	21.5
HSSCC	2401	2238	6.79	68.2	25.4	37.3	62.7	-	-	-	-
NSVC	2403	2222	7.53	44.0	9.4	21.4	78.6	44.0	29.0	65.8	34.2
HPVC	2516	2379	5.45	105.4	53.9	51.1	48.9	-	-	-	-

When concrete is exposed to a gradually increased temperature, the heat was transferred from the outer face of the concrete to its core, this process requiring less time for NSC and more time for HPC. The weak point of HSC classes is in that the heat was restricted by the dense microstructure, which leads to an explosion and spalling of concrete corners due to pore pressure, but the root cause of the failure was the cracking of concrete due to thermal tensile stresses, and the specimens with higher tensile strength can resist more pore pressure and spalling stresses. The relatively loose microstructure of NSC leads to absorption of heat to the concrete core and disintegrating its structure mainly due to the pore pressure build-up and the development of thermal stresses. Gravel particles can easily pull out and the burned paste is similar to dust, crushable with fingers, as shown in Figure 6, while in the case of HPC, the core of the cube is safer and the bond is still strong. The weight loss of 5–8% is recorded in this study, which is similar to that found by other authors. A mass loss of 4–6% was observed in [46] for 9 different mixes subjected to 1000 °C and last in the furnace for 90 min, and the mass loss of 2–9% has been reported in [19].

NSVC

HPSCC

Figure 6. The cleaned core of broken cubes heated to 700 °C.

3.12. Fire Resistance Test

Subjecting of concrete specimens directly to the fire is different from oven heating, regarding the distribution of the heat around concrete faces. Fire test results on +500 °C for 45 min were shown in Table 8. The loss of compressive strength for NSVC was 34% and it was 22% for HPSCC. Reduction in the strength of the cubes under fire was less than one-half when compared to the heating of the cubes in the oven; however, the main reason for these smaller reductions was the lower level of heating. Generally, the temperature which makes an NSC have a poor strength is in the range of 600 °C. The strength degradation is primarily ascribed to decomposition of hydration products, such as, calcium silicate hydrates, calcium hydroxide, and carbonates. Le et al. reported that HPC loses up to 50% of its ambient temperature strength at 600 °C [46]. The pore pressure development in HPC samples is much faster than in SCC samples. The moisture content, the dense microstructure, and the tensile strength are the main influencing factors that determine the spalling of HPSCC. Research showed that the critical pore diameter of SCC is bigger than HPC; therefore, SCC will have larger damage once exposed to fire. When exposed to the fire of 200 °C for 18 min, the highest pore pressure at 10 mm depth of HPC was 2.52 MPa; while in SCC it was 1.27 MPa [46].

3.13. Freezing and Thawing Cycles

The test results of the freeze-thaw cycles were evaluated through the mass loss of concrete and the residual compressive strength, as presented in Table 9. HPCs showed negligible mass loss of 0.02%, while NSVC exhibited a drastic loss of 83%. HPSCC had lost 9.7% in the compressive strength, whereas, NSVC was almost damaged by losing 86% in compressive strength. HSSCC had lost 6% of its weight and 37% of its strength (the mass loss of this type of specimen was primarily in the top surface, which had less relative density).

Table 9. Loss in mass and compressive strength for 100 mm concrete cubes due to freeze-thaw cycles, Mass losses of concrete specimens the scaling test, and Wearing of the concrete surface due to freeze-thaw cycles.

	Mix		HPSCC	HSSCC	NSVC	HPVC
Loss in mass and compressive strength for 100 mm concrete cubes due to freeze-thaw cycles	Mass loss (%) after	10 cycles	0	5.19	5.43	0
		25 cycles	0.02	5.67	22.94	0
		50 cycles	0.02	6.13	82.64	0.02
	Compressive strength (MPa)	Control cubes	109.8	90	56.1	130.2
		After 50 cycles	99.1	56.9	8.1	120.8
		Residual strength (%)	90.3	63.2	14.4	92.8
Mass losses of concrete specimens at the scaling test, and wearing of the concrete surface due to freeze-thaw cycles	Initial mass (kg)	0 cycles	12.591	12.048	11.621	13.026
	Mass loss (%) after	10 cycles	0	0.62	1.87	0
		25 cycles	0.01	0.87	6.83	0
		50 cycles	0.02	1.14	15.81	0.01
	50 cycles weight loss	kg/m^3	0.387	27.18	372.6	0.286
		kg/m^2	0.0194	1.359	18.631	0.014
	Scaling depth of exposed surface (mm)		$0.010 \approx 0$	$0.570 \approx 1$	$7.905 \approx 8$	$0.005 \approx 0$

The changes in concrete surface and the corresponding number of freeze-thaw cycles were shown in Figure 7. Deterioration processes typically begin when; aggressive fluids penetrate through capillary pore structure to the reaction sites where they trigger chemical or physical deterioration mechanisms [61]. When the test was running, in the first 10 cycles, the NSVC corners were subjected to the internal tensile stress. Later when the frozen salty water was causing volumetric internal pressure on the concrete surface, gravel particles started appearing and then got pulled out. If the pores are critically saturated, water will begin to flow to make room for the increased ice volume. The concrete will rupture if the hydraulic pressure exceeds its tensile strength. The cumulative effect of successive freeze-thaw cycles is the disruption of paste and aggregate eventually causing deterioration of the concrete. HPCs were resisting pore pressure due to their high tensile strength. In this test, loss in dimensions or lose of the concrete cover seems to be logically more acceptable when considering large structural members. The thickness loss was 25–30 mm in NSVC, 2–3 mm for HSSCC and the rest of the cubes almost had no thickness loss as in Figure 7. Similar deteriorations of NSVC and HSSCC concrete cubes have been observed in [30].

Figure 7. Freeze-thaw specimens after 25 and 50 cycles for NSVC and HSSCC and 50 cycles for HPSCC and HPVC.

Water absorption is a key parameter in the investigation of the durability of concrete. Because of its low w/b ratio of 0.20–0.45, it is widely believed that HPSCC should be highly resistant to both scaling and physical breakup due to freezing and thawing. Research showed that non-air-entrained HPC with w/b 0.22–0.31 could be extremely resistant to freeze-thaw damage and it was suggested that air-entrainment and supplementary cementitious materials are not needed. Among six mixtures tested; only the silica fume concrete with w/b 0.22 was frost resistant [8]. The weight change is an indication of the deterioration of the concrete specimen. Weight change of 0.3–5.3% recorded in [35] and 2.0–56.5% is recorded by [30] for 13 SCC mixes.

3.14. Scaling Test

The concrete specimens were exposed to 50 cycles of freezing and thawing. NSVC lose weight of 372.6 kg/m^3 and the aggregate particles appeared in early stages on the entire surface of the concrete; in HSSCC, initially, the first layer of concrete surface wore at earlier stages, but later, the degradation of the concrete surface was almost stopped or it was wearing very slowly so that the total weight loss after 50 cycles reached 27.2 kg/m^3; both HPSCC and HPVC mixes were durable, showed no scaling, and a negligible loss of weight by having 0.387 and 0.286 kg/m^3 respectively, as shown in Table 9. Mass loss of 0–0.5 kg/m^3 after 50 cycles is recommended for HPC. Rating of specimens was performed as in ASTM C672 [55]. Scale rating of 0–1 after 50 cycles is recommended for HPC. The rating results for important checkpoints were shown in Table 10. The test is qualitative, and the rating was decided with a visual examination based on the surface of two specimens. For HSSCC, For HPCs, the top thin paste layer or was resisting the wearing and was not removed until the end of the test; only several small dark spots appeared as shown in

Figure 8. Rating of the concrete surface can be evaluated by loss of concrete mass and visibility of gravel particles, whereas the interesting parameter in a practical point of view is the thickness of the deteriorated concrete, therefore it is better to determine the loss in thickness of the concrete which exposed to freezing and thawing cycles. Table 9 is also showing the concrete depth lost by the action of thermal stresses.

Table 10. Rating of concrete specimens in scaling test ASTM C672.

Mix	Rating/No. of Cycles for the Tested Specimens					Ranking-Surface Conditions According to ASTM C672
	0	5	10	15	25	50	
NSVC	0	1	2	3	4	5	(0) No scaling; (1) Very slight scaling (3 mm depth, max, no coarse aggregate visible); (2) Slight to moderate scaling; (3) Moderate scaling (some coarse aggregate visible); (4) Moderate to severe scaling; (5) Severe scaling (coarse aggregate visible over the entire surface)
HPVC	0	0	0	0	0	0	
HSSCC	0	1	2	2	2	3	
HPSCC	0	0	0	0	0	0	

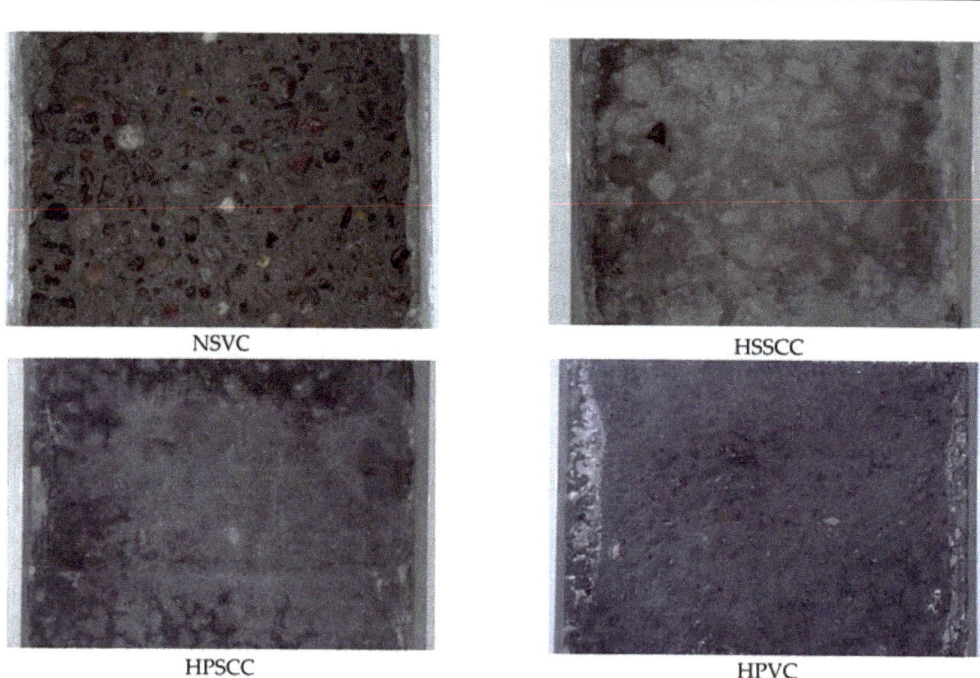

Figure 8. Scaling resistance test of concrete surface exposed to de-icing salts after 50 cycles.

When comparing the scaling test results and the results of freezing and thawing test cubes, NSVC had lost 15.8% of its weight in the scaling test, but the loss was 82.6% in concrete cubes. This difference can be justified by the that the cubes were entirely submerged in water and attacked all sides, but the scaling pans were exposed to freeze-thaw cycles only at the top surface. Gagne et al. tested 27 mixes using silica fume with w/b of 0.23, 0.26, and 0.30, and a wide range of air–void systems. All specimens performed exceptionally well in salt scaling, confirming the durability of HPC. Also in [8] the weight loss of 0.1–4.5 kg/m^2 is obtained after 40 cycles of scaling, for 3 concrete types with an air content of 2, 4, and 6% and w/c ratio of 0.25–0.50.

4. Hexagonal Model for Cost–Strength–Workability–Durability Relationship

An increased upfront cost must be expected for HPSCC, especially for better performance, enhanced quality control, along with better-quality formwork to withstand higher pressures. On the other hand, the incorporation of industry by-products increased productivity, reduced labor, and energy consumption, and fewer post-construction repairs can balance the final cost. The HPSCC is much more economical in certain applications on a basis of the original cost, and also in a point of view the durable upkeep, and more ecological than the usual concrete. Moreover, the lifespan of HPSCC is estimated at two or three times than on a usual concrete. The major obstacles that prevent a wider implementation of HPSCC in construction are its high cost (+25 to 50%), and the lack of knowledge of the properties. In Table 11 the major parameters (strength, workability, durability, and cost) are been considered simultaneously on a scale of 0–10. The strength and durability results were converted based on the highest value obtained among the four concrete types.

Table 11. Comparison scales and rating for the four concrete mixes.

Mix	Strength, 90 Days (MPa)		Durability, 90 Days (%)		Workability, (mm)	Cost ($)	MPAS		
	Compressive Strength	Tensile Strength	Oven Residual Strength	Residual Strength in F/T Cycles	Slump or Slump Flow	Cost-Effectiveness	Average Out of 10	Hexagon Area	Relative Area
HPSCC	100.2	6.21	48.2	90.3	810	135	-	-	-
S.N.	8	9	9	10	10	6	8.7	196.6	1
HSSCC	82.9	4.6	37.3	63.2	750	125	-	-	-
S.N.	7	6	7	7	10	6	6.1	132.1	0.67
NSVC	53.6	4.05	21.4	14.4	190	80	-	-	-
S.N.	4	6	4	2	6	10	5.3	69.3	0.35
HPVC	121.9	7.19	51.1	92.8	155	140	-	-	-
S.N.	10	10	10	10	4	6	8.3	173.2	0.88

S.N. standardized number.

For comparing the workability results, SCCs were considered as perfectly workable, while the vibrated concretes can achieve 6 points for a 200 mm slump. Decreasing the slump has to decrease in the scale by one point for each additional 25 mm. The cost is negatively influencing the concrete selection; therefore, the lowest price was divided by the cost of other concrete types and multiplied by 10. The shown values of the concrete cost are based on the local prices for the preparation of 1 m^3 of concrete. The average results of multi-parameter assessment scale (MPAS) are showing that the HPSCC is the best option (MPAS = 196.6) in selecting a mix among the tested classes (Figure 9).

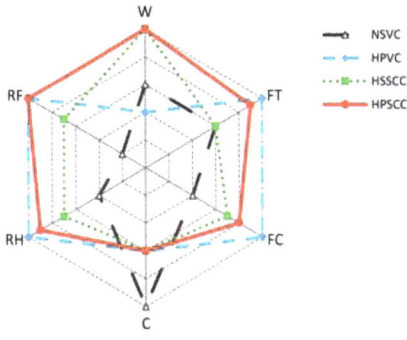

Figure 9. Hexagonal model for multi-parameter comparison of different concrete classes (W—workability measures; FT—tensile strength; FC—compressive strength; C—economic considerations; RH—residual strength after heating to +700 °C; RF—residual strength after 50 cycles of freezing and thawing).

5. Conclusions

The current study focuses on the assessment of HPSCC through an experimental program including 14 major performance tests that have mostly been used individually in the previous studies, and concluding improvement of some concrete properties, while some properties questionably remained if they were improved or worsened. A comparative and comprehensive study on the HPSCC can guarantee and answer that, what will happen to other properties. It was intended to apply the hexagonal model to some experimental studies in the literature, while the multi-parameter experiments (workability, cost, strength, and durability) can hardly or never be found. Therefore, it is of great interest to investigate such a type of comprehensive experiments. Based on the results and discussion, the following conclusions can be drawn:

i. HPSCC exhibited excellent workability and flowability compared to NSVC and HPVC mixes that need vibration or compaction effort, hence, HPSCC is eco-friendly and beneficial for the environment by using sustainable materials like micro-silica and fly ash.

ii. The density of the HPSCC was almost in the range of normal concretes, whereas the absorption of HPSCC was less than one half of both HSSCC and NSVC.

iii. HPSCC was an early strength-gaining concrete by obtaining 20.9 MPa of compressive strength in one day, and 100.2 MPa at 90 days, which is higher than that of NSVC and HSSCC by 86.9% and 20.9%, respectively. The tensile strength of HPSCC from both splitting tensile strength and flexural resistance was much higher than the two mentioned mixes. Thus, due to its strength HPSCC can be useful in decreasing structural section sizes, and thereby the amount of concrete and cement used in construction projects.

iv. HPSCC showed an elastic modulus of 42.7 GPa and had better resistance to the compressive strains and deformation, compared to those of NSVC and HSSCC. The relationship of the stress-strain curves was linear for HPSCC up to 80% of the ultimate load, while the curve for NSVC was starting deviation from linearity in an earlier stage of about 40% of the applied load.

v. When exposed to fire flame or high temperature of 700 °C for 2 h, HPSCC behaved as a durable concrete, and had a residual strength of 48.2%, while the residual strength of NSVC and HSSCC were 21.4% and 37.3%, respectively. So, during fire accidents, HPSCC can survive more, and the probability of demolishing after the fire and new construction is much less.

vi. Freeze-thaw 50 cycles were causing degradation in the compressive strength of HPSCC by only 9.7%, while the strength loss of NSVC and HSSCC was 85.6% and 36.8% respectively. In the scaling test, the average thickness losses for HPSCC were almost negligible, while the thickness loss of NSVC and HSSCC was 8 mm and 1 mm, respectively.

vii. The proposed hexagonal comparison model can successfully predict the most beneficial concrete mixes, considering workability, strength, cost, and durability.

Even though this study considered the main parameters of concrete such as technical properties and cost, further studies are still required on this path since the optimization (final outputs) was made based on 1 m^3 of concrete. Therefore, further studies on the mentioned parameters must be done by considering the real structural application such as beam and column.

Author Contributions: Conceptualization, G.H.A.; methodology, G.H.A., H.A., B.A. and R.A.; software, G.H.A., H.A., B.A. and R.A.; validation, G.H.A., H.A., B.A. and R.A.; formal analysis, G.H.A., H.A., B.A. and R.A.; investigation, G.H.A., H.A., B.A. and R.A.; resources, G.H.A., H.A., B.A. and R.A.; data curation, G.H.A.; writing—original draft preparation, G.H.A.; writing—review and editing, H.A., B.A. and R.A.; visualization, G.H.A., H.A., B.A. and R.A.; supervision, G.H.A., H.A., B.A. and R.A.; project administration, G.H.A., H.A., B.A. and R.A.; funding acquisition, R.A. All authors have read and agreed to the published version of the manuscript.

Funding: This research was funded by the Deanship of scientific research, Prince Sattam bin Abdulaziz University, grant number No. 2020/01/16810.

Institutional Review Board Statement: Not applicable.

Informed Consent Statement: Not applicable.

Data Availability Statement: No new data were created or analyzed in this study. Data sharing is not applicable to this article.

Acknowledgments: This publication was supported by the Deanship of Scientific Research, Prince Sattam bin Abdulaziz University under research project No. 2020/01/16810.

Conflicts of Interest: The authors declare no conflict of interest.

References

1. Łaźniewska-Piekarczyk, B. The influence of chemical admixtures on cement hydration and mixture properties of very high performance self-compacting concrete. *Constr. Build. Mater.* **2013**, *49*, 643–662. [CrossRef]
2. Pineaud, A.; Pimienta, P.; Rémond, S.; Carre, H. Mechanical properties of high performance self-compacting concretes at room and high temperature. *Constr. Build. Mater.* **2016**, *112*, 747–755. [CrossRef]
3. Łaźniewska-Piekarczyk, B. Effect of viscosity type modifying admixture on porosity, compressive strength and water penetration of high performance self-compacting concrete. *Constr. Build. Mater.* **2013**, *48*, 1035–1044. [CrossRef]
4. Farooq, F.; Rahman, S.K.U.; Akbar, A.; Khushnood, R.A.; Javed, M.F.; Alyousef, R.; Alabduljabbar, H.; Aslam, F. A comparative study on performance evaluation of hybrid GNPs/CNTs in conventional and self-compacting mortar. *Alex. Eng. J.* **2020**, *59*, 369–379. [CrossRef]
5. Noumowé, A.; Carre, H.; Daoud, A.; Toutanji, H. High-strength self-compacting concrete exposed to fire test. *J. Mater. Civ. Eng.* **2006**, *18*, 754–758. [CrossRef]
6. Ahmad, S.; Umar, A.; Masood, A. Properties of normal concrete, self-compacting concrete and glass fibre-reinforced self-compacting concrete: An experimental study. *Procedia Eng.* **2017**, *173*, 807–813. [CrossRef]
7. Kang, S.-H.; Hong, S.-G.; Moon, J. Shrinkage characteristics of heat-treated ultra-high performance concrete and its mitigation using superabsorbent polymer based internal curing method. *Cem. Concr. Compos.* **2018**, *89*, 130–138. [CrossRef]
8. Łaźniewska-Piekarczyk, B. Investigations on the relationship between porosity and strength of admixtures modified high performance self-compacting concrete. *J. Civ. Eng. Manag.* **2016**, *22*, 520–528. [CrossRef]
9. Wilson, M.; Kosmatka, S. *Design and Control of Concrete Mixtures*, 15th ed.; Portland Cement Association: Skokie, IL, USA, 2011.
10. Duran-Herrera, A.; De-León-Esquivel, J.; Bentz, D.; Valdez-Tamez, P. Self-compacting concretes using fly ash and fine limestone powder: Shrinkage and surface electrical resistivity of equivalent mortars. *Constr. Build. Mater.* **2019**, *199*, 50–62. [CrossRef]
11. Aggarwal, P.; Siddique, R.; Aggarwal, Y.; Gupta, S.M. Self-compacting concrete-procedure for mix design. *Leonardo Electron. J. Pract. Technol.* **2008**, *12*, 15–24.
12. Kumar, P.; Kumar, R.; Gupta, K. Study on Normally Vibrated Concrete to Self-Compacting Concrete. *J. Ceram. Concr. Technol.* **2016**, *1*, 1–18.
13. Boel, V.; Audenaert, K.; de Schutter, G. Gas permeability and capillary porosity of self-compacting concrete. *Mater. Struct.* **2008**, *41*, 1283–1290. [CrossRef]
14. Caldarone, M.A.; Taylor, P.C.; Detwiler, R.J.; Bhide, S.B. *Guide Specification for High Performance Concrete for Bridges*; Portland Cement Association: Skokie, IL, USA, 2005.
15. Zhutovsky, S.; Kovler, K. Influence of water to cement ratio on the efficiency of internal curing of high-performance concrete. *Constr. Build. Mater.* **2017**, *144*, 311–316. [CrossRef]
16. Aziz, O.Q.; Ahmed, G.H. Mechanical properties of ultra-high performance concrete (UHPC). In Proceedings of the 12th International Conference on Recent Advances in Concrete Technology and Sustainability Issues, Prague, Czech Republic, 31 October–2 November 2012; American Concrete Institute, ACI Special Publication: Prague, Czech Republic, 2012; pp. 1–16.
17. Wang, D.; Shi, C.; Wu, Z.; Xiao, J.; Huang, Z.; Fang, Z. A review on ultra high performance concrete: Part II. Hydration, microstructure and properties. *Constr. Build. Mater.* **2015**, *96*, 368–377. [CrossRef]
18. Afroughsabet, V.; Biolzi, L.; Ozbakkaloglu, T. Influence of double hooked-end steel fibers and slag on mechanical and durability properties of high performance recycled aggregate concrete. *Compos. Struct.* **2017**, *181*, 273–284. [CrossRef]
19. Nadeem, A.; Memon, S.A.; Lo, T.Y. The performance of fly ash and metakaolin concrete at elevated temperatures. *Constr. Build. Mater.* **2014**, *62*, 67–76. [CrossRef]
20. Kurda, R.; de Brito, J.; Silvestre, J.D. CONCRETop method: Optimization of concrete with various incorporation ratios of fly ash and recycled aggregates in terms of quality performance and life-cycle cost and environmental impacts. *J. Clean. Prod.* **2019**, *226*, 642–657. [CrossRef]
21. de Brito, J.; Kurda, R. The past and future of sustainable concrete: A critical review and new strategies on cement-based materials. *J. Clean. Prod.* **2021**, *281*, 123558. [CrossRef]
22. de Brito, J.; Kurda, R. Low Binder Concrete and Mortars. *Appl. Sci.* **2020**, *10*, 3866. [CrossRef]

23. Memon, S.A.; Shah, S.F.A.; Khushnood, R.A.; Baloch, W.L. Durability of sustainable concrete subjected to elevated temperature–A review. *Constr. Build. Mater.* **2019**, *199*, 435–455. [CrossRef]
24. Seleem, H.E.D.H.; Rashad, A.M.; Elsokary, T. Effect of elevated temperature on physico-mechanical properties of blended cement concrete. *Constr. Build. Mater.* **2011**, *25*, 1009–1017. [CrossRef]
25. Hamzah, A.F.; Ibrahim, M.W.; Jamaluddin, N.; Jaya, R.; Abidin, N.E.Z. Cementitious materials usage in self-compacting concrete: A review. *Adv. Mater. Res.* **2015**, *1113*, 153–160. [CrossRef]
26. Sfikas, P. Self-compacting concrete: History and current trends. *Concrete* **2017**, *51*, 12–16.
27. Dybeł, P.; Wałach, D.; Ostrowski, K. The top-bar effect in specimens with a single casting point at one edge in high-performance self-compacting concrete. *J. Adv. Concr. Technol.* **2018**, *16*, 282–292. [CrossRef]
28. Megid, W.A.; Khayat, K.H. Effect of concrete rheological properties on quality of formed surfaces cast with self-consolidating concrete and superworkable concrete. *Cem. Concr. Compos.* **2018**, *93*, 75–84. [CrossRef]
29. El-Chabib, H.; Syed, A. Properties of self-consolidating concrete made with high volumes of supplementary cementitious materials. *J. Mater. Civ. Eng.* **2013**, *25*, 1579–1586. [CrossRef]
30. Sideris, K.K.; Tassos, C.; Chatzopoulos, A.; Manita, P. Mechanical characteristics and durability of self compacting concretes produced with ladle furnace slag. *Constr. Build. Mater.* **2018**, *170*, 660–667. [CrossRef]
31. Kostrzanowska-Siedlarz, A.; Gołaszewski, J. Rheological properties of high performance self-compacting concrete: Effects of composition and time. *Constr. Build. Mater.* **2016**, *115*, 705–715. [CrossRef]
32. Huseien, G.F.; Sam, A.R.M.; Alyousef, R. Texture, morphology and strength performance of self-compacting alkali-activated concrete: Role of fly ash as GBFS replacement. *Constr. Build. Mater.* **2020**, 121368. [CrossRef]
33. Gholampour, A.; Ozbakkaloglu, T. Performance of sustainable concretes containing very high volume Class-F fly ash and ground granulated blast furnace slag. *J. Clean. Prod.* **2017**, *162*, 1407–1417. [CrossRef]
34. Tufail, M.; Shahzada, K.; Gencturk, B.; Wei, J. Effect of elevated temperature on mechanical properties of high strength concrete. *Int. J. Concr. Struct. Mater.* **2017**, *11*, 17–28. [CrossRef]
35. Olanike, A.O. Experimental Investigation into the Freeze-Thaw Resistance of Concrete Using Recycled Concrete Aggregates and Admixtures. *Ratio* **2014**, *85*, 120.
36. Alyousef, R.; Benjeddou, O.; Khadimallah, M.A.; Mohamed, A.M.; Soussi, C. Study of the Effects of Marble Powder Amount on the Self-Compacting Concretes Properties by Microstructure Analysis on Cement-Marble Powder Pastes. *Adv. Civ. Eng.* **2018**, *2018*, 6018613. [CrossRef]
37. Bogas, J.; Hawreen, A. Capillary Absorption and Oxygen Permeability of Concrete Reinforced with Carbon Nanotubes. *Adv. Civ. Eng. Mater.* **2019**, *8*, 1–13.
38. Hawreen, A.; Bogas, J.A.; Kurda, R. Mechanical Characterization of Concrete Reinforced with Different Types of Carbon Nanotubes. *Arab. J. Sci. Eng.* **2019**, *44*, 8361–8376. [CrossRef]
39. Abu Maraq, M.A.; Tayeh, B.A.; Ziara, M.M.; Alyousef, R. Flexural behavior of RC beams strengthened with steel wire mesh and self-compacting concrete jacketing—Experimental investigation and test results. *J. Mater. Res. Technol.* **2021**, *10*, 1002–1019. [CrossRef]
40. Li, Y.; Pimienta, P.; Pinoteau, N.; Tan, K.-H. Effect of aggregate size and inclusion of polypropylene and steel fibers on explosive spalling and pore pressure in ultra-high-performance concrete (UHPC) at elevated temperature. *Cem. Concr. Compos.* **2019**, *99*, 62–71. [CrossRef]
41. Alyousef, R.; Khadimallah, M.A.; Soussi, C.; Benjeddou, O.; Jedidi, M. Experimental and Theoretical Study of a New Technique for Mixing Self-Compacting Concrete with Marble Sludge Grout. *Adv. Civ. Eng.* **2018**, *2018*, 3283451. [CrossRef]
42. Alabduljabbar, H.; Alyousef, R.; Alrshoudi, F.; Alaskar, A.; Fathi, A.; Mohamed, A.M. Mechanical effect of steel fiber on the cement replacement materials of self-compacting concrete. *Fibers* **2019**, *7*, 36. [CrossRef]
43. Kannan, V.; Jerin, C.; Murali, D.K. A Review on Self Compacting Concrete. *Int. J. Adv. Res. Eng. Manag.* **2015**, *1*, 64–68.
44. Alyousef, R.; Benjeddou, O.; Soussi, C.; Khadimallah, M.A.; Mohamed, A.M. Effects of Incorporation of Marble Powder Obtained by Recycling Waste Sludge and Limestone Powder on Rheology, Compressive Strength, and Durability of Self-Compacting Concrete. *Adv. Mater. Sci. Eng.* **2019**, *2019*, 4609353. [CrossRef]
45. Celik, K.; Meral, C.; Gursel, A.P.; Mehta, P.K.; Horvath, A.; Monteiro, P.J. Mechanical properties, durability, and life-cycle assessment of self-consolidating concrete mixtures made with blended portland cements containing fly ash and limestone powder. *Cem. Concr. Compos.* **2015**, *56*, 59–72. [CrossRef]
46. Ye, G.; de Schutter, G.; Taerwe, L. Study of vapor pressure of high performance concrete and self-compacting concrete slabs subjected to standard fire conditions. In Proceedings of the 2nd International RILEM Workshop on Concrete Spalling due to Fire Exposure, Delft, The Netherlands, 5–7 October 2011; RILEM Publications SARL: Bagneux, France, 2011.
47. Phan, L.T. High-strength concrete at high temperature-an overview. In Proceedings of the 6th International Symposium on Utilization of High Strength/High Performance Concrete, Leipzig, Germany, 1 June 2002; pp. 501–518.
48. Ahmad, S.; Khushnood, R.A.; Jagdale, P.; Tulliani, J.-M.; Ferro, G.A. High performance self-consolidating cementitious composites by using micro carbonized bamboo particles. *Mater. Des.* **2015**, *76*, 223–229. [CrossRef]
49. Manjunath, R.; Narasimhan, M.C.; Kumar, S. Effects of fiber addition on performance of high-performance alkali activated slag concrete mixes: An experimental evaluation. *Eur. J. Environ. Civ. Eng.* **2020**, 1–16. [CrossRef]

50. Elemam, W.E.; Abdelraheem, A.H.; Mahdy, M.G.; Tahwia, A.M. Optimizing fresh properties and compressive strength of self-consolidating concrete. *Constr. Build. Mater.* **2020**, *249*, 118781. [CrossRef]
51. Ghezal, A.; Khayat, K.H. Optimizing self-consolidating concrete with limestone filler by using statistical factorial design methods. *Mater. J.* **2002**, *99*, 264–272.
52. Sharifi, E.; Sadjadi, S.J.; Aliha, M.; Moniri, A. Optimization of high-strength self-consolidating concrete mix design using an improved Taguchi optimization method. *Constr. Build. Mater.* **2020**, *236*, 117547. [CrossRef]
53. ASTM C1611/C1611M-18. *Standard Test Method for Slump Flow of Self-Consolidating Concrete*; ASTM International: West Conshohocken, PA, USA, 2018.
54. ASTM C666/C666M-15. *Standard Test Method for Resistance of Concrete to Rapid Freezing and Thawing*; ASTM International: West Conshohocken, PA, USA, 2015.
55. ASTM C672/C672M-12. *Standard Test Method for Scaling Resistance of Concrete Surfaces Exposed to Deicing Chemicals*; ASTM International: West Conshohocken, PA, USA, 2012.
56. Hisseine, O.A.; Basic, N.; Omran, A.F.; Tagnit-Hamou, A. Feasibility of using cellulose filaments as a viscosity modifying agent in self-consolidating concrete. *Cem. Concr. Compos.* **2018**, *94*, 327–340. [CrossRef]
57. ASTM C1621/C1621M-17. *Standard Test Method for Passing Ability of Self-Consolidating Concrete by J-Ring*; ASTM International: West Conshohocken, PA, USA, 2017.
58. Alhussainy, F.; Hasan, H.A.; Rogic, S.; Sheikh, M.N.; Hadi, M.N. Direct tensile testing of self-compacting concrete. *Constr. Build. Mater.* **2016**, *112*, 903–906. [CrossRef]
59. Craeye, B.; Van Itterbeeck, P.; Desnerck, P.; Boel, V.; De Schutter, G. Modulus of elasticity and tensile strength of self-compacting concrete: Survey of experimental data and structural design codes. *Cem. Concr. Compos.* **2014**, *54*, 53–61. [CrossRef]
60. Lura, P.; Terrasi, G.P. Reduction of fire spalling in high-performance concrete by means of superabsorbent polymers and polypropylene fibers: Small scale fire tests of carbon fiber reinforced plastic-prestressed self-compacting concrete. *Cem. Concr. Compos* **2014**, *49*, 36–42. [CrossRef]
61. Mirgozar Langaroudi, M.A.; Mohammadi, Y. Effect of nano-clay on the freeze-thaw resistance of self-compacting concrete containing mineral admixtures. *Eur. J. Environ. Civ. Eng.* **2019**, 1–20. [CrossRef]

Article

Effect of Varying Steel Fiber Content on Strength and Permeability Characteristics of High Strength Concrete with Micro Silica

Babar Ali [1], Rawaz Kurda [2,3,4,*], Bengin Herki [5,6], Rayed Alyousef [7], Rasheed Mustafa [8], Ahmed Mohammed [9], Ali Raza [10], Hawreen Ahmed [2,3,4] and Muhammad Fayyaz Ul-Haq [1]

1. Department of Civil Engineering, COMSATS University Islamabad—Sahiwal Campus, Sahiwal 57000, Pakistan; babar.ali@cuisahiwal.edu.pk (B.A.); m.fayyaz@cuisahiwal.edu.pk (M.F.U.-H.)
2. Department of Highway and Bridge Engineering, Technical Engineering College, Erbil Polytechnic University, Erbil 44001, Iraq; Hawreen.a@gmail.com
3. Scientific Research and Development Center, Nawroz University, Duhok 42001, Iraq
4. CERIS, Civil Engineering, Architecture and Georresources Department, Instituto Superior Técnico, Universidade de Lisboa, Av. Rovisco Pais, 1049-001 Lisbon, Portugal
5. Department of Civil Engineering, College of Science and Engineering, Bayan University, Erbil 44001, Iraq; bengin.awdel@bnu.edu.iq
6. Department of Civil Engineering, Faculty of Engineering, Soran University, Soran 44008, Iraq
7. Department of Civil Engineering, College of Engineering, Prince Sattam bin Abdulaziz University, Alkharj 16273, Saudi Arabia; r.alyousef@psau.edu.sa
8. Department of Environmental Engineering, College of Engineering, Knowledge University, Erbil 44001, Iraq; rasheed1954@yahoo.com
9. Civil Engineering Department, College of Engineering, University of Sulaimani, Sulaymaniyah 46001, Iraq; ahmed.mohammed@univsul.edu.iq
10. Department of Civil Engineering, Pakistan Institute of Engineering and Technology, Multan 66000, Pakistan; aliraza@piet.edu.pk
* Correspondence: rawaz.kurda@tecnico.ulisboa.pt; Tel.: +96-47505834949

Received: 26 November 2020; Accepted: 14 December 2020; Published: 16 December 2020

Abstract: For the efficient and durable design of concrete, the role of fiber-reinforcements with mineral admixtures needs to be properly investigated considering various factors such as contents of fibers and potential supplementary cementitious material. Interactive effects of fibers and mineral admixtures are also needed to be appropriately studied. In this paper, properties of concrete were investigated with individual and combined incorporation of steel fiber (SF) and micro-silica (MS). SF was used at six different levels i.e., low fiber volume (0.05% and 0.1%), medium fiber volume (0.25% and 0.5%) and high fiber volume (1% and 2%). Each volume fraction of SF was investigated with 0%, 5% and 10% MS as by volume of binder. All concrete mixtures were assessed based on the results of important mechanical and permeability tests. The results revealed that varying fiber dosage showed mixed effects on the compressive (compressive strength and elastic modulus) and permeability (water absorption and chloride ion penetration) properties of concrete. Generally, low to medium volume fractions of fibers were useful in advancing the compressive strength and elastic modulus of concrete, whereas high fiber fractions showed detrimental effects on compressive strength and permeability resistance. The addition of MS with SF is not only beneficial to boost the strength properties, but it also improves the interaction between fibers and binder matrix. MS minimizes the negative effects of high fiber doses on the properties of concrete.

Keywords: mechanical properties; fiber-reinforced concrete; permeability; durability; tensile strength; micro-silica/silica fume; steel fiber

1. Introduction

Plain cement concrete (PCC) is the most versatile construction material owing to its multiple benefits i.e., high compressive strength, cost-efficient, in-situ formability, thermal and electrical insulation, imperviousness, etc. Ingredients and mix design of PCC can be changed to obtain different types of concrete that suit different structural loadings and environments. To further raise the importance of conventional concrete, several performance-related issues need to be resolved. PCC generally performs weakly under tensile loadings. Its splitting tensile strength (STS) is very low compared to its compressive strength (CS). According to Ali and Qureshi [1,2] and Koushkbaghi et al. [3] PCC has a STS/CS ratio of about 7–9.5%. Zain et al. [4] showed that the STS/CS ratio decreases as the strength class of concrete is upgraded, therefore, high strength PCCs are more vulnerable to brittle failure than normal strength PCCs. PCC has low energy absorption capacity (or toughness) under both tensile and compressive loadings [5,6]. It undergoes a sudden failure after carrying the load beyond its peak capacity and it has very low residual strength (almost negligible compared to fiber-reinforced concrete) [7]. Mechanical performance of PCC undergoes significant degradation with time when subjected to environmental stresses (freeze-thaw) [8] and weathering actions of water and acid attack [9]. Due to the inherent weakness of PCC under tensile loadings, large structural dimensions cannot be avoided unless it is reinforced with some high strength material i.e., steel rebars, glass fiber-reinforced polymers (GFRP) bars, carbon fiber-reinforced polymer (CFRP) bars, etc.

Nowadays, fibers are being used as a discrete 3-dimensional reinforcement to overcome the deficiency of PCC in tensile strength. With the addition of proper fiber-type in concrete, initiation and proliferation cracks under both tensile and compressive loads can be controlled or delayed. Many types of reinforcements are available commercially that own their application-specific characteristics e.g., carbon fiber, steel fiber (SF), glass fiber, polypropylene fiber [10] organic fibers [11], carbon nano-tubes [12,13] etc. Tensile reinforcements disperse throughout the PCC matrix, bridge the microcracking [12,13]. Considering the mechanical performance of concrete, SF is by far more superior fiber compared to other industrial fibers [10]. SF has a very high tensile strength of over 1200 MPa and an elastic modulus of about 200 GPa. The literature confirms its vitality as a superior reinforcement material that ensures satisfying tensile, compressive, flexural and shear strength properties [14–17]. By improving the strength per unit quantity of material, SF-reinforced concrete (SFRC) shows lower economic and environmental impact compared to PCC [14].

There are several issues relevant to the underutilization of SF in fiber-reinforced concrete that must be addressed. According to Lee et al. [18], the primary reason for the failure of SF-Reinforced Concrete (SFRC) is the failure linked to the interface between fiber and binder matrix of the concrete. Two different types of failures are linked to SFRCs when the underutilized fiber is separated from the binder matrix [19]. The first type of failure occurs at the interface between the fiber and binder matrix, whereas another type of failure occurs in the adhering binder matrix. Both of these failures lead to under-utilization of matrix and full tensile strength potential of fibers, and consequently leading to cracking of concrete. Therefore, bond strength between fibers and binder matrix plays a significant role in defining the tensile capacity of fibrous composites.

The effect of SF on the properties of the fibrous composite also depends on the dosage of fiber. There is a consensus among researchers that the tensile and bending capacity of concrete improves the increasing fiber dosage (0 to 2% by volume fraction) [3,9,20–23], but the literature has shown the mixed effects of varying SF dosage on the compressive behavior of concrete. Some studies [23,24] report that SF induces porosity into concrete therefore, compressive strength and elastic modulus of concrete undergoes degradation with the rising fiber volume. Whereas, there are studies [3,9,21] which have shown positive effects of SF on compressive behavior of concrete. The improvements in compressive strength were attributed to the increased stiffness and confinement of concrete [3,9,21]. However, a consensus is found among findings of the researchers that SF is beneficial to compression toughness of concrete [5,23,25,26].

There are some issues related to application of SFRC that are vital to be understood and resolved. The reasons behind the mixed effects (positive, negative or inconsiderable) of SF dose on compressive strength of concrete are still needed to be understood for the effective use of fibers under compression loadings. Moreover, poor bond strength between fiber and binder matrix reduces the utilization of full potential of SF. It is essential to investigate the influence of such materials (i.e., MS) on SFRC that help in strengthening binder matrix and improve the dispersion of fibers [2,10]. This study is designed to evaluate the effects of varying SF dosage on the properties of a high strength concrete. To investigate the role of bond strength at interfacial zone between fiber and matrix, the effect of SF dosage was also explored with and without micro-silica (MS). SF was used at six different levels i.e., low fiber volume (Vf = 0.05% and 0.1%), medium fiber volume (Vf = 0.25% and 0.5%) and high fiber volume (Vf = 1% and 2%). Each dose of SF was investigated with 0%, 5% and 10% MS as by volume of binder. MS is an excellent consumer of portlandite CH (in pozzolanic reaction, that strengthens the binder matrix and improves the bond strength of fibers [27]. Concrete mixtures were assessed based on the results of basic mechanical and permeability tests. The results of this study provide a useful information on the selection of SF dose for the optimum mechanical and durability performance. Moreover, experimental results favor the use of MS to maximize the utilization of SF under compressive and tensile forces.

2. Materials and Methods

2.1. Materials

Conventional and supplementary materials used for the production of concrete mixtures are explained in this section. Type I general purpose cement (Bestway 53 Grade, Haripur, Pakistan) was used as the main binder as per specifications of ASTM C150 [28]. Micro-silica (also known as silica fume obtained from Jaza Minerals, Karachi, Pakistan) containing 90–94% silica was used as a partial cement substitute. Properties of cement and micro-silica can be assessed from the study of Ali et al. [2]. Siliceous sand from the Lawrancepur quarry (Attock, Pakistan) was used as fine aggregate. Dolomite sandstone of Kirana-hills of Sargodha was used as the coarse aggregate. The maximum aggregate size of coarse and fine aggregate is 12.5 mm and 4.75 mm. The distribution of different particle sizes in both coarse and fine aggregates is shown in Figure 1. Characteristics of aggregates are given in Table 1.

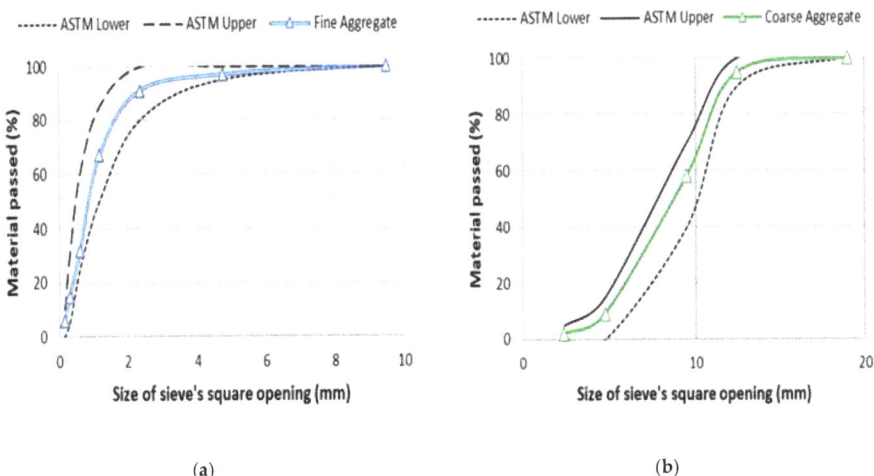

Figure 1. Gradation of (**a**) Siliceous sand/fine aggregate and (**b**) crushed dolomite limestone/coarse aggregate.

Table 1. Characteristics of aggregates.

Aggregate Type	Material	Dry Rodded Density	Water Absorption (%)	10% Fine Value (kN)	Specific Gravity	Maximum Aggregate Size
Fine aggregate	Siliceous sand	1615	0.97	-	2.67	4.75
Coarse Aggregate	Dolomite-sandstone	1519	0.65	148	2.68	12.5

Hooked steel fiber (SF) was studied as a discrete reinforcement in concrete. It has tensile strength of 750 MPa and a density of 7750 kg/m^3. It was sourced from Dramix© (Zwevegem, Belgium). SF has a length of 30 mm and a diameter of 0.38 mm. Its overview is shown in Figure 2. To control the loss of workability with a rising dose of SF, Viscorete 3110 (Sika Chemicals, Lahore, Pakistan) was used as an ultra-high range plasticizer. Tap water free from organic/inorganic impurities was used during the mixing and for curing as well.

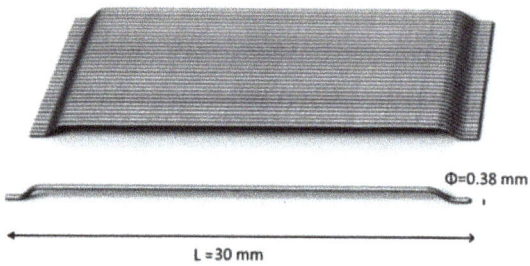

Figure 2. Overview of SF.

2.2. Design of Concrete Mixtures

A total of 21 concrete mixes were studied in this research. Six different doses (Vf = 0.05, 0.1, 0.25, 0.5, 1 and 2%) of SF were used to produce fiber-reinforced concretes. SF volumes were selected to observe the effects of a wide range of fiber doses on the properties of concrete. Fibers were used as by volume fraction of concrete (i.e., 1% Vf of SF = 78 kg of SF). Each fiber dosage is investigated with 0%, 5% and 10% micro-silica (MS). MS serves as the matrix-strengthening agent by producing calcium silicate hydrate gels from the free portlandite-CH a by-product from the hydration of cement. Plain concretes were also produced with and without MS and served as reference mixes. Composition and mix proportioning of all concrete mixtures are provided in Table 2.

Mixing of all concretes was done in a mechanical mixer having adjustable rotation speed. In the first stage, aggregates and binder were dry mixed for 3 min at speed of 40 rpm. In the second stage, half of the mixer and water reducer were added to the mix and blending continued for 3 min at speed of 40 rpm. In the third stage, the remaining water was added to the mix, and mixing was done a high speed of 80 rpm for 2 min. In the last stage, SF was added to the concrete mix, and blending continued for the next 4 min at 80 rpm. After mixing, a slump test was performed to check the desired workability of mixes (i.e., slump between 80 and 110 mm). Mixer continued to run at a slower speed of 20 rpm until the casting of specimens was completed.

Table 2. Design of concrete mixes.

Mix No.	Mix ID	MS (%)	SF (%)	Cement (kg/m³)	MS (kg/m³)	Siliceous Sand (kg/m³)	Crushed Limestone (kg/m³)	SF (kg/m³)	Water (kg/m³)	HWR (kg/m³)
1	MS0/SF0 (Control)	0	0.00	478	0	657	1077	0	185	2
2	MS0/SF0.05		0.05	478	0	656	1076	4	185	2
3	MS0/SF0.1		0.10	478	0	656	1076	8	185	2
4	MS0/SF0.25		0.25	478	0	654	1074	20	185	2
5	MS0/SF0.5		0.50	478	0	651	1071	39	185	3
6	MS0/SF1		1.00	478	0	644	1064	78	185	3
7	MS0/SF2		2.00	478	0	631	1051	156	185	3
8	MS5/SF0	5	0.00	454	18	657	1077	0	185	2
9	MS5/SF0.05		0.05	454	18	656	1076	4	185	2
10	MS5/SF0.1		0.10	454	18	656	1076	8	185	2
11	MS5/SF0.25		0.25	454	18	654	1074	20	185	2
12	MS5/SF0.5		0.50	454	18	651	1071	39	185	3
13	MS5/SF1		1.00	454	18	644	1064	78	185	3
14	MS5/SF2		2.00	454	18	631	1051	156	185	3
15	MS10/SF0	10	0.00	430	36	657	1077	0	185	2
16	MS10/SF0.05		0.05	430	36	656	1076	4	185	2
17	MS10/SF0.1		0.10	430	36	656	1076	8	185	2
18	MS10/SF0.25		0.25	430	36	654	1074	20	185	2
19	MS10/SF0.5		0.50	430	36	651	1071	39	185	3
20	MS10/SF1		1.00	430	36	644	1064	78	185	3
21	MS10/SF2		2.00	430	36	631	1051	156	185	3

MS: Micro-Silica; SF: Steel Fiber; HWR: High-range Water Reducer.

2.3. Sample Preparation and Testing Techniques

All specimens were cast in standard steel molds and protected with a waterproof membrane for 24 h setting immediately after casting. After setting, specimens were cured in tap water for 28-days at room temperature conditions. All mixes were tested for three important mechanical parameters i.e., compressive strength (CS), modulus of elasticity (MOE), and splitting tensile strength (STS). For CS, 100 φ mm × 200 mm cylindrical specimens were tested as per ASTM C39 [29]. To determine CS, specimens were tested under compressive-hydraulic press at the rate of 0.3 MPa/s. The static MOE of each mix was determined according to ASTM C469 [30]. MOE test was conducted on the specimens of 150 φ mm × 300 mm at the stress-rate of 0.15 MPa/s. The strain data (deformation characteristics) was recorded using compressometer-extensometer. To evaluate STS, 100 φ mm × 200 mm specimens were tested following ASTM C496 [31]. The splitting-load was applied at the stress rate of 0.015 MPa/s on the specimen to determine STS. All mechanical tests were performed in a controls compression testing machine with a loading capacity of 3000 kN. To understand the effects of varying SF and MS contents on the durability of concrete, two permeability-related durability indicators were evaluated i.e., water absorption and chloride ion penetration. To test for water absorption (WA) capacity, 100 φ mm × 50 mm concrete disc specimen of each mix was tested following ASTM C642 [32]. To determine chloride ion penetration (CIP) resistance of each mix, an immersion technique was adopted as explained by the authors [2]. For the CIP test, a 100 mm × 100 mm cylindrical specimen was first cured in normal water for 28 days. Then the specimen was immersed in a 10% NaCl solution for 56 days. After conditioning in chloride solution, the specimen was split, and the failed surface of the specimen was sprayed with 0.1 N AgNO$_3$ solution to observe the depth of CIP. The further detailed procedure for CIP testing can be assessed from studies [2,33]. All the results presented in this research are the mean values of the three results of each concrete mixture. The schedule of casting and testing is shown in Table 3.

Table 3. Overview of mechanical and permeability testing methods and schedule.

Property	Standard Followed	Size of Specimen	Age of Testing
Compressive Strength (MPa)	ASTM C39	100 φ mm × 200 mm cylinder	28 days
Modulus of Elasticity (MPa)	ASTM C469	150 φ mm × 300 mm cylinder	28 days
Splitting Tensile Strength (MPa)	ASTM C496	100 φ mm × 200 mm cylinder	28 days
Water Absorption (%)	ASTM C642	100 φ mm × 50 mm disc	28 days
Chloride Ion Penetration (mm)	Ali et al. [2]	100 φ mm × 100 mm cylinder	28 days curing + 56 days of condition in NaCl solution

3. Results and Discussion

3.1. Compressive Strength (CS)

Figure 3 shows the effect of varying SF dose on the CS of concrete. Figure 3b shows the net age change in CS with the varying SF dose. These results show a mixed effect of SF on CS at different doses. CS goes on increasing when the SF dose changed from 0 to 0.25%. Further increasing SF beyond 0.25%, CS starts reducing, and at 2% SF, CS of fibrous concrete is lesser than that of the plain concrete. Three different causes contribute to CS property due to the inclusion of fibers. The first cause is related to the confinement effect of fibers that increases the stiffness of concrete and it is known to positively affect the CS [10,34,35]. The second phenomenon is related to the entrainment of additional ITZs in concrete that has a detrimental effect on the CS. The introduction of a high number of ITZs contributes to porosity and permeable channels into the concrete and ITZs act as a weak link in the fibrous composite. The third phenomenon pertains to the resistance of cracking to the propagation of micro and macro-cracks; thus, it is known to improve the compressive stiffness of concrete. The first and third phenomenon prevails at 0.1–0.25% dose of SF, therefore, CS shows improvement due to fiber addition, whereas, at high fiber doses, a high number of ITZs introduction facilitate crack propagation and it adds to the total porosity of concrete.

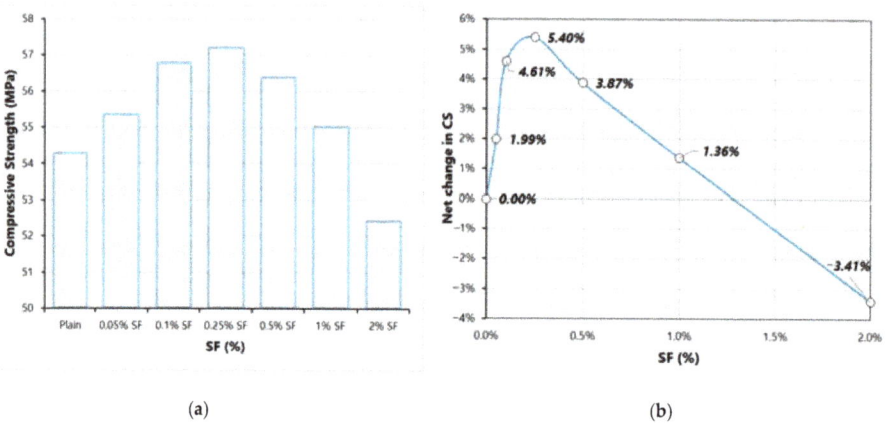

Figure 3. Compressive strength (CS) results (a) Variation of CS with SF dosage (b) Net change in CS with varying SF dosage.

Figure 4 shows the effect of MS on CS results of high strength concrete at different doses of SF. Figure 4b shows the effect of varying SF dose on CS at the levels of 0%, 5% and 10% MS. MS shows a positive effect on compressive strength concrete due to its ability to produce calcium silicate hydrate gels in pozzolanic reactive with free portlandite. The strengthening of the binder leads to improvement in the bond strength of fibers and matrix, that is why a clear difference (Figure 4b) between "net change" of SF mixes with and without MS. For example, at 0.25% SF the net changes in CS at 0%,

5% and 10% MS are 5.4%, 7.03% and 10.37%, respectively. MS also minimizes the negative effect of high fiber volumes (i.e., 2% SF) on CS. This can be credited to the strengthening of the bond at ITZ, which enhances the utilization of fibers in compression. It is confirmed from the results that the combined incorporation of 10%MS and 0.5%SF can increase the CS by more than 20%. It is verified by the literature that MS addition does not only contribute to the bond strength of fibers but it also improves the dispersion of fibers [27,36,37]. Therefore, it can be said that SF and MS have synergistic effects on the properties of concrete.

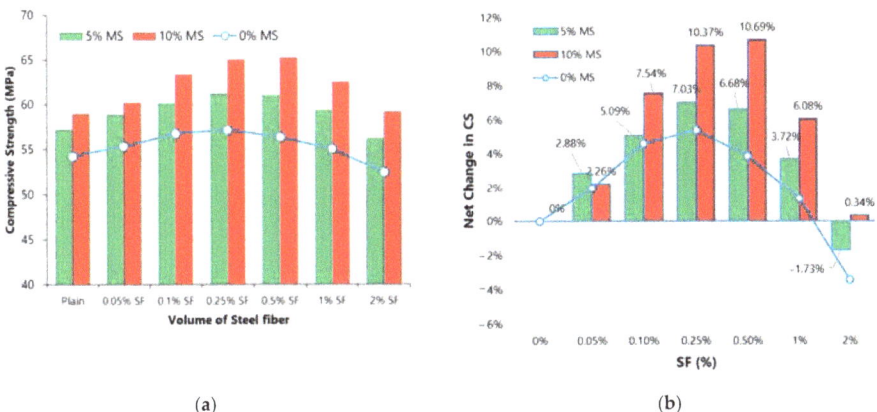

Figure 4. Compressive strength results (**a**) effect of MS on CS with varying SF dosage (**b**) effect of MS on the net change in CS with varying SF dosage.

3.2. Modulus of Elasticity (MOE)

Figure 5 shows the effect of SF on the MOE of concrete. MOE linearly increases when the SF dose changes from 0 to 0.5%. Figure 5b shows the effect of SF dose on the net change in MOE. The improvements in MOE at 0.05–0.5% can be ascribed to increment in the confinement of specimen under compression that helps in the utilization of the full potential of the concrete matrix. Beyond 0.5%SF, MOE starts degrading similar to CS. This shows that for given high-strength concrete the optimum dose of SF for optimum MOE is 0.5%. As already explained, high fiber doses can increase the number of ITZs in concrete which leads to the reduction in compression stiffness of concrete. A slight increase in porosity of concrete due to fibers (higher than 0.5%) can also damage the MOE considerably [38]. This finding is in line with the study of Xie et al. [24]. It was observed that during compression testing, mixes with high fiber doses showed more ductile failure before collapsing completely unlike the mixes with smaller doses. A linear increase in energy absorption capacity was observed with the rise in SF dose. Ou et al. [39] reported that the main role of SF is prominent in compression toughness of concrete (post-peak load behavior) because, before peak load in the determination of MOE, fibers are not activated.

In Figure 6, the effect of MS content is shown on the MOE of concrete. Figure 6b shows the net change in MOE due to varying dose of SF at 0, 5 and 10% replacement levels of MS. A clear improvement is noticed in the MOE of concrete due to MS addition. This is credited to (1) the improved packing density of binder particles and (2) the pozzolanic reaction that consumes free lime. MOE concrete with 10% MS is 11% higher than that of the plain concrete without MS. Figure 6b shows that MS enhances the utilization of fibers. Moreover, MS minimizes the negative effect of high SF volume on MOE. The combined incorporation of MS and SF shows synergistic behavior. For example, 0.5%SF and 10% MS individually leads to improvement of 3.4% and 7%, respectively. But simultaneous incorporation of both MS and SF improves the MOE by 14.8%. This is true for all mixes made with both MS and SF.

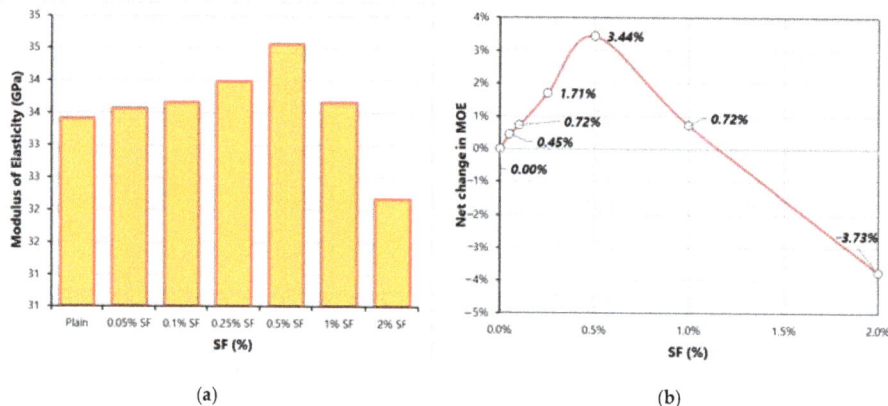

Figure 5. Modulus of elasticity (MOE) results (**a**) Variation of MOE with SF dosage (**b**) Net change in MOE with varying SF dosage.

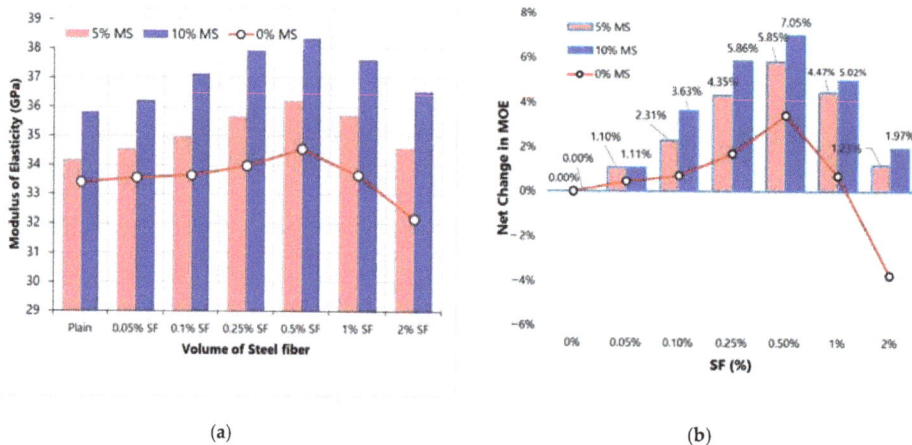

Figure 6. Modulus of elasticity (MOE) results (**a**) effect of MS on MOE with varying SF dosage (**b**) effect of MS on the net change in MOE with varying SF dosage.

From the results of CS and MOE, it is quite clear that both of these mechanical properties show a similar response to varying SF and MS contents. Therefore, both parameters can predict each with great accuracy. Since MOE is difficult to determine in the laboratory; therefore, it is predicted usually from CS. The relationship between MOE and half power of CS is shown in Figure 7. This relationship (Equation (1)) is drawn without considering the impact of SF or MS content:

$$MOE = 7007 \sqrt{CS} - 18500 \tag{1}$$

where MOE = modulus of elasticity (MPa); CS = compressive strength (MPa).

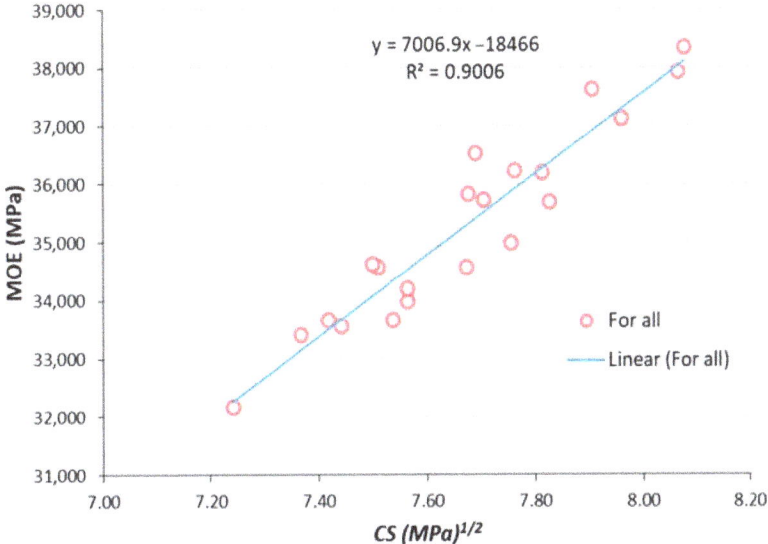

Figure 7. Relationship between MOE and $CS^{1/2}$.

3.3. Splitting Tensile Strength (STS)

STS in an estimate of true tensile strength of concrete. Due to the complexity of measuring the true tensile strength under the direct tension test, STS provides a simpler measurement of the tensile strength of cementitious materials. Figure 8 shows the effect of varying SF content on STS. Unlike results of CS and MOE, STS does not show a mixed response to increasing the dose of SF. This is because, under tensile load, fibers become active way before the failure at peak load; therefore, stretching action on concrete is resisted by both concrete matrix and fibers. Figure 8 shows that the net change in STS due to SF addition is very huge compared to that observed in the results of MOE and CS. STS achieves more than 3 times positive gain compared to CS and MOE at each dose of SF. This confirms that fibers are more useful in tensile stiffness than they are in the compressive stiffness of concrete.

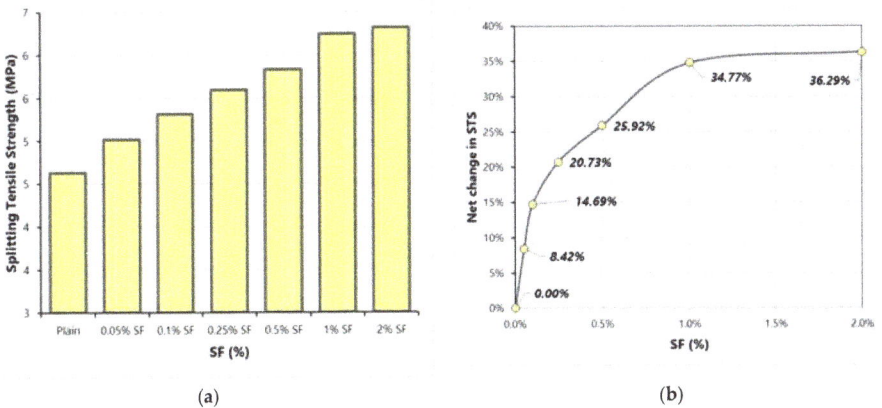

Figure 8. Splitting-tensile strength (STS) results (**a**) Variation of STS with SF dosage (**b**) Net change in STS with varying SF dosage.

MS addition provides a little advancement in the tensile strength, see Figure 9. Since MS strengthens the binder matrix, some small improvements can be anticipated in the STS. The filling effect of MS particles cannot contribute to STS, only the pozzolanic reaction between portlandite and silica strengthens the concrete matrix against tensile stresses [40]. A clear view of the synergistic effect of MS and SF on the STS can be seen in Figure 9. MS addition improves the net gain due to fibers by more than 30%. Densification of the matrix leads to an efficient transfer of tensile stresses to fiber-filaments; thus, MS addition improves the utilization of fibers. The results show that using MS along with SF can help in yielding 20% more STS than that could be achieved without MS. These results have important implications for fiber-reinforced concrete/composites. Since fibers are very expensive materials, their full utilization is very necessary to design cost and performance efficient structures. Therefore, MS and other high-performance mineral admixtures can help in enhancing the utilization of fibers.

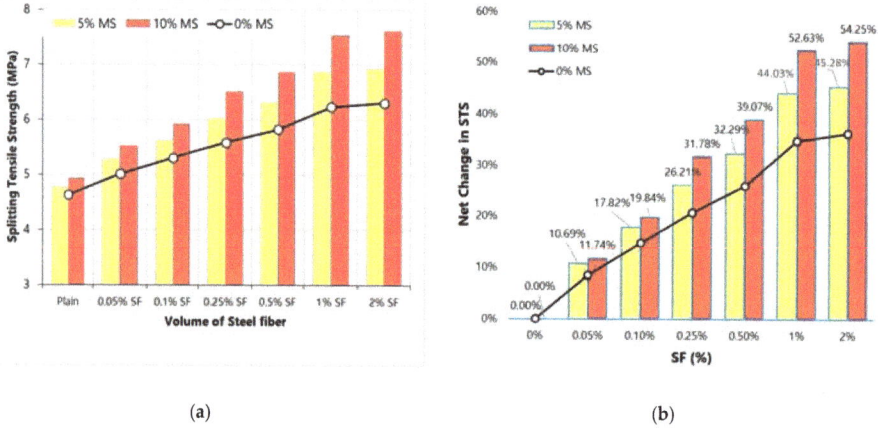

Figure 9. Splitting-tensile strength (STS) results (**a**) effect of MS on STS with varying SF dosage (**b**) effect of MS on the net change in STS with varying SF dosage.

For plain concrete, STS can be fairly correlated with CS or MOE. But for fibrous concrete STS cannot be correlated with CS or MOE, see Figure 10. As, activation of fibers during compression mostly starts near or after the peak load loads; therefore, fibers do not contribute a great deal towards the advancement of CS or MOE. Whereas, under tension, fibers activate way before peak load; therefore, concretes show a huge STS change with fiber addition. Under tension, fibers do not only contribute to the peak strength of concrete, but they are also useful in the post-peak load resistance. CS, MOE, and STS of each mix are correlated in Figure 10, without considering the role of SF dose. This surface plot shows a general trend that each mechanical parameter is directly proportional to each other but with a huge scatter (R2 < 0.6).

3.4. Water Absorption (WA)

WA capacity of concrete represents its water-permeable volume of voids. High WA generally indicates high porosity. The effect of SF on the WA capacity of each mix is shown in Figure 11. WA undergoes mixed changes with the rising dose of SF. Small fractions of SF cause minor reductions in the WA capacity of concrete, whereas, at high doses, SF, WA absorption of fibrous concrete is slightly higher than that of the plain concrete. Both positive [21,41] and negative [42] effects of SF on WA has been reported in the literature. No study in the literature has examined the permeability characteristics of SF-reinforced concretes considering a wide range of fiber dosage. Fibers can control micro-cracking during the evolution of cementitious compounds in concrete. These can restrict the plastic and temperature shrinkage cracking which ultimately improves the permeability resistance of concrete. At the same time, fiber addition increases the number of ITZs in concrete. Poor bond at ITZs favor permeability, hence it increases of WA capacity. Apparently, at low fiber volumes, controlled shrinkage leads to reduction in WA capacity and the role of ITZs is not very dominant at low fiber volumes. But as the fiber volume increases, the number of weak ITZs favor permeability and increase the WA. The minimum WA is observed at 0.1% SF, whereas maximum WA is noticed at 2% SF.

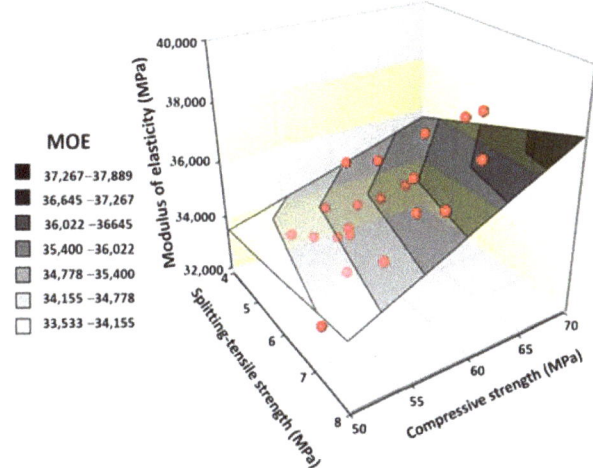

Figure 10. Correlation between mechanical properties (MOE, STS and CS).

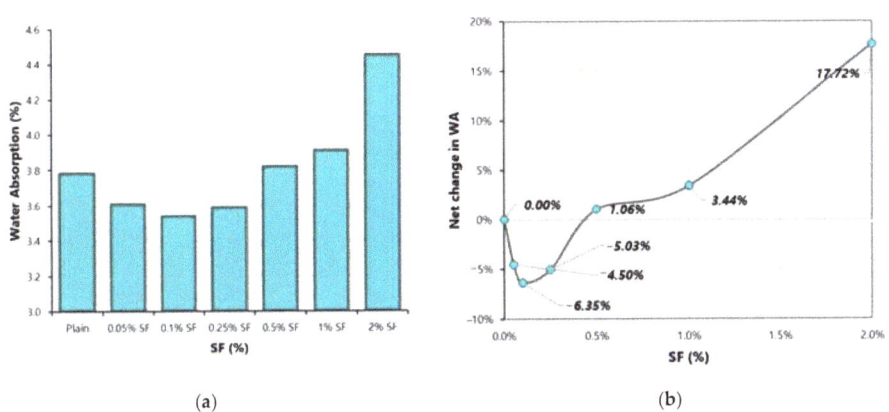

Figure 11. Water-absorption (WA) results (a) Variation of WA with SF dosage (b) Net change in WA with varying SF dosage.

Figure 12 shows the effect of MS on WA capacity of concrete. MS brings down the WA capacity of concrete significantly. As extremely fine particles of MS fill the gaps left between cement particles, the overall density of matrix undergoes improvement. MS can reduce the pore-size at the ITZ between fiber and matrix. MS can nullify the negative effect of SF on WA. These results implicate an important role of MS in fibrous concretes. Since, fibers at medium to high volumes (0.5–2%), increase the permeability which may favor the corrosion of SF. Corrosion of SF will significantly lower the performance of fibrous concrete over time. Therefore, the conjunctive use of fibers and MS can increase the durability life of fibrous concrete composites.

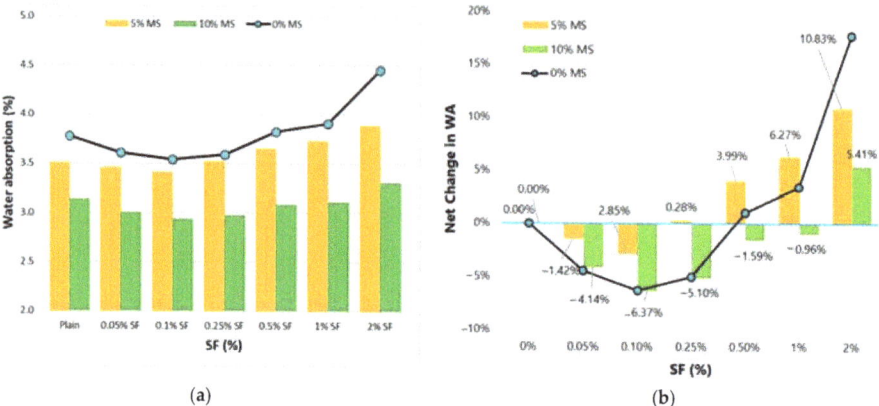

Figure 12. Water-absorption (WA) results (**a**) effect of MS on WA with varying SF dosage (**b**) effect of MS on the net change in WA with varying SF dosage.

3.5. Chloride Ion Penetration (CIP)

Figure 13 shows the effect of SF content on the CIP of concrete. CIP results also experience changes similar to WA with the variation of SF content. CIP undergoes reduction when fiber dose changes from 0 to 0.1%. Since there is no involvement of forced electrical transfer of chloride ions in the immersion technique high conductivity of SF does not play any role in determining the CIP resistance of concrete. CIP resistance improvement at low fiber volumes can be ascribed to a reduction in the WA capacity of concrete. On the other hand, reduction in CIP resistance at high fiber volumes (1% and 2%) can be blamed to an increase in porosity or absorption capacity of the matrix. At 2%SF, CIP of concrete is about 18% higher than that of the plain concrete. Since chloride-induced corrosion is usually experienced in most concrete structures, low chloride permeability resistance of fibrous concretes (especially with a high volume of fibers) can create durability issues which must be considered while designing a concrete mix.

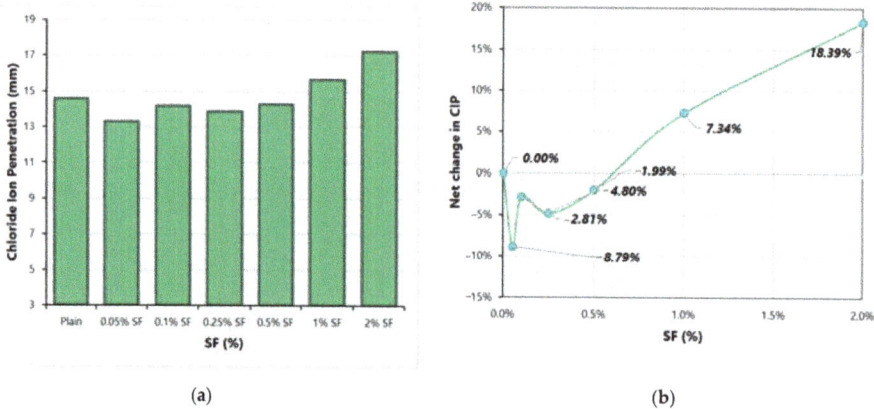

Figure 13. Chloride-ion penetration (CIP) results (**a**) Variation of CIP with SF dosage (**b**) Net change in CIP with varying SF dosage.

Figure 14 shows the effect of MS content on the CIP. The addition of 5%MS and 10%MS brings down the CIP by 23% and 33%, respectively w.r.t plain concrete (without MS). The behavior of WA and CIP with the addition of MS is very similar because MS substantially reduces the volume of permeable voids [2]. As fibrous concretes struggle with the issue of low CIP resistance at high fiber doses, MS can be a befitting addition to enhance the imperviousness of concrete. With 5 or 10% MS, high fiber volume concretes show lower CIP than control concrete (see Figure 14a).

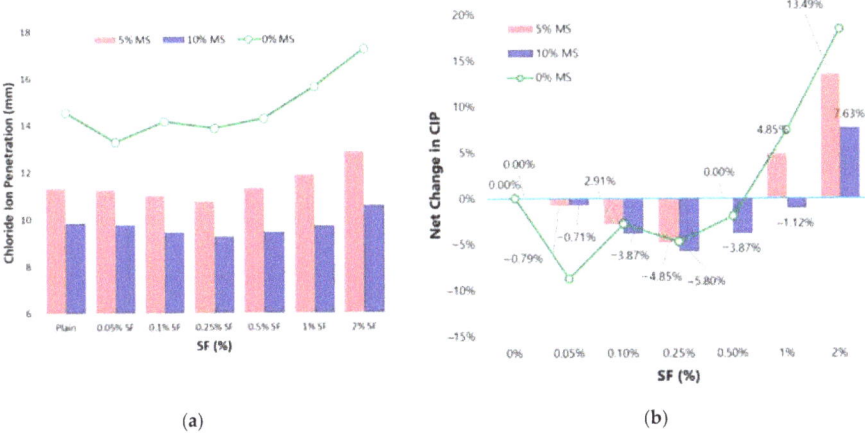

Figure 14. Chloride-ion penetration (CIP) results (**a**) effect of MS on CIP with varying SF dosage (**b**) effect of MS on the net change in CIP with varying SF dosage.

By constricting the microchannels across the ITZs at fibers, MS can efficiently minimize the degrading effect of fibers on CIP. Almost all engineering properties of concrete depend on the growth and density of microstructure i.e., strength and permeability characteristics. CS, CIP, and WA are correlated with each other in Figure 15. The surface plot shows a general trend that CS is inversely related to both WA and CIP. All data points in Figure 15, congregate near-surface plot which means CS, CIP and WA are strongly correlated ($R2 > 0.8$) and models developed to predict these parameters from each other can be formulated for design purposes.

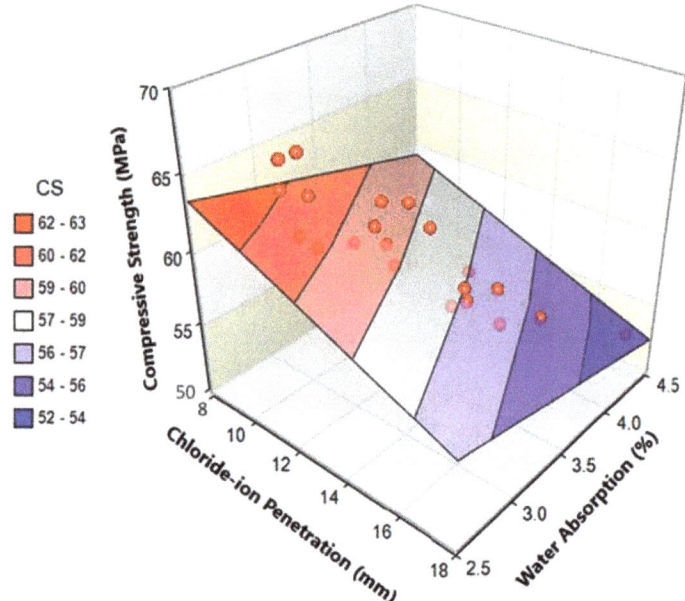

Figure 15. Correlation between CS, CIP and WA.

4. Conclusions

This study evaluates the influence of a wide range of SF doses on the basic engineering characteristics of high strength concrete. It also explores the modifications of SF-reinforced concrete properties with micro-silica (MS). Following important conclusions can be taken from this research:

(1) Fibers have mixed effects on compressive strength (CS). The positive effect of SF on CS is observed only at small doses. A high-volume dose of SF negatively affects CS. MS improves the utilization of SF in advancing the CS of concrete.

(2) Similar to CS results, modulus of elasticity (MOE) also shows mixed behavior with varying fiber dose. Low volumes (0.05–0.5%) of SF are beneficial to MOE, whereas high volumes (1–2%) are detrimental to MOE. MS shows synergistic effects with SF on MOE. SF doses of 0.5% produce optimum MOE and CS. The addition of MS is highly useful compared to SF addition if the increment in CS or MOE is desired.

(3) Splitting-tensile strength (STS) increases up to 36% with the rising dose of SF (0 to 2%). There is no significant achievement in STS when SF doses increased beyond 1%. STS experience more gain than CS and MOE at all doses of SF. MS improves the net gain in STS due to SF addition. The combined addition of 10%MS and 1%SF produces concrete with 60% more STS than plain concrete.

(4) Considering the combined behavior of CS, MOE, and STS, 1%SF can be taken as the optimum dose for high strength concrete.

(5) At low to medium fiber doses (0.05–0.25%), the WA of concrete was slightly lower than that of the plain concrete. Whereas high fiber doses (0.5–1%) are determinantal to the imperviousness of concrete. The positive effect of low fiber volumes is very negligible compared to that of the MS. MS can play a key role in downing the WA capacity of high fiber volume concretes.

(6) Similar to WA, chloride-ion penetration (CIP) experiences a small reduction of 0.05–0.25% SF. The detrimental effect of high fiber dose can be minimized by MS addition. With the help of 10%MS, 2%SF concrete shows a significant 25% lower CIP compared to plain concrete.

5. Future Research

Effect of micro-silica on the interfacial zones of fibers should be studied using scanning electron microscopy. Moreover, combined effect of MS and SF on different strength classes of concrete should also be investigated and compared. Effect of different MS contents on the freeze-thaw and corrosion resistance of SFRC can also be studied.

Author Contributions: Conceptualization, B.A. and B.H.; methodology, B.A. and B.H.; software, B.A. and B.H.; validation, R.K. and B.H.; formal analysis, B.A., R.K., B.H., R.A., A.R., A.M., R.M. and H.A.; investigation, B.A., B.H.; resources, B.A., R.K., B.H., R.A., A.R., A.M., R.M. and H.A.; data curation, B.A.; writing—original draft preparation, B.A. and B.H.; writing—review and editing, R.K., B.H., R.A., A.R., A.M., R.M., B.H., M.F.U.-H. and H.A.; visualization, R.K., B.H., R.A., A.R., A.M., R.M., M.F.U.-H. and H.A.; supervision, R.K.; project administration, B.A., and R.K.; funding acquisition, R.A. All authors have read and agreed to the published version of the manuscript.

Funding: This research was funded by Deanship of scientific research, Prince Sattam bin Abdulaziz University, grant number No. 2020/01/16810.

Acknowledgments: This publication was supported by Deanship of scientific research, Prince Sattam bin Abdulaziz University under research project No. 2020/01/16810.

Conflicts of Interest: The authors declare no conflict of interest.

Nomenclature

CIP	Chloride Ion Penetration
CS	Compressive Strength
MOE	Modulus of Elasticity
MS	Micro-Silica
SF	Steel Fiber
STS	Splitting Tensile Strength
WA	Water Absorption

References

1. Ali, B.; Qureshi, L.A. Influence of glass fibers on mechanical and durability performance of concrete with recycled aggregates. *Constr. Build. Mater.* **2019**, *228*, 116783. [CrossRef]
2. Ali, B.; Ahmed, H.H.; Qureshi, L.A.; Kurda, R.; Hafez, H.; Mohammed, H.; Raza, A. Enhancing the Hardened Properties of Recycled Concrete (RC) through Synergistic Incorporation of Fiber Reinforcement and Silica Fume. *Materials* **2020**, *13*, 4112. [CrossRef] [PubMed]
3. Koushkbaghi, M.; Kazemi, M.J.; Mosavi, H.; Mohseni, E. Acid resistance and durability properties of steel fiber-reinforced concrete incorporating rice husk ash and recycled aggregate. *Constr. Build. Mater.* **2019**, *202*, 266–275. [CrossRef]
4. Zain, M.; Mahmud, H.; Ilham, A.; Faizal, M. Prediction of splitting tensile strength of high-performance concrete. *Cem. Concr. Res.* **2002**, *32*, 1251–1258. [CrossRef]
5. Xie, J.; Li, J.; Lu, Z.; Li, Z.; Fang, C.; Huang, L.; Li, L. Combination effects of rubber and silica fume on the fracture behaviour of steel-fibre recycled aggregate concrete. *Constr. Build. Mater.* **2019**, *203*, 164–173. [CrossRef]
6. Cao, S.; Xue, G.; Yilmaz, E. Flexural Behavior of Fiber Reinforced Cemented Tailings Backfill under Three-Point Bending. *IEEE Access* **2019**, *7*, 139317–139328. [CrossRef]
7. Ali, B.; Qureshi, L.A.; Khan, S.U. Flexural behavior of glass fiber-reinforced recycled aggregate concrete and its impact on the cost and carbon footprint of concrete pavement. *Constr. Build. Mater.* **2020**, *262*, 120820. [CrossRef]
8. Wang, J.; Dai, Q.; Si, R.; Ma, Y.; Guo, S. Fresh and mechanical performance and freeze-thaw durability of steel fiber-reinforced rubber self-compacting concrete (SRSCC). *J. Clean. Prod.* **2020**, *277*, 123180. [CrossRef]
9. Qureshi, L.A.; Ali, B.; Ali, A. Combined effects of supplementary cementitious materials (silica fume, GGBS, fly ash and rice husk ash) and steel fiber on the hardened properties of recycled aggregate concrete. *Constr. Build. Mater.* **2020**, *263*, 120636. [CrossRef]

10. Hussain, I.; Ali, B.; Akhtar, T.; Jameel, M.S.; Raza, S.S. Comparison of mechanical properties of concrete and design thickness of pavement with different types of fiber-reinforcements (steel, glass, and polypropylene). *Case Stud. Constr. Mater.* **2020**, *13*, e00429. [CrossRef]
11. Ahmad, W.; Farooq, S.H.; Usman, M.; Khan, M.; Ahmad, A.; Aslam, F.; Alyouef, R.; Alabduljabbar, H.; Sufian, M. Effect of Coconut Fiber Length and Content on Properties of High Strength Concrete. *Materials* **2020**, *13*, 1075. [CrossRef] [PubMed]
12. Ahmed, H.H.; Bogas, J.A. Influence of carbon nanotubes on steel–concrete bond strength. *Mater. Struct.* **2018**, *51*, 155. [CrossRef]
13. Hawreen, A.; Bogas, J.A.; Kurda, R.; Hawreen, A.; Bogas, J.A.; Kurda, R. Mechanical Characterization of Concrete Reinforced with Different Types of Carbon Nanotubes. *Arab. J. Sci. Eng.* **2019**, *44*, 8361–8376. [CrossRef]
14. Ali, B.; Qureshi, L.A.; Kurda, R. Environmental and economic benefits of steel, glass, and polypropylene fiber reinforced cement composite application in jointed plain concrete pavement. *Compos. Commun.* **2020**, *22*, 100437. [CrossRef]
15. Wu, Z.; Shi, C.; He, W.; Wu, L. Effects of steel fiber content and shape on mechanical properties of ultra high performance concrete. *Constr. Build. Mater.* **2016**, *103*, 8–14. [CrossRef]
16. Raza, S.S.; Qureshi, L.A.; Ali, B.; Raza, A.; Khan, M.M. Effect of different fibers (steel fibers, glass fibers, and carbon fibers) on mechanical properties of reactive powder concrete. *Struct. Concr.* **2020**. [CrossRef]
17. Alabduljabbar, H.; Alyouef, R.; Alrshoudi, F.; Alaskar, A.; Fathi, A.; Mohamed, A.M. Mechanical Effect of Steel Fiber on the Cement Replacement Materials of Self-Compacting Concrete. *Fibers* **2019**, *7*, 36. [CrossRef]
18. Lee, S.F.; Jacobsen, S. Study of interfacial microstructure, fracture energy, compressive energy and debonding load of steel fiber-reinforced mortar. *Mater. Struct.* **2011**, *44*, 1451–1465. [CrossRef]
19. Li, V.C.; Wu, H.-C.; Chan, Y.-W. Effect of Plasma Treatment of Polyethylene Fibers on Interface and ementitious Composite Properties. *J. Am. Ceram. Soc.* **2005**, *79*, 700–704. [CrossRef]
20. Teng, S.; Afroughsabet, V.; Ostertag, C.P. Flexural behavior and durability properties of high performance hybrid-fiber-reinforced concrete. *Constr. Build. Mater.* **2018**, *182*, 504–515. [CrossRef]
21. Afroughsabet, V.; Biolzi, L.; Ozbakkaloglu, T. Influence of double hooked-end steel fibers and slag on mechanical and durability properties of high performance recycled aggregate concrete. *Compos. Struct.* **2017**, *181*, 273–284. [CrossRef]
22. Afroughsabet, V.; Ozbakkaloglu, T. Mechanical and durability properties of high-strength concrete containing steel and polypropylene fibers. *Constr. Build. Mater.* **2015**, *94*, 73–82. [CrossRef]
23. Xie, J.; Zhang, Z.; Lu, Z.; Sun, M. Coupling effects of silica fume and steel-fiber on the compressive behaviour of recycled aggregate concrete after exposure to elevated temperature. *Constr. Build. Mater.* **2018**, *184*, 752–764. [CrossRef]
24. Wang, J.; Xie, J.; He, J.; Sun, M.; Yang, J.; Li, L. Combined use of silica fume and steel fibre to improve fracture properties of recycled aggregate concrete exposed to elevated temperature. *J. Mater. Cycles Waste Manag.* **2020**, *22*, 1–16. [CrossRef]
25. Carneiro, J.A.; Lima, P.R.L.; Leite, M.B.; Filho, R.D.T. Compressive stress–strain behavior of steel fiber reinforced-recycled aggregate concrete. *Cem. Concr. Compos.* **2014**, *46*, 65–72. [CrossRef]
26. Song, W.; Yin, J. Hybrid effect evaluation of steel fiber and carbon fiber on the performance of the fiber reinforced concrete. *Materials* **2016**, *9*, 704. [CrossRef]
27. Wu, Z.; Khayat, K.H.; Khayat, K. Influence of silica fume content on microstructure development and bond to steel fiber in ultra-high strength cement-based materials (UHSC). *Cem. Concr. Compos.* **2016**, *71*, 97–109. [CrossRef]
28. C01 Committee. *Specification for Portland Cement*; ASTM International: West Conshohocken, PA, USA, 2018.
29. ASTM International. *ASTM-C39, Standard Test Method for Compressive Strength of Cylindrical Concrete Specimens*; ASTM International: West Conshohocken, PA, USA, 2015.
30. ASTM International. *ASTM-C469, Standard Test Method for Static Modulus of Elasticity and Poisson's Ratio of Concrete in Compression*; ASTM International: West Conshohocken, PA, USA, 2014.
31. ASTM International. *ASTM-C496, Standard Test Method for Splitting Tensile Strength of Cylindrical Concrete Specimens*; ASTM International: West Conshohocken, PA, USA, 2017.
32. ASTM International. *ASTM-C642-13, Standard Test Method for Density, Absorption, and Voids in Hardened Concrete*; ASTM International: West Conshohocken, PA, USA, 2013.

33. Ali, B.; Qureshi, L.A. Durability of recycled aggregate concrete modified with sugarcane molasses. *Constr. Build. Mater.* **2019**, *229*, 116913. [CrossRef]
34. Das, C.S.; Dey, T.; Dandapat, R.; Mukharjee, B.B.; Kumar, J. Performance evaluation of polypropylene fibre reinforced recycled aggregate concrete. *Constr. Build. Mater.* **2018**, *189*, 649–659. [CrossRef]
35. Chan, R.; Santana, M.A.; Oda, A.M.; Paniguel, R.C.; Vieira, L.B.; Figueiredo, A.D.; Galobardes, I. Analysis of potential use of fibre reinforced recycled aggregate concrete for sustainable pavements. *J. Clean. Prod.* **2019**, *218*, 183–191. [CrossRef]
36. Ali, B.; Raza, S.S.; Hussain, I.; Iqbal, M. Influence of different fibers on mechanical and durability performance of concrete with silica fume. *Struct. Concr.* **2020**, 201900422. [CrossRef]
37. Chan, Y.-W.; Chu, S.-H. Effect of silica fume on steel fiber bond characteristics in reactive powder concrete. *Cem. Concr. Res.* **2004**, *34*, 1167–1172. [CrossRef]
38. Lee, S.-C.; Oh, J.-H.; Cho, J.-Y. Compressive Behavior of Fiber-Reinforced Concrete with End-Hooked Steel Fibers. *Materials* **2015**, *8*, 1442–1458. [CrossRef] [PubMed]
39. Ou, Y.-C.; Tsai, M.-S.; Liu, K.-Y.; Chang, K.-C. Compressive Behavior of Steel-Fiber-Reinforced Concrete with a High Reinforcing Index. *J. Mater. Civ. Eng.* **2012**, *24*, 207–215. [CrossRef]
40. Dinh-Cong, D.; Keykhosravi, M.H.; Alyousef, R.; Salih, M.N.A.; Nguyen, H.; Alabduljabbar, H.; Alaskar, A.; Alrshoudi, F.; Poi-Ngian, S. The effect of wollastonite powder with pozzolan micro silica in conventional concrete containing recycled aggregate. *Smart Struct. Syst.* **2019**, *24*, 541–552.
41. Afroughsabet, V.; Biolzi, L.; Monteiro, P.J. The effect of steel and polypropylene fibers on the chloride diffusivity and drying shrinkage of high-strength concrete. *Compos. Part B Eng.* **2018**, *139*, 84–96. [CrossRef]
42. Li, D.; Liu, S. The influence of steel fiber on water permeability of concrete under sustained compressive load. *Constr. Build. Mater.* **2020**, *242*, 118058. [CrossRef]

Publisher's Note: MDPI stays neutral with regard to jurisdictional claims in published maps and institutional affiliations.

© 2020 by the authors. Licensee MDPI, Basel, Switzerland. This article is an open access article distributed under the terms and conditions of the Creative Commons Attribution (CC BY) license (http://creativecommons.org/licenses/by/4.0/).

Article

Steatite Powder Additives in Wood-Cement Drywall Particleboards

Viet-Anh Vu [1], Alain Cloutier [1,*], Benoît Bissonnette [2], Pierre Blanchet [1] and Christian Dagenais [1,3]

[1] Department of Wood and Forest Sciences, Université Laval, Quebec, QC G1V 0A6, Canada; viet-anh.vu.1@ulaval.ca (V.-A.V.); pierre.blanchet@sbf.ulaval.ca (P.B.); christian.dagenais@fpinnovations.ca (C.D.)
[2] Department of Civil and Water Engineering, Université Laval, Quebec, QC G1V 0A6, Canada; benoit.bissonnette@gci.ulaval.ca
[3] FPInnovations, Quebec, QC G1V 4C7, Canada
* Correspondence: alain.cloutier@sbf.ulaval.ca; Tel.: +1-418-656-5851

Received: 23 September 2020; Accepted: 23 October 2020; Published: 29 October 2020

Abstract: The objective of this study was to develop a new drywall wood-based particleboard as an alternative to gypsum board. Various development iterations have led to the use of wood particles, steatite powder and Portland cement. The resulting outcome shows that screw withdrawal resistance was improved by 37% and bending properties by 69% compared to gypsum board of a similar density (0.68–0.70). The raw surface of the boards is of good quality and comparable to the paper-faced surface of gypsum board. Furthermore, the reaction to fire was evaluated through bench-scale test with a cone calorimeter. The investigated particleboard did not reveal visual signs of combustion after 20 min when exposed to a radiant heat of 50 kW/m^2, while burning of the overlay paper of gypsum board occurred at about 57 s, suggesting that wood-cement-steatite powder particleboard could be classified as a quasi non-combustible material.

Keywords: steatite; wood particles; Portland cement; fire performance

---

1. Introduction

Steatite (also known as soapstone or soap rock) is a type of metamorphic rock. It is primarily composed of mineral talc, rich in magnesium. Its main component is hydrated magnesium silicate:$Mg_3Si_4O_{10}(OH)_2$. As it is relatively soft because of its high talc content, it has been used as carving material for thousands of years. This stone is soft, dense, heat-resistant and has a high specific heat capacity [1]. Steatite can be pressed into complex shapes before heating. It is also used in the paint industry, particularly in marine paints and protective coatings for ceramics due to its high electrical resistivity [2]. Due to its electrical characteristics, steatite is mostly used in electrotechnics. In the world market, steatite with more than 92% brightness, less than 1.5% $CaCO_3$ and less than 1% Fe_2O_3 is preferred for exports [3].

Many studies have been carried out to evaluate the performance and applications of wood-cement composites because of their low cost and important contribution in mitigating the housing problem in developing countries [4]. Indeed, many studies have shown that wood-cement boards, could be used for ceilings or walls covering [5,6]. The most important advantages of wood–cement boards are their high resistance to insect, fungi, decay, acoustic waves and fire [6,7]. In fact, the sugars present in wood can inhibit cement setting. Therefore, the main problem in wood-cement board design is the compatibility between wood and cement [6]. The effect of wood on cement setting depends on several factors, among which harvesting season and wood species have the higher impacts [8]. Several special cement-based mortar containing additions of fine powder such as steatite [9], glass [10] and wood ash [11] have emerged.

The replacement of cement with steatite powder (SP) decreases setting time of cement and increases mortar cube compressive strength, but the consistency of the binding material increases [2]. The replacement of cement with SP was reported to result in improvements of the mortar microstructure, up to maximum replacement rates of the order of 20% by weight [12].

Gypsum boards (GB) are widely used in North America building construction for interior partitioning. Gypsum boards consist of calcium sulphate in the form of dihydrate crystals with overlay paper on both sides. The board core is a non-combustible material. It contains nearly 21% chemically combined water which is slowly released as steam when submitted to high levels of heat. Because steam does not exceed 100 °C at normal atmospheric pressure, it effectively retards the transfer of heat and the spread of fire [13,14]. Even after complete calcination, when all the water has been released from its core, GB continue to serve as heat-insulating barriers. When installed in combination with other materials such as walls and ceiling assemblies, GB serve to protect building elements from fire effectively for prescribed durations. While GB fails the flaming criteria for determining the non-combustibility of materials due to the paper overlay [15], it is typically an accepted material for non-combustible construction in most building codes due to its good fire performance. However, the paper overlay plays a vital role in the mechanical resistance of GB [16]. Besides, it appears that construction wastes from this material are a problem [17], which is aggravated by its extensive use. Economic pressures and environmental concerns are some of the driving forces of today's industrial development. Hence, many research projects are being conducted for increasing the utilization of waste materials in order to decrease threats to the environment and to streamline existing waste disposal and recycling methods by making them more affordable [17]. On the market, several alternatives to gypsum have been used such as plastic panels, plywood, fiberboard and veneer plaster.

The aim of the present study was to evaluate the mechanical, physical and thermal properties and reaction to fire of wood-cement particleboards incorporating SP as a supplementary cementing material, intended as an eco-responsible alternative to the GB. In this study, two in three of the raw materials used for particleboard production, wood particles and SP, are secondary low-cost products from lumber production and mineral extraction of steatite.

2. Materials and Methods

2.1. Material

The primary binder used was Portland cement type 10 (GU, General Use), an ordinary CSA (Canadian Standards Association).

The SP selected for this research project was provided by Polycore Inc, Quebec, Canada.

The wood-cement mixtures were prepared with air-dried wood particles obtained from white spruce (*Picea glauca*) trees harvested at the Petawawa Research Forest in Mattawa (ON), Canada. The wood chips were refined with a Pallmann PSKM8-400 ring refiner (Ludwig Pallmann K.G, Zweibrücken, Germany). Then, the wood particles were screened using nine sieve sizes: 1.19, 1.40, 1.70, 2.38, 2.80, 3.35, 4.00, 4.46 and 5.00 mm.

The regular GB used in the study for comparison purposes were 12.7 mm [1/2 in] in thickness. They are commercialized by Georgia Pacific under the trade name ToughRock®. They were typical regular drywall boards used for interior partitioning in building construction.

2.2. Material Characterisation

2.2.1. Wood Particles

Figure 1 shows the wood particles size distribution by mass. According to the results, all of the particles are smaller than 5 mm in size and the highest volume fraction (37%) is the particles with a diameter of 1.7 mm. In the study of Vu et al. [11], the size of the wood particles was less than 3 mm and the highest volume fraction was 1.7 mm. Wood particles size reaches a maximum of 5 mm for the purpose of increasing the mechanical strength of the particleboard.

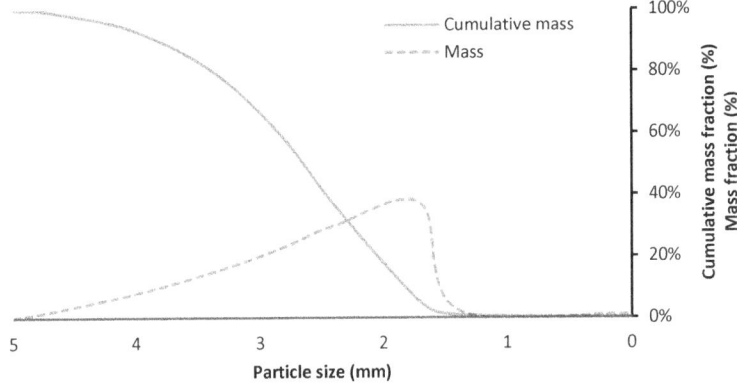

Figure 1. Wood particles size distribution.

2.2.2. Steatite Powder

Chemical Composition

Table 1 shows the results of the chemical analysis of SP. The combined content of aluminum oxide (Al_2O_3 = 0.7%), iron oxide (Fe_2O_3 = 6.32%), and silicon dioxide (SiO_2 = 38.3%) reaches 45.32%, while the minimum value required for the material to qualify as a pozzolan is 70%. The relative mass loss during combustion observed at 950 °C was 20.4%, which is considerably more than the maximum requirement for pozzolans set at 12%. The alkali content recorded (%Na_2O + 0.658 × %K_2O) was less than 0.23%, which is lower than the maximum alkali content of 1.5% required for pozzolans [18]. Therefore, SP does not qualify as a pozzolan. The specific gravity of SP was found to be 2.91. This is lower than the specific gravity of Portland cement (3.15), but larger than for mineral aggregates typically used in cementitious materials (limestone, granite, quartzite).

Table 1. Composition and properties of steatite powder.

Chemical Composition (%)		Property	Value
SiO_2	38.3	Density	2.91 g/cm^3
Al_2O_3	0.70	Blaine fineness	6505 cm^2/g
Fe_2O_3	6.32	pH	9.4
MgO	33.9		
CaO	0.77		
Na_2O	0.22		
K_2O	<0.01		
TiO_2	0.02		
MnO	0.09		
P_2O_5	<0.01		
Cr_2O_3	0.34		
V_2O_5	<0.01		
ZrO_2	<0.02		
ZnO	<0.01		
PAF	20.4		

Particle Size Analysis

The most commonly used metrics when describing particles size distributions are D-Values (D10, D50 and D90) which are the intercepts for 10, 50 and 90% of the cumulative mass [19]. According to the results shown in Figure 2, D10, D50, and D90 values of the SP were 3.9 μm, 18.5 μm and 52.3 μm, respectively. The D90 value of the SP was smaller than the corresponding values (114.1 μm) recorded for wood ash in the study of Vu et al. [11]. Moreover, the tested material contained 14% of ultrafine particles ($\phi < 5$ μm). Therefore, SP is suitable for use as a filler to reduce the porosity in the particleboard.

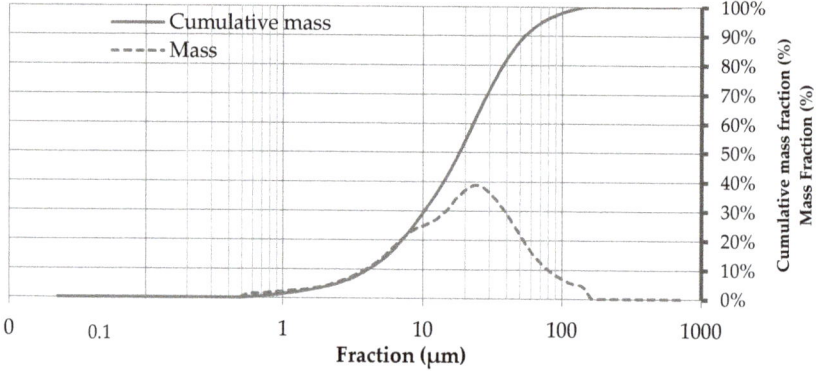

Figure 2. Particle size analysis of SP.

Material Preparation

The wood-cement steatite powder (WCSP) mixtures tested in this project were all prepared with the same ratio by weight of wood-binder and SP-binder, where the binder phase is the sum of cement and SP. The wood-binder ratio and SP-binder ratio selected were 0.35 and 0.15 (Table 3—P3). After mixing the materials in the mortar mixer, each particleboard was cast using the same $450 \times 330 \times 14$ mm^3 wooden mold. The wet mixture was poured into the mold, the surface was then levelled off with a wood screed, and in the end a wooden lid was secured on top of the mold with C-clamps. The particleboard thickness was reduced to 13 mm due to the pressure of the lid. The particleboards were unmoulded at the age of 3 days and stored in a conditioning chamber at 23 °C and 60% R.H. The various test specimens were sawn from the particleboard using a 5 mm thick saw blade at the age of 28 days (Figure 3). Particleboards nos. 1, 2, 3, 6, 7 and 8 were tested for bending modulus of rupture (MOR) and modulus of elasticity (MOE), and screw-withdrawal later. Thermal properties and water absorption tests were carried out on particleboards nos. 4 and 5. The reaction to fire was determined on particleboards nos. 9 and 10.

Due to the settling of the SP at the bottom of the panels, this face of the WSCP which was in contact with the mold had a less porous, denser microstructure than at the top. This face is the smoothest and is called front face. The top face of the panel in the mold which is the roughest is referred to as back face throughout this paper (Figure 4) and should be used against the structure when mounting a wall. In Section 2, the front face will be used for reaction to fire testing and nail pull resistance testing, while the three-point bending test is applied on both faces of the WCSPs.

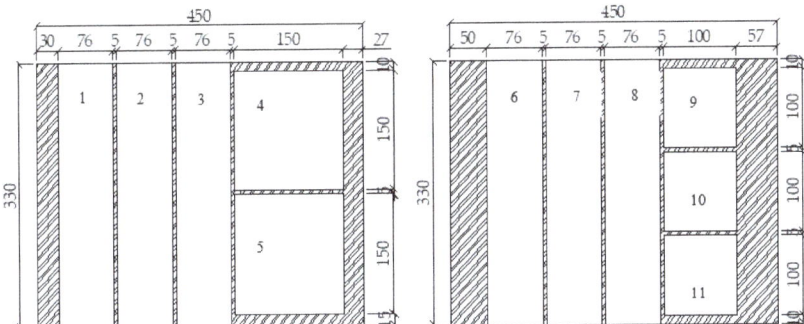

Figure 3. Sketch of samples cutting for WCSP (all measurements in mm).

Figure 4. Edges and faces of a wood–cement steatite powder particleboard cut with a saw.

2.3. Test Methods

In this study, the mechanical properties of the investigated particleboards and GB were determined in accordance with ASTM D1037-12 Standard test methods for evaluating the properties of wood-based fiber and particle panel materials [20]. Beside, the nail pull resistance test were determined in accordance with ASTM C473-17 Standard test methods for physical testing of gypsum panel products [21]. In both method, MOR and MOE, screw withdrawal resistance and nail pull resistance were determined using an MTS QTest-5 Universal Test Frame (MTS systems corporation, Eden Prairie, MN, USA) featuring the Elite Modular Control System. All experiments on WCSP test specimens were conducted at the age of 28 days. As shown in Figure 4, the molded WCSP samples have the shape of a panel. Therefore, the determination of density was based on the weight and the average dimensions of the samples.

Water absorption was determined in accordance with ASTM D1037-12. The reaction to fire was tested following the ISO 5660 [22] using a cone calorimeter (Fire testing technology Limited, West Sussex, UK). Thermal capacity, specific heat and thermal conductivity were determined with a FOX 314 Heat Flow Meter (TA instruments-LaserComp Inc., Wakefield, MA, USA) following the ASTM C518 [23]. The sample was placed between the two plates of the heat flow meter at a controlled temperature. The flux meter was attached on each side of sample. The temperature and heat flux could therefore be measured at the board surface. The bottom face of WSCP (in the mold) is the exposed face in the test. The bottom face was exposed directly to the heat flux and spark igniter. The four parameters (two temperatures and two heat fluxes) can then be used to calculate heat capacity and thermal conductivity of the sample.

Finally, solid samples were observed under a Scanning Electron Microscope in order to analyse its microstructure by the JEOL JSM-840A (JEOL USA Inc, Peabody, MA, USA) equipped with an energy dispersive X-ray analysis system (EDS). The specimens were placed on double-sides adhesive tape and coated with a thin alloy of Au-Pd. The operating conditions were set at 15 kV.

2.4. Preliminary Work

A preliminary test program was conducted to evaluate the effect of SP when used in partial replacement of cement in a mixture of wood particles and cement. Seven mixtures were investigated, the variable being the fraction of cement replaced by SP. The mixing sequence used with a mortar mixer (HOBART A-120, Hobart Canada Inc, Don Mills, ON, Canada) is presented in Table 2.

Table 2. Mixing Sequence.

Step	Mixer Rotor Speed (rpm)	Cumulative Time (s)
Addition of cement and wood ashes	140	0
Addition of water	140	60
Addition of wood particles	140	120
	285	180
End of mixing	0	270

Unsurprisingly, the presence of steatite was found to increase the amount of water necessary to produce mixtures with adequate workability. The quantity of water required was estimated according to ASTM C1437 [24] to make sure that all mixture have the same workability value as P1 (Table 3). Assessing the workability and bending strength of mixtures with different percentages of SP was intended to determine the maximum amount of SP that could be used in the mixture without affecting negatively the mechanical properties of the particleboard in comparison with those of the reference wood-cement particleboard and GB. Only cement and wood particles were selected to prepare the control mixture (P1), while six other mixtures were prepared by incorporating SP at replacement rates of 10, 15, 20, 30, 40 and 50% respectively (P2 to P7).

Table 3. Mass ratio of steatite powder, cement and water used for the seven mixtures considered.

Mass Ratio	P1	P2	P3	P4	P5	P6	P7
Steatite powder/Cement	0.00	0.10	0.15	0.20	0.30	0.40	0.50
Wood/(Steatite powder + cement)	0.35	0.35	0.35	0.35	0.35	0.35	0.35
Water/(Steatite powder + cement)	1.00	1.15	1.24	1.32	1.45	1.56	1.65

Preliminary mechanical results have shown that the replacement of cement by SP in WCSP has a significant impact. The three-point bending test results at 3, 7, 14 and 28 days of moist curing show that the bending strength of the sample particleboards increases with the curing time as expected for Portland cement-based systems, although it does not increase much beyond the age of 14 days. A density change test revealed that the weight of all particleboards was stable after 14 days of curing. The study of Vu et al. [8] has also shown that the difference of bending resistance between the particleboard cement-wood-wood ash at 7 and 28 days of curing time, was not significant (4.2% max.). In the freshly consolidated particleboard, the heavier SP particles tend to settle in the bottom, yielding non-uniform characteristics across the thickness of the board. This segregation results in non-isotropic particleboards with different bending MOR depending on which side is subjected to tension stress during the test. These preliminary results have shown that particleboards with 15% of the cement replaced by SP (P3) is optimum, with the best mechanical properties obtained among the six tested mixtures. Indeed, the study of P. Kumar et al. [12] shown that the replacement of SP should be maintained below 20%.

3. Results

3.1. Change in Density

The variation of particleboard density was determined by the recording of weight of all particleboard at the begin and at the end of the curing period in the wood mold (from 0 to 3 days). A reduction of weight of about 5% during that period occurred due to water evaporation through the mold. After removal from the mold, the particleboard mass typically reached a plateau at about 6 days, meaning that most of the free water in the mortar had evaporated in the conditioning chamber at 23 °C and 60% R.H by then. At 14 days, the particleboards had a specific gravity ranging between 0.68 and 0.70.

3.2. Scanning Electron Microscopy

According to the results shown in Figure 5, both materials show a low porosity and pore sizes smaller than 10 µm. Based on SEM examination, the difference of microstructure of a mixture of neat cement-wood and a mixture containing 15% of SP in replacement of cement is not significant. Both materials show a uniform and dense microstructure.

(a)

(b)

Figure 5. Scanning electron microscope images of cement-wood particles (**a**) and cement +15% replacement of powder steatite+ wood particles (**b**).

3.3. Bending Properties

Figure 6 presents the evaluation of bending properties obtained for WCSP and GB. Three replicates per products were tested. The average values of elastic modulus and bending strength for each tested specimen are shown in Table 4.

Table 4. Bending strength of WCSP and GB (s = standard deviation) accordance with ASTM D1037-12.

Property		GB	WCSP
Specific gravity		0.7 (s = 0.02)	0.68 (s = 0.2)
Sample parallel to paper fiber direction	MOR (MPa)	5.4 (s = 0.08)	
	MOE (GPa)	1.9 (s = 0.03)	
Sample perpendicular to paper fiber direction	MOR (MPa)	1.6 (s = 0.08)	
	MOE (GPa)	1.3 (s = 0.04)	
Load on front face	MOR (MPa)		2.7 (s = 0.2)
	MOE (GPa)		1.7 (s = 0.24)
Load on back face	MOR (MPa)		5.1 (s = 0.12)
	MOE (GPa)		2.1 (s = 0.09)

As mentioned previously, 28 days after casting, the particleboards were tested in static bending using in each case six specimens. Separate test series were carried out with the front or back face subjected to bending test. Typical load-deflection curves obtained for these different test configurations are displayed in Figure 6. WCSP were tested with the load being applied on the front or on the back face of the samples. As explained before, due to the settling of the SP, the mortar at the front face of the WCSP has a less porous, denser microstructure than at the back face. Therefore, the bending strength when the load is applied on the front face of the sample is lower than on the back face of the WCSP. The WCSP with the load on the back face exhibited a bending strength of 5.1 MPa, which is over 1.9 times more than it is on the front face. The analysis of the stress–displacement curves indicates that there are three material behavior stages in the course of the test that corresponds to the mechanical behavior in the constituent composite materials (wood-cement-steatite). Each experimental curve included a linear period at the beginning of the test and a non-linear period later. The linear period represents the elastic behavior of the material. The tangent elastoplastic modulus decreases in the second period, which corresponds to a non-linear plastic behavior, the third period corresponds to the last part of the curve when the WCSP began to fracture. The MOE of WCSP is about 1.7 GPa with the load applied on the front face and about 2.1 GPa with the load applied on the back face.

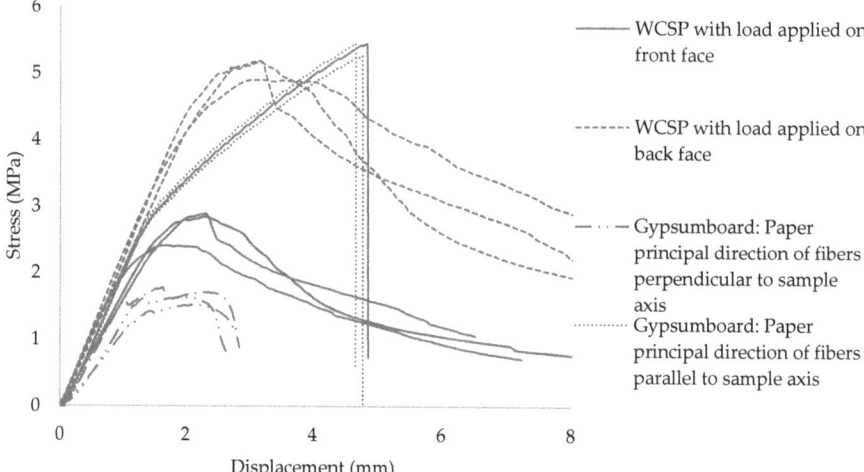

Figure 6. Characteristic stress–displacement curve for a three-point bending test of WCSP and GB in accordance with ASTM D1037-12.

For the GB, separate bending tests were conducted with the samples oriented perpendicular and along the direction of the overlay paper fibers (Figure 6). The respective stress–displacement curves in bending were entirely different. In the overlay paper fiber direction, the GB exhibit a fragile behavior, while it is more ductile in the perpendicular direction. This is due to the different tensile properties of the overlay paper in the two orthogonal directions. Therefore, for samples oriented perpendicular to the paper fiber direction, the paper failed at the beginning of the test and did not have a significant effect on the bending test. For samples oriented in the paper fiber direction, the overlay paper had a significant contribution to the mechanical properties and played the role of a reinforcement. Table 4 shows that the MOR in the overlay paper fiber direction is 5.4 MPa, which is 3.4 times higher than in the perpendicular direction. It is approximately equal to the MOR of WCSP in the case of a load applied on the back face and two times higher than in the case of a load applied on the front face. In fact, the mechanical quality of the GB depends significantly on the gypsum core properties. Therefore, the whole GB failed as the reinforcement failed. That is why the behavior is fragile (Figure 6). This mechanical comportment of GB was also noticed in the study of P. Tittelein et al. [6]. The GB bending MOE in the overlay paper fiber direction is 1.9 GPa, which is 1.5 times higher than it is in the perpendicular direction, 1.1 times lower than the bending MOE of WCSP in the case of loading on the back face and 1.1 times higher than it is in the case of loading on the front face. The results reveal that the MOR and MOE of WCSP in the case of loading on the front face are lower than in the GB overlay paper fiber direction and higher than across the GB overlay paper direction. However, the WCSP still could replace GB when adjusting the distance of the studs in the wall composition. A good bending strength of WCSP in the case of loading on the back face is an advantage for transportation and installation.

3.4. Screw Withdrawal and Nail Pull Test

The results of the screw withdrawal and nail pull resistance tests are shown in Table 5. According to these results, WCSP has better resistance to screw and nail withdrawal than GB. The recorded screw withdrawal resistance and nail pull resistance of WCSP are respectively 37 and 11% higher than the corresponding values recorded for the GB. The resistance values of GB show less variation, as it is a more homogeneous material than WCSP.

Table 5. Screw withdrawal and nail pull resistance of WCSP and GB (s = standard deviation).

PROPERTY		GB	WCSP
Screw withdrawal resistance test (N)	ASTM C473-15	374 (s = 8)	415 (s = 20)
Nails pull resistance (N)	ASTM 1037-12	328 (s = 7)	450 (s = 38)

3.5. Water Absorption

The water absorption test results are shown in Table 6. While the data show that water absorption of WCSP is just slightly lower than that of GB, the difference with respect to swelling is considerable. In fact, WCSP shows virtually no swelling, whereas for almost the same water uptake, GB undergoes a 5% expansion. In the case of WCSP, most of the free water in the mortar had evaporated about 6 days after removal from the mold based on the samples weight. Therefore, the cement was not fully cured leading to shrinkage in water. In addition, the negative value may be caused by the erosion of the specimen due to the flow of water during the test.

Table 6. Moisture absorption characteristics of WSCP and GB according to ASTM D1037-12.

Property	GB	WCSP
Water absorption (%)	54	51
Thickness swelling (%)	5	−1

3.6. Thermal Properties

Table 7 shows the results obtained for thermal capacity, specific heat, and thermal conductivity of WCSP and GB using the test method described in Section 2.2. It is worth noting that WCSP has a thermal conductivity almost three times lower than that of GB. This low thermal conductivity results from the high porosity of the WCSP compared to GB. The thermal capacity and specific heat of WCSP is 1.4 times higher than that of GB. This indicates that WCSP's ability to store thermal energy is higher than for GB. This is an important characteristic for building applications, namely those where fire-resistance rated components are required.

Table 7. Thermal properties of WCSP and GB according to ASTM C1784-13.

Property	GB	WCSP
Thermal capacity (kJ/kg·K)	970	1338
Specific heat (kJ/m^3·K)	679	910
Thermal conductivity (W/m·K)	0.32	0.12

3.7. Reaction to Fire

The thermal properties and results of the cone calorimeter tests are presented in Table 8 and Figure 7.

Table 8. Results of the test calorimeter cone for WCSP and GB in accordance with ASTM E1354-17 (from start of test +15 min).

Property/Characteristic	GB	WCSP
Average ignition time (s)	57	none
Average peak rate of heat release (kW/m^2) for 15 min	109.87	15.03
Average time to peak rate of heat release (s)	65	423
Average total smoke production (m^2) for 15 min	0.19	0.28
Average total heat release (MJ/m^2) for 15 min	5.89	7.88

The WCSP with a SP replacement rate of 15% was tested for comparison with GB. Test results in Figure 7a indicate that the heat release rate (HRR) of GB increases very quickly and reaches a maximum at about 65 s, due to burning of the overlay paper occurring at about 57 s (Table 8). The WCSP showed very low HRR throughout the 20 min duration. The peak of heat release is not obvious as no combustion took place (Table 8). The HRR of WCSP varied from −2 to 16 kW/m^2 and also varied from −4 to 12 kW/m^2 in the case of GB after the ignition. Such variations are related to the accuracy of the cone calorimeter for materials exhibiting low combustibility characteristics, such as a THR less than 15 MJ/m^2.

The National Building Code of Canada (NBCC) states that a material can be used in non-combustible construction provided that, when tested in accordance with CAN/ULC-S135 [25] at a heat flux of 50 kW/m^2, the total heat release is no more than 3 MJ/m^2 and the total smoke extinction area is no more than 1.0 m^2.

Figure 7b shows that WCSP behaves similarly to GB with respect to mass loss. Indeed, the remaining mass of GB after 15 min is 81% while the WCSP is 78%. The average total smoke production for 15 min recorded for both materials are less than 1 m^2 (Table 8). Their average total heat release (THR) exceed in each case 3 MJ/m^2 (Figure 7 and Table 8). While CAN/ULC-S135 slightly differs from ISO 5660, the results suggest that WCSP would most likely fail these requirements due to its average THR exceeding 3 MJ/m^2. Given the low value threshold of 3 MJ/m^2, very few products consisting in whole or in part of combustible materials will pass this test [26]. It is unclear from the NBCC as to how this threshold value was determined.

In Japan, based on the performance during a cone calorimeter test when subjected to an irradiance level of 50 kW/m^2, the reaction to fire of interior finishing materials are classified as being non-combustible, quasi-non-combustible or fire retardant [27]. Neither WCSP nor GB actually meet the Japanese criteria of a non-combustible material: THR ≤ 8 MJ/m^2 and peak rate of heat release ≤200 kW/m^2 after 20 min of exposure. It is noted that the Japanese THR threshold of 8 MJ/m^2 is much less severe than the Canadian value of 3 MJ/m^2. They would however both meet the criteria for a quasi-non-combustible material: THR ≤ 8 MJ/m^2 and peak rate of heat release ≤200 kW/m^2 after 10 min of exposure.

Flame-spread, which is used to describe the surface burning characteristics of building materials, is one of the most commonly tested fire performance characteristics for limiting fire growth in the early stage of fire development. The results generated in this study show that the surface of WCSP is still not burnt after 20 min. The surface color barely changed and neither fractures nor burnt surface could be observed in Figure 7d). The side face of the specimen became dark due to direct contact with the metallic specimen holder and the aluminum wrap, while the surface of the GB burned completely (Figure 7e). The side face of the sample shows that the GB burned and became dark (Figure 7f) across the thickness, while the 3 mm surface layer of the WCSP almost did not change color (Figure 7g) and did not burn. However, the wood particles on the inside and back face of the WCSP became dark due to the high temperature and large porosity (Figure 7g,h). It may explain why the HRR of WCSP (−2 to 16 kW/m^2) was slightly higher than that of GB (−4 to 12 kW/m^2) after the end of ignition of GB (Figure 7a).

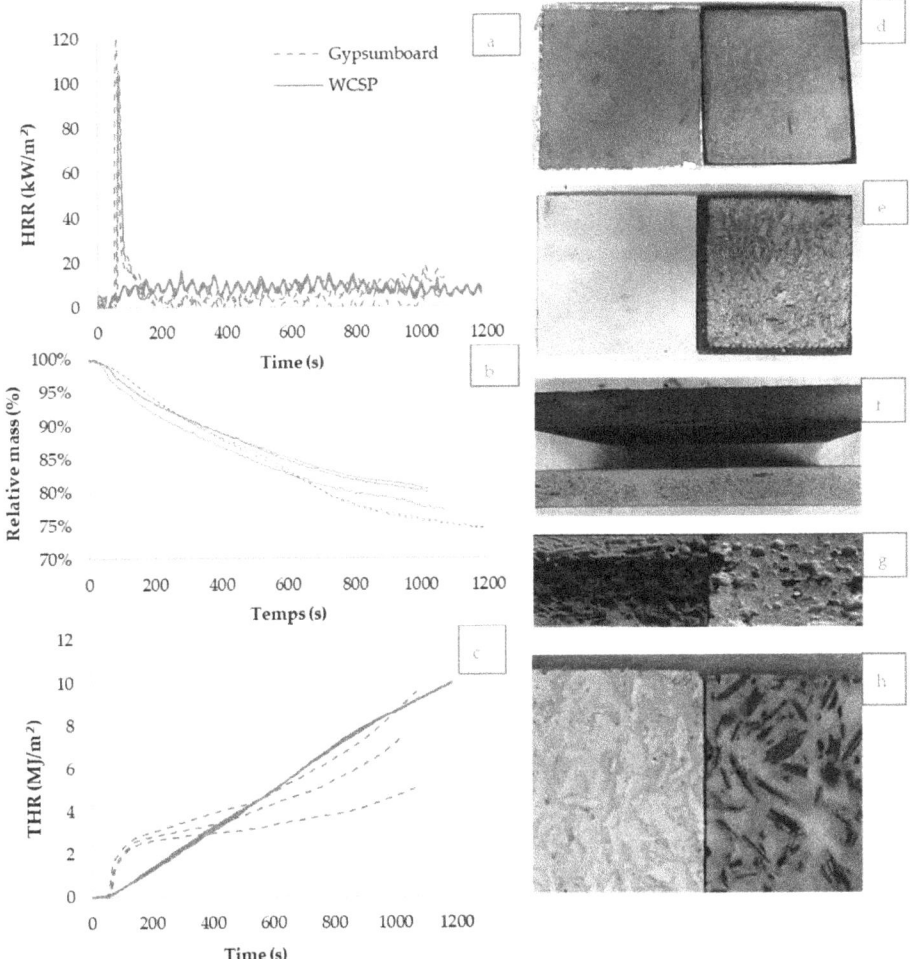

Figure 7. Cone calorimeter measurements: (**a**) Heat release rate (HRR). (**b**) Relative mass. (**c**) Total heat release (THR). (**d**) Front face of WCSP before (**left**) and after (**right**) testing. (**e**) Front face of GB before (**left**) and after (**right**) testing. (**f**) Side face of gypsum specimen before and after testing. (**g**) Side face of WCSP before (**left**) and after (**right**) testing. (**h**) Back face (**left**) before and after testing (**right**) of WCSP.

4. Conclusions

Due to the settling of steatite powder, the formed surface (bottom face) of a wood-cement steatite powder (WCSP) board was of good quality even without paper overlay. It compares favorably to the surface of paper-faced gypsum boards. Besides, the ASTM D 1037-12 screw withdrawal resistance and ASTM C473-15 nail pull resistance of wood-cement-steatite powder boards were found to be 37% and 11% higher, respectively. When the load was applied on the front face, their bending strength is 69% higher. These panels also exhibit better water-resistance and better reaction to fire than those of gypsum boards. Indeed, with regards to reaction to fire, no ignition was observed for the WSCP, and the remaining mass of both type of boards after 15 min from start of the test was similar. The test results obtained in the present study actually show that wood-cement-steatite powder boards could be classified as a quasi-non-combustible material. While the replacement of cement with steatite powder at a rate of 15% improved the mechanical and thermal properties of the panel, it could also contribute to reduce CO_2 emissions caused by cement production. Two-thirds of the raw materials used for wood-cement-steatite powder board production are low cost secondary products from mineral extraction of steatite and lumber production. The above results show that replacing gypsum boards by such an engineered material may be a worthy choice for buildings of the future.

Author Contributions: Conceptualization V.-A.V., A.C., B.B., P.B. and C.D.; Formal analysis V.-A.V., A.C., B.B., P.B. and C.D.; Funding acquisition A.C. and P.B.; Investigation V.-A.V., A.C., B.B. and P.B.; Methodology V.-A.V., A.C., B.B., P.B. and C.D.; Project administration A.C. and P.B.; Resources A.C., B.B., P.B. and C.D.; Supervision A.C. and B.B.; Validation V.-A.V., A.C., B.B. and C.D.; Writing—original draft V.-A.V.; Writing—review & editing A.C., B.B., P.B. and C.D. All authors have read and agreed to the published version of the manuscript.

Funding: This work is part of the research program of the Natural Sciences and Engineering Research Council of Canada (NSERC) Industrial Research Chair on Eco-Construction in Wood (CIRCERB) through programs IRC (IRCPJ 461745-12) and CRD (RDCPJ 445200-12).

Acknowledgments: The authors are also grateful to the industrial partners of the NSERC Industrial Chair on Eco-Responsible Wood Construction (CIRCERB).

Conflicts of Interest: The authors declare no conflict of interest.

References

1. Hobart, M.K. Soapstone What is Soapstone? How does it Form? How is it Used? Geoscience New and Information. Available online: https://geology.com/rocks/soapstone.shtml (accessed on 30 April 2020).
2. Sudalaimani, K.; Shanmugasundaram, M. Influence of Ultrafine Natural Steatite Powder on Setting Time and Strength Development of Cement. *Adv. Mater. Sci. Eng.* **2014**, *2014*, 532746. [CrossRef]
3. Anonymous. *Talc, Soapstone and Steatite in Indian Minerals Yearbook*; Government of India, Ministry of Mines, Indian Bureau of Mines: Nagpur, India, 2013.
4. Ramirez-Coretti, A.; Eckelman, C.; Wolfe, R. Inorganic-bonded composite wood panel systems for low-cost housing: A Central American perspective. *Forest Prod. J.* **1998**, *48*, 62–68.
5. Berger, F.; Gauvin, F.; Brouwers, H.J.H. The recycling potential of wood waste into wood-wool/cement composite. *Constr. Build. Mater.* **2020**, *260*, 119786. [CrossRef]
6. Tittelein, P.; Cloutier, A.; Bissonnette, B. Design of a low-density wood–cement particleboard for interior wall finish. *Cem. Concr. Compos.* **2012**, *34*, 218–222. [CrossRef]
7. Němec, M.; Igaz, R.; Gerge, T.; Danihelová, A.; Ondrejka, V.; Krišťák, L.; Gejdoš, M.; Kminiak, R. Acoustic and thermophysical properties of insulation materials based on wood wool. *Akustika* **2019**, *33*, 115–123.
8. Govin, A.; Peschard, A.; Guyonnet, R. Modification of cement hydration at early ages by natural and heated wood. *Cem. Concr. Compos.* **2006**, *28*, 12–20. [CrossRef]
9. Panzera, T.; Hallak, T. Cement-steatite composites reinforced with carbon fibres: An alternative for restoration of brazilian historical buildings. *Mater. Res.* **2011**, *14*, 118–123. [CrossRef]
10. Mahsa, K.; Ali, G. Effect of glass powders on the mechanical and durability properties of cementitious materials. *Constr. Build. Mater.* **2015**, *98*, 407–416.
11. Vu, V.-A.; Cloutier, A.; Bissonnette, B.; Blanchet, P.; Duchesne, J. The effect of wood ash as a partial cement replacement material for making wood-cement panels. *Materials* **2019**, *17*, 2766. [CrossRef] [PubMed]
12. KumarK, P.; Sudalaimani, M.; Shanmugasundaram, K. An Investigation on Self-Compacting Concrete Using Ultrafine Natural Steatite Powder as Replacement to Cement. *Adv. Mater. Sci. Eng.* **2017**, *6*, 1–8. [CrossRef]
13. Weber, B. Heat transfer mechanisms and models for a gypsum board exposed to fire. *Int. J. Heat Mass Transf.* **2012**, *55*, 1661–1678. [CrossRef]
14. Anonymous. *Gypsum Board for Walls and Ceilings*; Gypsum Association: Dallas, TX, USA, 1980.
15. Anonymous. *Fire Safety Design in Buildings*; Canadian Wood Council: Ottawa, ON, Canada, 1996.
16. Cramer, S.; Friday, O.; White, R.; Sriprutkiat, G. Mechanical properties of gypsum board at elevated temperatures. In Proceedings of the Fire and Materials 2003: 8th International Conference, San Francisco, CA, USA, 27–28 January 2003.

17. Blanchet, P. Gypsum Replacement Material—A Biobased Proposal. In Proceedings of the 3rd International Conference on Bio-Based Building Materials, Belfast, UK, 26–28 June 2019.
18. ASTM C618. *Standard Specification for Fly Ash and Raw or Calcined Natural Pozzolans for Use Portland Cement Concrete*; American Standard Test of Materials; ASTM International: Montgomery, PA, USA, 2019.
19. Baig, Z.; Mamat, O.; Mustapha, M.; Mumtaz, A.; Sarfraz, S.M. An efficient approach to address issues of graphene nanoplatelets (GNPs) incorporation in aluminium powders and their compaction behavior. *Metal* **2018**, *8*, 90. [CrossRef]
20. ASTM C473. *Standard Test Methods for Physical Testing of Gypsum Panel Products*; American Standard Test of Materials; ASTM International: Montgomery, PA, USA, 2017.
21. ASTMD1037. *Standard Test Method for Evaluating the Properties of Wood-Base Fiber and Particle Panel Materials*; American Standard Test of Materials; ASTM International: Montgomery, PA, USA, 2012.
22. ISO 5660. *Reaction-to-Fire Tests-Heat Release, Smoke Production and Mass Loss Rate-Part 1: Heat Release Rate (Cone Calorimeter Method) and Smoke Production Rate (Dynamic Measurement)*; International Organization for Standardization: Geneva, Switzerland, 2015.
23. ASTM C518. *Standard Test Method for Steady-State Thermal Transmission Properties by Means of the Heat Flow Meter Apparatus*; American Standard Test of Materials; ASTM International: Montgomery, PA, USA, 2017.
24. ASTM C1437. *Standard Test Method for Flow of Hydraulic Cement Mortar*; Standard Test Methods for Physical Testing of Gypsum Panel Products; ASTM International: Montgomery, PA, USA, 2020.
25. CAN/ULC-S135-92. *Standard Method of Test for Determination of Degrees of Combustibility of Building Materials Using an Oxygen Consumption Calorimeter (Cone Calorimeter)*; Standards Council of Canada: Ottawa, ON, Canada, 1998.
26. Mehaffey, J.; Dagenais, C. *Assessing the Flammability of Mass Timber Components: A Review*; FPInnovations: Pointe-Claire, QC, Canada, 2014.
27. Mehaffey, J.R. *Fire Performance of Interior Finishes, Room Linings and Structural Panel Products*; Project Report No. 3638; Forintek Canada Corp: Ottawa, ON, Canada, 2006.

Publisher's Note: MDPI stays neutral with regard to jurisdictional claims in published maps and institutional affiliations.

© 2020 by the authors. Licensee MDPI, Basel, Switzerland. This article is an open access article distributed under the terms and conditions of the Creative Commons Attribution (CC BY) license (http://creativecommons.org/licenses/by/4.0/).

Review

Calcined Clay as Supplementary Cementitious Material

Roman Jaskulski [1,*], Daria Jóźwiak-Niedźwiedzka [2] and Yaroslav Yakymechko [1]

1 Faculty of Civil Engineering, Mechanics and Petrochemistry, Warsaw University of Technology, Łukasiewicza 14, 09-400 Płock, Poland; Yaroslav.Yakymechko@pw.edu.pl
2 Institute of Fundamental Technological Research, Polish Academy of Sciences, Pawińskiego 5b, 02-106 Warsaw, Poland; djozwiak@ippt.pan.pl
* Correspondence: Roman.Jaskulski@pw.edu.pl

Received: 3 October 2020; Accepted: 21 October 2020; Published: 23 October 2020

Abstract: Calcined clays are the only potential materials available in large quantities to meet the requirements of eco-efficient cement-based materials by reducing the clinker content in blended cements or reducing the cement content in concrete. More than 200 recent research papers on the idea of replacing Portland cement with large amounts of calcined clay are presented and discussed in detail. First, the fundamental information about the properties and structure of clay minerals is described. Then, the process of activation and hydration of clays is discussed, including the methods of pozzolanic activity assessment. Additionally, various testing methods of clays from different worldwide deposits are presented. The application of calcined clay in cement and concrete technology is then introduced. A separate chapter is devoted to lime calcined clay cement. Then an influence of calcined clay on durability of concrete is summarized. Finally, conclusions are formulated.

Keywords: calcined clay; binder; supplementary cementitious materials; cement-based materials

1. Introduction

Sustainable development, understood as progress that meet the needs of the present without compromising the ability of future generations to meet their needs, is now the basic idea behind the current paradigm of technological progress. The greatest threat to the achievement of the objectives of sustainable development is the excessive emission of greenhouse gases, including, above all, carbon dioxide, which results in global warming with many serious negative consequences for future development and its prospects.

The building industry makes a significant contribution to increasing the carbon dioxide content in the atmosphere, with cement production accounting for 5% of global carbon dioxide emissions [1] being its largest source. This is due, on the one hand, to the high energy intensity of the Portland clinker production process. The production of one ton of clinker requires the supply of 3.1–3.8 GJ of heat, whereas in the older generation wet method kilns the demand may reach even 6 GJ/t [1]. On the other hand, during cement production, carbon dioxide is emitted from raw materials, mainly limestone. The amount of this emission is about 0.53 kg/kg of clinker [1].

The possibilities of reducing the energy intensity of cement production are limited, although it is already possible to reduce heat consumption to 2.9 GJ/t of clinker. The use of alternative fuels further reduces carbon dioxide emissions from fossil fuels. However, the use of limestone in clinker production cannot be eliminated or reduced. Therefore, solutions to further reduce CO_2 emissions to the atmosphere by the cement and concrete industry are the production of blended cements that reduce the amount of clinker and the use of supplementary cementitious materials (SCM), mainly pozzolans, in the concrete technology. A pozzolan, according to ASTM C125, is "a siliceous and aluminous material which, in itself, possesses little or no cementitious value but which will, in finely divided form in the presence of moisture, react chemically with calcium hydroxide at ordinary temperature to form compounds possessing cementitious properties" [2].

The most commonly used SCM so far is fly ash from coal power plants. However, this additive, which is used in the production of both cement and concrete, is slowly losing its primary importance due to the progressive decommissioning of coal and lignite power plants. As a result, the supply of fly ash will be significantly reduced in the near future. Another cement additive used to reduce clinker content is a granulated blast furnace slag. However, it has no potential to replace fly ash, not least because its global supply is far below potential demand. Additionally, it is only available in countries where the steel industry exists. However, even where it is available, its share in blended cements production is not large. In India, for example, Portland slag cement accounts for only about 8% of total cement production, while ordinary Portland cement accounts for 24% of production, and Portland pozzolana cement accounts for 65%, where the main addition is fly ash [3,4].

The decreasing supply of fly ash from the power industry is encouraging the search for new sources of pozzolan additives for cement and concrete production. One of the directions of this search is ashes obtained from biomass combustion, [5]. Another is the use of natural and artificial pozzolana. The former include, among others, volcanic tuffs already known in ancient times or zeolites. The best known representative of the second group is the metakaolin formed by the calcination of kaolinite from clays with its significant content.

Metakaolin was shown to be the best clay raw material for SCMs production, but in its pure form it is only found in a limited number of deposits, so its availability is not sufficient to meet the needs of the building materials industry. It is also in the focus of interest of other industries [6,7]. For this reason, there has been interest in the possibility of producing SCMs from other locally available natural clays containing, in addition to kaolinite, other minerals which have the potential to develop pozzolanic activity upon appropriate activation. Research conducted in this direction led to the separation of a new group of pozzolanic materials—calcined clay.

Clay is a widely spread material in the world, cheap and easily accessible [8]. At the same time, it is a material with a great diversity in terms of mineralogical composition, hence numerous literature items devoted to the analysis of the possibility of using clays from specific deposits for the production of SCM in the calcination process [9–11].

In general, the literature on the activation of clay minerals and their use in civil engineering is very rich and covers a wide range of issues. It would be very difficult to discuss this vast area of research in one review article, so certain assumptions on the scope of this review were made First of all, it focuses on calcined clays used as SCM in binders, where cement is the dominant component, and in the cement-based concrete. Outside the scope of interest remain cement-free binders based on calcined clays, which are the basis of geopolymers, as well as other alkaline-activated binders (including clay minerals). Issues that do not fall within the assumed topics of the article are mentioned where it was indicated for various reasons.

The paper is divided into eight thematic sections. The first section describes a fundamental knowledge about clay minerals and publications on clays from various deposits in the world. In the next section the topic of clay minerals activation with emphasis on calcination process and research of hydration mechanisms and pozzolanic activity with factors influencing them are presented. A separate section is devoted to the role of calcined clays in concrete technology (as SCM) and in blended cements production. Due to the wealth of literature devoted to lime calcined clay cement (LC^3), a separate section is also devoted to it. The next chapter discusses the results of research on the impact of calcined clays on concrete durability. Numerous publications on calcined clay, which cannot be clearly classified as one of the above thematic areas, have been placed in a separate section devoted to suitable aspects of the application of these materials. The summary presents the conclusion of the literature review

2. Characteristic of Clay Minerals

Clay minerals, that are the subject of this review, belong to two main groups, which are often referred to in the literature as 1:1 and 2:1 minerals. These two terms result from their structure, which consists of repetitive tetrahedral (T) and octahedral (O) layers [12]. The T layer consists of corner-contacting tetrahedral Si^{4+}, Al^{3+}, and Fe^{3+} cations, while the O layer consists of edge-contacting octahedral Al^{3+}, Fe^{3+}, Mg^{2+}, and Fe^{2+} cations, which are arranged alternately in *cis* and *trans* configurations (see Figures 1 and 2).

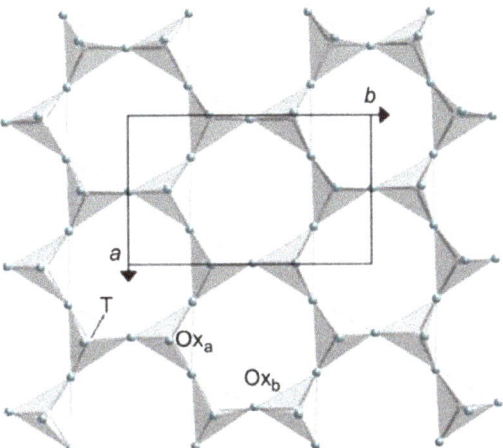

Figure 1. The tetrahedral sheet. T, tetrahedral cations; O*xa*, apical oxygen atoms; O*xb*, basal oxygen atoms. *a* and *b* refer to unit cell parameters. Reprinted from Brigatti, M.F.; Galán, E.; Theng, B.K.G. Structure and Mineralogy of Clay Minerals. In *Handbook of Clay Science*; Bergaya, F., Lagaly, G., Eds.; Elsevier Ltd: Amsterdam, The Netherlands, 2013; pp. 21–81. License number 4912071405833.

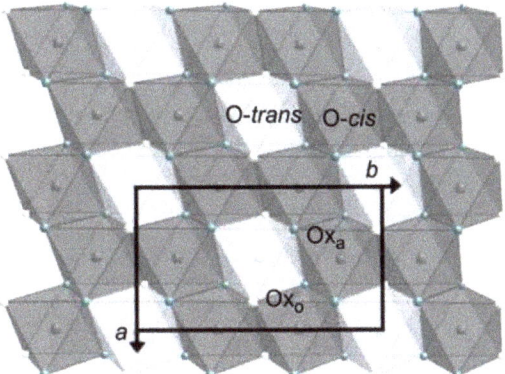

Figure 2. The octahedral sheet. O-*trans*, *trans*-oriented octahedra; O-*cis*, *cis*-oriented octahedra; O_{xa}, apical oxygen atoms; O_{xo}, OH, F, Cl octahedral anion. *a* and *b* refer to unit cell parameters. Reprinted from Brigatti, M.F.; Galán, E.; Theng, B.K.G. Structure and Mineralogy of Clay Minerals. In *Handbook of Clay Science*; Bergaya, F., Lagaly, G., Eds.; Elsevier Ltd.: Amsterdam, The Netherlands, 2013; pp. 21–81. License number 4912071405833.

At the contact of T and O layers there is a layer of oxygen atoms (so-called apical oxygen). Additionally, the O layer contains oxygen atoms (non-apical oxygen) forming OH^- groups and other anions (F⁻, Cl⁻), which are located in the corners of octahedrons, but are not shared with the T layer [12].

The division of minerals into 1:1 (TO) or 2:1 (TOT) groups results from the repetitive arrangement of layers of the one or the other type (see Figure 3). This division is far from covering the rich diversity of clay minerals, at least because 2:1 minerals may additionally contain anhydrous interlayer cations, hydrated interlayer cations or octahedral interlayer sheet. In addition, clays also contain complex minerals in which TO and TOT layers co-exist and can be distributed regularly or randomly.

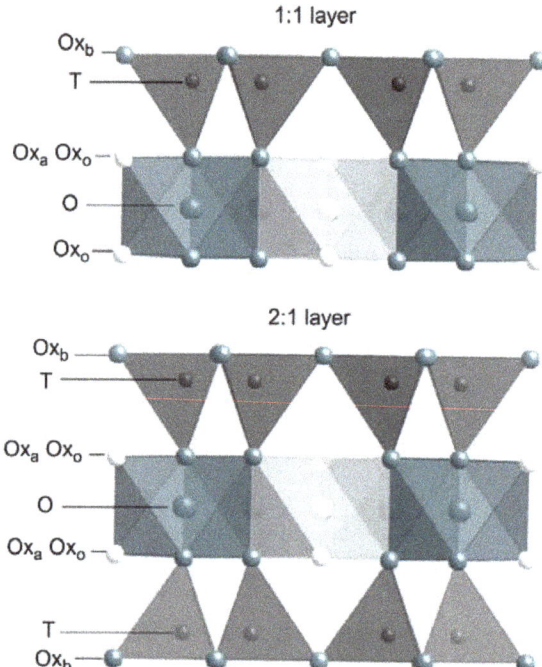

Figure 3. Models of 1:1 and in 2:1 layer structures. Ox_b, basal oxygen atoms; T, tetrahedral cations; O, octahedral cations; Ox_a, apical oxygen atoms; Ox_o, octahedral anions (OH, F, Cl). Reprinted from Brigatti, M.F.; Galán, E.; Theng, B.K.G. Structure and Mineralogy of Clay Minerals. In *Handbook of Clay Science*; Bergaya, F., Lagaly, G., Eds.; Elsevier Ltd.: Asmetdam, The Netherlands, 2013; pp. 21–81. License number 4912071405833.

Minerals 1:1 include two groups: the kaolin group and the serpentine group. The first group includes: kaolinite, dickite, nicrite, halloysite, hisingerite. To the second group belong, among others, lizardite, antigorite, chrysotile, caryopilite, pyrosmalite, polygonal serpentines and polyhedral serpentines.

Minerals 2:1 form a large group of which the following should be mentioned: pyrophyllite and talc both of ideally layered structure; mica, more than 200 variants [13], of which muscovite and illites will be discussed further on; and smectites, of which montmorillonite is discussed further on. The 2:1 structure also has vermiculite, chrorites and, finally, minerals with mixed layer structure, such as illite-smectite [14,15], chlorite-smectite [16] or illitic-chlorite. The calcination of the former is devoted, among others, in the publication of Garg and Skibsted [15].

A more detailed discussion of the properties and structure of clay minerals is beyond the scope of this review. In the following section, the minerals which are part of calcined clays and which determine their properties are discussed in more detail. The discussion is focused on their properties, which are important in the activation process and determine their properties as pozzolan materials. The article describes also factors influencing the optimization of activation process.

The most important clay minerals, which are subject to temperature activation and, therefore, are discussed in more detail in this paper, are kaolinite and montmorillonite, which also occurs in the form of variants called Ca-montmorillonite and Na-montmorillonite [17–19]. Additionally, illite, being a poorly crystallised mica, and muscovite belonging to the same family, can be subjected to the calcination process, although both show low pozzolan activity even after heat treatment (illite, however, slightly higher than muscovite). Nevertheless, both are also included in this review of clay minerals. Relatively little attention is paid to sepiolite, halloysite and mixed-layer minerals in the research on pozzolan activity, therefore, due to a small number of publications, these minerals are discussed more briefly in this paper.

Kaolinite and the product of its calcination, i.e., metakaolinite, are the most thoroughly tested of the clay minerals that are part of the clays subjected to calcination in order to use them as SCMs due to their pozzolanic activity. Part of this literature can be found in [20,21]. Other minerals are much less popular so far, so Garg and Skibsted's publications on montmorillonite are worth mentioning here [22,23]. Garg and Skibsted [22] presented the result of investigation on the activity of calcined pure montmorillonite due to the changes in the structure under the heating. The authors traced the changes in the structure of montmorillonite heated up to 1100 °C using the NMR technique. The results indicate that montmorillonite exhibits pozzolanic activity both in the raw state as well as heated up to any temperature value it was subjected to in the mentioned studies, although above 800 °C a clear decrease in this activity is visible.

It is not only montmorillonite that exhibits pozzolanic activity in its raw state. Among such minerals we can also include illite, kaolinite and muscovite [24–26]. However, it is montmorillonite that is most active in the raw state. The least active of them is muscovite due to a small proportion of amorphous phase, which is additionally accompanied by very high water content [27]. Nevertheless, it is commonly accepted that the most active mineral of the pozzolanic is metakaolinite. The study of He et al. [17] revealed that Ca-montmorillonite may show even higher pozzolanic activity after calcination. It is associated with a high content of amorphous silica. Additionally, in [28] two types of bentonite (containing mainly montmorillonite) were tested and obtained better results than most of the five kaolinites, although it should be added that such results were obtained only in lime consumption and electric conductivity tests.

To the group of kaolinites, i.e., minerals 1:1 belongs halloysite—a lesser known close "cousin" of kaolinite. Both these minerals have identical chemical structure with the only one difference that halloysite may contain two additional water molecules in the interlayer space. Halloysite is, therefore, sometimes referred to as hydrated kaolinite. This small difference has significant consequences for the structure of this mineral, which forms different forms such as tubes, spheres and laths [12,29]. Water in the interlayer space can be easily removed by thermal treatment and such dehydrated halloysite is sometimes referred to as metahalloysite. The results of a study on the pozzolanic activity of clays rich in halloysite are presented in papers of Tironi et al. [30,31]. Obtained results confirmed the high pozzolanic activity of calcined clays containing halloysite, confirming their usefulness as a raw material for SCM production.

Muscovite is also included among the clay minerals that show pozzolanic activity after appropriate heat treatment, although in this case the results are ambiguous. Ambroise [24], by examining the possibility of activation at 750 °C of, among others, muscovite and phlogopite, obtained results indicating low effectiveness of this form of treatment. Samples of calcined muscovite blended with calcium hydroxide after seven days of curing in the moulds disintegrated after placing them in water, where they were to undergo further curing. On the other hand, [8] showed that muscovite increases the hydration heat of the cement and its effect varies depending on the temperature of thermal treatment (the raw material was tested, as well as heated at various temperatures from 500–950 °C). Additionally, in [27] the calcined muscovite pozzolanic activity at 800 °C was shown, although it was relatively low compared to calcined illite and kaolinite.

Another clay mineral which has been studied due to its potentially pozzolanic properties is sepiolite [17,32]. It is a 2:1 group mineral with an unusual structure due to discontinuity of TOT layers [12]. Studies have shown that without heat treatment sepiolite is not a material that is active as a pozzolan. Its additional disadvantage, from a technological point of view, is high water demand. The use of sepiolite in raw form as a partial substitute for cement led to a significant decrease in the strength of the mortar used to assess the hydraulic properties of potentially pozzolanic materials. On the other hand, calcination carried out at the temperature of 830 °C allows to obtain a material with distinct, although relatively low pozzolanic activity and significantly reduced water demand [32]. The low pozzolanic activity of sepiolite may result from low Al content (about 1%), which, together with Si, is responsible for pozzolanic activity of clay minerals.

Illit is one of the minerals which show practically no pozzolanic properties in their raw state. Heat treatment carried out at high temperatures (930–950 °C) both pure mineral with small admixtures [33] as well as clays with a dominant share of illite [34] leads to a material with moderate pozzolanic activity. The dehydroxylation process itself, occurring at a lower temperature, does not lead to activation of the illite and formation of a material with a positive value of hydraulic index whereas, at the temperature that causes the destruction of its structure, a partial recrystallization of the resulting amorphous phase takes place at the same time, which reduces the potential pozzolanic activity of this mineral.

Mixed-layer minerals, which are characterised by the presence of different types of layers, are very common in clays. According to Weaver [35], it is estimated that out of 6000 tested samples of clay minerals from all over the USA as much as 70% of them contain different varieties of mixed-layer minerals. Such widespread minerals could not be omitted in studies on pozzolanic activity of clays, although the number of publications on this topic is small. He et al. [14], investigated the pozzolanic activity of synthetic clay consisting of layers of mica and smectite, used as a catalyst [36]. The share of illite (i.e., mica) in relation to montmorillonite (smectite) according to various studies of this material is in the range from 3:1 to 2:1. His studies showed that this mineral, after calcination at 960 °C, showed good pozzolanic properties, achieving in the test mortar strength of 113% in relation to mortar consisting only of OPC.

The study of natural mixed-layer mineral of similar composition (illite/smectite in the ratio 70/30) was conducted by Garg and Skibsted [15]. In this case, the optimal calcination temperature was 900 °C and its exceeding led to crystallization of a part of amorphous phase and decrease of pozzolanic activity. Apart from a slightly different mutual ratio of mica and smectite, the mineral studied by Garg and Skibsted did not contain NH^{4+} ions, which may partially explain the discrepancy in the value of the optimal calcination temperature in relation to the findings of He et al.

Lemma et al. [16] study of mixed-layer clay consisting mainly of illite (45%) and chlorite (14%) containing also non-clay minerals, such as quartz (30%) and feldspar (10%), revealed high optimal calcination temperature of the material. The highest value of strength activity index (SAI) of this clay was observed after calcination at 1100 °C. More detailed analysis of the obtained results indicated that an increase in the pozzolanic activity at this temperature, in relation to that observed at 900 °C after total dehydroxylation of the illite, was caused by the transition of feldspar from a crystalline to an amorphous form. Although this mineral is not typical for clay rocks, its occurrence in the composition of clay intended for calcination may be a premise for carrying out this process at a higher temperature.

3. Activation and Hydration of Clays

3.1. Activation Process and Its Conditionalities

Raw clays usually have a moderate or low pozzolanic activity. To increase it, they need to be activated. As a result of activation, clay minerals undergo processes of dehydroxylation and amorphisation and the accompanying change in coordination of Al ions. Those processes lead to, among other things, greater solubility of Al and Si ions and their greater reactivity, which is a basic condition for the demonstration of pozzolanic activity by clay minerals [37]. The activation process can be carried out mechanically (by grinding) [38–40] or thermally by heating to a temperature high enough to destroy the structure of the clay minerals, but low enough to avoid recrystallization and the formation of chemically inert phases. Some publications indicate that kaolinitic clays should be ground before they undergo calcination [41,42], i.e., their activation should have both mechanical and thermal component. Processes occurring during thermal or mechanical activation can also be induced chemically to some extent [37,43]. Sanchez et al. [44] studied the effects of thermal activation of kaolin supported by an activator in the form of 1% ZnO admixture. The efficacy of the activation process of acid treatment of clay before its calcination was also studied [45]. Due to the subject matter of this article, only thermal activation, i.e., calcination, will be described in more detail.

The effectiveness of the thermal activation process, and consequently pozzolanic activity of the obtained material, depends on many factors. These include: calcination temperature, particle size and shape, time and others [24]. The most attention is paid to the analysis of the temperature influence. According to Fernández et al. [18], the exposure of clay to too low or too high a temperature can significantly affect the activation process. In the former case, due to incomplete dehydroxylation, in the latter case, due to the melting of minerals and their subsequent recrystallization leading to the formation of phases which do not react with the cement hydration products and do not exhibit any pozzolanic activity.

The process of dehydroxylation, i.e., tearing off OH^- groups [17], leads to serious damage in the crystal structure of clay minerals. As a result, among other things, it leads to increased exposure of Al ions on the surface of the mineral grains and increased solubility [18]. This effect is more pronounced in the case of kaolinite than in the case of group 2:1 minerals, which is reflected in the differentiation of pozzolanic activity of clay minerals. The structure of the minerals also affects the dehydroxylation temperature, which occurs in a wide range from 350 °C to 900 °C, and for its full effect in most cases the mineral must be heated to temperatures between 600 and 800 °C [46,47]. At temperatures below 600 °C, this reaction takes place only with kaolinite [48,49]. Montmorillonites undergo this process at temperatures between 550–850 °C. Illite is dehydroxylated in the temperature range 600–900 °C [18]. Neißer-Deiters et al. [8] have studied the influence of temperature on the properties of calcined mica consisting of muscovite with a small admixture of phengite. The investigated material was subjected to calcination at different temperatures ranging from 500 °C to 950 °C. As the raw muscovite already exhibits pozzolanic properties, the aim of the study was to determine whether and how the calcination influences its pozzolanic activity and water demand. The results indicate that the calcination process, regardless of the temperature, leads to slight changes in the above-mentioned parameters.

Another variable of importance in the thermal activation process is time. Its impact has been studied by Chakchouk et al. [50]. They determined the optimal parameters of mortar with the use of calcined clay. For this purpose a model with three variables was developed. In addition to the calcination time, the temperature of the calcination and the exchange rate of cement to calcined clay in the subsequently tested blended binder were also taken into account. In 23 individual experiments, the time varied from 1.32 h to 4.68 h. In most cases, it was 3 h. They showed, that a longer calcination time was beneficial in increasing the strength of the tested mortar if the temperature was lower than 700 °C, and above this temperature the effect was the opposite.

The duration of the thermal activation process is usually counted in hours or minutes when using fluidized bed reactors [51]. This type of calcination is referred to as soak-calcination. This process is most commonly used and most of the publications are devoted to it, however, in these publications the time is sometimes given for the whole process, sometimes it is only the time of keeping the material at the highest temperature, and sometimes it is not specified at all. This makes it difficult an attempt to summarise and make clear recommendations. Especially since this time also depends on the degree of material fragmentation, its pre-treatment (e.g., drying to constant mass) or the type of device in which the activation is carried out (rotary kiln or fluidized bed reactor).

However, regardless of whether the process of soak-calcination takes several minutes or several hours, the activation process known as flash-calcination, which lasts from fractions of a second to several seconds, can be clearly distinguished from it [51–55]. The shortening of the process results from a very large fineness of the processed material and from the high temperature in the calciner reaching 1100 °C. The combination of high pulverization and high temperature causes the clay subjected to calcination to be heated very quickly in its entire mass. The rise of temperature reaches 0.5–1.5×10^4 °C/s [52,54]. Such a rapid temperature increasing leads to a material with slightly different characteristics from clay obtained from the soak-calcination process. In such clays in particular, the dehydroxylation process is sometimes not complete. Moreover, due to the rapid process, the obtained material has a lower density and higher porosity. This usually also translates into increased reactivity [51,53,54]. Most of the research on flash-calcination concerns clays with high kaolinite content, therefore, the work of Rasmussen et al. [53] is worth mentioning. They carried out a study using clays with a composition dominated by minerals from the 2:1 group. They concluded that, also in the case of this type of clays, flash-calcination can lead to the production of clays with a high-level of pozzolanic activity, and the advantage as compared to soak-calcination procedure in this case lies in the lower risk of melting and recrystallization of minerals, such as feldspar or spinel [53].

It is not only time, temperature or particle size that determine the success of the calcination process. Bich et al. [56] studied the conditions of the dehydroxylation process depending on the degree of ordering of the mineral structure. They used three clays with high kaolinite content and different degree of order of this mineral defined by values of P_0 coefficient (it is defined further in the article). They showed, that in case of kaolinite with disordered structure ($P_0 = 0.68$), that it was enough to heat the material at 650 °C for 45 min for full dehydroxylation whereas, in the case of kaolinite with well-ordered structure ($P_0 = 1.4$), such time allowed dehydroxylation of 71–95% of kaolinite. The relationship between kaolinite dehydroxylation degree and lime consumption from CaO saturated solution after 28 days was also shown. Although the level of linear regression adjustment ($R^2 = 0.73$) indicates that the kaolinite dehydroxylation level is an important but not the only factor shaping the calcined clay pozzolan activity even if it contains more than 75% of kaolinite.

The calcination process, due to its complexity resulting from a number of parameters regulating it, as well as from various possible variants of its execution, has also gained mathematical models describing it. Teklay et al. [54,57] developed two mathematical models of the calcination process. The first one [57], with reference to the material itself, models the calcined mineral in the form of spherical particles with specific physical parameters, which are affected by temperature, as a result of which some changes in its properties occur. This model is independent of the form of the process (flash- or soak-calcination). The other model [54] is an attempt to mathematically describe the flash-calcination process itself, allowing to determine its key parameters in relation to specific input data referring to the material subjected to heat treatment.

Among publications devoted to the activation of clays and clay minerals, there are also those concerning the conduct of this process with the use of additional substrates aimed at changing certain characteristics of the finished material. The primary purpose of such additives or admixtures is to increase pozzolanic activity [58]. As an example of research on the influence of admixtures on calcined clay, Taylor-Lange et al. [59,60] can be mentioned. The admixture in this case was zinc oxide, which was added to clays containing kaolinite, montmorillonite and illite both before and after calcination. The results were promising only in case of the clays containing kaolinite. The presence of calcined clay significantly reduced the delaying effect of zinc oxide on the cement hydration process.

The increase in kaolinite pozzolanic activity was also obtained by Ghorbel and Samet [61], who added iron nitrate solution to clay rich in this mineral. As a result of calcination of such additionally enriched clay, hematite and goethite were formed. The authors came to the conclusion that the presence of the former increases the kaolinite pozzolanic activity as long as it does not exceed

2.7%. The presence of iron ions in clay subjected to calcination also has its less favourable effects. They are responsible for its reddish hue, which is visible in cements blended with use of such clay. The reddish shades of cement are sometimes improperly interpreted as a sign of poor quality of the material, hence the idea to study the calcination process in a reducing atmosphere [62]. In this process, ground petroleum coke was also added to the raw ground clay. As a result of such calcination, a product was obtained whose red colouring was much less intense. The durability of the obtained effect remains an issue to be investigated.

3.2. Activation of Various Clays and Clay Minerals

It is assumed that the largest pozzolanic activity characterizes kaolinite, which also has the lowest activation temperature range. According to He et al. [17], calcination of this mineral can be successfully carried out even at 450 °C, although full completion of the dehydroxylation process requires a temperature of about 650 °C [63]. The high pozzolanic activity of calcined kaolinite results from the high content of hydroxyl groups and their location, which favours the exposure of Al groups on the surface of material grains after the dehydroxylation process [18]. Illite and montmorillonite also lose hydroxyl groups under the influence of temperature, but Al atoms remain in the structure of these minerals in places more difficult to access for cement hydration products, which may react with them [18]. As a result, their pozzolanic activity is lower than that of calcined kaolinite [24].

Metakaolinite, which is a product of kaolinite calcination, is a transition phase during the thermal transformation of kaolinite into mullite, which is an inactive pozzolanic mineral. Sperinck et al. [63] presents the process of kaolinite calcination modelled using the molecular dynamics (MD) simulations. The simulations were based on "heating" the simulated structure of the mineral to 1000 K, removing 10% of the initial number of hydroxyl groups and then quickly "cooling" it to 300 K in order to analyse the obtained results. The next steps of the simulation were carried out until all OH^- groups were removed. As a result of the simulation, a disordered mineral structure was obtained, in which the silicon layers showed little disorder, while the aluminium layers were significantly disordered and about 20% of the Al ions gained 5-fold coordination. It is less stable than the original six-fold coordination and four-fold coordination, which was adopted by most Al ions after calcination. This results in greater solubility and, as a consequence, also pozzolanic activity. Interestingly, the Al ions only moved within their own layer, as they were held back from further migration by a less affected silicon layer.

Minerals of the 2:1 group are characterised by a fairly wide range of temperatures at which dehydroxylation occurs. As far as illite is concerned, the temperature required for this process is 650 °C, but further heating is possible (up to 930 °C), which leads to an increase in its pozzolanic activity, probably due to the progressive amorphisation of the structure. According to He et al. [17], Ca-montmorillonite and Na-montmorillonite undergo total dehydroxylation at temperatures of 730 °C and 740 °C, respectively. In the case of both these minerals, the pozzolanic activity increases also after heating at higher temperature, but not exceeding 930 °C. Differences in decarboxylation temperature between individual minerals of group 2:1, and even for the same mineral, were explained by Drits et al. in [64]. They indicated the role of *cis*-vacant (cv) and *trans*-vacant (tv) modified layers in determining this temperature. Both these variants of layers differ in the distance between the nearest hydroxyl groups. In tv 2:1 layers the bond length is 2.45 Å ± 0.05 Å, whereas in cv 2:1 layers it is considerably longer and is 2.85–2.88 Å. Hydrogen, which in the process of dehydroxylation passes from one group OH^- to another creating a water molecule, needs a little more energy in the latter case. This explains well the difference in temperature required for full hydroxylation between the 2:1 group minerals.

Studies on the process of montmorillonite calcination, using the NMR technique, were conducted by Garg and Skibsted [23]. They showed that up to the temperature of 200 °C the dehydration of the mineral takes place, then up to the temperature of 500 °C no significant changes occur and then between 500 °C and 600 °C the dehydroxylation process begins. This process lasts until the material reaches a temperature of 900 °C and is connected with the progressive amorphisation of the mineral structure. Starting from the temperature of 1000 °C, another process of changing the structure of the heated material begins, resulting in its recrystallization as a result of the final destruction of the original layered mineral structure. In addition to NMR studies, Gard and Skibsted [23] also studied the mineral's pozzolanic activity by preparing mortars containing Portland cement and calcined clay at a ratio of 70:30 and analysing the products of hydration at specific intervals up to 1 year. The results indicated that 800 °C was the upper limit of the optimal calcination temperature range. The lower limit of this range was 750 °C. Interestingly, montmorillonite heated to a temperature above 850 °C showed significantly lower pozzolanic activity than in the raw state. The presented results are consistent with the results of Al and Si solubility tests carried out by the same authors [23]. Samples of calcined clay at two temperatures: 800 °C and 900 °C were subjected to 24-h treatment with 0.1 M NaOH. The solubility of Si and Al, which is a good indicator of pozzolanic activity, was about 4 times lower for clay activated at the higher of the given temperatures.

The less frequently occurring clay minerals, and thus slightly less popular among researchers, include pyrophylite, whose dehydroxylation processes and effects of thermal treatment are presented in papers [65,66]. Additionally, halloysite is not the subject of numerous publications [30,31], even though it is a clay mineral showing high pozzolanic activity after calcination. Moreover, it is interesting due to its differentiated structure. Its structure can form tubes or spheres, and these differences manifest themselves to a large extent in the pozzolanic activity of the calcined mineral. Halloysite in the spheroidal form is mainly active in the initial period due to more easily available reactive alumina. In the tubular form alumina cations are not so easy available, therefore, this form exhibits delayed activity [30].

Among the publications devoted to the study of clay activation, paper [67] should be mentioned, which covers the subject of activation of pure clays, among others kaolinite, illite, etc. Activation of clays containing smectites (bentonites) and kaolinite was studied by the authors of papers [68–70]. The analysis of activation process of clays containing illite and smectites (montmorillonite) in comparison to those containing kaolinite is included in [71]. Heat treatment of clays containing about 40% kaolinite as well as about 40% illite and montmorillonite is described in [49].

3.3. Tests of Eligibility of Clays for Activation Procedure

Since deposits of pure clays containing only one type of clay mineral are not rich enough to meet the growing demand of the construction industry for calcined clays as supplementary cementitious materials, it is important to assess the clays from existing deposits in terms of the possibility of using their resources as raw materials for the production of SCMs. The pozzolanic activity of clay is not a simple sum of the activity of its individual constituent minerals. The co-occurrence of some of them may give a synergy effect, but in other configurations the potential of any of the components may not be fully exploited. Hence the need to develop a methodology for the assessment of specific deposits arises in terms of their suitability for activation.

Diaz et al. [72] presented an idea for a method of initial assessment of clay based on its chemical composition. This method is primarily intended for the assessment of clays containing minerals from the kaolinite group, but its application is also possible for clays containing 2:1 minerals. In order to assess the suitability of clays from a particular deposit using this method, first of all their chemical composition should be determined, with the Al_2O_3 content, which should be more than 18%, then the ratio of Al_2O_3 to SiO_2 content, which should be higher than 0.3 and, finally, the loss on ignition, which should not be less than 7%. Due to the deleterious effects of calcite and pyrite, the authors have added two additional conditions: $CaO < 3\%$ and $SO_3 < 3\%$. The authors are of the opinion that first of all, the exploitation should be carried out with the use of aluminium deposits containing over 40% of kaolinite. However, in order to take into account the pozzolanic activity of 2:1 group clays, they proposed a parameter defined as kaolinite equivalent calculated according to Equation (1):

$$\%KEQ = \{[m(350\ °C) - m(850\ °C)]/[m(200\ °C) \times 0.1396]\} \times 100, \qquad (1)$$

in which $m(x)$ is the mass of the mineral after heat treatment at a given temperature x.

A promising tool for a preliminary assessment of potential clay minerals' pozzolanic activity may be cation exchange capacity, but not yet at this stage, as the author himself admits in his paper [73]. The development of an effective and simple methodology for the assessment of clay deposits for the cost-effectiveness of their exploitation as raw materials for SCMs production still remains an issue to be resolved.

3.4. Methods of Assessment of Pozzolanic Activity

The pozzolan reaction occurs between Al_2O_3-$2SiO_2$ and $Ca(OH)_2$, which in the case of concrete comes from the hydration of cement. The presence of water is also necessary for this reaction, as the reaction products are hydrates. As a result of the pozzolan reaction, CSH gel ($CaO\cdot SiO_2\cdot H_2O$) and hydrated calcium aluminates (e.g., C_4AH_{13}, C_2AH_8, C_3AH_6) are produced, as well as hydrated calcium aluminosilicates of the hydrogehlenite type C_2ASH and hydrogarnets (C_3AS_3-C_3AH_6). Carboaluminates (e.g., calcium hemicarboaluminate hydroxide C_4ACH_{11} which more detailed formula is $C_3A\cdot 0.5CaCO_3\cdot 0.5Ca(OH)_2\cdot 10.5H_2O$) may also be formed in the presence of carbon dioxide or limestone filler [47,74–77].

There are many different methods of determining pozzolanic activity. They are described among others in [28,58,78,79]. According to [78] they can be divided into direct and indirect ones. Direct methods include methods based on the analysis of $Ca(OH)_2$ content and products of its reaction and changes in the content of these compounds over time. These are methods using XRD, TGA and DTA techniques, as well as methods based on solution titration, i.e., the Fratini test and its simplification, also referred to as the saturated lime (SL) [78] test or lime consumption (LC) test [28]. Among the indirect methods, the authors [78] include those that allow the determination of the pozzolanic activity of the material on the basis of the study of the characteristic which is affected by this activity. These features can be, e.g., the compressive strength of mortar specimens, the electrical conductivity of a saturated solution of $Ca(OH)_2$ in which the tested material is placed [80–83], or the determination of the amount of heat released in calorimetric tests [84,85].

A wide review of methods of testing the pozzolanic activity is presented in paper of Tkaczewska [79]. The methods described therein were divided into chemical and physical. Although some of them come from currently withdrawn standards, it is worth quoting them here to show the richness and diversity of approaches to this topic, which even with this review will not be exhausted.

One of the methods is the measurement of portlandite consumption by SCM using thermal analysis methods, i.e., TGA in combination with DTA [86]. In this method, the measure of pozzolanicity is the amount of calcium hydroxide bound by the tested material. It is defined as the difference between the initial $Ca(OH)_2$ mass and the remaining unbound mass. The latter is determined in a thermal test from the loss in mass of the sample in the temperature range 490–510 °C, which corresponds to the decomposition of portlandite. This method has several weaknesses which must be taken into account when using it. First of all, the temperature of calcium hydroxide decomposition may vary depending on the alkali content or even grain size [87]. Another factor is the possibility of loss of portlandite in the sample due to carbonation and mass loss of other hydration products in the temperature range corresponding to calcium hydroxide. These problems were identified by Kim and Olek [88] and they proposed an appropriate test procedure.

TGA/DTA testing requires expensive equipment, so it is not a method that can be used in many laboratories that are less well equipped. To address such situations, Mendoza and Tobón [89] have carried out a comparative study of the results of weight loss measurement of calcium hydroxide mixture with potentially pozzolanic material using a moisture analyser. They succeeded in demonstrating that if a sufficiently high sample heating temperature in the moisture analyser is ensured (in the quoted article it was 230 °C), the mass loss at this temperature obtained using both methods is characterised by a high correlation ($R^2 = 0.971$), while the mass loss in the moisture analyser is slightly higher. With appropriate assumptions, it is possible to test the pozzolanic activity of the materials by determining the amount of water lost by the hydration products, which increases with the progress of the pozzolanic reaction.

Fratini is identified with at least two different methods to test the activity of pozzolan. The first one [90–92] consists in the determination of the amount of calcium ions bound by the tested material mixed with cement. A sample of the ground test material is placed in a $Ca(OH)_2$ suspension at 40 °C for eight days [93]. During this time, the sample is shaken from time to time. The suspension is then filtrated and the total alkalinity of the filtrate is determined by titration with 0.1 mol HCl solution. The next step is to neutralise the filtrate with ammonia and to determine the amount of CaO in it by complexometric method. The concentration of OH^- ions (total alkalinity) and Ca^{2+} ions determined in the test is compared with the calcium hydroxide solubility isotherm curve. The location of the result below the curve means that the material is active pozzolan, the greater its activity, the greater the distance of the result beneath the curve.

The second Fratini method is the strength method. It consists in testing the strength of two series of mortar samples prepared with the use of cement and the tested material in the amount of 30–50% of the hot water. One series of samples is stored for seven days in water at a temperature of 20 °C, and the other only three days, after which it is transferred to water at 50 °C for another four days. The difference in strength of both series of samples proves pozzolanic activity of the tested material [94].

The method according to the PN-EN 450-1:2012 standard [95] is based on a similar methodology as the Fratini strength method. It is designed to test the activity of fly ashes, however, it can also be applied to other materials showing the pozzolanic activity. It is based on the determination of the pozzolanic activity index (IAP), defined as the ratio, expressed as a percentage, of the compressive strength of a mortar with a fixed proportion of binder, water and sand of a specific grain size made in two versions. In the first version, 25% of cement is replaced by fly ash (or other pozzolanic material), while in the second version, 100% cement is used. The standard requires that it is Portland cement CEM I 42.5R. Pozzolanic activity index is calculated on the basis of the strength test after 28 and 90 days. After 28 days it should reach ≥75% and after 90 days it should be ≥85% for the test material to be considered as active pozzolan. The method of measuring the pozzolanic activity index gives results that closely correlate with the determination of lime concentration in saturated solution with the test material [78,80,81,96]. This method is often referred to in the literature as SAI (strength activity index) and varies, depending on the standard on which it is based, in the composition of the tested mortar (the amount of cement exchanged for the tested material, e.g., 20% [97] or 30% [98] instead of 25%) and the required minimum SAI value (e.g., it is 80% after 28 days [98] or 75% after seven and 28 days [97]).

Apart from testing the strength of mortars containing cement, there are also methods based on testing the strength of mortars prepared from slaked lime and the tested material. One such method is contained in the Serbian standard SRPS B.C1.018:2015 [99]. A mixture of slaked lime, pozzolana, sand and water with a mass ratio of 1:2:9:1.8 is prepared for testing. From this mixture samples of 40 × 40 × 160 mm are formed, which are stored for 24 h in temperature 20 ± 2 °C with a relative humidity of 90%. Then they are placed in thermostat chamber at temperature 55 ± 2 °C for five days. Before testing, specimens are stored again 24 h in temperature 20 ± 2 °C and relative humidity of at least 90%. Compression and flexural strength of the specimens is the basis for the division into three classes of pozzolanic activity. The requirements for each class are as follows: class I—bending strength ≥ 4.0 MPa, compressive strength ≥ 15.0 MPa; class II—bending strength > 3.0 MPa, compressive strength ≥ 10.0 MPa; class III—bending strength ≥ 2.0 MPa, compressive strength ≥ 5.0 MPa.

The method based on the study of the influence of pozzolanic material on the strength of cement mortar is also Graff's method [79]. In this method three series of mortars differing in binder are prepared. In the first series it is Portland cement and in the second series it is a mixture of Portland cement and the tested material, with the content of the latter in the samples: 15, 30, 50, 70 and 90%. The third series is prepared just like the second series, but replacing the test material in the binder with quartz sand in the same proportions. After seven, 28 and 90 days from the preparation the sample is subjected to strength tests according to PN-EN 196-1:2006 [100] and the pozzolanity number is calculated according to Equation (2):

$$P = (a - c)/(b - c), \qquad (2)$$

in which one: P is the pozzolanity number, a is the compressive strength after 28 days of the second series of samples, b is the compressive strength after 28 days of the first series of samples and c is the compressive strength after 28 days of the third series of specimens.

The method contained in the withdrawn Russian standard GOST 6269-54 [101] is similar to the first of the Fratini methods described above. According to it, the measure of pozzolanic activity is the total amount of calcium hydroxide bound by pozzolana during 30 days. A sample of the investigated pozzolana (2 g) is placed in a container with 100 mL of saturated calcium hydroxide solution, which is sealed and occasionally shaken vigorously. After two days, 50 mL of the solution is taken and is titrated with HCl in the presence of methyl orange. Then 50 mL of saturated calcium hydroxide solution is poured into the cylinder. This procedure is repeated every two days and the whole test takes 30 days. A modification of this method is using barium hydroxide instead of calcium hydroxide and determining its loss in time [79]. Barium hydroxide dissolves better in water but simultaneously carbonates faster [102].

Chemical methods include the method contained in the withdrawn ASTM C379-65T [103]. It assumes the determination of the pozzolanic activity of the material on the basis of the determination of the amounts of soluble SiO_2 and Al_2O_3 in it. The test material should be placed for 1.5 h in a 1-molar solution of sodium hydroxide at 80 °C. The total content of dissolved SiO_2 and Al_2O_3 above 20% indicates that the material shows pozzolanic activity. The same standard also includes a method based on testing the strength of the lime - pozzolanic mortar. Samples of such mortar should be stored for seven days at 65 °C in a humid atmosphere, then cooled to 23 °C and kept at this temperature for another 21 days, also in a humid atmosphere. The material exhibits adequate pozzolanic activity if the samples reach compressive strength ≥4.1 MPa. ASTM C379-65T was withdrawn in 1966 and replaced by ASTM C593-95 [104], which only contains the latter method.

Chapelle test consists in boiling the pozzolana sample in $Ca(OH)_2$ solution for 16 h and then determining the calcium hydroxide content in the filtrate [105]. A measure of the pozzolanic activity of the material is the difference in $Ca(OH)_2$ content in the initial solution and in the filtrate. Nino et al. [106] have modified this method for their research. They used 300 mL of saturated calcium hydroxide solution at 20 °C, when the solubility of this compound in water is 1.65 g/L. They placed 0.5 g of kaolin, almost as much as $Ca(OH)_2$ in the solution. After heating the solution to 100 °C, part of the calcium hydroxide was precipitated, but during the test, as part of $Ca(OH)_2$ reacted with kaolin, the precipitate dissolved, keeping the solution saturated. The test was extended to 24 h and the solution was mixed at 150 rpm.

Another chemical method is the Feret–Florentin method, about which Feret mentioned in [107] and which he credited to Florentin [108]. The ground sample of test material is shaken for 10 min in 100 mL of HCl (30%) and then the amounts of SiO_2, Al_2O_3 and Fe_2O_3 that have entered the solution are determined. Then a further aliquot of the test material is mixed 1:1 with $Ca(OH)_2$ adding enough water to give the prepared mixture a plastic consistency. The prepared mixture is stored for three days in a humid atmosphere and then it is placed in water at 15 °C. After one, four and 26 weeks from the preparation of the mixture, part of the mixture is taken, dried, ground and the SiO_2, Al_2O_3 and Fe_2O_3 content is determined in the same way as in the case of the test sample itself (shaking in 30% HCl). Pozzolan activity is measured by the increase in the amount of leached oxides in a sample of test material with $Ca(OH)_2$ compared to a reference sample containing the test material itself.

Feret himself is the author of the method in which pozzolanic activity is tested firstly by treating the material with a 30% solution of HCl and then with a 25% solution of NaOH. The measure of pozzolanicity is the amount of material tested, expressed as a percentage, which has migrated to the solution. In order to consider that the material has a high pozzolanic activity, at least 60% of its mass should be dissolved. The simplification of the Feret method is the Jarrige and Decreux method [109].

It consists in treating the tested material with HCl for five and 30 min. The measure of pozzolanic activity is an increase in the amount of dissolved material with an increase in the leaching time. Material characterized by good pozzolanic activity should have a difference in solubility of more than 10%.

The Guillaume method [110] consists in comparing the content of insoluble parts of two blended cements, one of which contains 20% by weight of the tested material and the other 20% by weight of quartz sand. Both cements are treated with HCl and the amount of insoluble residue is then determined. In the next step the cements samples are heated at 1000 °C for 1 h, after which they are treated with HCl as previously unheated samples. The pozzolan activity is determined by the increase in solubility of the cements after heating. If active SiO_2 is present in the test material, it reacts with calcium oxide or alite to form HCl-soluble calcium silicates. Cement with added quartz sand does not change significantly its solubility in HCl.

The method proposed by Battaglino and Schipp [111] is based on two criteria: the determination of specific surface area of hardened cement paste prepared with the addition of 30% of tested material and the determination of free $Ca(OH)_2$ content. The tests are carried out after different hardening periods. A sample of the paste should be ground and then the 30–60 µm fraction should be separated. The specific surface area is determined by the BET method and the free $Ca(OH)_2$ content by the ethylacetylacetate method (Franke method) [112]. A material with good pozzolanic activity is considered as one which reacts quite slowly with calcium hydroxide and leads to a large specific surface area.

Another suggestion for the test of pozzolanic activity is the R^3 method (rapid, relevant and reliable). A detailed description of it and the underlying research can be found in [113,114]. This method has two variants. The first one is based on a heat release test in an isothermal calorimeter, in which the mixture is placed. It consists of the material tested, calcium hydroxide (additionally calcium carbonate in the case of testing mixtures of LC^3) and potassium hydroxide as well as potassium sulphate both added in such quantities that the reaction environment has a similar composition to a pore liquid in concrete. Detailed compositions of the mixtures are given in the above mentioned publications. At the method development stage the prepared mixture was tested in a calorimeter at 20 °C and 40 °C for seven days. Since the amount of heat released at the higher of these temperatures after 1 day is equal to the amount released at the lower one after six days, further analyses were based on the results obtained after one day at 40 °C. The second variant of this method starts with the preparation of a mixture with the same composition as the first variant. This mixture is placed in a sealed container and treated at 40 °C for one day. The sample is then dried at 110 °C until its mass change per day becomes less than 0.5%. The sample so dried is exposed to a temperature of 400 °C for two hours and then cooled to 110 °C. The sample is weighed before and after heating at 400 °C and the quantity of water bound in the hydration products is determined from Equation (3):

$$W_b = (m_s - m_c)/m_s\ [\%], \qquad (3)$$

In Equation (3) W_b is the quantity of water bound by the hydration products, m_s is the mass of the sample after drying to a constant mass but before exposure to 400 °C and m_c is the mass of the sample after heating and then cooling to 110 °C. According to [113], the results obtained show a high correlation with the results of the strength tests and the Chapelle test. A quick test of pozzolanic activity based on the conductivity test was proposed by Luxán et al. [115], developing an idea presented by Raasek and Bhaskar [116]. The idea of this method is to test the conductivity of a saturated solution of $Ca(OH)_2$ kept at a constant temperature of 40 ± 1 °C, in which a sample of the tested material, dried earlier at 105 ± 5 °C, is placed. Pozzolan activity is measured by the difference of the compensated conductivity of the solution at the start of the test and after 120 s. According to the classification proposal of [115], a material is considered as non-pozzolanic if the difference in compensated conductivity is less than 0.4 mS/cm, and if it is greater than 1.2 mS/cm, the material is considered as having good pozzolanicity. Between these two results there are materials which show variable pozzolanicity. In its original form this test was not resistant to interference from soluble salts increasing the conductivity of the solution. Payá et al. [80] proposed several modifications to this test, such as extending its time, carrying it out at different temperatures or using unsaturated calcium hydroxide solution. The most important amendment to the test methodology was the proposal to additionally measure the conductivity of distilled water, in which the material sample was placed. Both curves obtained in this way (conductivity of calcium hydroxide solution and distilled water) should be subtracted from each other and analysed.

Each method has its strengths and weaknesses, so it is best to use more than one method to obtain as complete a picture as possible of pozzolanic activity of the tested material. Danatello et al. [78] comparing the saturated lime method, Fratini test and SAI concluded that the first method is the most questionable. It did not show any correlation with two other methods, between which a high correlation was found. Therefore, they recommend using the Fratini test together with SAI, and also one of the direct methods. There is also a number of voices that suggest only direct methods should be used in the determination of pozzolanic activity [117].

4. The Results of Investigations of Clays from Various Worldwide Deposits

Maier et al. [10] examined 11 clays from various deposits in Germany. The calcination temperature of each clay was determined from the DTG results by adding 100 °C to the main peak temperature of the dehydroxylation process. The mineralogical composition of the raw materials and the proportion of amorphous phase after the calcination process were determined. Reactivity was investigated using R^3 calorimetry according to [114] (described in the previous section) and solubility of Al and Si ions. The results showed a high agreement between the results of the last two tests. The authors also came to a conclusion that in order to properly determine the suitability of clay as a raw material for SCM production it is not enough to know the chemical composition, especially in case of clays with low kaolinite content. In those cases other minerals play an important role, therefore, it is necessary to determine the complete mineralogical composition. The R^3 calorimetric test can also be useful in the evaluation of this type of clays.

Tironi et al. [11,118] have studied five kaolinitic clays from various deposits located in Argentina with kaolinite content between 16% and 94%. These clays were tested by X-ray diffraction (XRD) and Fourier transformed infrared spectrometers (FTIR) before activation. These tests determined the phase composition of the clays and the order or disorder of the kaolinite they contain. The latter determination was performed in accordance with the methodology proposed by Bich [56], which consists in calculating the P_0 coefficient, which is the quotient of the intensity bands at 3620 cm^{-1} and 3700 cm^{-1} obtained in the FTIR test. If $P_0 > 1$, then according to [56], kaolin has well-organised structure and $P_0 < 1$ indicates disordered structure. In the case of three clays with intermediate kaolinite content the structure of this mineral was disordered. The clays were heated to 700 °C where they stayed for 5 min (and the whole heating process took 1 h). An XRD study after calcination showed that the time and temperature were sufficient to transform the kaolinite into metakaolinite. After calcination, pozzolanic activity was also determined by the Fratini test and the electrical conductivity test (previously described). The compressive strength of mortars in which 30% of cement was replaced with ground calcined clay was also tested. On the basis of the results of the latter test, a model was proposed in which the strength of the mortars (after seven, 28 and 90 days), in which 30% OPC was replaced with ground calcined clay, depends on the content of kaolinite in the clay, on the specific surface defined by the Blaine method and on the inverse of the P_0 coefficient. The authors proved that the results obtained from the application of the proposed model are highly consistent with the test results. Thus, they showed that the calcined clays' pozzolanic activity depends not only on kaolinite content but also on the degree of order of its structure and specific surface of the clays after calcination and grinding.

Huenger et al. [119] selected three clays from Lower Lusatia deposits and tested their pozzolanic activity after burning. Among the selected clays, one was characterized by a high content of quartz (about 60%) co-occurring with smaller amounts of kaolinite and illite, the second one consisted mostly of kaolinite (about 60%) and quartz and a small amount of illite, and in the third one more than half of the composition consisted of quartz, and the other clay minerals mentioned earlier were present in smaller, similar amounts. Each clay was calcined separately, as well as their mixtures. Three temperatures of heat treatment were applied: 600 °C, 650 °C and 700 °C. The study of mineralogical composition of the calcined clays showed a slight decrease in the illite content and a significant decrease in kaolinite content accompanied by a proportional increase in the amorphous phase. The studies of pozzolanic activity allowed to formulate a conclusion that the mixture of kaolinite-rich clay with quartz-rich clay in the proportion 60:40 allowed to obtain a material with satisfactory parameters.

Beuntner and Thienel [120,121] presented the results of research on clays with clay content of about 70% (with a predominance of illite) from deposits located in southern Germany. In [120] the authors compared the material obtained by calcination of clay with kaolinite content of 25% in two technological processes. One was carried out on a laboratory scale and the other on an industrial scale. The comparison of research results showed that calcination in industrial conditions allowed to obtain material with slightly lower pozzolanic activity, but still at a satisfactory level. It was also shown that the mineralogical composition of clay subjected to heat treatment varies according to grading. Whereas [121] was devoted to the study of selected technological, mechanical and durability properties of concrete prepared from nine different types of cement with different exchange rate of the cement with calcined clay, which was used as a Type II addition. Mixtures of clay with cement in most cases showed slower development of strength in the initial period. However, later on, the strength increase accelerated and in the case of mixes with 20–25% clay, the strength exceeded the values obtained for concrete with sole cement. Moreover, it was found that the replacement of cement parts with calcined clay reduces consistency, bleeding and shrinkage.

Papers [122–124] are devoted to research on the possibility of using kaolinite waste fraction from Amazonian deposits for the production of calcined clays, which were an intermediate product for further zeolite production. The research was carried out on waste fractions of kaolin from deposits of the Brazilian Amazon calcined at various temperatures in the range 550–800 °C for 2 h. Results of the tests showed that the wastes were composed primarily of coarse-grained kaolinite, which can be an excellent starting material for pozzolana and zeolite production.

Akasha [125] presented the results of research on the use of clay obtained from five locations in Libya and calcined as a partial substitute for cement. The studied clays contained from 30% to 90% of kaolinite and, besides that, mainly quartz (and in one case additionally a small amount of illite). The clays were calcined at 800 °C for two hours. Subsequently blended cements were prepared in which the share of calcined clays was 10%, 15% and 20%. The obtained results do not show a clear correlation between pozzolanic activity and kaolinite content in the raw material. In comparison with the control mortar, in which pure cement was used, only the mortar containing calcined clay containing initially 30% kaolinite (the least) and 70% quartz showed higher strength at all test dates and at all exchange rates.

Chakchouk et al. [126] have selected six clays from five Tunisian deposits located in different parts of the country. Samples of clays have been tested for basic properties such as plastic limit, liquid limit and plasticity index. Blue methylene test and calcimetry were also carried out on them. The chemical composition and qualitative mineralogical composition with XRD was determined. The chemical composition was used to verify the criterion of pozzolan activity according to two standards: ASTM C618 [97] and an unspecified Indian standard. The results showed that with the exception of one of the clays, which did not meet the Indian standard of CaO < 10%, the others should be a good material for the production of pozzolanic active materials. These five clays were subjected to calcination at 600 °C, 700 °C or 800 °C, and then their pozzolan activity was determined by examining the compressive strength of clay mixed with calcium hydroxide in two proportions CC:CH = 1 and CC:CH = 3. The highest strength was obtained in case of two clays with the highest content of kaolinite and at three times the mass prevalence of calcined clay to calcium hydroxide. Their optimal calcination temperatures are 700 °C and 800 °C. In the case of two of the remaining clays, the measured pozzolanic activity was moderate and the optimal temperature (among the applied ones) was 800 °C. The fifth of the examined clays showed weak pozzolanic activity, which the authors attribute to high quartz content.

The reported results of the research indicate that it is necessary and advisable to develop a research protocol which could be used as a standard for examining clays from unexplored local deposits. The diversity of clays from different deposits in the same country indicates the need to take into account the specific mineral composition of a particular clay in developing and optimising its calcination technology.

5. Calcined Clay in Cement and Concrete Technology

The issue of calcined clays as SCM cannot be considered in isolation from the technological determinants related to the use of this active pozzolanic material in the production of blended cements and concrete. In [41] Samet et al. applied the response surface methodology (RSM) [127,128] to determine the value of: calcination temperature, specific surface of calcined clay and its share in the blended cement to obtain optimal parameters of the blended cement. Normal consistency, setting time, stability of expansion, mechanical properties and the specific surface of the calcined clay were taken into account. The presented methodology is universal and can be used to determine the optimal composition of cement blends with different types of calcined clay.

The issue of optimization of the composition of cement blends with calcined clays was also considered by Pierkes et al. [129]. Due to a large number of variable parameters, they planned their research using the DOE (design of experiments) with statistic tool MINITAB®. Four types of Portland clinker and three types of clays (illitic, kaolinitic and chloritic) were used in the research. The clay was added to the clinker in amounts of 20% and 40%. Additionally, the anhydrite was added in a different amount (2% or 4%). The research included compressive strength after two, seven and 28 days and studies of hydration products after 3 h and two days. The authors came to the conclusion that there is space to improve the strength parameters of calcined clay cements by selecting both the appropriate chemical and mineralogical composition of the raw materials and adjusting the amount of sulphates accordingly.

Siddique and Klaus [130] have reviewed widely the literature on the influence of metakaolin on the properties of mortars and concrete. The analysis included fresh properties, properties of hardened materials and durability properties of concrete. They stated, that the use of metakaolin had a positive impact on both early age and long-term strength properties of cement paste, mortar cement and concrete. It also reduced water absorption of concrete by capillary suction (i.e., its sorptivity [131]), as well as other parameters that may reduce the durability of the material (e.g., permeability) due to the refinement of pore structure. Moreover, mortars and concrete performed with metakaolin replacing cement in the amount of 10% and 15% showed very good chemical resistance, including sulphate corrosion.

Paiva et al. [132] have studied properties of fresh and hardened concrete prepared with metakaolin with the pre-set workability. The influence of metakaolin particles dispersion achieved with water and polycarboxylic acid based HRWRA on concrete structure was researched and discussed. The results clearly favoured the use of HRWRA to control the workability of mixture with such fine additions as metakaolin. The use of water alone had resulted in increased porosity and the formation of agglomerates of calcined kaolin, which has had a negative impact on the strength parameters of the hardened material.

Research on the compatibility of superplasticisers with cement mixes containing metakaolin and limestone was undertaken by Zaribaf et al. [133]. They used cement type IL according to ASTM C595 [134], containing limestone and commercially available metakaolin. Four admixtures each with different chemical base were used as superplasticisers. The results, which were discussed, included flowability measured by the minislump size, compressive strength, hydration heat measured in isothermal calorimeter and setting time. On the basis of the results obtained, the authors came to the conclusion that admixtures based on polycarboxylate ether (PCE) and polymelamine sulfonate (PMS) were the most compatible.

An extended analysis of the efficiency of superplasticizers depending on the mineralogical composition of calcined clay and the type of cement (OPC or PLC) was conducted by Sposito et al. [135].

Apart from the two types of cement mentioned above, they used four types of calcined clay in their research, one of which was a typical clay with different mineralogical composition, one was commercially available metakaolin and the other two contained in its raw form more than 90% of one of the minerals: illite or muscovite. Quartz powder was also used to evaluate its impact on the rheology. Three different admixtures were used as superplasticisers. The tests were carried out on both clay and cement mixes as well as on clinker-free suspensions. The results and their analysis allowed the conclusion that zeta potential, total surface area and water demand were reliable predictors of superplasticizer demand with the exception of mixtures with calcined muscovite. However, the latter reservation should be alleviated in the case of clays with different mineralogical composition, in which muscovite content is a medium. The share of quartz powder in the tested systems for the variety significantly improved workability.

Nied et al. [136] investigated the synergistic effect of the use of metakaolin and limestone on the mechanical properties and phase assemblage of the blends composed in this way. For this purpose they prepared 12 blends with a fixed OPC content of 60% and a variable content (10%, 20%, 30% and 40%) of metakaolin of two qualities, limestone (10–40%) and quartz instead of limestone (20%) in reference blends. The authors concluded that part of metakaolin in binary blends with OPC could be replaced with limestone, which positively affects both workability and 28-day compression strength, with the exchange ratio depending on the quality of metakaolin. Moreover, it was observed that the addition of limestone stabilises ettringite, refines pore structure and alters the hydration of OPC/metakaolin blend at early age.

Beuntner et al. [137] presented the results of a study on the possibility of using calcined clay in high-performance concrete. It is worth emphasizing that they used not only metakaolin, but also mixed layer clays, which on the one hand occur in much larger quantities in world clay deposits, but on the other hand are treated as "poorer sisters" of kaolinite. The results of the research, which included, among other things, compressive strength, modulus of elasticity and porosity, showed that calcined clays, including those obtained from raw mixed layer ones, can successfully replace silica fume in HPC, although this may require doses of superplasticisers larger than those currently recommended by their manufacturers.

The use of calcined clays as binder components can be combined with the simultaneous use of other types of pozzolan. Mechti et al. [138] presented results of a study in which finely ground sand was used as an additional component which, in spite of its crystalline structure, showed the pozzolan activity. The aim of the study was to determine the optimal proportions of cement blend components and, as additional parameters, fineness of ground sand as well as temperature of calcination of clay. The plan of the experiment included 28 mixtures characterised by different values of the optimised parameters. The analysis of the results showed that the optimum composition under the assumed assumptions was the one containing 80% of cement, 15% of sand with fineness lower than 40 μm and 5% of clay calcined at temperature 750 °C.

Two different pozzolana (calcined clay and fly ash) in cement mixes were also used by Ng and Østnor [139]. This combination had its justification in the different characteristics of both pozzolana materials. Fly ash is a material which very slowly increases the strength of cement-based composites, while calcined clays, especially when used with limestone, are significantly faster in the process of pozzolan reaction. In the presented studies, 20% of OPC was exchanged for fly ash, clay or mixtures of these two pozzolana in different proportions. The results showed that the simultaneous application of fly ash and calcined clay improved the workability of the mixture, which is usually impaired when using only calcined clays as a substitute for cement. It has also been shown that the combination of these two pozzolanic materials allows strength values to be achieved after 28 days at the level of mixtures using only OPC.

Workability and hydration of blended cements with calcined illitic clay have been studied by Marchetti et al. [140]. They used two types of ground, calcined illitic clays and OPC, the content of which was reduced in favour of each clay by 15%, 25% and 35% by weight. Cement mixtures were tested for their packing density (according to [141]), heat release, flow, compressive strength and some additional parameters related to the hydration course. The obtained results allowed to state that with the increase of the share of calcined illitic clay in the mixture, its workability decreased. As far as heat release is concerned, the clay caused a decrease in the amount of heat generated during the first 48 h, but at the same time it did not cause any significant delay in the hydration process.

Machner et al. [142] have studied the possibility of using dolomite to replace limestone as a third component in mixtures with OPC and metakaolin. The test mixtures were composed of Portland clinker, dolomite, limestone, metakaolin and gypsum. A total of six blends were prepared, which together with the cement gave seven different series. In all mixtures the ratio of OPC to metakaolin was constant at 6:1 and the addition of limestone or dolomite was 5%, 10% or 20%. The results showed that if the 90-day strength was taken as a determinant, there was no significant difference between limestone and dolomite. The latter mineral, however, showed less reactivity, which was particularly evident at low curing temperature.

As it can be seen that the addition of calcined clay reduces the early strength of cement composites due to lower reactivity of pozzolana. Paper [143] presents the results of research on the possibility of acceleration of cement strength development in the initial period by using CSH seeds. The clay obtained from a deposit in southern Germany, which after calcination was mixed with OPC in the proportion 20:80, was used in the study. Thus, the prepared cement, corresponding to type CEM II-A/Q, was enriched with 1%, 2% and 3% of CSH seeds admixture. In another series 5%, 10% and 20% of the cement mass was replaced with the microlimestone. There were performed calorimetric tests, TGA and SEM investigations as well as flow and compressive strength tests on mortars. The results allowed to recommend the addition of 3% of CSH seeds admixture or 10% microlimestone to optimize the early strength development of cement with 20% addition of calcined clay.

There is a consensus among researchers that the replacement of part of cement with metakaolin (up to 15%) reduces drying shrinkage of concrete compared to material without metakaolin or with silica fume [144–146]. However, as far as autogenous shrinkage is concerned, the results are divergent. Gleize et al. [147] investigated the influence of metakaolin on the autogenous shrinkage of cement paste. They used CEM I 52.5 cement, which was replaced by metakaolin at 0%, 5%, 10%, 15% and 20% rates. Mixtures were prepared with two water to binder ratios: 0.5 and 0.3. In the latter case a superplasticizer was used. The results allowed to formulate a conclusion that long-term autogenous shrinkage decreased with increasing proportion of metakaolin in the binder. The majority of this effect was caused by the pozzolanic activity of the applied metakaolin, and not by the effect of dilution of the cement with the addition.

A review of the studies on the use of calcined clays in cement and concrete technology shows an overwhelming predominance of articles devoted to various aspects of application of metakaolin. It is the best-researched calcined clay mineral, although even in its case the results obtained are far from being generalised. Although kaolin is a grateful object both for testing and for subsequent use after calcination, due to its limited availability, more extensive research into the properties and application of clays with different mineralogical compositions is necessary.

In addition, many of the work presented is based on results obtained using pastes or mortars cement. Concrete testing results are in a clear minority. This is probably due to the need to produce larger amounts of calcined clay with homogeneous parameters, which can be difficult in laboratory conditions. Nevertheless, this is an important direction of research that deserves more attention.

6. Lime Calcined Clay Cement (LC^3)

The cement consisting of Portland clinker, calcined clay (preferably rich in kaolinite), calcium carbonate and gypsum is described in literature as LC^3 [148]. It is a solution for a demand for cement that is more environmentally friendly, the production of which takes place with lower CO_2 emissions to the atmosphere, and which at the same time is not inferior to ordinary Portland cement with a clinker content of at least 90%. The abundance of raw clays and limestone is also an important factor in the development of LC^3 production, in contrast to the shrinking resources of good quality fly ash or even their unavailability in some countries.

The most common composition for the LC^3 is 50% of ground Portland clinker, 30% of ground calcined clay, 15% of ground limestone and 5% of ground gypsum. Such a mixture is sometimes referred to in the literature as LC^3-50. Other proportions of components are also possible [149], but it is known that cement prepared according to the above composition, according to the research [113,150], reaches mechanical parameters corresponding to OPC already after seven days of hydration, provided that the clay contains at least 40% kaolin.

An important aspect of LC^3 production is its cost and profitability. Papers [151,152] present the results of economic analysis of production of this type of ternary blend in India. The authors concluded that the production of LC^3 is economically viable if the following conditions are met: the cost of fly ash will be high, the quality of fly ash will be low, the acquisition of fly ash will require longer transport than the acquisition of clay and the quality of the extracted clay will reduce the amount of clinker in cement. Most of these conditions are already met, for example, in countries which do not have sources of good quality fly ash. Given the restrictions on the use of fossil fuels in energy production, these conditions may soon be met automatically.

Apart from the economic aspect, the durability and performance of concrete with LC^3 is no less important. Several publications have been devoted to this issue. In papers [153,154] the authors assessed the pore structure in concrete made of three types of cement: OPC, Portland pozzolana cement with 30% of Class F fly ash and LC^3. The results of the study showed a much finer pore structure in concrete with LC^3, which was proved in the mercury intrusion porosimetry study. Additionally, the conductivity of the concrete obtained in this way turned out to be lower, which allows one to assume that it will have higher resistance to the penetration of harmful ions into its structure. In short, its durability can be predicted to be significantly higher than that of other concrete series compared in the work.

Khan et al. [155] discussed carbonation of LC^3. The study covered investigation of concrete with cement, in which 15%, 30% and 45% of the mass was replaced by a mixture of calcined clay (containing about 50% metakaolin and 50% quartz) and limestone in a ratio of 2:1. For comparison, two series of concrete with OPC were used. One of them had the same aggregate, water and binder ratio as the LC^3 concrete series, and the other modified proportions and quantities. The results indicate that concrete with LC^3 and 15% cement exchange rate showed higher carbonation resistance than concrete with OPC. At 30% a slight advantage was gained by concrete with OPC, and at a 45% cement exchange rate to a mixture of calcined clay and limestone, concrete with LC^3 showed a high carbonation rate. The same authors [156] presented the results of the same concretes according to their fresh properties, strength, porosity and drying shrinkage. They showed that the concrete with LC^3 was characterized by worse workability. As far as strength was concerned, the highest was reached by concrete with a 15% exchange rate of cement, which also had the lowest porosity. As far as drying shrinkage is concerned, all series of LC^3 concrete were lower than OPC concrete.

An interesting approach to the issue of carbonation is presented in paper of Joseph et al. [157]. The authors have been tempted to create the protocol of testing the durability of LC^3 (although not only) with respect to carbonation. This protocol included the determination of indicators of the carbonation process, such as diffusivity of CO_2, hydration and carbonation products, total possible carbonation, and pH change due to the carbonation and rate of carbonation. The paper also contains a suggestion to determine the above mentioned elements and the possibility to apply the presented approach to other issues concerning concrete durability.

Nguyen and Castel [158] evaluated the resistance of concrete with LC^3 to chloride diffusion. Three series of concrete were made for research purposes. One contained only general purpose cement corresponding to Type I cement according to ASTM C150/C150M [159]. In the remaining two 15% and 20% of the cement was replaced by a mixture of calcined clay and limestone in the ratio 2:1. The results indicate that the use of LC^3 in concrete significantly increased its resistance to chloride diffusion.

The continuation of the research on chloride diffusion resistance Nguyen et al. presented in [160], where they compared concrete made with OPC with concrete containing calcined clay obtained in two technological processes: flash calcination and calcination in rotary kiln. Due to differences in activity of both types of calcined clay, the authors decided to apply different proportions of general purpose cement (as per Australian Standard AS 3972 [161]) to calcined clay (20% for flash calcined clay and 44% for clay calcined in rotary kiln). This variation was due to the intention to obtain the appropriate compressive strength of the hardened concrete (>45 MPa). The results indicated four times higher chloride diffusion resistance of concrete with LC^3. The difference between the results of concrete made with different types of calcined clay turned out to be insignificant, which led the authors to conclude that it is crucial in this case to properly select the proportion of cement exchanged for calcined clay to meet the requirements of specific strength of concrete exposed to chloride ingressives.

A different approach to chloride diffusion research was applied by Yang et al. [162]; they discussed the results of simulations of this phenomenon in LC^3 concrete and concrete with fly ash compared to it. The authors conducted extensive simulations taking into account many elements affecting the course of chloride diffusion in hardened concrete. The results showed that both concretes had comparable service life even though the LC^3 concrete contained less clinker. The conclusions also indicated important parameters that should be introduced to the simulation in order for the results to be reliable.

Nguyen and Castel [163] presented the results of research on corrosion of reinforcing steel in concrete made using LC^3. This issue is important because of the lower alkalinity of this type of concrete due to the lower amount of portlandite, which is consumed by calcined clay in the pozzolanic reaction. The research, carried out over 500 days, allowed the conclusion to be drawn that performance of reinforced concrete prepared with LC^3 was comparable to that for concrete with OPC in the propagation phase.

Rengaraju et al. [164] also investigated corrosion of steel in concrete with LC^3. They presented the results of corrosion testing of steel in three concrete mixes prepared with the use of: OPC, a blend of 70% OPC with 30% of fly ash and a blend of 50% OPC with a 50% mixture of limestone and calcined clay. In their previous work [165] they carried out tests according to AASHTO T 358 standard [166], which showed that concrete with LC^3 showed very high resistivity, which is a good predictor of resistance to chloride ion penetration and, consequently, to corrosion of reinforcing steel. As a continuation of the tests, corrosion tests of steel in concrete were carried out using the method included in ASTM G109 [167] and the impressed current corrosion test method. In the conclusions, apart from the statement that LC^3 cement composites showed higher resistance to chloride ingress and better protection of steel against corrosion, the authors also claimed that the corrosion products of steel in concrete with LC^3 were less expansive and thus less destructive to concrete.

The study of corrosion of reinforcement in concrete made with LC^3 was also conducted by Pillai et al. [168]. They determined such concrete parameters as chloride diffusion coefficient, ageing coefficient and chloride threshold for seven mixtures containing OPC and blends of OPC with pulverised fuel ash (PFA) or with calcined clay and limestone (i.e., LC^3). Using these parameters they determined the service life of the two construction elements. The results showed that a construction made of LC^3 or blended cement with PFA will had a significantly longer service life with a much smaller carbon footprint.

The issue of carbon footprint is focusing our attention on the eco-efficiency of LC^3. In [169] the authors have undertaken the assessment of the sustainability of LC^3 using two methods: life cycle analysis and eco-efficiency. The first method was used to assess the environmental impact of LC^3 production compared to OPC and Portland pozzolana cement (with 20% zeolite content). The second method assessed how the use of LC^3 in the construction of a model residential building will increase its eco-efficiency. The calculations led to the conclusion that the use of LC^3 may lead to a reduction of cement production costs by 4–40% and CO_2 emissions by 15–30%.

No less important are the technological aspects related to LC^3 production. The authors of paper [170] undertook the assessment of suitability of four clays from deposits located in Cuba for LC^3

production. The results turned out to be promising as they indicated that all examined clays had the potential to be used in ternary blend cement production. Additionally, the authors showed that the pozzolan activity was directly proportional to kaolinite equivalent (KEQ), and that the specific surface area of the obtained cements depended mainly on the calcination temperature of the clays and their mineralogical composition.

Nair et al. [171] presented the results of rheological properties test of cement paste and mortar prepared using LC^3 in comparison with OPC and Portland-fly ash cement (containing 30% Class F fly ash). The effectiveness of five superplasticising admixtures based on polycarboxylic ethers (PCE) differing in chemical composition and one admixture based on sulphonated naphthalene formaldehyde (SNF) were also compared. The studies carried out have led to the conclusion that larger quantities of superplasticisers were required for blends with LC^3, with PCE providing a lower viscosity than SNF and the latter additive had to be dosed in larger quantities to not allow to reduce the w/b ratio below 0.4.

Li, X. and Scrivener, K.L [172] devoted their paper to the comparison of three methods of determination of reacted metakaolin in LC^3. The authors compared: mass balance [173], thermodynamic modelling with Gibbs Free Energy Minimization Software [174] and the Partial or not Known Crystalline Structure method (PONKCS) [175]. Of the methods analysed, the mass balance approach yielded the most reliable results, although the use of GEMS software was, on the other hand, less time-consuming and required less labour-intensive input. The PONKCS method proved to be reliable in the case of LC^3-50 mixtures containing clays with metakaolin content above 60%.

Papers [176,177] present the results of two approaches to the pilot production of LC^3 in India. The composition of blends in both cases was almost the same: 50% Portland clinker, 30% or 31% calcined clay, 15% crushed limestone and 4% or 5% gypsum. In the first approach [176] several problems of technological nature emerged, such as too little fineness of the cement due to the use of ground limestone and calcination of only part of the used clay. This resulted in a large scattering of results and lower strength values of LC^3 concretes. In the second approach [177], the authors did not report any technological complications concluding that it is possible to produce good quality LC^3 even without the need to change technologies in existing cement plants.

The influence of the degree of fineness of LC^3 components on its selected physical and mechanical parameters was examined and presented in [178]. It was assumed that each of the three binder components (Portland cement—55%, calcined clay—30% and limestone—15%) can be ground to two degrees: coarse and fine. In addition, in part of the series, clay or limestone was replaced by finely ground quartz. The results indicate that the greatest influence on the strength parameters of concrete with LC^3 had the degree of clinker fineness, and a slightly lesser calcined clay fineness. Limestone fineness was important only in the initial period of concrete strength development (up to three days).

Examples of practical application of LC^3 were presented by Maity et al. in [179]. Four different LC^3 blends with two types of clay and two types of limestone and, for comparison, Portland Pozzolan Cement were used to produce concrete and various structural elements. All these products have undergone quality control and have been incorporated into the construction of residential buildings. The obtained results showed that the mechanical parameters of LC^3 concrete can be even higher than those of PPC concrete. The resulting buildings are tangible proof that LC^3 had moved from the concept and research phase to the first practical application tests.

Dhandapani and Santhanam [180] analysed the impact of the mutual ratio of limestone and calcined clay on the binding and strength development of LC^3 and blends in which the calcined clay was replaced by Class F fly ash. Binary blends without added limestone were used as a reference. The research showed a clear influence of limestone on the setting of the concrete and much less on the further development of its strength. In addition, concrete using calcined clay achieved a very significantly higher strength after seven and 28 days of hardening.

Ferreiro et al. [181] analysed the influence of the clay calcination temperature in the flash calcination process and the fineness of raw clay on the workability and strength performance of LC^3. They used two types of clay, which were calcined in two different installations (gas suspension calciner and flash calciner) and at two different temperatures. The clays used consisted mainly of minerals of group 2:1, which was an additional value of the work, as most of the research on LC^3 was based mainly on clays containing kaolin. The results obtained allowed to conclude that both the calcination temperature and the degree of fragmentation of the material subjected to this process had a significant impact on the workability of fresh LC^3 mortars and their strength development.

The influence of the method of constituent grinding on rheology and early strength of LC^3 tested on mortars was examined by Perez et al. [182]. The constituent were ground both separately and jointly. Clinker, limestone and calcined clay ground separately were divided into three fractions, from which LC^3 was then composed using different combinations of the degree of grinding of the components. In the case of components ground jointly the grinding time varied (25, 45 and 65 min). As far as the latter case is concerned, the results indicate that the longer the grinding time, the higher the strength of the LC^3 later reached, while the change of mini slump radius over 45 min of grinding time was no longer significant. In the case of separated milling, the degree of grinding of the clinker was crucial and the degree of grinding of limestone was of secondary importance. The influence of the degree of milling calcined clay was not studied by the authors.

The suitability of LC^3 as a material for 3D printing process has been investigated by Chen et al. [183]. They prepared four mixtures containing different proportions of two calcined clay with lower (49%) and higher (90%) metakaolinite content, and then conducted extrudability and earlier strength tests on them. The results showed that as the amount of metakaolinite in LC^3 blend increased, its extrudability decreased but at the same time the early strength increased. Therefore, it is crucial to find the optimal content of metakaolinite in LC^3 used for blends in 3D printing. A referral to the same topic is in [184], in which the authors analyse the influence of viscosity modifying admixture on extrudability of LC^3 based materials.

Avet and Scrivener [185,186] focused on the issue of hydration of LC^3 depending on the content of calcined kaolinite in the blend. The research was carried out on LC^3-50 blends with different metakaolinite content. The results showed that with calcined kaolinite above 65% hydration of clinker was slowed down after three days due to refinement of the pore structure. However, despite this, the strength was still increasing, which is connected with further reaction of metakaolinite and increase of the C-A-S-H amount.

Zunino and Scrivener [187] dealt with reactivity and mechanical performance of three mixtures containing Portland cement, Portland cement-limestone blend and LC^3. The distinction of this publication is that the prepared mixtures were cured at a lowered temperature of 10 °C. The same mixtures cured at 20 °C were used as a reference. The research included determination of compressive strength after 1, 7 and 28 days, isothermal calorimetry as well as XRD phase assemblage assessment and MIP pore refinement assessment. They found that LC^3 cured at lower temperature achieved significantly higher compressive strength than the same cement paste cured at higher temperature. The authors claim that this was an effect of, among others, a slightly different course of hydration of this cement.

In the summary of this chapter, it would be appropriate to repeat the concluding statements of the previous section which indicate the need to extend the interest in clays which may be active after calcination and which do not contain kaolinite or contain small amounts of kaolinite. Another interesting aspect that comes to mind after reading the above article on the practical application of LC3 is the possibility to test the suitability of this type of binder as a base for concrete for road construction.

7. Influence of Calcined Clay on the Durability of Concrete

Durability issues are important for each of the cement-based materials used nowadays. A studies related to the durability of cement-based composites prepared with calcined clays are presented in this chapter.

Trümer and Ludwig [188] put forward the thesis that one of the obstacles to the wider use of calcined clays in cement and concrete technology is the lack of information on the long-time behaviour of the concrete. They used clays of various mineralogical composition, which they subjected to the process of thermal activation and used in concrete mixtures as a substitute for 30% of the cement. The investigated concretes focused on the resistance to sulphate attack, alkali-silica reaction, chloride ingress as well as freezing-thawing resistance and carbonation. They revealed that calcined clays in concrete showed significantly lower pozzolanic activity than in other cement systems, which had a direct impact on the durability of concrete. As far as resistance to chloride ingress is concerned, all clays have passed the test. At the other extreme is freeze-thaw resistance, where only concrete with metakaolin showed satisfactory performance. Carbonation of concrete with calcined clay, due to the reduction of $Ca(OH)_2$ consumed in the pozzolanic reaction, progressed faster than in concrete with OPC. The confirmation of pozzolanic activity should not be synonymous with the recognition that the tested clay is suitable for the production of cement or concrete, because depending on its composition and quality its influence on durability parameters may be diametrically different.

Slightly more optimistic conclusions about the durability of concrete with cement blended with calcined clay were formulated by Pierkes et al. [189]. They prepared a series of concretes using cements with 20% (CEM II/A-Q) and 40% (CEM IV/A-Q) of various calcined clays and determined their durability parameters, i.e., resistance to carbonation, chloride migration, freezing-thawing and frost-deicing salts. The first three tests were carried out on non-air-entrained concrete and the fourth on air-entrained concrete. The results indicated that concrete with the addition of calcined clays was able to achieve comparable values of durability parameters as concrete with the addition of other pozzolanic additives, such as fly ash or silica fume.

Shah et al. [190] prepared a short literature review on the durability of concrete with low clinker content, which is replaced by various mineral additives: fly ash, calcined clay, limestone and slag. This review covered chloride migration/penetration, carbonation, sulphate attack and alkali-silica reaction. The authors concluded that significant chloride resistance can be achieved by using a combination of mineral additives and, after that, the resistance was higher than blended cements. The influence of mineral additions on carbonation was summarised by the statement that above a certain level of exchange rate resistance of cement with mineral additions this resistance started to decrease. Calcined clays were found to cause the faster front of carbonation in the material. ASR (alkali-silica reaction) could be mitigated by SCMs, of which metakaolin is directly named. When it comes to resistance to sulphate attack, the authors are of the opinion that a combination of calcined clay and limestone appears to deteriorate in a sulphate environment.

Castillo Lara et al. [191] have conducted tests of physical-mechanical and durability properties on micro-concrete. From this work only the part concerning the tests of durability, which in this case were water absorption and sorptivity, was discussed. The tests were carried out on a cement system defined as micro-concrete, whose characteristic feature was a maximum aggregate grain size of 5 mm, i.e., more than for mortars and less than for ordinary concrete. Two kinds of calcined clay obtained from raw clay soil were used as a substitute for 30% cement. The results showed a decrease in both water absorption and sorptivity of the micro-concrete tested after using a cement blend with calcined clay compared to the material with OPC only.

The issue of carbonation of binders with metakaolin has been investigated by Bucher et al. [192]. The authors have carried out the research using the cements of type CEM I, CEM II/A-LL and CEM II/A-V as a reference. The depth of carbonation was determined on concrete specimens made with cements without additives and with partial replacement of cement with metakaolin. The results confirmed that the use of calcined clay accelerated the carbonation of concrete except when the component of cement blend was limestone (i.e., in the case of CEM II/A-L cement). Concrete with this blend containing 15% of metakaolin showed higher resistance to carbonation than concrete with Portland cement itself.

Studies [193,194] were devoted to resistance of cement mortars, containing, e.g., calcined clays, to chloride ingress. The first one discusses the results of tests carried out on mortar specimens prepared using ten different binders. These binders were composed by replacing 35% of the cement with pozzolanic active materials, including calcined clay. The research concerned concretes subjected to an artificial marine environment for 270 days. The use of alternative binders allowed the obtaining of a material with adequate resistance to chloride ingress while reducing CO_2 emissions by 15%. In the other paper the authors investigated, among others, the resistance to chloride ingress of mortar prepared from cement with 35% content of calcined clay. In one of the mortar series, water was replaced with 0.5 M Na_2SO_4 solution, which served as a chemical activator and precursor of ettringite formation. The results of strength and resistance to chloride ingress revealed that the use of such a chemical activator positively affected both parameters.

Calcined clay is a pozzolanic material which makes it possible to reduce the negative effects of alkali-silica reaction, which has been demonstrated in a number of studies [188,195,196]. This is one of the methods of preventing ASR, besides the use of low-reactivity aggregate [197]. Paper [195] presents results of tests of mortars made with high alkali Portland limestone cement, in which 0%–30% of cement binder was replaced with calcined clay. Cement used in the research was characterized by 4.32% Na_2O_{eq}. As the amount of calcined clay in cement increased, the expansion of mortar bars decreased. The lowest expansion value was achieved by mortar with 25% content. The authors attributed this to the formation of more $CaSi_2O_5$ and the associated reduction of sodium silicate, which is the main product of the ASR reaction responsible for the damage caused by it. In [196] the authors presented the results of research on effectiveness of calcined clays in ASR mitigation. Mortars containing highly reactive aggregate and four different types of calcined clays (replacing cement in the amount of 5%, 10%, 15% and 20%) were used in the research. The research showed that the chemical and mineral composition of the clay was crucial for the effectiveness of ASR mitigation. Trümer and Ludwig [188] tested the resistance of concrete with cement mixed with calcined clay to both ASR and sulphate attack. Three calcined clays containing as basic minerals, respectively, kaolinite, illite and montmorillonite after mixing with CEM I cement in a ratio of 30:70 significantly reduced the expansion of mortars and increased their resistance to sulphate attack.

Studies [198–200] were devoted to research on the resistance of cement composites to sulphate attack. Aramburo et al. [198] tested cement with a high content of calcined clay to show that it was possible to produce CEM IV/A-SR and CEM IV/B-SR type cements, which complied with the requirements of the current European standard UNE EN 197-1:2011 [201]. The specificity of this study was the very high degree of cement exchange for calcined clay, which was 40%, 50%, 60% and 70%. Apart from calcined clay, two different Portland cements were used in cement mixes, one with high C_3A content and the other with high C_3S content. The research showed the required increase in resistance to sulphate cements with calcined clay was accompanied by a decrease in compressive strength. Al-Akhras [199] presented the results of a wide range of research, in which the influence of the exchange of 5%, 10% and 15% of cement to metakaolin on the durability of concrete to sulphate attack was analysed. Apart from various degrees of cement exchange to calcined kaolin, the w/c ratio, initial moist curing period, curing type and air content were also variable parameters. The results allowed to formulate a conclusion that 10% and 15% of the exchange of cement to metakaolin allowed to achieve excellent durability to sulphate attack. Concrete with a lower w/c ratio achieved better durability performances, the time of moist curing proved to be insignificant, autoclaving increased the resistance to moist curing ratio as well as offer a higher air content in concrete. Shi et al. [200] presented the results of a study on the influence of 35% of cement exchange on calcined clay (metakaolin or calcined montmorillonite) or calcined clay and limestone in different proportions on sulphate resistance of mortars. White Portland cement and ordinary Portland cement were used as base cements. The results showed that all mixtures in which the ratio of calcined clay to the sum of calcined clay and limestone was greater than 0.5 showed excellent resistance to sulphate attack.

Quite a large group of papers on durability are focused on cement systems with calcined kaolin. Saillio et al. [202], as well as Yazıcı [203], have analysed various aspects of the durability of concrete and mortars produced with metakaolin addition. Tafraoui et al. [204] gave a brief literature review on the durability of ultra-high performance concrete containing metakaolin. Badogiannis and Tsivilis [205] dedicated their article to the study of the effect of Greek low kaolinite on the durability of concrete, and Shekarchi et al. [206] examined the transport properties (i.e., water penetration, gas permeability, water absorption, electrical resistivity and chloride ingress) of metakaolin-blended cement.

Shi et al. [207] analysed the results of a durability study of Portland cement blends containing, apart from cement, the following additives: pure limestone, pure metakaolin, metakaolin and limestone in 3:1 mass proportion, metakaolin and silica fume, and the three additions simultaneously. The obtained results allowed to state that mortar with pure limestone showed the worst durability parameters in all tests. Mortar with pure cement had the highest resistance to carbonation, but it did not perform well in resistance tests to sulphate attack and chloride ingress, and mortar with pure metakaolinite obtained exactly the opposite resistance parameters.

The issues of concrete durability with calcined clay have also been addressed in two review articles on calcined clays [47] and SCMs in general [58].

A review of studies on the durability of concrete with calcined clay makes it clear that the subject is by no means exhausted. There are relatively few studies on frost resistance and resistance to surface scaling. These issues are important for countries where, during the colder seasons, the temperature can repeatedly pass through zero, while relatively fewer articles come from these countries. When studying these durability properties, it also seems appropriate to determine the compatibility of binders containing calcined clay with air-entraining agents.

Due to the confirmed effectiveness of the addition of calcined clay in limiting the alkali-silica reaction, it is worth considering the simultaneous use of these co-binders with waste glass, either as an aggregate in concrete or as an additional material exhibiting pozzolanic activity. The use of waste glass has so far been limited by concerns about the durability of concrete. Among the studies on the durability aspects of concrete with calcined clays, it is also difficult to find those concerning air permeability. It can be assumed that in this aspect of concrete with the addition of calcined clay will be superior to concrete based on traditional cements, but there is a great deal of room for research between the conjecture and hard evidence.

8. Conclusions

This review paper presents the idea of the replacement of Portland cement by addition of calcined clays. Such application of this kind of pozzolanic materials, as clinker/cement replacement or as supplementary cementitious materials, can be an ecological and economically justified way to meet the global need to reduce CO_2 emissions in concrete technology. Although clay is a material with a very diverse mineralogical composition, its low price, availability and, above all, its distribution in the world make it a valuable supplementary cementitious material in concrete technology.

Several studies have shown that the thermal treatment is necessary to activate the clay's minerals or to increase their pozzolanic activity. To determine the conditions of success of the calcination process, the effects of time, temperature and particle size were investigated.

It was shown that mixing various clays in appropriate proportions with cement allowed to obtain a concrete with satisfactory mechanical parameters, much better than the reference one without calcined clay. It was described that the use of calcined clay increased both early age and long-term mechanical properties of cement paste, mortar and concrete. Moreover, it was found that the replacement of cement content by calcined clay influenced the reduction of the bleeding and shrinkage. The simultaneous use of calcined clays as binder components with other types of pozzolan was also shown.

Research on various aspects of a new binder type—lime calcined clay cement, LC^3—was thoroughly described. Its composition (Portland clinker, calcined clay, calcium carbonate and gypsum) and as well as fresh mix properties and concrete resistance against aggressive liquid and gaseous media were presented.

It is confirmed that cement systems containing calcined clay, also with the addition of limestone, showed better durability and increased resistance to most of the aggressive actions to which concrete was exposed. The exception was carbonation, but satisfactory results have also been achieved in this area.

Most of the discussed papers were carried out on mortar or cement paste, and clearly less on concrete. It seems that the application of calcined clay as a supplementary cementitious material in concrete technology requires more research, as not all the findings performed on a smaller scale (i.e., on mortars and cement paste) can be directly transferred to the parameters and performance of concrete.

Future research aimed at improving the long-term durability of cement-based composites containing calcined clays is needed. It is suggested to optimize the composition of calcined clay cements by selecting the appropriate chemical and mineral components of raw materials to obtain the required mechanical parameters. Additionally, more research should be carried out on the practical application in the cement and concrete production of calcined clays of group 2:1. While there is a great deal of basic research on clays and minerals themselves, application research in this field is dominated by clays containing mainly kaolinite.

Author Contributions: Conceptualization: R.J. and D.J.-N.; methodology: R.J., D.J.-N. and Y.Y.; formal analysis: R.J. and D.J.-N.; resources: R.J. and Y.Y.; writing—original draft preparation: R.J. and D.J.-N.; writing—review and editing: R.J. and D.J.-N.; supervision: D.J.-N. and R.J. All authors have read and agreed to the published version of the manuscript.

Funding: This research received no external funding.

Conflicts of Interest: The authors declare no conflict of interest.

References

1. Damtoft, J.S.; Lukasik, J.; Herfort, D.; Sorrentino, D.M.; Gartner, E.M. Sustainable development and climate change initiatives. *Cem. Concr. Res.* **2008**, *38*, 115–127. [CrossRef]
2. ASTM. *ASTM C125-20 Standard Terminology Relating to Concrete and Concrete Aggregates*; ASTM Standard; ASTM International: West Conshohocken, PA, USA, 2020; p. 9.
3. Emmanuel, A.C.; Parashar, A.; Bishnoi, S. The Role of Calcined Clay Cement vis a vis Construction Practices in India and Their Effects on Sustainability. In *Calcined Clays for Sustainable Concrete*; RILEM Bookseries; Springer Science and Business Media LLC: Dordrecht, The Netherlands, 2015; pp. 411–417. ISBN 9789401799393.
4. WBCSD-IEA. *Low-Carbon Technology for the Indian Cement Industry*; IEA Technology Roadmaps; OECD: Paris, France, 2013; ISBN 9789264197008.
5. Kosior-Kazberuk, M.; Józwiak-Niedzwiedzka, D. Influence of Fly Ash From Co-Combustion of Coal and Biomass on Scaling Resistance of Concrete. *Arch. Civ. Eng.* **2010**, *56*, 239–254. [CrossRef]
6. Murray, H.H. Industrial Applications of Kaolin. *Clays Clay Miner.* **1961**, *10*, 291–298. [CrossRef]
7. Murray, H.H. Traditional and new applications for kaolin, smectite, and palygorskite: A general overview. *Appl. Clay Sci.* **2000**, *17*, 207–221. [CrossRef]
8. Neißer-Deiters, A.; Scherb, S.; Beuntner, N.; Thienel, K.-C. Influence of the calcination temperature on the properties of a mica mineral as a suitability study for the use as SCM. *Appl. Clay Sci.* **2019**, *179*, 105168. [CrossRef]
9. Nawel, S.; Mounir, L.; Hedi, H. Effect of temperature on pozzolanic reaction of Tunisian clays calcined in laboratory. *SN Appl. Sci.* **2020**, *2*, 157. [CrossRef]
10. Maier, M.; Beuntner, N.; Thienel, K.-C. An Approach for the Evaluation of Local Raw Material Potential for Calcined Clay as SCM, Based on Geological and Mineralogical Data: Examples from German Clay Deposits. In *Calcined Clays for Sustainable Concrete*; RILEM Bookseries; Springer: Dordrecht, The Netherlands, 2020; Volume 25, pp. 37–47.
11. Tironi, A.; Trezza, M.A.; Scian, A.N.; Irassar, E.F. Kaolinitic calcined clays: Factors affecting its performance as pozzolans. *Constr. Build. Mater.* **2012**, *28*, 276–281. [CrossRef]

12. Brigatti, M.F.; Galán, E.; Theng, B.K.G. Structure and Mineralogy of Clay Minerals. In *Handbook of Clay Science*; Bergaya, F., Lagaly, G., Eds.; Elsevier Ltd.: Amsterdam, The Netherlands, 2013; pp. 21–81.
13. Brigatti, M.F.; Guggenheim, S. Mica Crystal Chemistry and the Influence of Pressure, Temperature, and Solid Solution on Atomistic Models. *Rev. Mineral. Geochem.* **2002**, *46*, 1–97. [CrossRef]
14. He, C.; Makovicky, E.; Øsbæck, B. Thermal stability and pozzolanic activity of raw and calcined mixed-layer mica/smectite. *Appl. Clay Sci.* **2000**, *17*, 141–161. [CrossRef]
15. Garg, N.; Skibsted, J. Pozzolanic reactivity of a calcined interstratified illite/smectite (70/30) clay. *Cem. Concr. Res.* **2016**, *79*, 101–111. [CrossRef]
16. Lemma, R.; Castellano, C.C.; Bonavetti, V.L.; Trezza, M.A.; Rahhal, V.F.; Irassar, E.F. Thermal Transformation of Illitic-Chlorite Clay and Its Pozzolanic Activity. In *Calciner Clays for Sustainable Concrete*; RILEM Bookseries; Springer: Dordrecht, The Netherlands, 2018; Volume 16, pp. 266–272. ISBN 9789402412062.
17. He, C.; Øsbæck, B.; Makovicky, E. Pozzolanic reactions of six principal clay minerals: Activation, reactivity assessments and technological effects. *Cem. Concr. Res.* **1995**, *25*, 1691–1702. [CrossRef]
18. Fernández, R.; Martirena-Hernández, J.F.; Scrivener, K.L. The origin of the pozzolanic activity of calcined clay minerals: A comparison between kaolinite, illite and montmorillonite. *Cem. Concr. Res.* **2011**, *41*, 113–122. [CrossRef]
19. Brown, I.W.M.; MacKenzie, K.J.D.; Meinhold, R.H. The thermal reactions of montmorillonite studied by high-resolution solid-state 29Si and 27Al NMR. *J. Mater. Sci.* **1987**, *22*, 3265–3275. [CrossRef]
20. Rashad, A.M. Metakaolin as cementitious material: History, scours, production and composition—A comprehensive overview. *Constr. Build. Mater.* **2013**, *41*, 303–318. [CrossRef]
21. Garg, N.; Wang, K. Comparing the performance of different commercial clays in fly ash-modified mortars. *J. Sustain. Cem.Based Mater.* **2012**, *1*, 111–125. [CrossRef]
22. Garg, N.; Skibsted, J. Thermal Activation of a Pure Montmorillonite Clay and Its Reactivity in Cementitious Systems. *J. Phys. Chem. C* **2014**, *118*, 11464–11477. [CrossRef]
23. Garg, N.; Skibsted, J. Heated Montmorillonite: Structure, Reactivity, and Dissolution. In *Calcined Clays for Sustainable Concrete*; RILEM Bookseries; Springer: Dordrecht, The Netherland, 2015; pp. 117–124. ISBN 9789401799393.
24. Ambroise, J.; Murât, M.M.; Péra, J. Hydration reaction and hardening of calcined clays and related minerals V. Extension of the research and general conclusions. *Cem. Concr. Res.* **1985**, *15*, 261–268. [CrossRef]
25. He, C.; Makovicky, E.; Øsbæck, B. Thermal treatment and pozzolanic activity of Na- and Ca-montmorillonite. *Appl. Clay Sci.* **1996**, *10*, 351–368. [CrossRef]
26. Pomakhina, E.; Deneele, D.; Gaillot, A.-C.; Paris, M.; Ouvrard, G. 29Si solid state NMR investigation of pozzolanic reaction occurring in lime-treated Ca-bentonite. *Cem. Concr. Res.* **2012**, *42*, 626–632. [CrossRef]
27. Scherb, S.; Beuntner, N.; Thienel, K.-C. Reaction Kinetics of Basic Clay Components Present in Natural Mixed Clays. In *Calcined Clays for Sustainable Concrete*; RILEM Bookseries; Springer: Dordrecht, The Netherland, 2018; Volume 16, pp. 427–433. ISBN 9789402412062.
28. Tironi, A.; Trezza, M.A.; Scian, A.N.; Irassar, E.F. Assessment of pozzolanic activity of different calcined clays. *Cem. Concr. Compos.* **2013**, *37*, 319–327. [CrossRef]
29. Joussein, E.; Petit, S.; Churchman, J.; Theng, B.; Righi, D.; Delvaux, B. Halloysite clay minerals—A review. *Clay Miner.* **2005**, *40*, 383–426. [CrossRef]
30. Tironi, A.; Cravero, F.; Scian, A.N.; Irassar, E.F. Pozzolanic activity of calcined halloysite-rich kaolinitic clays. *Appl. Clay Sci.* **2017**, *147*, 11–18. [CrossRef]
31. Tironi, A.; Cravero, F.; Scian, A.N.; Irassar, E.F. Hydration of Blended Cement with Halloysite Calcined Clay. In *Calcined Clays for Sustainable Concrete*; RILEM Bookseries; Springer: Dordrecht, The Netherland, 2018; Volume 16, pp. 455–460. ISBN 9789402412062.
32. He, C.; Makovicky, E.; Øsbæck, B. Thermal treatment and pozzolanic activity of sepiolite. *Appl. Clay Sci.* **1996**, *10*, 337–349. [CrossRef]
33. He, C.; Makovicky, E.; Øsbæck, B. Thermal stability and pozzolanic activity of calcined illite. *Appl. Clay Sci.* **1995**, *9*, 337–354. [CrossRef]
34. Lemma, R.; Irassar, E.F.; Rahhal, V. Calcined Illitic Clays as Portland Cement Replacements. In *Calcined Clays for Sustainable Concrete*; Springer: Dordrecht, The Netherland, 2015; pp. 269–276. ISBN 9789401799393.
35. Weaver, C.E. The Distribution and Identification of Mixed-layer Clays in Sedimentary Rocks. *Am. Mineral.* **1956**, *41*, 202–221.

36. Wright, A. Catalysis by layer lattice silicates. I. The structure and thermal modification of a synthetic ammonium dioctahedral clay. *J. Catal.* **1972**, *25*, 65–80. [CrossRef]
37. Khalifa, A.Z.; Cizer, Ö.; Pontikes, Y.; Heath, A.; Patureau, P.; Bernal, S.A.; Marsh, A.T.M. Advances in alkali-activation of clay minerals. *Cem. Concr. Res.* **2020**, *132*, 106050. [CrossRef]
38. Frost, R.L.; Makó, É.; Kristóf, J.; Horváth, E.; Kloprogge, J.T. Mechanochemical Treatment of Kaolinite. *J. Colloid Interface Sci.* **2001**, *239*, 458–466. [CrossRef]
39. Frost, R.L.; Makó, É.; Kristóf, J.; Horváth, E.; Kloprogge, J.T. Modification of Kaolinite Surfaces by Mechanochemical Treatment. *Langmuir* **2001**, *17*, 4731–4738. [CrossRef]
40. Vizcayno, C.; de Gutiérrez, R.M.; Castello, R.; Rodriguez, E.; Guerrero, C.E. Pozzolan obtained by mechanochemical and thermal treatments of kaolin. *Appl. Clay Sci.* **2010**, *49*, 405–413. [CrossRef]
41. Samet, B.; Mnif, T.; Chaabouni, M. Use of a kaolinitic clay as a pozzolanic material for cements: Formulation of blended cement. *Cem. Concr. Compos.* **2007**, *29*, 741–749. [CrossRef]
42. Morsy, M.S.; El-Enein, S.A.A.; Hanna, G.B. Microstructure and hydration characteristics of artificial pozzolana-cement pastes containing burnt kaolinite clay. *Cem. Concr. Res.* **1997**, *27*, 1307–1312. [CrossRef]
43. Komadel, P. Chemically modified smectites. *Clay Miner.* **2003**, *38*, 127–138. [CrossRef]
44. Sánchez, I.; de Soto, I.S.; Casas, M.; Vigil de la Villa, R.; García-Giménez, R. Evolution of Metakaolin Thermal and Chemical Activation from Natural Kaolin. *Minerals* **2020**, *10*, 534. [CrossRef]
45. Belviso, C.; Cavalcante, F.; Niceforo, G.; Lettino, A. Sodalite, faujasite and A-type zeolite from 2:1 dioctahedral and 2:1:1 trioctahedral clay minerals. A singular review of synthesis methods through laboratory trials at a low incubation temperature. *Powder Technol.* **2017**, *320*, 483–497. [CrossRef]
46. Ambroise, J.; Murât, M.M.; Péra, J. Investigations on synthetic binders obtained by middle-temperature thermal dissociation of clay minerals. *Silic. Ind.* **1996**, *7*, 99–107.
47. Sabir, B.B.; Wild, S.; Bai, J. Metakaolin and calcined clays as pozzolans for concrete: A review. *Cem. Concr. Compos.* **2001**, *23*, 441–454. [CrossRef]
48. Todor, D.N. *Thermal Analysis of Minerals*; Abacus Press: Tunbridge Wells, UK, 1976; ISBN 978-0856261015.
49. Alujas Díaz, A.; Martirena-Hernández, J.F. Influence of Calcination Temperature on the Pozzolanic Reactivity of a Low Grade Kaolinitic Clay. In *Calcined Clays for Sustainable Concrete*; RILEM Bookseries 10; Scrivener, K.L., Favier, A.R., Eds.; RILEM: Dordrecht, The Netherlands, 2015; pp. 331–338.
50. Chakchouk, A.; Trifi, L.; Samet, B.; Bouaziz, S. Formulation of blended cement: Effect of process variables on clay pozzolanic activity. *Constr. Build. Mater.* **2009**, *23*, 1365–1373. [CrossRef]
51. Salvador, S. Pozzolanic properties of flash-calcined kaolinite: A comparative study with soak-calcined products. *Cem. Concr. Res.* **1995**, *25*, 102–112. [CrossRef]
52. Slade, R.C.T.; Davies, T.W.; Atakül, H.; Hooper, R.M.; Jones, D.J. Flash calcines of kaolinite: Effect of process variables on physical characteristics. *J. Mater. Sci.* **1992**, *27*, 2490–2500. [CrossRef]
53. Rasmussen, K.E.; Moesgaard, M.; Køhler, L.L.; Tran, T.T.; Skibsted, J. Comparison of the Pozzolanic Reactivity for Flash and Soak Calcined Clays in Portland Cement Blends. In *Calcined Clays for Sustainable Concrete*; RILEM Bookseries; Springer: Dordrecht, The Netherland, 2015; pp. 151–157. ISBN 9789401799393.
54. Teklay, A.; Yin, C.; Rosendahl, L.; Køhler, L.L. Experimental and modeling study of flash calcination of kaolinite rich clay particles in a gas suspension calciner. *Appl. Clay Sci.* **2015**, *103*, 10–19. [CrossRef]
55. Salvador, S.; Davies, T.W. Modelling of combined heating and dehydroxylation of kaolinite particles during flash calcination; production of metakaolin. *Process. Adv. Mater.* **1994**, *4*, 128–135.
56. Bich, C.; Ambroise, J.; Péra, J. Influence of degree of dehydroxylation on the pozzolanic activity of metakaolin. *Appl. Clay Sci.* **2009**, *44*, 194–200. [CrossRef]
57. Teklay, A.; Yin, C.; Rosendahl, L.; Bøjer, M. Calcination of kaolinite clay particles for cement production: A modeling study. *Cem. Concr. Res.* **2014**, *61–62*, 11–19. [CrossRef]
58. Juenger, M.C.G.; Siddique, R. Recent advances in understanding the role of supplementary cementitious materials in concrete. *Cem. Concr. Res.* **2015**, *78*, 71–80. [CrossRef]
59. Taylor-Lange, S.C.; Riding, K.A.; Juenger, M.C.G. Increasing the reactivity of metakaolin-cement blends using zinc oxide. *Cem. Concr. Compos.* **2012**, *34*, 835–847. [CrossRef]
60. Taylor-Lange, S.C.; Rajabali, F.; Holsomback, N.A.; Riding, K.; Juenger, M.C.G. The effect of zinc oxide additions on the performance of calcined sodium montmorillonite and illite shale supplementary cementitious materials. *Cem. Concr. Compos.* **2014**, *53*, 127–135. [CrossRef]

61. Ghorbel, H.; Samet, B. Effect of iron on pozzolanic activity of kaolin. *Constr. Build. Mater.* **2013**, *44*, 185–191. [CrossRef]
62. Chotoli, F.F.; Quarcioni, V.A.; Lima, S.S.; Ferreira, J.C.; Ferreira, G.M. Clay Activation and Color Modification in Reducing Calcination Process: Development in Lab and Industrial Scale. In *Calcined Clays for Sustainable Concrete*; RILEM Bookseries; Springer: Dordrecht, The Netherland, 2015; pp. 479–486. ISBN 9789401799393.
63. Sperinck, S.; Raiteri, P.; Marks, N.; Wright, K. Dehydroxylation of kaolinite to metakaolin—A molecular dynamics study. *J. Mater. Chem.* **2011**, *21*, 2118–2125. [CrossRef]
64. Drits, V.A. An Improved Model for Structural Transformations of Heat-Treated Aluminous Dioctahedral 2:1 Layer Silicates. *Clays Clay Miner.* **1995**, *43*, 718–731. [CrossRef]
65. Fitzgerald, J.J.; Hamza, A.I.; Dec, S.F.; Bronnimann, C.E. Solid-State 27 Al and 29 Si NMR and 1 H CRAMPS Studies of the Thermal Transformations of the 2:1 Phyllosilicate Pyrophyllite. *J. Phys. Chem.* **1996**, *100*, 17351–17360. [CrossRef]
66. Frost, R.L.; Barron, P.F. Solid-state silicon-29 and aluminum-27 nuclear magnetic resonance investigation of the dehydroxylation of pyrophyllite. *J. Phys. Chem.* **1984**, *88*, 6206–6209. [CrossRef]
67. Hollanders, S.; Adriaens, R.; Skibsted, J.; Cizer, Ö.; Elsen, J. Pozzolanic reactivity of pure calcined clays. *Appl. Clay Sci.* **2016**, *132*, 552–560. [CrossRef]
68. Muñoz-Santiburcio, D.; Kosa, M.; Hernández-Laguna, A.; Sainz-Díaz, C.I.; Parrinello, M. Ab Initio Molecular Dynamics Study of the Dehydroxylation Reaction in a Smectite Model. *J. Phys. Chem. C* **2012**, *116*, 12203–12211. [CrossRef]
69. Drachman, S.R.; Roch, G.E.; Smith, M.E. Solid state NMR characterisation of the thermal transformation of Fuller's Earth. *Solid State Nucl. Magn. Reson.* **1997**, *9*, 257–267. [CrossRef]
70. Taylor-Lange, S.C.; Lamon, E.L.; Riding, K.A.; Juenger, M.C.G. Calcined kaolinite–bentonite clay blends as supplementary cementitious materials. *Appl. Clay Sci.* **2015**, *108*, 84–93. [CrossRef]
71. McConville, C.J.; Lee, W.E. Microstructural Development on Firing Illite and Smectite Clays Compared with that in Kaolinite. *J. Am. Ceram. Soc.* **2005**, *88*, 2267–2276. [CrossRef]
72. Alujas Díaz, A.; Almenares Reyes, R.S.; Carratalá, F.A.; Martirena-Hernández, J.F. Proposal of a Methodology for the Preliminary Assessment of Kaolinitic Clay Deposits as a Source of SCMs. In *Calcined Clays for Sustainable Concrete*; RILEM Bookseries; Springer: Dordrecht, The Netherland, 2018; Volume 16, pp. 29–34. ISBN 9789402412062.
73. Chatterjee, A.K. Pozzolanicity of Calcined Clay. In *Calcined Clays for Sustainable Concrete*; RILEM Bookseries; Springer: Dordrecht, The Netherland, 2015; pp. 83–89. ISBN 9789401799393.
74. de Silva, P.S.; Glasser, F.P. Hydration of cements based on metakaolin: Thermochemistry. *Adv. Cem. Res.* **1990**, *3*, 167–177. [CrossRef]
75. Dunster, A.M.; Parsonage, J.R.; Thomas, M.J.K. The pozzolanic reaction of metakaolinite and its effects on Portland cement hydration. *J. Mater. Sci.* **1993**, *28*, 1345–1350. [CrossRef]
76. Dron, R. L'activite Pouzzolanique. Bulletin de liaison des laboratoires des ponts et chaussées. *Spécial* **1978**, *93*, 66–69.
77. Bonavetti, V.; Donza, H.; Rahhal, V.F.; Irassar, E.F. Influence of initial curing on the properties of concrete containing limestone blended cement. *Cem. Concr. Res* **2000**, *30*, 703–708. [CrossRef]
78. Donatello, S.; Tyrer, M.; Cheeseman, C.R. Comparison of test methods to assess pozzolanic activity. *Cem. Concr. Compos.* **2010**, *32*, 121–127. [CrossRef]
79. Tkaczewska, E. Methods of testing pozzolanic actvity of mineral additives. *Mater. Ceram. Ceram. Mater.* **2011**, *63*, 536–541. (In Polish)
80. Payá, J.; Borrachero, M.; Monzó, J.; Peris-Mora, E.; Amahjour, F. Enhanced conductivity measurement techniques for evaluation of fly ash pozzolanic activity. *Cem. Concr. Res* **2001**, *31*, 41–49.
81. McCarter, W.J.; Tran, D. Monitoring pozzolanic activity by direct activation with calcium hydroxide. *Constr. Build. Mater.* **1996**, *10*, 179–184. [CrossRef]
82. Sinthaworn, S.; Nimityongskul, P. Effects of temperature and alkaline solution on electrical conductivity measurements of pozzolanic activity. *Cem. Concr. Compos.* **2011**, *33*, 622–627. [CrossRef]
83. Velázquez, S.; Monzó, J.; Borrachero, M.; Payá, J. Assessment of Pozzolanic Activity Using Methods Based on the Measurement of Electrical Conductivity of Suspensions of Portland Cement and Pozzolan. *Materials* **2014**, *7*, 7533–7547. [CrossRef]

84. Mostafa, N.Y.; El-Hemaly, S.A.S.; Al-Wakeel, E.I.; El-Korashy, S.A.; Brown, P.W. Activity of silica fume and dealuminated kaolin at different temperatures. *Cem. Concr. Res.* **2001**, *31*, 905–911. [CrossRef]
85. Mostafa, N.Y.; Brown, P.W. Heat of hydration of high reactive pozzolans in blended cements: Isothermal conduction calorimetry. *Thermochim. Acta* **2005**, *435*, 162–167. [CrossRef]
86. Tironi, A.; Trezza, M.A.; Scian, A.N.; Irassar, E.F. Thermal analysis to assess pozzolanic activity of calcined kaolinitic clays. *J. Therm. Anal. Calorim.* **2014**, *117*, 547–556. [CrossRef]
87. Federico, L.M.; Chidiac, S.E.; Raki, L. Reactivity of cement mixtures containing waste glass using thermal analysis. *J. Therm. Anal. Calorim.* **2011**, *104*, 849–858. [CrossRef]
88. Kim, T.; Olek, J. Effects of Sample Preparation and Interpretation of Thermogravimetric Curves on Calcium Hydroxide in Hydrated Pastes and Mortars. *Transp. Res. Rec. J. Transp. Res. Board* **2012**, *2290*, 10–18. [CrossRef]
89. Mendoza, O.; Tobón, J.I. An alternative thermal method for identification of pozzolanic activity in Ca(OH)2/pozzolan pastes. *J. Therm. Anal. Calorim.* **2013**, *114*, 589–596. [CrossRef]
90. Fratini, N. Hydrated lime solubility in the presence of potassium hydroxide and sodium hydroxide. *Ann. Chim. Appl.* **1949**, *39*, 616. (In Italian)
91. Fratini, N. Research of hydrolysis lime in cement pastes. Note II. Proposal test for the chemical evaluate of pozzolanic cement. *Ann. Chim. Appl.* **1950**, *40*, 461–469. (In Italian)
92. Rio, A.; Fratini, N. Researches on the control and manufacture of pozzolan cements. *Ann. Chim. Appl.* **1952**, *42*, 526. (In Italian)
93. *ISO 863:2008 Cement—Test Methods—Pozzolanicity test for Pozzolanic Cements*; ISO Standard; International Organization for Standardization: Geneva, Switzerland, 2008; p. 9.
94. Odler, I.; Skalny, J. Hydration of tricalcium silicate at elevated temperatures. *J. Appl. Chem. Biotechnol.* **2007**, *23*, 661–667. [CrossRef]
95. *PN-EN 450-1:2012 Fly Ash for Concrete*; Definition, Specifications and Conformity Criteria; Polish Standard; Polish Committee for Standardization: Warsaw, Poland, 2012; p. 30.
96. Frías, M.; Villar-Cociña, E.; Sánchez de Rojas, M.I.; Valencia-Morales, E. The effect that different pozzolanic activity methods has on the kinetic constants of the pozzolanic reaction in sugar cane straw-clay ash/lime systems: Application of a kinetic–diffusive model. *Cem. Concr. Res.* **2005**, *35*, 2137–2142. [CrossRef]
97. *ASTM C618-19 Standard Specification for Coal Fly Ash and Raw or Calcined Natural Pozzolan for Use in Concrete*; ASTM Standand; ASTM International: West Conshohocken, PA, USA, 2019; p. 5.
98. *BS 3892:1997 Pulverised-Fuel Ash*; Part 1: Specification forpulverised fuel ash for use with Portland cement; British Standard; ASTM International: West Conshohocken, PA, USA, 1997; p. 23.
99. *SRPS B.C1.018:2015 Non-Metallic Mineral Raws-Puzzolanic Materials-Constituents for Cement Production-Classification, Technical Conditions and Test Methods*; Serbian; Standard; Institute for Standardization of Serbia: Beograd, Serbia, 2015.
100. *PN-EN 196-1:2016-07 Methods of Testing Cement*; Determination of Strength; Polish Standard; Polish Committee for Standardization: Warsaw, Poland, 2016; p. 35.
101. *GOST Active Mineral Additives to Binders*; Russian Standard; Federal Agency on Technical Regulating and Metrology: Moscow, Russia, 1963. (In Russian)
102. Slížková, Z.; Drdácký, M.; Viani, A. Consolidation of weak lime mortars by means of saturated solution of calcium hydroxide or barium hydroxide. *J. Cult. Herit.* **2015**, *16*, 452–460. [CrossRef]
103. *ASTM C379-65T Specification for Fly Ash for Use as a Pozzolanic Material with Lime*; ASTM Standard; ASTM International: West Conshohocken, PA, USA, 1965.
104. *ASTM C593-19 Standard Specification for Fly Ash and Other Pozzolans for Use With Lime for Soil Stabilization*; ASTM Standand; ASTM International: West Conshohocken, PA, USA, 2019; p. 5.
105. Chapelle, J. Sulpho-calcic attack of slags and pozzolans. *Rev. Matériaux Constr. Trav. Publics. Ed. C Chaux Cim. Plâtre Agglomérés* **1958**, *511–513*, 87–100, 136–151, 159–170, 514–515. (In French)
106. Ninov, J.; Donchev, I.; Dimova, L. On the kinetics of pozzolanic reaction in the system kaolin–lime–water. *J. Therm. Anal. Calorim.* **2010**, *101*, 107–112. [CrossRef]
107. Feret, R. Researches into the Nature of Pozzolanic Reaction and Materials. *La Rev. Matériaux Constr. Trav. Publics* **1933**, *281*, 41–44, 85–92.

108. Moran, W.; Gilliland, J. Summary of Methods for Determining Pozzolanic Activity. In *Symposium on Use of Pozzolanic Materials in Mortars and Concretes*; ASTM International: West Conshohocken, PA, USA, 1950; pp. 109–130.
109. Jarrige, A.; Ducreux, R. Quelques résultats d'expériences de laboratoire sur les cendres volantes et les ciments aux cendres. *Rev. Matériaux Constr. Trav. Publics* **1958**, *518*, 300–305.
110. Guillaume, L. L'activité pouzzolanique des cendres volantes dans les ciments portland et les ciments au laitier. *Silic. Ind.* **1962**, *28*, 297–300.
111. Battaglino, G.; Schippa, G. Su un nuovo metodo di valutazione dei materiali pozzolanicici. *L'Industria Ital. Cem.* **1968**, *38*, 175–178.
112. Pressler, E.E.; Brunauer, S.; Kantro, D.L. Investigation of Franke Method of Determining Free Calcium Hydroxide and Free Calcium Oxide. *Anal. Chem.* **1956**, *28*, 896–902. [CrossRef]
113. Avet, F.; Snellings, R.; Alujas Díaz, A.; Ben Haha, M.; Scrivener, K.L. Development of a new rapid, relevant and reliable (R3) test method to evaluate the pozzolanic reactivity of calcined kaolinitic clays. *Cem. Concr. Res.* **2016**, *85*, 1–11. [CrossRef]
114. Li, X.; Snellings, R.; Antoni, M.G.; Alderete, N.M.; Ben Haha, M.; Bishnoi, S.; Cizer, Ö.; Cyr, M.; De Weerdt, K.; Dhandapani, Y.; et al. Reactivity tests for supplementary cementitious materials: RILEM TC 267-TRM phase 1. *Mater. Struct.* **2018**, *51*, 151. [CrossRef]
115. Luxán, M.P.; Madruga, F.; Saavedra, J. Rapid evaluation of pozzolanic activity of natural products by conductivity measurement. *Cem. Concr. Res.* **1989**, *19*, 63–68. [CrossRef]
116. Raask, E.; Bhaskar, M.C. Pozzolanic activity of pulverized fuel ash. *Cem. Concr. Res.* **1975**, *5*, 363–375. [CrossRef]
117. Cara, S.; Carcangiu, G.; Massidda, L.; Meloni, P.; Sanna, U.; Tamanini, M. Assessment of pozzolanic potential in lime–water systems of raw and calcined kaolinic clays from the Donnigazza Mine (Sardinia–Italy). *Appl. Clay Sci.* **2006**, *33*, 66–72. [CrossRef]
118. Tironi, A.; Trezza, M.A.; Scian, A.N.; Irassar, E.F. Potential use of Argentine kaolinitic clays as pozzolanic material. *Appl. Clay Sci.* **2014**, *101*, 468–476. [CrossRef]
119. Huenger, K.-J.; Gerasch, R.; Sander, I.; Brigzinsky, M. On the Reactivity of Calcined Clays from Lower Lusatia for the Production of Durable Concrete Structures. In *Calcined Clays for Sustainable Concrete*; RILEM Bookseries; Springer: Dordrecht, The Netherland, 2018; Volume 16, pp. 205–211. ISBN 9789402412062.
120. Beuntner, N.; Thienel, K.-C. Properties of Calcined Lias Delta Clay—Technological Effects, Physical Characteristics and Reactivity in Cement. In *Calcined Clays for Sustainable Concrete*; RILEM Bookseries; Springer: Dordrecht, The Netherland, 2015; pp. 43–50. ISBN 9789401799393.
121. Beuntner, N.; Thienel, K.-C. Effects of Calcined Clay as Low Carbon Cementing Materials on the Properties of Concrete. In Proceedings of the Concrete in the Low Carbon Era, Dundee, UK, 9–11 July 2012; pp. 504–517.
122. Pöllmann, H.; da Costa, M.L.; Angélica, R.S. Sustainable Secondary Resources from Brazilian Kaolin Deposits for the Production of Calcined Clays. In *Calcined Clays for Sustainable Concrete*; RILEM Bookseries; Springer: Dordrecht, The Netherland, 2015; pp. 21–26. ISBN 9789401799393.
123. Maia, A.Á.B.; Saldanha, E.; Angélica, R.S.; Souza, C.A.G.; Neves, R.F. Utilização de rejeito de caulim da Amazônia na síntese da zeólita A. *Cerâmica* **2007**, *53*, 319–324. [CrossRef]
124. Maia, A.Á.B.; Angélica, R.S.; de Freitas Neves, R.; Pöllmann, H.; Straub, C.; Saalwächter, K. Use of 29Si and 27Al MAS NMR to study thermal activation of kaolinites from Brazilian Amazon kaolin wastes. *Appl. Clay Sci.* **2014**, *87*, 189–196. [CrossRef]
125. Akasha, A.M. Using of Libyan Calcined Clay in Concrete. In *Calcined Clays for Sustainable Concrete*; RILEM Bookseries; Springer: Dordrecht, The Netherland, 2015; pp. 555–561. ISBN 9789401799393.
126. Chakchouk, A.; Samet, B.; Mnif, T. Study on the potential use of Tunisian clays as pozzolanic material. *Appl. Clay Sci.* **2006**, *33*, 79–88. [CrossRef]
127. Box, G.E.P.; Hunter, J.S.; Hunter, W.J. *Statistics for Experimenters: Design, Innovation, and Discovery*, 2nd ed.; John Wiley & Sons, Inc.: Hoboken, NJ, USA, 2005.
128. Montgomery, D.C. *Design and Analysis of Experiments*, 9th ed.; John Wiley & Sons, Inc.: Hoboken, NJ, USA, 2017; ISBN 9781119113478.
129. Pierkes, R.; Schulze, S.E.; Rickert, J. Optimization of Cements with Calcined Clays as Supplementary Cementitious Materials. In *Calcined Clays for Sustainable Concrete*; RILEM Bookseries; Springer: Dordrecht, The Netherland, 2015; pp. 59–66. ISBN 9789401799393.

130. Siddique, R.; Klaus, J. Influence of metakaolin on the properties of mortar and concrete: A review. *Appl. Clay Sci.* **2009**, *43*, 392–400. [CrossRef]
131. Kubissa, W.; Jaskulski, R. Measuring and Time Variability of The Sorptivity of Concrete. *Procedia Eng.* **2013**, *57*, 634–641. [CrossRef]
132. Paiva, H.; Velosa, A.; Cachim, P.; Ferreira, V.M. Effect of metakaolin dispersion on the fresh and hardened state properties of concrete. *Cem. Concr. Res.* **2012**, *42*, 607–612. [CrossRef]
133. Zaribaf, B.H.; Uzal, B.; Kurtis, K.E. Compatibility of Superplasticizers with Limestone-Metakaolin Blended Cementitious System. In *Calcined Clays for Sustainable Concrete*; RILEM Bookseries; Springer: Dordrecht, The Netherland, 2015; pp. 427–434. ISBN 9789401799393.
134. *ASTM C595/C595M-20 Standard Specification for Blended Hydraulic Cements*; ASTM Standard; ASTM International: West Conshohocken, PA, USA, 2020; p. 8.
135. Sposito, R.; Beuntner, N.; Thienel, K.-C. Characteristics of components in calcined clays and their influence on the efficiency of superplasticizers. *Cem. Concr. Compos.* **2020**, *110*, 103594. [CrossRef]
136. Nied, D.; Stabler, C.; Zajac, M. Assessing the Synergistic Effect of Limestone and Metakaolin. In *Calcined Clays for Sustainable Concrete*; RILEM Bookseries; Springer: Dordrecht, The Netherland, 2015; pp. 245–251. ISBN 9789401799393.
137. Beuntner, N.; Kustermann, A.; Thienel, K.-C. Pozzolanic Potential of Calcined Clay in High- Performance Concrete. In Proceedings of the International Conference on Sustainable Materials, Systems and Structures (SMSS 2019) New Generation of Construction Materials, Rovinj, Croatia, 20–22 March 2019; pp. 470–477.
138. Mechti, W.; Mnif, T.; Chaabouni, M.; Rouis, J. Formulation of blended cement by the combination of two pozzolans: Calcined clay and finely ground sand. *Constr. Build. Mater.* **2014**, *50*, 609–616. [CrossRef]
139. Ng, S.; Østnor, T. Ternary cement blends with Fly ash-Calcined clay-OPC: An evaluation on their early age and mechanical properties as binders. In Proceedings of the XXIII Nordic Concrete Research Symposium, Aalborg, Danmark, 21–23 August 2017; pp. 19–22.
140. Marchetti, G.; Pokorny, J.; Tironi, A.; Trezza, M.A.; Rahhal, V.F.; Pavlík, Z.; Černý, R.; Irassar, E.F. Blended Cements with Calcined Illitic Clay: Workability and Hydration. In *Calcined Clays for Sustainable Concrete*; RILEM Bookseries; Springer: Dordrecht, The Netherland, 2018; Volume 16, pp. 310–317. ISBN 9789402412062.
141. Wong, H.H.C.; Kwan, A.K.H. Packing density of cementitious materials: Part 1-measurement using a wet packing method. *Mater. Struct.* **2008**, *41*, 689–701. [CrossRef]
142. Machner, A.; Zajac, M.; Ben Haha, M.; Kjellsen, K.O.; Geiker, M.R.; De Weerdt, K. Portland metakaolin cement containing dolomite or limestone–Similarities and differences in phase assemblage and compressive strength. *Constr. Build. Mater.* **2017**, *157*, 214–225. [CrossRef]
143. Ouellet-Plamondon, C.; Scherb, S.; Köberl, M.; Thienel, K.-C. Acceleration of cement blended with calcined clays. *Constr. Build. Mater.* **2020**, *245*, 118439. [CrossRef]
144. Ding, J.; Li, Z. Effects of Metakaolin and Silica Fume on Properties of Concrete. *ACI Mater. J.* **2002**, *99*, 393–398.
145. Zhang, M.H.; Malhotra, V.M. Characteristics of a thermally activated alumino-silicate pozzolanic material and its use in concrete. *Cem. Concr. Res.* **1995**, *25*, 1713–1725. [CrossRef]
146. Brooks, J.; Megat Johari, M. Effect of metakaolin on creep and shrinkage of concrete. *Cem. Concr. Compos.* **2001**, *23*, 495–502. [CrossRef]
147. Gleize, P.J.P.; Cyr, M.; Escadeillas, G. Effects of metakaolin on autogenous shrinkage of cement pastes. *Cem. Concr. Compos.* **2007**, *29*, 80–87. [CrossRef]
148. Scrivener, K.L.; Martirena-Hernández, J.F.; Bishnoi, S.; Maity, S. Calcined clay limestone cements (LC3). *Cem. Concr. Res.* **2018**, *114*, 49–56. [CrossRef]
149. Krishnan, S.; Emmanuel, A.C.; Bishnoi, S. Effective Clinker Replacement Using SCM in Low Clinker Cements. In *Calcined Clays for Sustainable Concrete*; RILEM Bookseries; Springer: Dordrecht, The Netherland, 2015; pp. 517–521. ISBN 9789401799393.
150. Alujas Díaz, A.; Fernández, R.; Quintana, R.; Scrivener, K.L.; Martirena-Hernández, J.F. Pozzolanic reactivity of low grade kaolinitic clays: Influence of calcination temperature and impact of calcination products on OPC hydration. *Appl. Clay Sci.* **2015**, *108*, 94–101. [CrossRef]
151. Joseph, S.; Bishnoi, S.; Maity, S. An economic analysis of the production of limestone calcined clay cement in India. *Indian Concr. J.* **2016**, *90*, 22–27.

152. Joseph, S.; Joseph, A.M.; Bishnoi, S. Economic Implications of Limestone Clinker Calcined Clay Cement (LC3) in India. In *Calcined Clays for Sustainable Concrete*; RILEM Bookseries; Springer: Dordrecht, The Netherland, 2015; pp. 501–507. ISBN 9789401799393.
153. Dhandapani, Y.; Santhanam, M. Assessment of pore structure evolution in the limestone calcined clay cementitious system and its implications for performance. *Cem. Concr. Compos.* **2017**, *84*, 36–47. [CrossRef]
154. Dhandapani, Y.; Vignesh, K.; Raja, T.; Santhanam, M. Development of the Microstructure in LC3 Systems and Its Effect on Concrete Properties. In *Calcined Clays for Sustainable Concrete*; RILEM Bookseries; Springer: Dordrecht, The Netherland, 2018; pp. 131–140. ISBN 9789402412062.
155. Khan, M.S.H.; Nguyen, Q.D.; Castel, A. Carbonation of Limestone Calcined Clay Cement Concrete. In *Calcined Clays for Sustainable Concrete*; RILEM Bookseries; Springer: Dordrecht, The Netherland, 2018; Volume 16, pp. 238–243. ISBN 9789402412062.
156. Nguyen, Q.D.; Khan, M.S.H.; Castel, A. Engineering Properties of Limestone Calcined Clay Concrete. *J. Adv. Concr. Technol.* **2018**, *16*, 343–357. [CrossRef]
157. Joseph, A.M.; Shah, V.; Bishnoi, S. Protocol for Prediction of Durability of New Cements: Application to LC3. In *Calcined Clays for Sustainable Concrete*; RILEM Bookseries; Springer: Dordrecht, The Netherland, 2015; pp. 403–409. ISBN 9789401799393.
158. Nguyen, Q.D.; Castel, A. Chloride Diffusion Resistance of Limestone Calcined Clay Cement (LC3) Concrete. In Proceedings of the SynerCrete'18: Interdisciplinary Approaches for Cement-based Materials and Structural Concrete: Synergizing Expertise and Bridging Scales of Space and Time, Funchal, Portugal, 24–26 October 2018; pp. 129–134.
159. *ASTM C150/C150M-20 Standard Specification for Portland Cement*; ASTM Standard; ASTM International: West Conshohocken, PA, USA, 2020; p. 9.
160. Nguyen, Q.D.; Afroz, S.; Castel, A. Influence of Calcined Clay Reactivity on the Mechanical Properties and Chloride Diffusion Resistance of Limestone Calcined Clay Cement (LC3) Concrete. *J. Mar. Sci. Eng.* **2020**, *8*, 301. [CrossRef]
161. *AS 3972-2010 General Purpose and Blended Cements*; Standards Australia: Sydney, Australia, 2010; p. 19.
162. Yang, P.; Dhandapani, Y.; Santhanam, M.; Neithalath, N. Simulation of chloride diffusion in fly ash and limestone-calcined clay cement (LC3) concretes and the influence of damage on service-life. *Cem. Concr. Res.* **2020**, *130*, 106010. [CrossRef]
163. Nguyen, Q.D.; Castel, A. Reinforcement corrosion in limestone flash calcined clay cement-based concrete. *Cem. Concr. Res.* **2020**, *132*, 106051. [CrossRef]
164. Rengaraju, S.; Pillai, R.G.; Neelakantan, L.; Gettu, R.; Santhanam, M. Chloride-Induced Corrosion Resistance of Steel Embedded in Limestone Calcined Clay Cement Systems. In *Calcined Clays for Sustainable Concrete*; RILEM Bookseries; Springer: Singapore, 2020; Volume 25, pp. 613–619.
165. Rengaraju, S.; Neelakantan, L.; Pillai, R.G. Investigation on the polarization resistance of steel embedded in highly resistive cementitious systems–An attempt and challenges. *Electrochim. Acta* **2019**, *308*, 131–141. [CrossRef]
166. *AASHTO T 358 Standard Method of Test for Surface Resistivity Indication of Concrete's Ability to Resist Chloride Ion Penetration*; AASHTO Standard; American Association of State Highway and Transportation Officials: Washingtion, DC, USA, 2019; p. 10.
167. *ASTM G109-07(2013) Standard Test Method for Determining Effects of Chemical Admixtures on Corrosion of Embedded Steel Reinforcement in Concrete Exposed to Chloride Environments*; ASTM Standard; ASTM International: West Conshohocken, PA, USA, 2013; p. 6.
168. Pillai, R.G.; Gettu, R.; Santhanam, M.; Rengaraju, S.; Dhandapani, Y.; Rathnarajan, S.; Basavaraj, A.S. Service life and life cycle assessment of reinforced concrete systems with limestone calcined clay cement (LC3). *Cem. Concr. Res* **2019**, *118*, 111–119. [CrossRef]
169. Berriel, S.S.; Díaz, Y.C.; Martirena-Hernández, J.F.; Habert, G. Assessment of Sustainability of Low Carbon Cement in Cuba. Cement Pilot Production and Prospective Case. In *Calcined Clays for Sustainable Concrete*; RILEM Bookseries; Springer: Dordrecht, The Netherland, 2015; pp. 189–194. ISBN 9789401799393.
170. Almenares Reyes, R.S.; Alujas Díaz, A.; Rodríguez, S.B.; Rodríguez, C.A.L.; Martirena-Hernández, J.F. Assessment of Cuban Kaolinitic Clays as Source of Supplementary Cementitious Materials to Production of Cement Based on Clinker–Calcined Clay–Limestone. In *Calcined Clays for Sustainable Concrete*; RILEM Bookseries; Springer: Dordrecht, The Netherland, 2018; Volume 16, pp. 21–28. ISBN 9789402412062.

171. Nair, N.; Mohammed Haneefa, K.; Santhanam, M.; Gettu, R. A study on fresh properties of limestone calcined clay blended cementitious systems. *Constr. Build. Mater.* **2020**, *254*, 119326. [CrossRef]
172. Li, X.; Scrivener, K.L. Determination of the amount of reacted metakaolin in calcined clay blends. *Cem. Concr. Res.* **2018**, *106*, 40–48.
173. Durdziński, P.T.; Ben Haha, M.; Zajac, M.; Scrivener, K.L. Phase assemblage of composite cements. *Cem. Concr. Res.* **2017**, *99*, 172–182. [CrossRef]
174. GEMS Development Team. GEM Software (GEMS) Home. Available online: http://gems.web.psi.ch/ (accessed on 20 September 2020).
175. Snellings, R.; Salze, A.; Scrivener, K.L. Use of X-ray diffraction to quantify amorphous supplementary cementitious materials in anhydrous and hydrated blended cements. *Cem. Concr. Res.* **2014**, *64*, 89–98. [CrossRef]
176. Bishnoi, S.; Maity, S.; Mallik, A.; Joseph, S.; Krishnan, S. Pilot scale manufacture of limestone calcined clay cement: The Indian experience. *Indian Concr. J.* **2014**, *88*, 22–28.
177. Emmanuel, A.C.; Haldar, P.; Maity, S.; Bishnoi, S. Second pilot production of limestone calcined clay cement in India: The experience. *Indian Concr. J.* **2016**, *90*, 57–63.
178. Vizcaíno-Andrés, L.M.; Antoni, M.G.; Alujas Díaz, A.; Martirena-Hernández, J.F.; Scrivener, K.L. Effect of fineness in clinker-calcined clays-limestone cements. *Adv. Cem. Res.* **2015**, *27*, 546–556. [CrossRef]
179. Maity, S.; Bishnoi, S.; Kumar, A. Field Application of Limestone-Calcined Clay Cement in India. In *Calcined Clays for Sustainable Concrete*; RILEM Bookseries; Springer: Dordrecht, The Netherlands, 2015; pp. 435–441. ISBN 9789401799393.
180. Dhandapani, Y.; Santhanam, M. Influence of Calcined Clay-Limestone Ratio on Properties of Concrete with Limestone Calcined Clay Cement (LC3). In *Calcined Clays for Sustainable Concrete*; RILEMBookseries 25; Springer: Singapore, 2020; pp. 731–738. ISBN 9789811528064.
181. Ferreiro, S.; Canut, M.M.C.; Lund, J.; Herfort, D. Influence of fineness of raw clay and calcination temperature on the performance of calcined clay-limestone blended cements. *Appl. Clay Sci.* **2019**, *169*, 81–90. [CrossRef]
182. Perez, A.; Favier, A.R.; Martirena-Hernández, J.F.; Scrivener, K.L. Influence of the Manufacturing Process on the Performance of Low Clinker, Calcined Clay-Limestone Portland Cement. In *Calcined Clays for Sustainable Concrete*; RILEM Bookseries; Springer: Dordrecht, The Netherland, 2015; pp. 283–289. ISBN 9789401799393.
183. Chen, Y.; Li, Z.; Chaves Figueiredo, S.; Çopuroğlu, O.; Veer, F.; Schlangen, E. Limestone and Calcined Clay-Based Sustainable Cementitious Materials for 3D Concrete Printing: A Fundamental Study of Extrudability and Early-Age Strength Development. *Appl. Sci.* **2019**, *9*, 1809. [CrossRef]
184. Chen, Y.; Chaves Figueiredo, S.; Yalçinkaya, Ç.; Çopuroğlu, O.; Veer, F.; Schlangen, E. The Effect of Viscosity-Modifying Admixture on the Extrudability of Limestone and Calcined Clay-Based Cementitious Material for Extrusion-Based 3D Concrete Printing. *Materials* **2019**, *12*, 1374. [CrossRef]
185. Avet, F.; Scrivener, K.L. Hydration Study of Limestone Calcined Clay Cement (LC3) Using Various Grades of Calcined Kaolinitic Clays. In *Calcined Clays for Sustainable Concrete*; RILEM Bookseries; Springer: Dordrecht, The Netherland, 2018; Volume 16, pp. 35–40. ISBN 9789402412062.
186. Avet, F.; Scrivener, K.L. Investigation of the calcined kaolinite content on the hydration of Limestone Calcined Clay Cement (LC3). *Cem. Concr. Res.* **2018**, *107*, 124–135. [CrossRef]
187. Zunino Sommariva, F.A.; Scrivener, K.L. Reactivity and Performance of Limestone Calcined-Clay Cement (LC3) Cured at Low Temperature. In *Calcined Clays for Sustainable Concrete*; RILEM Bookseries; Springer: Dordrecht, The Netherland, 2018; Volume 16, pp. 514–520. ISBN 9789402412062.
188. Trümer, A.; Ludwig, H.-M. Sulphate and ASR Resistance of Concrete Made with Calcined Clay Blended Cements. In *Calcined Clays for Sustainable Concrete*; RILEM Bookseries; Springer: Dordrecht, The Netherland, 2015; pp. 3–9. ISBN 9789401799393.
189. Pierkes, R.; Schulze, S.E.; Rickert, J. Durability of Concretes Made with Calcined Clay Composite Cements. In *Calcined Clays for Sustainable Concrete*; RILEM Bookseries; Springer: Dordrecht, The Netherland, 2018; Volume 16, pp. 366–371. ISBN 9789402412062.
190. Shah, V.; Joseph, A.M.; Bishnoi, S. Durability Characteristics of Sustainable Low Clinker Cements: A Review. In *Calcined Clays for Sustainable Concrete*; RILEM Bookseries; Springer: Dordrecht, The Netherland, 2015; pp. 523–530. ISBN 9789401799393.

191. Castillo Lara, R.; Antoni, M.G.; Alujas Díaz, A.; Scrivener, K.L.; Martirena-Hernández, J.F. Study of the addition of calcined clays in the durability of concrete Rancés. *Rev. Ing. Construcción* **2011**, *26*, 25–40. [CrossRef]
192. Bucher, R.; Cyr, M.; Escadeillas, G. Carbonation of Blended Binders Containing Metakaolin. In *Calcined Clays for Sustainable Concrete*; RILEM Bookseries; Springer: Dordrecht, The Netherland, 2015; pp. 27–33. ISBN 9789401799393.
193. Geiker, M.R.; De Weerdt, K.; Garzón, S.F.; Jensen, M.M.; Johannesson, B.; Michel, A. Durability testing of low clinker blends-chloride ingress in similar strength mortar exposed to seawater. In Proceedings of the XXIII Nordic Concrete Research Symposium, Aalborg, Danmark, 21–23 August 2017; pp. 97–100.
194. Marangu, J.M.; Thiong'o, J.K.; Wachira, J.M. Chloride Ingress in Chemically Activated Calcined Clay-Based Cement. *J. Chem.* **2018**, *2018*, 1–8. [CrossRef]
195. Sarfo-Ansah, J.; Atiemo, E.; Boakye, K.A.; Adjei, D.; Adjaottor, A.A. Calcined Clay Pozzolan as an Admixture to Mitigate the Alkali-Silica Reaction in Concrete. *J. Mater. Sci. Chem. Eng.* **2014**, *2*, 20–26. [CrossRef]
196. Li, C.; Ideker, J.H.; Drimalas, T. The Efficacy of Calcined Clays on Mitigating Alakli-Silica Reaction (ASR) in Mortar and Its Influence on Microstructure. In *Calcined Clays for Sustainable Concrete*; RILEM Bookseries; Springer: Dordrecht, The Netherland, 2015; pp. 211–217. ISBN 9789401799393.
197. Jóźwiak-Niedźwiedzka, D.; Jaskulski, R.; Glinicki, M.A. Application of Image Analysis to Identify Quartz Grains in Heavy Aggregates Susceptible to ASR in Radiation Shielding Concrete. *Materials* **2016**, *9*, 224. [CrossRef]
198. Aramburo, C.H.; Pedrajas, C.; Talero, R. Portland Cements with High Content of Calcined Clay: Mechanical Strength Behaviour and Sulfate Durability. *Materials* **2020**, *13*, 4206. [CrossRef]
199. Al-Akhras, N.M. Durability of metakaolin concrete to sulfate attack. *Cem. Concr. Res.* **2006**, *36*, 1727–1734. [CrossRef]
200. Shi, Z.; Ferreiro, S.; Lothenbach, B.; Geiker, M.R.; Kunther, W.; Kaufmann, J.; Herfort, D.; Skibsted, J. Sulfate resistance of calcined clay–Limestone–Portland cements. *Cem. Concr. Res.* **2019**, *116*, 238–251. [CrossRef]
201. EN 197-1. *Cement-Part 1: Composition, Specifications and Conformity Criteria for Common Cements*; European Committee for Standardization: Brussels, Belgium, 2002; pp. 1–25.
202. Saillio, M.; Baroghel-Bouny, V.; Pradelle, S. Various Durability Aspects of Calcined Kaolin-Blended Portland Cement Pastes and Concretes. In *Calcined Clays for Sustainable Concrete*; RILEM Bookseries; Springer: Dordrecht, The Netherland, 2015; pp. 491–499. ISBN 9789401799393.
203. Yazıcı, Ş.; Arel, H.Ş.; Anuk, D. Influences of Metakaolin on the Durability and Mechanical Properties of Mortars. *Arab. J. Sci. Eng.* **2014**, *39*, 8585–8592. [CrossRef]
204. Tafraoui, A.; Escadeillas, G.; Vidal, T. Durability of the Ultra High Performances Concrete containing metakaolin. *Constr. Build. Mater.* **2016**, *112*, 980–987. [CrossRef]
205. Badogiannis, E.; Tsivilis, S. Exploitation of poor Greek kaolins: Durability of metakaolin concrete. *Cem. Concr. Compos.* **2009**, *31*, 128–133. [CrossRef]
206. Shekarchi, M.; Bonakdar, A.; Bakhshi, M.; Mirdamadi, A.; Mobasher, B. Transport properties in metakaolin blended concrete. *Constr. Build. Mater.* **2010**, *24*, 2217–2223. [CrossRef]
207. Shi, Z.; Geiker, M.R.; De Weerdt, K.; Lothenbach, B.; Kaufmann, J.; Kunther, W.; Ferreiro, S.; Herfort, D.; Skibsted, J. Durability of Portland Cement Blends Including Calcined Clay and Limestone: Interactions with Sulfate, Chloride and Carbonate Ions. In *Calcined Clays for Sustainable Concrete*; RILEM Bookseries; Springer: Dordrecht, The Netherland, 2015; pp. 133–141. ISBN 9789401799393.

Publisher's Note: MDPI stays neutral with regard to jurisdictional claims in published maps and institutional affiliations.

© 2020 by the authors. Licensee MDPI, Basel, Switzerland. This article is an open access article distributed under the terms and conditions of the Creative Commons Attribution (CC BY) license (http://creativecommons.org/licenses/by/4.0/).

Article

Characteristics of CO_2 and Energy-Saving Concrete with Porous Feldspar

Jung-Geun Han [1], Jin-Woo Cho [2,*], Sung-Wook Kim [3], Yun-Suk Park [1] and Jong-Young Lee [1,*]

- [1] School of Civil and Environmental Engineering, Urban Design and Study, Chung-Ang University, Seoul 06974, Korea; jghan@cau.ac.kr (J.-G.H.); dbstjr9653@naver.com (Y.-S.P.)
- [2] Construction Automation Research Center, Korea Institute of Civil Engineering and Building Technology (KICT), Goyang 10223, Korea
- [3] Geo Information Research Group Co., LTD., Busan 47598, Korea; suwokim@chol.com
- * Correspondence: jinucho@kict.re.kr (J.-W.C.); geoljy@cau.ac.kr (J.-Y.L.)

Received: 24 August 2020; Accepted: 16 September 2020; Published: 21 September 2020

Abstract: In this study, to reduce the use of cement and sand, porous feldspar with excellent economic efficiency was used as a substitute in the heat storage concrete layer. When porous feldspar and four other silicate minerals were used as substitute materials for sand in cement mortar, the specimen with the porous feldspar exhibited approximately 16–63% higher compressive strength, thereby exhibiting a higher reactivity with cement compared to the other minerals. To compensate for the reduction in strength owing to the decreased cement content, mechanical and chemical activation methods were employed. When the specific surface area of porous feldspar was increased, the unit weight was reduced by approximately 30% and the compressive strength was increased by up to 90%. In addition, the results of the thermal diffusion test confirmed that thermal diffusion increased owing to a reduction in the unit weight; the heat storage characteristics improved owing to the better porosity of feldspar. When chemical activation was performed after reducing the cement content by 5% and replacing the sand with porous feldspar, the compressive strength was found to be approximately twice that of an ordinary cement mortar. In a large-scale model experiment, the heat storage layer containing the porous feldspar exhibited better heat conduction and heat storage characteristics than the heat storage layer composed of ordinary cement mortar. Additionally, energy savings of 57% were observed.

Keywords: porous feldspar; activation; compressive strength; substitute material; energy saving concrete

---

1. Introduction

One of the characteristics of South Korea's housing culture is that hot water is circulated through the piping underneath the floor. Fossil fuels, such as LNG (liquefied natural gas), LPG (liquefied petroleum gas), and coal, are mainly used as the heat sources for the hot water supply, and cement is mostly used as a flooring material. Cement, which is based on carbonate minerals, has continuously generated debates about human health risks, such as sick house syndrome and atopy, owing to heavy metal release and high pH [1]. In South Korea, the energy target management system and the emissions trading scheme were introduced based on the 21st United Nations Framework Convention on Climate Change, and efforts are being made to reduce greenhouse gas emissions [2]. Therefore, a method for reducing the use of cement, which emits 700 kg of carbon dioxide per ton, is required [3]. In construction, fly ash, which is the residue of the thermal power generation process, is used as a substitute for cement. The mixture of fly ash and cement has been used as roadbed and fill material [4–8]. When substitute materials are mixed with cement, compressive strength characteristics vary depending on the mixing ratio. To prevent the reduction in strength owing to the decreased cement content, studies have been conducted on activation methods for increasing the reactivity of substitute materials. Furthermore, studies on eco-friendly materials and the reduction in the use of cement have been conducted of late [9–13].

With the development of nanotechnology, porous materials, for which cavities represent more than 30% of the total volume, have recently been developed. Representative porous materials include active carbon and zeolite, and studies on their use as construction materials have been reported [14–16]. However, active carbon increases environmental hazards, such as fine dust, owing to logging and heating. In the case of zeolite, only mordenite and clonoptilolite are functional among the entire zeolite mineral groups, but their reserves in South Korea are small [17].

Feldspar, a representative aluminosilicate mineral, is a commonly found mineral as it accounts for 60% of Earth's crust. It is used for the manufacture of glass in addition to various potteries and ceramics, and is also used for non-functional purposes, such as land reclamation [18,19]. In South Korea, feldspar is mainly extracted from granite and quartz bedrock, and its reserves are abundant; thus, it is available at low cost.

For feldspar, the mineral composition and surface structure are changed by the weathering process. Cavities are observed on the surface of weathered feldspar porphyry, showing a porous structure. Especially, in feldspar phenocrysts, tens of thousands to hundreds of thousands of cavities are observed. The formation of cavities is related to the specific surface area and the cation exchange capacity [20]. Therefore, feldspar with a porous structure is expected to improve the physical and chemical characteristics, such as adhesion to cement, heat transfer, and preservation capabilities.

In the floor structure of a typical Korean house, cushioning or insulation materials (more than 20 mm) and lightweight foamed concrete (more than 40 mm) are placed on the concrete slab and then hot water pipes are installed, as shown in Figure 1. The heat storage layer (more than 40 mm) composed of sand and cement is then constructed on the top [21]. In this study, a certain proportion of cement was replaced with porous feldspar in the heat storage layer to increase the thermal efficiency of floor heating. Thermal, mechanical, and chemical methods were used for the activation of natural feldspar, and changes in the density, strength, and surface structure were observed. For the utilization of porous feldspar as a flooring material, it was mixed with cement and the strength characteristics were evaluated according to the mixing ratio. In addition, the thermal conductivity and heat storage characteristics were monitored through the test construction to evaluate porous feldspar as a substitute for cement.

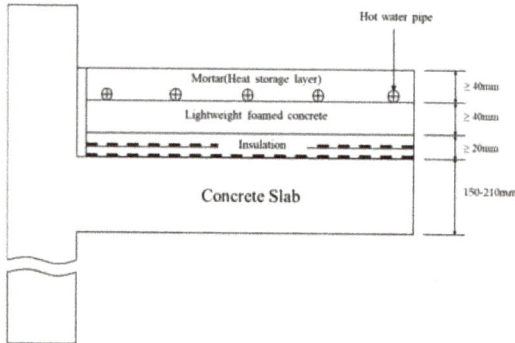

Figure 1. Construction standard of the bottom layer in Korea.

2. Materials and Methods

2.1. Materials

Table 1 shows the mineral and chemical compositions of the porous feldspar used in this study. X-ray fluorescence (XRF) analysis was conducted with samples from three areas in South Korea.

The analysis results showed that the content of two components, i.e., SiO_2 and Al_2O_3, accounted for more than 80%. Figure 2 shows the scanning electron microscope (SEM, VEGA3 SBH, TESCAN, Brno, Czech Republic) image of weathered feldspar. It can be observed that irregular cavities are present on the surface and they are connected to each other. When the pore distribution of porous feldspar was measured, a high specific surface area of 334.5 m^2/g was obtained. The measurement was performed using TriStar TM as an analyzer (TriStarTM II 3020, Micromeritics, GA, USA) and the Brunauer–Emmett–Teller (BET) method. It has been reported that materials with porous structures have excellent physical and chemical characteristics owing to the high specific surface area and the cation exchange capacity. Therefore, it was judged that the pore characteristics of the feldspar used in this study satisfied the above characteristics. As for the materials used in the experiment, rocks with developed feldspar phenocrysts were collected from the granite and diorite rocks in the Chung-ju area, and were used in their powder form.

Table 1. Characteristics of experimental materials (porous feldspar).

I. Mineral Composition of Feldspar Porphyry			
Albite ($NaAlSi_3O_8$)	Quartz (SiO_2)	Orthoclase ($KAlSi_3O_8$)	Chlorite
(%)			
38.8	26.5	22.7	8.3

II. Chemical Composition of Feldspar									
Location	SiO_2	Al_2O_3	K_2O	Na_2O	CaO	Fe_2O	MgO	TiO_2	LOI	Other
	(%)									
Chung-ju	69.59	13.07	2.7	4.53	2.56	2.49	1.61	0.49	2.46	0.50
Muan	67.10	15.26	5.02	3.84	1.88	3.34	0.73	0.40	1.59	0.84
Namwon	68.95	14.59	4.63	3.39	1.29	3.03	0.67	0.37	2.71	0.37

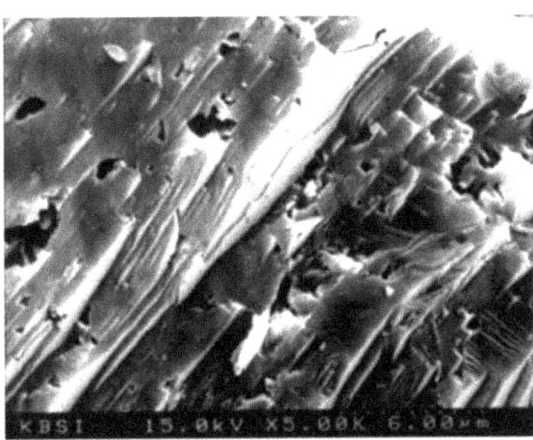

Figure 2. SEM image of weathered feldspar.

2.2. Experimental Conditions

To examine the reactivity of porous feldspar with cement, the uniaxial compressive strength according to the mixing ratio was measured first and the results were compared with the compressive strengths of other substitute materials. Furthermore, when cement is replaced with porous feldspar, a reduction in strength is expected. To compensate for the reduction, mechanical activation for reducing the particle size of materials and chemical activation for improving chemical reactions by mixing a solidifying agent were employed. In addition, mixing tests were conducted to evaluate the applicability of porous feldspar to the heat storage layer by replacing cement and sand. The purpose and method of each test are as follows.

2.2.1. Characteristics of Strength

The purpose of this test was to investigate the strength characteristics according to the weight ratio of porous feldspar powder. The mixing ratio of cement and porous feldspar was varied, and three specimens were fabricated for each mixing ratio in accordance with KS L ISO 679 [22]. The specimens were cured at room temperature (20–24 °C) and humidity (50–60%) for 7 days. The average compressive strength of the three specimens was used as the compressive strength under each condition (EXP-R1 to EXP-R10). Table 2 shows the reactivity test of cement and feldspar.

Table 2. Test conditions for reactivity of cement and feldspar.

Mixed Ratio (Cement: Feldspar)										Feldspar Size	Curing Time	W/C Ratio
EXP-R1	EXP-R2	EXP-R3	EXP-R4	EXP-R5	EXP-R6	EXP-R7	EXP-R8	EXP-R9	EXP-R10	20–500 μm	7 day	0.5
10:0	9:1	8:2	7:3	6:4	5:5	4:6	3:7	2:8	1:9			

Industrial minerals that can be widely utilized as construction materials are clay minerals generated from weathered silicate minerals [23]. Among silicate minerals, silica fume, metakaolin, illite, and dolomite, which are representative pozzolanic materials containing a large amount of silica and alumina and are highly reactive with cement, were selected as substitutes for cement. When cement was replaced with porous feldspar, the compressive strength was compared with those of these materials (EXP-RM to EXP-RF). Five specimens were fabricated under each condition in accordance with KS L ISO 679, and the average compressive strength was used. After the fabricated specimens were cured at room temperature for three days, the uniaxial compressive strength was measured. Table 3 shows the mixing ratios of the materials under the above experimental conditions.

Table 3. Test conditions for reactivity of cement and silicate minerals.

Mixed Ratio (Cement: Clay Mineral)	Metakaolin	Silica Fume	Ilite	Dolomite	Feldspar	Curing Time (Day)	W/C Ratio
7:3	EXP-RM	EXP-RS	EXP-RL	EXP-RD	EXP-RF	3	0.5

2.2.2. Method of Activation and Experimental

To compensate for the strength reduction when porous feldspar was used as a substitute for cement, mechanical and chemical activation methods were used. The particle size of the material causes changes in the physical characteristics, such as the unit weight and compressive strength. The unit weight of a material can change its thermal diffusion and heat storage characteristics. In this study, the unit weight of materials for each particle size was measured in accordance with ASTM C 128 (KS F 2504) for mechanical activation [24,25]. In this instance, all the tests were conducted five times to improve objectivity. The compressive strength test was then conducted on mortar, in which porous feldspar with various particle sizes was substituted for sand. The specimens used in the experiment were fabricated and the compressive strength test was conducted in accordance with KS L ISO 679. Section 1 in Table 4 shows the test conditions for mechanical activation.

Table 4. Test conditions for activation methods (mechanical and chemical activation).

	I. Mechanical Activation								
Test Methods	Feldspar Size (μm)						Mixed Ratio (PC:FS) *	Curing Time (day)	W/C Ratio
Unit weight test	20	38	48	80	100	500	0:10	-	-
Compressive strength test	30	50	80	100	150	200	70:30	3	0.5
	II. Chemical Activation							Curing time (day)	W/C Ratio
Test No.	EXP-A1		EXP-A2		EXP-A3		Feldspar size (μm)		
Mixed ratio (PC:FS:S) *	100:0:0		100:0:0.1		30:70:0.1		80	28	0.5

* PC: Portland cement (%), FS: Feldspar (%), S: Solidifying agent of 0.1% by weight of cement (%).

When feldspar was used as a material to replace cement, a chemical activation method involving mixing a solidifying agent was used for preserving strength. A developed liquid-type inorganic product was used as a solidifying agent in the test. Section 2 in Table 4 shows the test conditions for chemical activation. The fabrication of specimens and the compressive strength test were conducted in accordance with KS L ISO 679, and changes in the surface structure were observed using SEM imaging (EXP-A1 to EXP-A3). The particle size of the feldspar used for specimen fabrication was 80 μm, which was selected based on the strength change results obtained via mechanical activation.

2.2.3. Strength Test of Feldspar and Mortar

This test was conducted to investigate the strength characteristics when cement and sand were replaced with porous feldspar. Specimens in which the ratio of ordinary cement to sand was 1:3 (EXP-FM1) and other specimens in which sand was replaced with feldspar smaller than 1 mm and feldspar powder smaller than 40 μm (EXP-FM2) were fabricated in a cubic form (side length: 50 mm). Three specimens were fabricated under each condition. After they were cured in water for 3–28 days, the compressive strength was measured in accordance with the ASTM C109/C109M (KS L 5105) method [26,27]. Table 5 shows the material mix and experimental conditions.

Table 5. Test condition of substitute materials with feldspar.

Test Method	Mixed Ratio		Curing Time (Day)	W/C Ratio
	EXP-FM1 (PC:AG) *	EXP-FM2 (PC:AGF:PS:S) *	3, 7, 14, 28	0.5
Compressive strength test	25:75	20:40:4%:0.1		

* PC: Portland cement (%), AG: sand (%), AGF: Feldspar ≤ 1000 μm (%), PF: Feldspar ≤ 40 μm (%), S: Solidifying agent of 0.1% by weight of cement (%).

2.2.4. Thermal Diffusion and Heat Storage Test

In the thermal diffusion and heat storage test, a temperature sensor was embedded at the center of 50 × 50 × 50 mm specimens, which were then subjected to water curing for 28 days [28]. The mixing proportions of the specimens were the same as those in EXP-FM1 and EXP-FM2, as shown in Table 5. The fabricated specimens were installed on top of a 400 × 400 mm hot plate, as shown in Figure 3. Their temperatures were measured every minute using a data logger. The test was conducted under two conditions. First, the temperature of the hot plate was set to 100 °C after installing the two specimens on top of the hot plate. Subsequently, the specimens were separated from the hot plate after a 60 min heating period. They were then cooled at room temperature (22–24 °C) to investigate their thermal diffusion effects and heat storage characteristics. In the second method, the thermal diffusion characteristics were investigated by repeating the heating and cooling periods to simulate conditions similar to those of the actual floor heating. Heating and cooling periods were repeated for 300 min at 30 min intervals.

Floor heating circulates water heated to a high temperature through a pipe embedded in the heat storage concrete. In this instance, hot water is repeatedly supplied according to the set temperature. The second method is similar to this process.

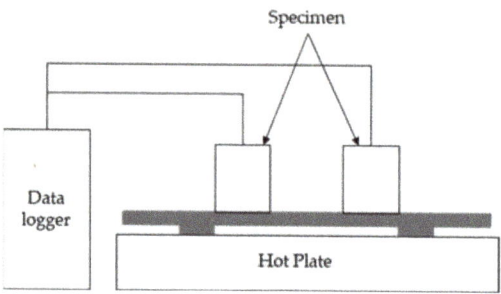

Figure 3. Thermal diffusion and heat storage test device.

2.3. Pilot Test

To evaluate the thermal conductivity, heat storage characteristics, and energy efficiency of porous feldspar, a large-scale single-story experimental building composed of temporary structures was installed. The outer wall of the experimental building was insulated to reduce the influence of the external temperature. Inside this experimental building, two temporary houses each of dimensions 3000 (L) × 4000 (W) × 3000 (H) mm were constructed. Figure 4 shows the design drawings of the heat storage experiment. Based on the floor layer construction standard in Figure 1, a heat storage layer composed of typical concrete mortar (PT-1) was constructed in a temporary house and a heat storage layer in which sand was replaced with feldspar (PT-2) was installed in the other temporary house. EXP-FM1 and EXP-FM2 were applied as the concrete mixing ratios of the heat storage layers. In addition, the temporary houses were separated from the ground by 50 cm to minimize the influence of the ground temperature. Construction and measurement for the two conditions were simultaneously performed to minimize the influence of external environmental factors. After installing a 2 kW electric boiler in each temporary house for hot water supply, a watt-hour meter was installed to determine the power consumption according to the experimental conditions. The temperature of the heat storage layer was measured using an infrared thermal-imaging camera (FLIR A615, temperature range: 20–150 °C, measurement error: 1 °C) (FLIR Systems Srl, Milan, Italy). Temperature sensors were installed at intervals of 80 cm in the heat storage layer to measure the temperature change and power consumption owing to the boiler operation. The measured data were transmitted to the Internet via a wireless router and stored in a cloud service to apply a remote measurement method [29]. The power consumption during the boiler operation was calculated from the images obtained by the CCTVs installed in the watt-hour meters [30].

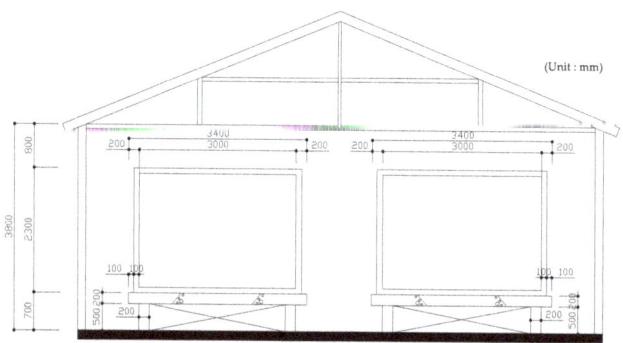

Figure 4. Design drawings of heat storage experiment.

3. Results and Discussion

3.1. Response Characteristics

In the development of substitute materials for cement, the cement replacement rate is generally determined by the uniaxial compressive strength. The strength decreases if the mixing proportions of substitute materials are excessive, and the cement content increases if they are too low. In other words, appropriate mixing of cement and substitute materials is important because increasing the cement content is beneficial for strength but not for the environment.

Figure 5 shows the uniaxial compressive strength according to the mixing ratio of cement and porous feldspar powder. The compressive strength of the specimen with only cement was 11.43 MPa, and the compressive strength decreased as the cement content decreased. The compressive strength linearly decreased for EXP-R10 to EXP-R4 in which the cement content was reduced to 40%, and it rapidly changed for EXP-R3 to EXP-R1 in which the cement content was less than 30%. This indicates that the proper mixing proportion of porous feldspar powder is less than 70% for a modest reduction in strength.

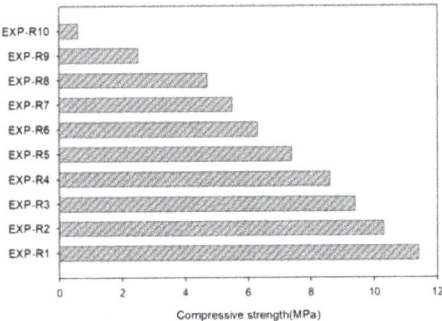

Figure 5. Compressive strength according to the ratio of cement and porous feldspar.

Figure 6 shows the uniaxial compressive strength according to the silicate mineral type. When feldspar powder was used, the strength was approximately 16–63% higher compared with that when other silicate minerals were used, indicating that porous feldspar can be used as a substitute for cement. As for the characteristics of aluminosilicate minerals, SiO_2 and Al_2O_3 are representative pozzolanic components. It appears that porous feldspar increased the strength through the reaction with $Ca(OH_2)$ in the cement hydration process because approximately 80% of its content is accounted for by these two components.

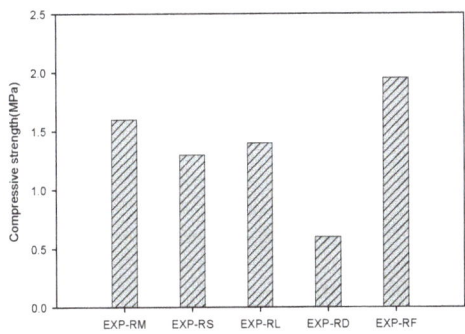

Figure 6. Compressive strength according to the type of silicate minerals.

3.2. Mechanical Activation

Figure 7 shows the unit weight and compressive strength according to the particle size of porous feldspar. The unit weight decreased as the particle size of the feldspar decreased. The unit weight for the particle size of 20 μm was 1.06 g/cm³, which was approximately 31% lower than that for 500 μm. As for the compressive strength according to the particle size, the lowest strength was observed for the largest particle size of 200 μm. As the particle size decreased, the strength slowly increased up to approximately 90%, confirming that the physical characteristics were improved by reducing the particle size through mechanical activation.

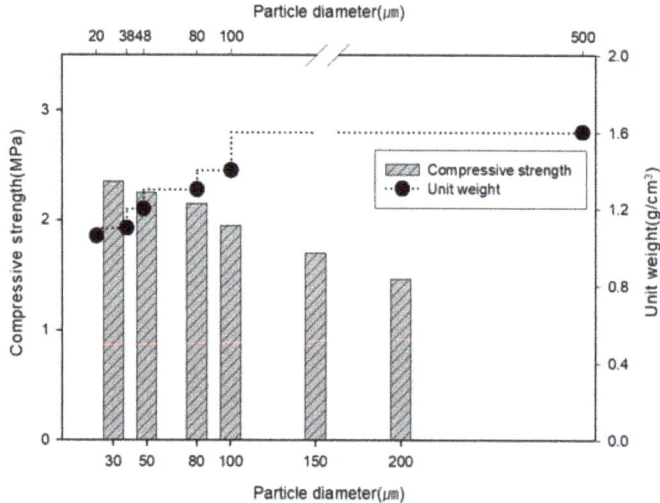

Figure 7. Relationship between the unit weight and compressive strength of feldspar according to the particle size.

3.3. Chemical Activation

Figure 8 shows the compressive strength according to the mixing condition of porous feldspar. The compressive strength of EXP-A2 in which 100% cement was mixed with solidifying agent corresponding to 0.1% of the cement weight ranged from 15 to 19 MPa, showing that the compressive strength was improved by approximately 33% compared with that of EXP-A1 in which only 100% cement was used. The compressive strength of EXP-A3 in which 70% of the cement weight was replaced with porous feldspar and solidifying agent corresponding to 0.1% of the cement weight added ranged from 15 to 18 MPa, which was approximately 30% higher than that of EXP-A1 in which only cement was used. Figure 9 shows the surface structures of the specimens analyzed using SEM. For EXP-A3 in which cement was replaced with porous feldspar, reaction products of the chemical reactions of the inorganic solidifying agent and porous feldspar were observed. Table 6 shows the results of analyzing (SEM-EDS, TESCAN VEGA3 SBH) the chemical compositions of the specimens used in the tests. Na and Cl, which are the major components of the solidifying agent, were detected in EXP-A2 and EXP-A3, in which the solidifying agent and porous feldspar were added. In EXP-A3, the Si and Al contents were 29.6% and 7.2%, respectively, values two to three times higher than those of the other samples. They appear to have increased the strength through the reaction with $Ca(OH)_2$ generated from the cement hydration process.

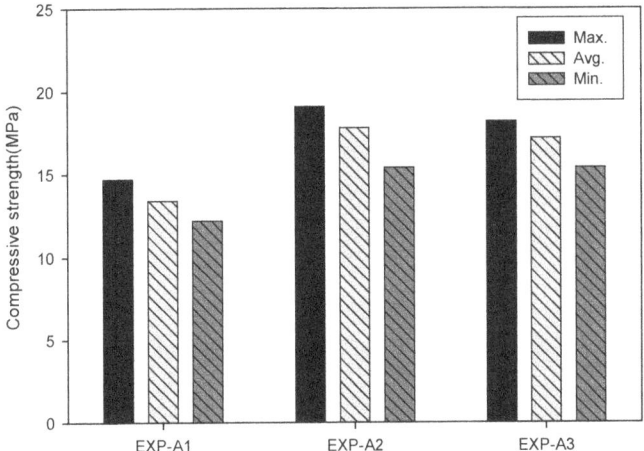

Figure 8. Compressive strength of cement and feldspar mixture.

Figure 9. SEM image of specimens. (**a**) EXP-A1, (**b**) EXP-A2, and (**c**) EXP-A3.

Table 6. SEM-EDS elemental composition of specimens (EXP-A1, EXP-A2, EXP-A3).

Component Sample No.	Si	Al	Ca	Na	Cl	Mg	K	S	Fe
				(%)					
EXP-A1	14.1	2.9	73.8	N.D *	N.D *	1.9	1.2	3.5	2.6
EXP-A2	11.2	3.1	73.6	0.8	0.9	1.7	3.2	2.7	2.8
EXP-A3	29.6	7.2	43.9	1.6	1.3	3.1	4.7	2.5	6.1

* N.D: Non-detection.

3.4. Evaluation of Substitute Materials

The compressive strength test was conducted to evaluate the applicability of porous feldspar to the heat storage layer in the heating floor layer. For a relative comparison, a specimen was fabricated in the same manner by mixing cement mortar and sand in a ratio of 1:3. The mixing proportion of porous feldspar less than 70% was proposed in Section 3.1, and the inorganic solidifying agent corresponding to 0.1% of the cement weight was added based on the experiment results in Section 3.3 to prevent the rapid reduction in strength and to increase the addition of porous feldspar.

Figure 10 shows the results of the compressive strength test. As for the strength characteristics according to the curing time, the strength showed a tendency to increase over time for both materials. Especially, for EXP-FM2 with porous feldspar, the compressive strength at 3 days was 10.53 MPa even though the cement content was reduced by 5%. This result indicates that the strength was improved by approximately 43% compared with that of EXP-FM1. The compressive strengths of EXP-FM1 and EXP-FM2 at seven days were 7.3 and 14.94 MPa, respectively, showing that the strength of EXP-FM2 was two times higher. Both compressive strengths satisfied the quality criterion of South Korea (strength at seven days: 7 MPa) [31]. The compressive strengths of EXP-FM1 and EXP-FM2 at 28 days were 14.14 and 18.97 MPa, respectively, confirming that the strength of EXP-FM2 with porous feldspar was approximately 35% higher than that of EXP-FM1 even though its strength increment slightly decreased.

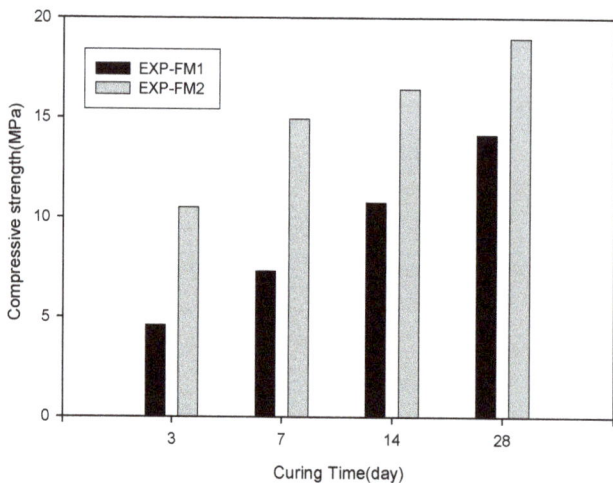

Figure 10. Compressive strength for different mixing ratios and curing times.

3.5. Characteristics of Thermal Diffusion and Heat Storage

Figure 11 shows the temperature changes in the EXP-FM1 and EXP-FM2 specimens when they were heated on the heating plate for 60 min and then cooled at room temperature (22–24 °C) for 150 min. The maximum temperature of the EXP-FM2 specimen, which replaced the sand with porous feldspar, was 66.4 °C. This was 7 °C higher than the maximum temperature (59.4 °C) of

EXP-FM1, a comparison target, confirming the high thermal diffusivity of the specimen containing the porous feldspar. In contrast, when cooled at room temperature after heating for 60 min, EXP-FM2 exhibited a sharp decrease in temperature at the beginning of the cooling period and maintained an approximately 1.3 °C higher temperature than EXP-FM1. This could be because the heat loss in EXP-FM2, which exhibited relatively higher temperatures, was more, owing to the equilibrium between the temperature inside the specimen and the outside temperature during the cooling period at room temperature.

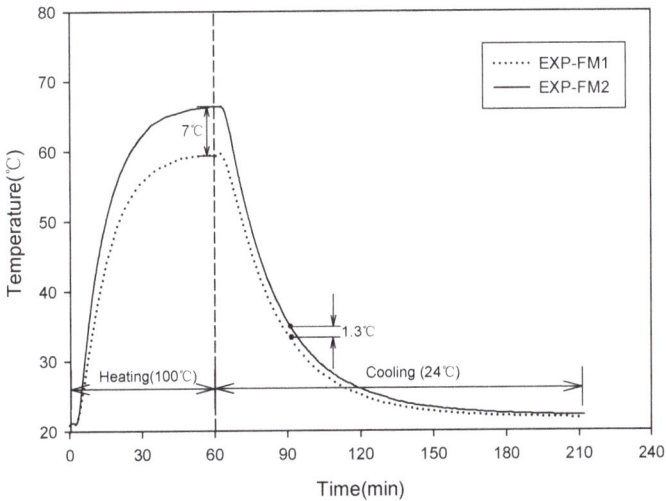

Figure 11. Result of heat storage (heating for 60 min).

Figure 12 shows the results of the test in which the heating and cooling periods were repeated at 30 min intervals. The maximum temperatures during the heating period were approximately 59 °C for EXP-FM1 and 65 °C for EXP-FM2, resulting in a difference of approximately 6 °C. Contrarily, the minimum temperatures during the cooling period were approximately 45 °C for EXP-FM1 and 48 °C for EXP-FM2, resulting in a difference of approximately 3 °C. As the test was repeated, the maximum and minimum temperatures exhibited similar values.

In the cooling process, EXP-FM2 had higher temperatures than EXP-FM1 unlike the results shown in Figure 11. This is because a method similar to the actual heating process was selected in the cooling process without separating the specimens from the heating plate. A point to be noted from these results is whether the created mortar, in which the sand was replaced with porous feldspar and the cement content was decreased, is suitable as a heat storage layer in flooring material. In Section 3.4, it was confirmed that the specimen containing the porous feldspar satisfied the strength criterion for the heat storage layer. In addition, the results of the thermal diffusion and the heat storage test confirmed that EXP-FM2, which replaced the sand with porous feldspar, had faster thermal diffusion and better heat storage characteristics than EXP-FM1, a comparison target. In general, low density of concrete causes high thermal diffusivity but leads to a low specific heat capacity. In this test, however, EXP-FM2, which had a lower density, was superior to EXP-FM1 in terms of both heat transfer and heat storage. It appears that the heat storage effect was high owing to the porosity characteristics of feldspar, a substitute material for sand.

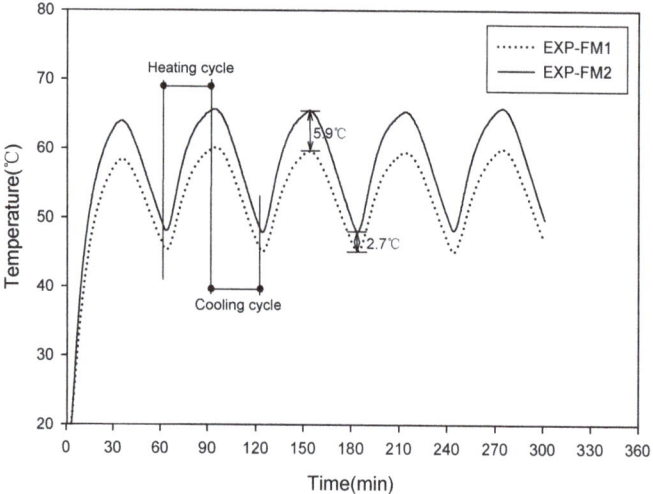

Figure 12. Characteristics of heat diffusion and storage (repeated heating).

3.6. Pilot Test

3.6.1. Characteristics of Thermal Conductivity and Heat Storage

Figure 13 shows the temperature distribution of the heat storage layer measured using the thermal-imaging camera during the heating time after hot water was supplied to the floor. The temperature of the supplied hot water was 55 °C. At 10 min after the boiler operation, the heat storage layer mixed with porous feldspar (PT-2) exhibited a temperature approximately 2 °C higher than that of the heat storage layer composed of ordinary concrete (PT-1). The temperature difference increased to approximately 3 °C at 2 h after the boiler operation, indicating that PT-2 had better heat transfer than that of PT-1. Figure 14 shows the temperature variation during the cooling period after the boiler operation was stopped. PT-1 exhibited a fast cooling speed and no thermal image could be obtained after 90 min. The average temperature was 21 °C after 120 min and 18 °C after 180 min, which was identical to the temperature before the boiler operation. In contrast, PT-2 with porous feldspar exhibited temperatures approximately 2 °C higher than those of PT-1. The temperature remained at over 23 °C even after 180 min of cooling, confirming the excellent heat storage characteristics.

These results can be divided into the influence of the material properties of the heat storage layer and that of the bottom storage layer structure (Figure 1). First, in terms of material properties, the concrete containing the porous feldspar (PT-2) was favorable for thermal diffusion because its density was lower than that of PT-1, as shown in the thermal diffusion and heat storage test in Section 3.5. In contrast, PT-1 exhibited slow thermal diffusion because the density of its heat storage layer was higher compared to that of PT-2. Owing to the high thermal resistance, more heat flow occurred in the bottom direction (lightweight foamed concrete + slab), thereby causing relatively low thermal diffusion to the heat storage layer.

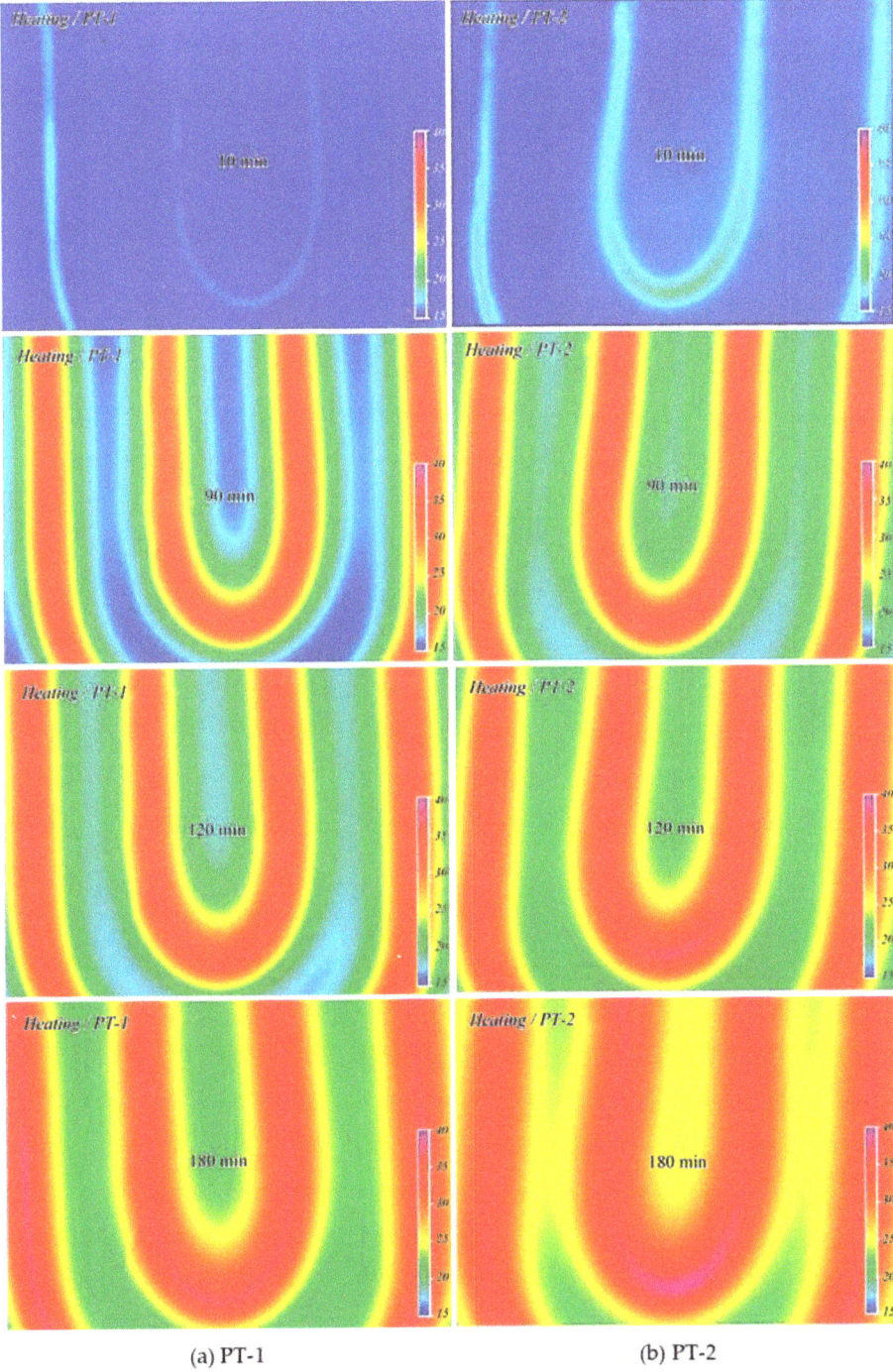

(a) PT-1 (b) PT-2

Figure 13. Temperature change of heat storage layer (heating cycles).

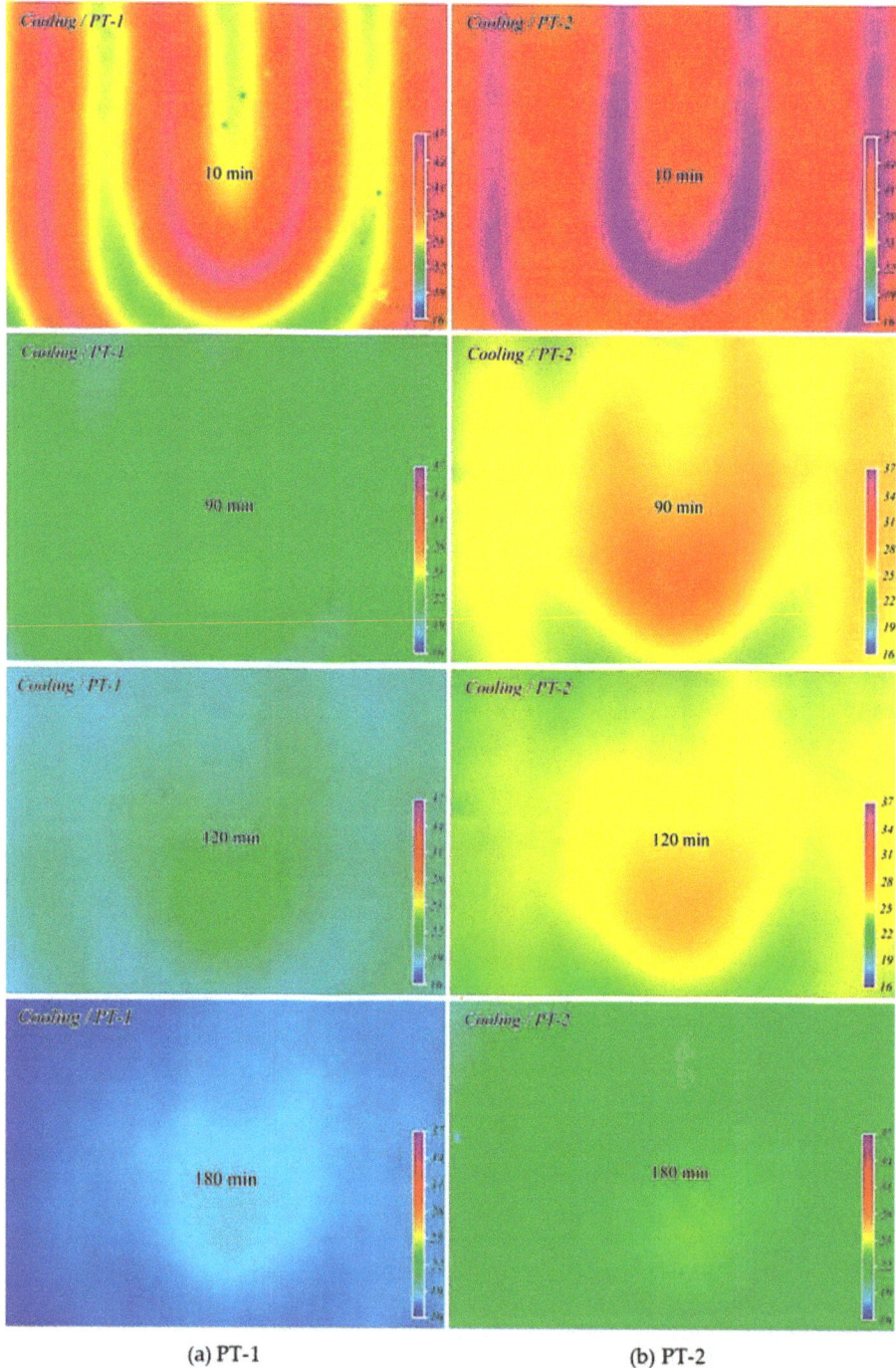

(a) PT-1 (b) PT-2

Figure 14. Temperature change of heat storage layer (cooling cycles).

3.6.2. Energy Efficiency

Figure 15 shows the heat storage layer temperature and power consumption during the experiment period under the two conditions. PT-2 exhibited a higher temperature increase rate than that of PT-1 until 100 min after heating, and the increase rate decreased after 30 °C. PT-1 showed a temperature increase rate similar to that of PT-2 until 20 °C, but its temperature increase rate became lower than that of PT-2. During 180 min of heating, the maximum temperature was 34.4 °C for PT-2 and 31.6 °C for PT-1, showing that the value for PT-2, which used porous feldspar, was 2.7 °C higher. When the target temperature of the heat storage layer was set to 25 °C, the boiler operation was stopped after approximately 40 min and the power consumption was 1.84 kWh for PT-2. For PT-1, the boiler operation was stopped after approximately 70 min and the power consumption was 3.22 kWh. In other words, the boiler was operated 30 min longer for PT-1 than for PT-2, and PT-1 required the additional power of 1.38 kWh, confirming that PT-2 could save approximately 57% power. If the boiler restart temperature is set to 20 °C after the cooling process, it is expected that the boiler will restart after 380 min for PT-2 and after 310 min for PT-1; thus, PT-2 will delay the boiler operation by approximately 70 min. Through the experiment, the rapid temperature increase effect in the heating process owing to the high thermal conductivity caused by the specific surface area characteristics of porous feldspar and the energy-saving effect in the cooling process owing to the delay effect caused by the heat storage characteristics of porous feldspar could be confirmed.

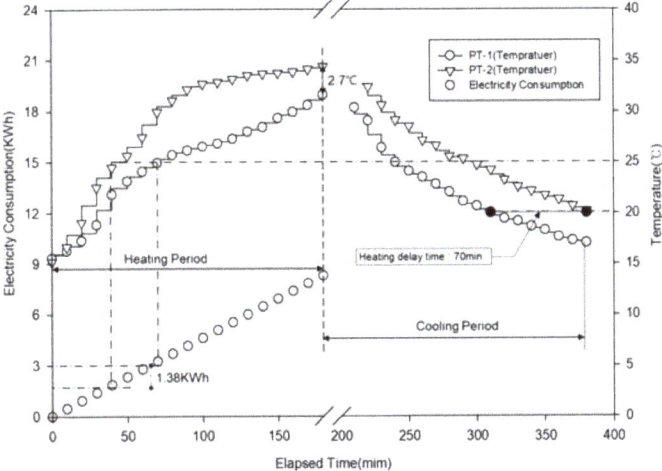

Figure 15. Change of temperature and electric consumption.

4. Conclusions

In this study, feldspar, which is found in South Korea in large quantities and has a porous structure owing to weathering, was used as a substitute for sand in cement. When sand was replaced with porous feldspar and four other silicate minerals in the cement mortar, the specimen that used the porous feldspar exhibited approximately 16–63% higher compressive strength, thereby confirming a higher reactivity with cement than other minerals.

When the particle size was reduced via mechanical activation to increase the specific surface area of porous feldspar, the unit weight decreased by approximately 30%, but the uniaxial compressive strength increased by up to 90%, confirming that the physical characteristics were improved.

A solidifying agent was mixed in to compensate for the strength reduction caused by the addition of porous feldspar. When 70% of the sand weight was replaced with porous feldspar and the solidifying agent was mixed in, the compressive strength was improved by approximately 30% compared with when only cement was used.

When chemical activation (2:8) was performed after reducing the cement content by 5% and replacing the sand with porous feldspar, the compressive strength at 3 days improved by approximately 43% compared to that of the specimen in which cement and sand were mixed in a ratio of 1:3. The compressive strength at 7 days was approximately two times higher.

As for the thermal diffusion, the mortar in which the sand was replaced with the porous feldspar exhibited approximately 6–7 °C higher temperatures than that of the standard concrete in the heating process. In addition, it maintained approximately 1–3 °C higher temperatures in the cooling process. This was because the mechanical activation increased the thermal diffusion by reducing the density of the porous feldspar mortar and the heat storage effect of the feldspar was relatively better owing to its porosity.

In a large-scale model experiment, porous feldspar exhibited excellent thermal diffusion and heat storage characteristics in the heat storage layer as well as an approximately 57% energy saving effect, thereby confirming its high applicability as a heat storage layer material.

The use of porous feldspar as a substitute material for aggregate can reduce the cement content, thereby decreasing the CO_2 emissions. In addition, porous feldspar is an economical option owing to its easy availability and inexpensive characteristics.

Author Contributions: J.-G.H. and J.-W.C. provided the idea and applied for funding to support this paper. J.-W.C., S.-W.K., and Y.-S.P. performed these experiments. J.-W.C. and J.-Y.L. contributed to the analysis of experiment data. J.-Y.L. and J.-W.C. wrote this paper. J.-G.H., J.-W.C., and J.-Y.L. revised and put forward opinions for this paper. All authors have read and agreed to the published version of the manuscript.

Funding: This research was supported by the MSIT (Ministry of Science and ICT), Korea, under the ITRC (Information Technology Research Center) support program (IITP-2020-2020-0-01655) and the MSIP (NRF-2019R1A2C2088962), the X-mind Corps program (2017H1D8A1030599) from the National Research Foundation (NRF) of Korea, the Human Resources Development (No.20204030200090), and the of the Korea Institute of Energy Technology Evaluation and Planning (KETEP) grant funded by the Korean government.

Conflicts of Interest: The authors declare no conflict of interest.

References

1. Mehraj, S.S.; Bhat, G.A.; Balkhi, H.M. Cement factories and human health. *Int. J. Cur. Res. Rev.* **2013**, *5*, 47–54.
2. Greenhouse gas inventory and research center of Korea. *2018 Korean Emissions Trading System (K-ETS) Summary Report*; Greenhouse gas inventory and research center of Korea: Seoul, Korea, 2020.
3. Kewalramani, M.A.; Syed, Z.I. Application of nanomaterials to enhance microstructure and mechanical properties of concrete. *Int. J. Civ. Eng. Technol.* **2018**, *9*, 115–129. [CrossRef]
4. Islam, A.; Alengaram, U.J.; Jumaat, M.Z.; Bashar, I.I. The development of compressive strength of ground granulated blast furnace slag-palm oil fuel ash-fly ash based geopolymer mortar. *Mater. Des.* **2004**, *56*, 833–841. [CrossRef]
5. Bernal, S.A.; Rodríguez, E.D.; Mejía de Gutiérrez, R.; Gordillo, M.; Provis, J.L. Mechanical and thermal characterisation of geopolymers based on silicate-activated metakaolin/slag blends. *J. Mater. Sci.* **2001**, *42*, 5477–5486. [CrossRef]
6. Deb, P.S.; Nath, P.; Sarker, P.K. Strength and Permeation Properties of Slag Blended Fly Ash Based Geopolymer Concrete. *Adv. Mater. Res.* **2013**, *651*, 168–173.
7. Malhotra, S.K.; Dave, N.G. Investigation into the effect of addition of fly-ash and burnt clay pozzolana on certain engineering properties of cement composites. *Cem. Concr. Comp.* **1999**, *21*, 285–291. [CrossRef]
8. Indraratna, B.; Nutalaya, P.; Koo, K.S.; Kuganenthira, N. Engineering behaviour of a low carbon, pozzolanic fly ash and its potential as a construction fill. *J. Can. Geotech.* **1991**, *28*, 542–555. [CrossRef]
9. Idawati, I.; Susan, A.B.; John, L.P.; Rackel, S.N.; Sinin, H.; Jannie, S.J. Modification of phase evolution in alkali-activated blast furnace slag by the incorporation of fly ash. *Cem. Concr. Comp.* **2014**, *45*, 125–135.
10. Thanongsak, N.; Watcharapong, W.; Arnon, C. Utilization of fly ash with silica fume and properties of Portland cement–fly ash–silica fume concrete. *Fuel* **2010**, *89*, 768–774.

11. Xinghua, F.; Wenping, H.; Chunxia, Y.; Dongxue, L.; Xuequan, W. Study on high-strength slag and fly ash compound cement. *Cem. Concr. Res.* **2000**, *30*, 1239–1243.
12. Li, D.; Shen, J.; Chen, Y.; Cheng, L.; Xuequan, W. Study of properties on fly ash-slag complex cement. *Cem. Concr. Res.* **2000**, *30*, 1381–1387. [CrossRef]
13. Roy, D.M.; Silsbee, M.R. Alkali activated cementilious materials, an overview. *Mater. Res. Soc.* **1992**, *245*, 153–164. [CrossRef]
14. Arkaiusz, D.; Wojciech, F.; Halina, W.N.; Adriana, C.Z. Textural properties vs. CEC and EGME retention of Na–X zeolite prepared from fly ash at room temperature. *Int. J. Miner. Process.* **2007**, *82*, 57–68.
15. Malik, R.; Ramteke, D.S.; Wate, S.R. Adsorption of malachite green on groundnut shell waste based powdered activated carbon. *Waste Manag.* **2007**, *27*, 1129–1138. [CrossRef] [PubMed]
16. Jha, B.; Singh, D.N. A review on synthesis, characterization and industrial application of fly ash zeolites. *J. Mater. Educ.* **2011**, *33*, 65–132.
17. Cho, K.W. Engineering properties of weathered alumino-silicate minerals as an eco-friendly construction materials. Ph.D. Thesis, Chung-Ang University, Seoul, Korea, 2018.
18. Kauffman, R.A.; Van Dyk, D. *Feldspars: In Industrial Minerals and Rocks*, 6th ed.; Carr, D.D., Ed.; Society for Mining, Metallurgy, and Exploration Inc.: Littleton, CO, USA, 1994; pp. 473–481.
19. Potter, M.J. Feldspar and nepheline syenite. In *Minerals Yearbook*; US Dept. of Interior, US Geological Survey: Reston, VA, USA, 1996.
20. Cecen, F. *Activated Carbon, Kirk-Othmer Encyclopedia of Chemical Technology*; John Wiley and Sons: New York, NY, USA, 2014.
21. Ministry of Land, Infrastructure and Transport. *Standard Floor Finishing and Structures for Interlayer Noise Prevention*; Ministry of Land, Infrastructure and Transport: Governing City, Korea, 2015.
22. KS L ISO 679. *Methods of Testing Cement-Determination of Strength*; Korea standard association: Seoul, Korea, 2016.
23. Brooks, J.J.; Megat Johari, M.A. Effect of metakaolin on creep and shrinkage of concrete. *Cem. Concr. Comp.* **2001**, *23*, 495–502. [CrossRef]
24. ASTM C128. *Standard Test. Method for Relative Density (Specific Gravity) and Absorption of Fine Aggregate*; ASTM International: West Conshohocken, PA, USA, 2015.
25. KS F 2504. *Standard Test. Method for Density and Absorption of Fine Aggregates*; Korea standard association: Seoul, Korea, 2019.
26. ASTM C109/C109M-16a. *Standard Test. Method for Compressive Strength of Hydraulic Cement Mortars (Using 2-in. or [50-mm] Cube Specimens]*; ASTM International: West Conshohocken, PA, USA, 2016.
27. KS L 5105. *Testing Method for Compressive Strength of Hydraulic Cement Mortar*; Korea standard Association: Seoul, Korea, 2017.
28. Park, Y.S. Experimental Study on the Mortar using Alumino-Silicate Minerals as Aggregate. Master's Thesis, Chung-Ang University, Seoul, Korea, 2020.
29. Ferdoush, S.; Li, X. Wireless Sensor Network System Design Using Raspberry Pi and Arduino for Environmental Monitoring Applications. *Procedia Computer Sci.* **2014**, *34*, 103–110. [CrossRef]
30. Choi, E.J.; Lee, W.J.; Lee, K.H.; Kim, J.K.; Kim, J.H. Real-time pedestrian recognition at night based on far infrared image sensor. In Proceedings of the 2nd International Conference on Communication and Information Processing, Singapore, 26–29 November 2016; pp. 115–119.
31. KS L 5220. *Dry Ready Mixed Cement Mortar*; Korea standard Association: Seoul, Korea, 2018.

© 2020 by the authors. Licensee MDPI, Basel, Switzerland. This article is an open access article distributed under the terms and conditions of the Creative Commons Attribution (CC BY) license (http://creativecommons.org/licenses/by/4.0/).

Review

End-of-Life Materials Used as Supplementary Cementitious Materials in the Concrete Industry

Adrian Ionut Nicoara [1,2], Alexandra Elena Stoica [1,2], Mirijam Vrabec [3], Nastja Šmuc Rogan [3], Saso Sturm [4], Cleva Ow-Yang [5,6], Mehmet Ali Gulgun [5,6], Zeynep Basaran Bundur [7], Ion Ciuca [8] and Bogdan Stefan Vasile [1,2,*]

1. Department of Science and Engineering of Oxide Materials and Nanomaterials, Faculty of Applied Chemistry and Materials Science, University Politehnica of Bucharest, 011061 Bucharest, Romania; adrian.nicoara@upb.ro (A.I.N.); elena_oprea_93@yahoo.co.uk (A.E.S.)
2. National Research Center for Micro and Nanomaterials, Faculty of Applied Chemistry and Materials Science, University Politehnica of Bucharest, 060042 Bucharest, Romania
3. Department of Geology, Faculty of Natural Sciences and Engineering, University of Ljubljana, Aškerčeva 12, 1000 Ljubljana, Slovenia; mirijam.vrabec@geo.ntf.uni-lj.si (M.V.); nastja.rogan@guest.arnes.si (N.Š.R.)
4. Department for Nanostructured Materials, Jozef Stefan Institute, Jamova cesta 39, SI-1000 Ljubljana, Slovenia; saso.sturm@ijs.si
5. Materials Science and Nano-Engineering Program, Sabanci University, Orta Mahalle, Üniversite Caddesi No:27, 34956 Tuzla–İstanbul, Turkey; cleva.ow-yang@sabanciuniv.edu (C.O.-Y.); m-gulgun@sabanciuniv.edu (M.A.G.)
6. Nanotechnology Application Center (SUNUM), Sabanci University, Orta Mahalle, Üniversite Caddesi No:27, 34956 Tuzla–İstanbul, Turkey
7. Department of Civil Engineering, Ozyegin University, Nişantepe District, Orman Street, Çekmeköy, 34794 Istanbul, Turkey; zeynep.basaran@ozyegin.edu.tr
8. Faculty Materials Science and Engineering, University Politehnica of Bucharest, 060042 Bucharest, Romania; ion.ciuca11@gmail.com
* Correspondence: bogdan.vasile@upb.ro

Received: 22 March 2020; Accepted: 19 April 2020; Published: 22 April 2020

Abstract: A sustainable solution for the global construction industry can be partial substitution of Ordinary Portland Cement (OPC) by use of supplementary cementitious materials (SCMs) sourced from industrial end-of-life (EOL) products that contain calcareous, siliceous and aluminous materials. Candidate EOL materials include fly ash (FA), silica fume (SF), natural pozzolanic materials like sugarcane bagasse ash (SBA), palm oil fuel ash (POFA), rice husk ash (RHA), mine tailings, marble dust, construction and demolition debris (CDD). Studies have revealed these materials to be cementitious and/or pozzolanic in nature. Their use as SCMs would decrease the amount of cement used in the production of concrete, decreasing carbon emissions associated with cement production. In addition to cement substitution, EOL products as SCMs have also served as coarse and also fine aggregates in the production of eco-friendly concretes.

Keywords: construction debris; cement; recycling; circular economy; eco-friendly concretes; fly ash (FA); silica fume (SF); palm oil fuel ash (POFA); rice husk ash (RHA); sewage sludge ash (SSA) and sugarcane bagasse ash (SBA); mine tailings; marble dust; construction and demolition debris (CDD)

1. Introduction

In the face of rapidly expanding urbanization, environmental sustainability represents a serious challenge for the construction industry, whose consumption of concrete requires a significant quantity of natural reserves worldwide and necessitates the development of alternative materials and sources. The fabrication of concrete consumes around 27 billion tonnes of feedstock, representing 4 tonnes of concrete for each person every year. By 2050, concrete production will be four times higher than in 1990. Aggregates and binder (i.e., cement) represent around 60%–80% and 10%–15% of the total weight of concrete, respectively [1]. Along with processing of substantial amount of aggregates and around 2.8 billion tonnes of cement products per year, concrete generates approximately 5%–7% of the global total carbon dioxide emissions. By 2025, around 3.5 billion tonnes of carbon dioxide is foreseen to be released to the atmosphere during cement production. One solution for more sustainable production can be harvesting locally available end-of-life (EOL) and/or recyclable materials [1,2].

Global quarry practices to obtain coarse aggregates have substantially modified the ecological equilibrium. Figure 1 shows the amount, in tonnes of aggregates produced per capita in 39 countries. For the sustainable future of our planet, it is essential to find substitutes for virgin materials to harvest for producing binder and aggregates necessitated by the construction industry. Meanwhile, a lot of EOL materials are disposed of in open fields as landfill. One example of this kind of waste is construction and demolition debris with enormous potential for recycling as a profitable recycled concrete aggregate (RA) [2,3].

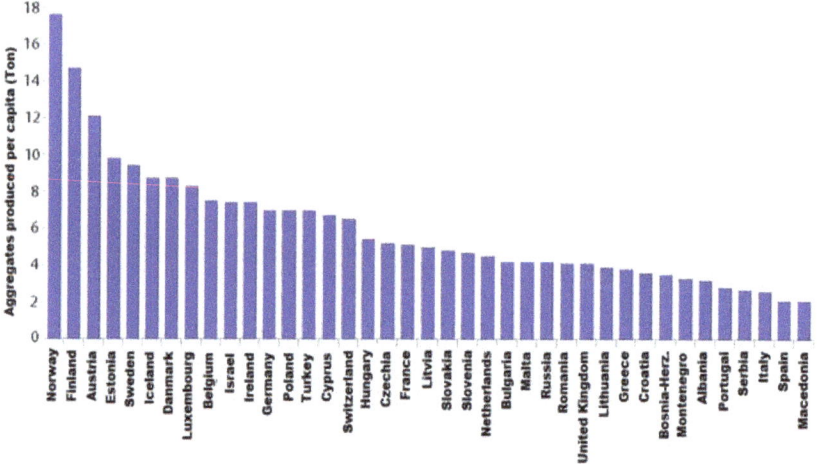

Figure 1. Aggregates production, in tonnes per capita, in 39 countries [3].

RA can be found in almost all developed and developing countries as a result of the demolition of older buildings and structures. Moreover, in war-afflicted regions in other parts of the world, a large number of buildings have been destroyed by bomb attacks. Such buildings have become impractical and are of no-value, but offer significant potential as material sources for reconstruction projects. Currently, a significant volume of RA is being disposed of as zero-value debris. Therefore, RA can play an important role in sustainable development. RA has recently become an important domain in construction as substitutes for natural aggregate raw materials. Several studies [2,4–9] claimed that the incorporation of RA in concrete production would alter the hardened properties. Several studies demonstrated that 100% substitution of aggregates with RA in concrete is unacceptable, due to a significant decrease in the hardened strength varying from 15 to 25% [2].

In addition, concrete has been produced with traditional supplementary cementitious materials (SCMs) that possess a high pozzolanic activity [10], such as fly ash (FA) [11–14], silica fume (SF) [15–17] and ground granulated blast slag (GGBS) [2,18–21] that yielded notable improvement in strength and durability. Numerous industrial solid by-products containing calcareous siliceous, and aluminous materials (fly ash, ultrafine fly ash, silica fume, etc.), along with some natural pozzolanic materials [22] (volcanic tuffs, diatomaceous earth, sugarcane bagasse ash, palm oil fuel ash, rice husk ash, mine tailings, etc.) can be used as SCMs, because they possess cementitious and/or pozzolanic properties [23].

The abundance of these classes of materials and their broad diversity in chemical and physical composition compel the development of a common strategy for their application in concrete production industry (see Figure 2) [24,25].

Figure 2. Most common industrial by-products used as substitutes [26].

Although traditional SCMs have been attractive due to their superior long-term durability [27], sustainable SCMs can also be developed to decrease the quantity of cement required for concrete production with lower ecological impact. One example is the incorporation of industrial EOL by-products into conventional cement. Concrete containing up to 30% of cement substituted by SCMs has been regarded as environmentally-friendly concrete [2,10,28]. In recent years, the use of SCMs and/or natural pozzolans has increased in the concrete industry, due to their superior long-term performance. Consequently, there is a strong interest in activating large amounts EOL materials to replace traditional SCMs as different sustainable resources that are otherwise deemed as of-zero-value [2,10].

2. Supplementary Cementitious Materials (SCMs)

SCMs play a pivotal role in concrete performance across many civilizations. They encompass a wide spectrum of alumino-silicious materials, including natural or processed pozzolans and industrial by-products, like GGBS, FA/UFFA and SF [29]. In spite of broad variations in properties across the various types of SCMs, they share in common the capacity to react chemically in concrete and form cementitious binders replacing those obtained by OPC hydration [30]. The key feature of SCMs is their pozzolanicity, i.e., their capability to react with calcium hydroxide (portlandite, CH) aqueous solutions to form calcium silicate hydrate (C–S–H) [31].

In the right proportion, SCMs can improve the fresh and hardened properties of concrete, especially the long-term durability. The use of SCMs in concrete composition is an ancient technique [32–34], as evidenced by the widespread use of natural pozzolans, like volcanic ash, in Greek and Roman civilizations. The testament to their efficacy is the persistence of an important number of constructions built using pozzolanic materials that are still standing today [30].

The most frequently employed SCMs in the cement industry are discussed below: fly ash (FA) and/or ultrafine fly ash (UFFA) [35], silica fume (SF) [36,37] and natural pozzolanic materials like rice husk ash (RHA) [38–40], sugarcane bagasse ash (SBA) [41–43], sewage sludge ash (SSA) [44,45], palm oil fuel ash (POFA) [46–49], mine tailings [50–53], marble dust [54–58], construction and demolition debris (CDD) [59–61].

2.1. Silica Fume (SF)

Silica fume (SF) used to be EOL products harvested from industrial processes. However, lately the demand for the silica fume for high-performance concrete has increased so strongly that there are now foundries dedicated to producing SF for ultra-high-performance concrete. SF may be in the form micro-silica, condensed silica fume or volatilized silica [26]. It is a fine powder produced in silicon foundries, where it is ultimately condensed from the vapour phase upon cooling [10]. As a consequence, SF is composed almost entirely of very small round particles of amorphous SiO_2, whose fineness contributes to a relatively high pozzolanic activity [62]. The size distribution of regular SF particles is on the order of 0.1 μm, where the majority of the particles—more than 95%—should be smaller than 1 μm. They are about 100 times smaller than cement particles, with a specific surface of SF is around 20,000 m^2/kg, therefore their specific surface is 10 to 20 times larger than that of other pozzolanic materials. Considering its characteristics, SF is a very reactive pozzolanic material [63]. The use of SF can notably increase mechanical properties of concrete due to effective filling and pozzolanic properties [64].

SF's most important effect on concrete is on the short- and long-term strength and long-term durability [26]. SF is claimed to be capable of increasing the bonding between cement paste and aggregates at the interfacial transition zone (ITZ). With a small particle size, SF not only improves the packing in the ITZ but also uses the localized CH in the ITZ during pozzolanic reaction, in combination with additional C–S–H. The net effect is creating better adhesion between the cement paste and aggregate. The high surface area of SF also provides a large reaction surface for enhancing the hydration process, resulting in improved density [26]. SF's influence on the improved density, or reduced macro porosity, also stems from its fluidized bed effect. The perfectly round fine particles of SF facilitate denser packing in concrete. Improved flow properties of concrete in the fresh state reduce bleeding and segregation, leading to superior performance of fresh and hardened concrete. Additional pozzolanic activity helps fill the excess water volume during further hydration [26,65].

A study by Seong Soo Kim et al. [66] demonstrated the improvement in compressive strength due to SF incorporation. Three types of coarse aggregates (basalt, quartzite and granite) were used to produce different concrete samples. The main binder used was OPC (according to ASTM C150), while SF was used to substitute 10 wt. % of cement. Figure 3, from this study, demonstrates the increase in compressive strength increased for all coarse aggregates types (i.e., basalt, quartzite and granite), when SF was added (see Figure 3).

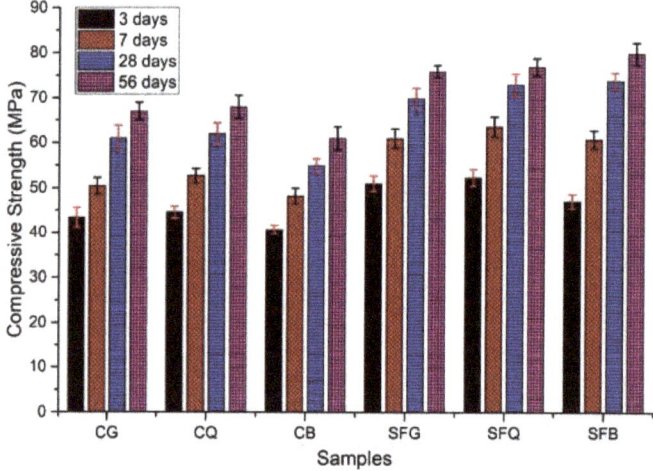

Figure 3. Compressive strength test results (the specimens made with only cement as binder were noted with the acronyms CG, CQ and CB for concrete with granite, with quartz and with basalt aggregates, respectively). Those containing silica fume were called SFG, SFQ and SFB, accordingly [66].

Tensile performance was investigated in another study, in which 3 different series of concrete samples (see Table 1) with 0 wt.%, 12 wt.% and 16 wt.% SF as a cement substitute were produced and tested [67]. The results indicated that SF added to concrete yielded superior mechanical properties under dynamic tensile loading. Moreover, the dynamic tensile strength of samples rose with the silica fume (SF) amount, as well as with strain rate. In addition, the strain rate sensitivity of concrete augmented with SF appeared to have increased in comparison to concrete without SF (see Figure 4) [67].

Table 1. The proportion of concrete constituents with different SF levels by weight [67].

Series	Mass of Concrete Constituents (kg/m^3)				
	Water	Cement	SF	Fine Aggregate	Aggregate
I	210.00	389.00	–	614	1141
II	210.00	340.80	48.20	614	1141
III	210.00	326.28	62.72	614	1141

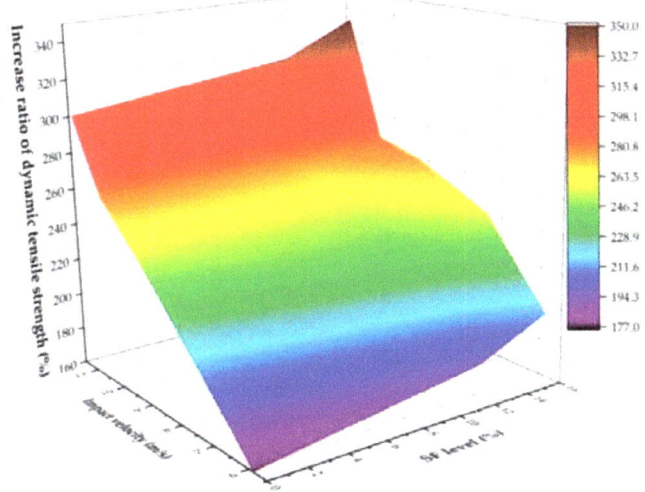

Figure 4. Dynamic tensile strength under different impact velocities and silica fume levels [67].

SF additions to concrete improve several other physical and chemical properties besides the mechanical strength: lower adverse environmental impact, decreased permeability, increased corrosion protection for steel bars and improved resistance against the sulphate and chemicals attacks [68,69].

2.2. Fly Ash (FA)

Fly ash (FA) is a by-product generated from coal-fired power plants. It is a fine, volatile powder emitted from chimneys alongside flue gases [10]. There are three distinct types of FA, including Class N, Class F and Class C, produced by burning black coal or brown coal, respectively. Class C and Class F are used in production of building materials, like lightweight aggregate, concrete, bricks, etc. (see Tables 2–4) [26,70]. There have been numerous studies, in which fly ash to be used as SCM in concrete have been analysed and classified according to types and characteristics. The American Society for Testing and Materials (ASTM) produced the initial report on standards for applying FA more than 40 years ago, ASTM C 618 [26].

Table 2. Type of fly ash as per American Society for Testing and Materials [26,70].

Type	CaO Amount, [%]	Properties
class C	≥10	cementitious, pozzolanic, hydraulic
class F	≤10	Pozzolanic
class N	is not commonly used in construction because of the existence of clay and shale	

Coal-fired power plants produce significant solid wastes that contained FA. About 700 million tons of FA every year throughout the world can be used in cement and/or concrete production, thanks to its pozzolanic activity. Replacing with FA in cement or concrete results in gross energy requirement (GER), carbon dioxide emissions (Global Warming Potential: GWP) and natural resource consumption [71,72]. Despite these advantages, the development of optimal composition of concrete with a high volume fraction of fly ash (HVFA) remains a challenge, due to the broad variation in chemical composition (lime, sulphates, alkalis and organics), fineness and mineralogy of fly-ash [71].

Table 3. Type of fly ash as per S 3812-1981.

Type	$SiO_2+Al_2O_3+Fe_2O_3$ Fraction, [%]
Grade I	≥70
Grade II	≥50

Table 4. Type of fly ash based on boiler operations.

Type	Short Name	Forming Temperature
Low temperature fly ash	LT	≤900 °C
High temperature fly ash	HT	≤1000 °C

In recent years, the integration of fly ash as a partial substitute of cement in concrete is a common procedure. The amount of FA to substitute the cement for regular applications is restricted to 15%–35% (depending on the type of FA) by mass of the total amount of cementitious material. However, even at these levels, the use of FA has overcome some critical issues focused on sustainable construction. FA's pozzolanic activity is related to the presence of amorphous SiO_2 and Al_2O_3 in its composition [73]. An active FA reacts with $Ca(OH)_2$ during the cement hydration process and forms supplementary C–S–H and calcium aluminate hydrate (C–A–H) [73,74]. These hydration products allow formation of a denser matrix, resulting in better strength and superior durability (see Figure 5) [73,74].

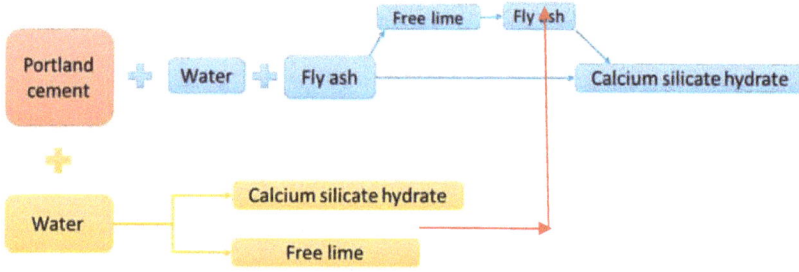

Figure 5. Reaction of fly ash (FA) in cement [70].

Ashish Kumer Saha [75] studied the use of Class F fly ash as a partial substitute of binder in concrete. A set of five samples was cast—a control sample without FA and 4 samples consist of 10, 20, 30 and 40% of FA as a substitute for cement. The water-to-cement ratio and the volume of superplasticizer were maintained constant for all 5 samples. The compressive strength (see Figure 6) of the FA mixtures showed lower early compressive strength than the ones of the control samples. The positive influence of FA additions could be observed at the longer hydration times, i.e., longer than 28 days of hydration. The small sized FA particles with high surface area and high percentage of amorphous silica phase provided the pozzolanic activity, leading to enhanced strength for longer periods of hydration. In addition, the spherical shape of FA particles improves the fresh state properties by increasing the workability of the concrete mix.

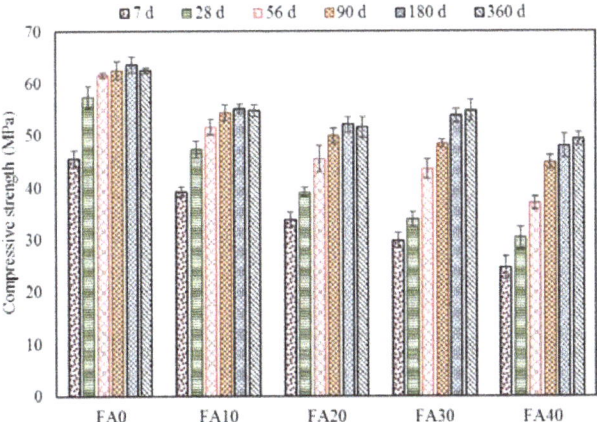

Figure 6. Compressive strength development [75].

Bingqian Yan et al. [76] studied the changes of mechanical properties of cement mortar with an admixture of FA and lime. As brine water is used in the preparation of filling slurry of Sanshandao Gold Mine, the chloride ions in the slurry have a great negative effect on the strength of the backfill. Figure 7 shows that the uniaxial compressive strength of the cement mortar sample with an amount of FA of 5% is 0.2 MPa better than the one without FA. When the amount continued to increase, the uniaxial compressive strength of cement mortar declined with increasing FA content. Analysis revealed that the activity of FA under the excitation of lime and brine water was restricted and large amounts of FA use presented a negative effect on the strength development and on cementing properties of binder [76].

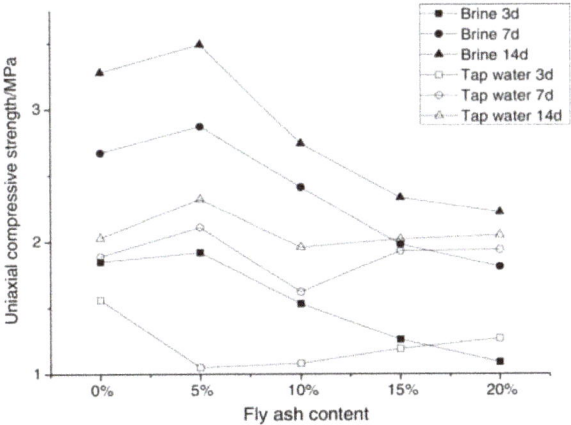

Figure 7. The compressive strength of the specimen and the quantity FA content [76].

FA is in the form of spherical particles composed of many phases—amorphous and also crystalline compounds, mostly silicon, calcium, aluminium and iron oxides. The versatility of fly ash for production of different types of cements is attributed to its physical features, chemical properties and phase composition [77–79].

2.3. Other Pozzolanic Ashes

Rice Husk Ash (RHA). An agricultural by-product that is suitable for cement replacement in rice-growing regions is Rice Husk Ash (RHA) [38]. The cementitious characteristics of RHA unappreciated before the 1970s [26]. RHA is the combustion residue from rice husks, which are the stiff outer layer that accumulates during de-husking of paddy rice. Every tonne of paddy rice can yield around 200 kg of husk, which produces about 40 kg of ash after combustion. It is known that rice plants consume H_4SiO_4 from underground water that exists in saturated zones beneath the earth surface. H_4SiO_4 at this point is polymerized and leads to development of amorphous silica in the husks [80]. During combustion of the organic compounds, CO_2 is produced along with the silica remaining in the ash leftovers. The researchers have demonstrated that the principal chemical composition of rice husk ash consists of biomass-driven silicon dioxide (SiO_2). Table 5 summarizes its typical chemical composition and some of its properties. As a result that the nature of silica in rice husk ash is sensitive to processing conditions (see Table 5), the ash obtained through open-field burning or uncontrolled combustion in furnaces generally includes a high percentage of crystalline silica minerals, like tridymite or cristobalite, with inferior reactivity. The highest amount of amorphous silica is obtained when RHA is burnt at temperatures ranging from 500 to 700 °C [81]. The superior reactivity of RHA is due to its large amount of amorphous silica, which has high surface area due to the porous architecture of the host material. RHA can be used as a substitute in Portland cement (acceptable up to 15%), thanks to its pozzolanic activity. Fine RHA can increase the compressive strength of cement paste and can lead to preparation of mortars with low porosity [26,39].

Table 5. Typical chemical and physical properties of Rice Husk Ash (RHA) [82].

Chemical Composition *, [%]					
SiO_2	Al_2O_3	Fe_2O_3	CaO	MgO	K_2O
93.4	0.05	0.06	0.31	0.35	1.4
Physical Properties					
Fineness—median particle size (μm)			8.6		
Specific gravity			2.05		
Pozzolanic activity index (%)			99		
Water absorption (%)			104		

* Minor Constituents Not Given.

As a cement substitute, the application of RHA in concrete production has advantages and disadvantages [39]. Improved compressive strength of concrete is one of the essential advantages of using RHA as substitute. Recent studies have highlighted important benefits of replacing cement with RHA in small percentages [26]. In the context of durability, the use of RHA as a substitute in concrete production can lead to notable improved water permeability resistance, Cl^- penetration and sulphate deterioration [83].

Weiting Xu et al. [83] compared the pozzolanic impact of SF and ground RHA as SCMs on the properties of composite cement pastes and concretes. The authors evaluated mechanical property, workability, durability and microstructure. In the composite cement pastes, the binder (OPC) was substituted with SF and finely ground RHA (named FRHA) at 5%, 10%, 15%, 20%, 25% and 30%, respectively by mass. They reported that the optimal substitution percentage of SF and RHA were 10%

by weight of cement in pastes and concretes. Compressive strength for this composition was evaluated (see Figure 8) [83].

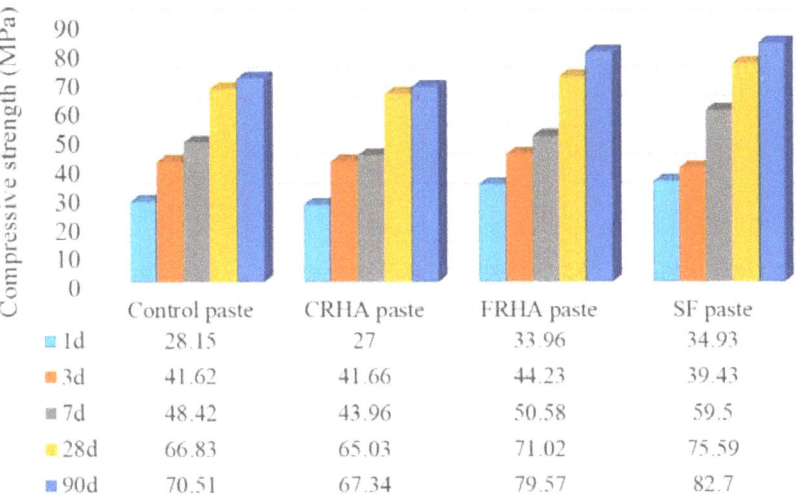

* Note: CRHA: RHA with 5 min grinding; and FRHA: RHA with 30 min grinding.

Figure 8. Compressive strength of studied pastes [83].

It can be observed that the sample with coarse rice husk ash (CRHA) had the lowest compressive strength at all curing ages, which could be because of the bigger particle size and small surface area of coarse RHA particles. Tests results showed that addition of FRHA to the paste led to an increased compressive strength compared to the control concrete, as a result of the growing specific surface area and pozzolanic activity of RHA. In addition, the morphological dissimilarity may also be implicated by the differences in compressive strength between the concrete specimens (see Figures 9 and 10) [83].

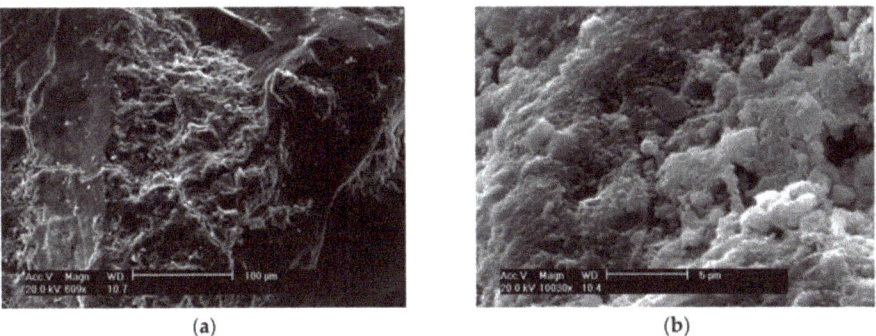

Figure 9. SEM images of concrete containing SF [83]. (**a**) Global Image—100 µm (**b**) Detailed image 5 µm.

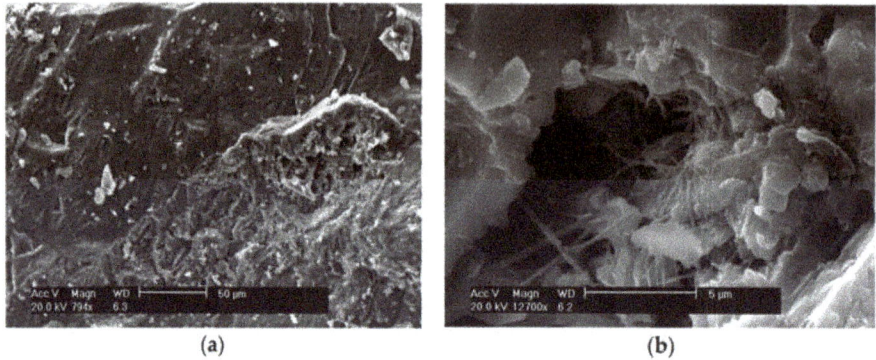

Figure 10. SEM images of concrete containing SF [83]. (**a**) Global Image—100 μm (**b**) Detailed image 5 μm. In view of the empirical results [83], it appeared that FRHA exhibits similar pozzolanic and rheological activity to SF and can lead to notable improvement in the properties of a cementitious system.

Sugarcane bagasse ash (SBA) is a by-product of producing juice from sugar cane by crushing the stalks of the plants (see Table 6). The addition of SBA in concrete production can decrease the hydration temperature up to 33%, when 30% of OPC is substituted by SBA [42]. Furthermore, water permeability considerably decreases when compared to control concrete samples. With the aim of superior compressive strength, OPC was substituted in the range from 15% to 30%. It was also claimed that SBA aids the reduction of ASR expansion in concrete, by binding alkalis [26]. Table 6 lists the properties of SBA.

Table 6. Usual chemical and physical properties of sugarcane bagasse ash (SBA) [82].

Chemical Composition *, [%]					
SiO_2	Al_2O_3	Fe_2O_3	CaO	MgO	K_2O
65.3	6.9	3.7	4.0	1.1	2.0
Physical Properties					
Fineness—median particle size (μm)			5.1		
Specific gravity			1.8		
Blaine fineness (m^2/kg)			900		

* Minor constituents not given.

Use of SBA as limited cement substitute reduces hydration heat, compared to that of the control reference. The decrease of semi-adiabatic temperature rise (°C) is proportional to the percentage of SBA replaced (see Figure 11). Therefore, SBA may be used to control the temperature in mass concrete pouring [82].

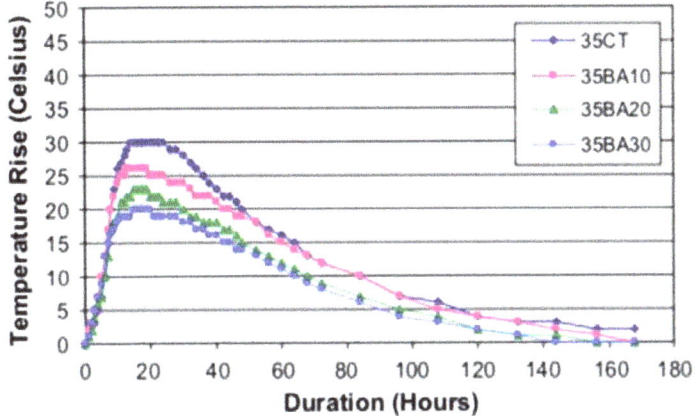

* a number at the end indicates the percentage of sugarcane bagasse ash

Figure 11. Semi-adiabatic temperature rise (°C) in concrete containing SBA as substitute [82].

SBA incorporation improved concrete durability. A composite concrete with additions of different amounts of SBA was studied in the context of chloride attack (see Figure 12), as well as gas and water permeability [41,82].

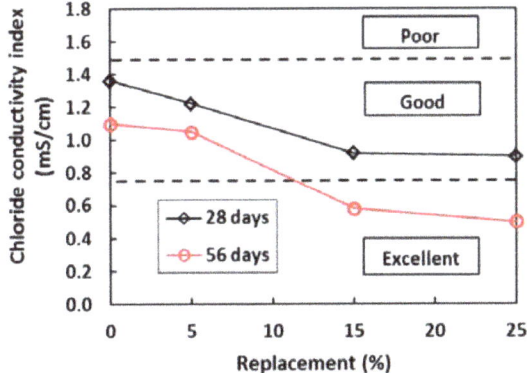

Figure 12. Effect of SBA on the chloride conductivity index of concrete [82].

Wastes of different sources have been investigated for their possibility in re-use, to reduce their environmental impact, in landfill volume and decomposition by-products [84]. Sewage sludge ash (SSA) is an urban waste that may be used as fertilizer, as well as a cement substitute [26,85–87]. SSA was not only considered as SCM in blended cements but also in a large scale of building materials like pave-stones, tiles, bricks, light aggregates production. The impact of SSA in mortar was a decrease in the compressive strength, when SSA was applied as partial cement substitute. Therefore, use of SSA as an SCM was shown to be limited, in the construction domain. The cement community does not include SSA in the group of pozzolanic materials [88].

Palm oil fuel ash (POFA) is an important cash-crop in tropical countries, especially in Malaysia and Indonesia [82]. For every 100 t of fresh fruit bunches handled, there will be about 20 t of nut shells, 7 t of fibres and 25 t of empty bunches released from the mills. POFA can be used in concrete either as aggregates, SCM or as filler material [89,90]. Comparable to RHA and SBA, the amorphous SiO_2 (around 76%) content of POFA offers relatively high pozzolanic activity, when used as binder in concrete production. Even though a few performance parameters of concrete (especially setting time and strength) are negatively influenced by POFA, several studies claimed that palm oil fuel ash may be appropriate in different applications [91,92]. It may be an important resource in developing countries, although more studies are certainly needed, to support the use of POFA in structural applications [82,93,94].

Mine tailings. The amount of mine tailings has grown excessively with an ever increasing demand for metal and mineral resources [95]. Mining wastes are produced during mineral extraction by the mining industry and is at present one of the largest waste flows worldwide [25,95,96]. Mine tailings are finely ground in the time of mineral processing and separation of minerals of interest [96]. A significant part of milling processes and separation procedure uses water as the transport medium. Therefore, tailings obtained by mining enterprises usually consist of small particle slurries with high water content, flow ability and poor mechanical stability [95]. Mine tailings are deposited in dammed ponds along with industrial wastewater or as thickened pastes in piles close to the extraction sites [96]. At present, their principal use is as backfill in mined-out underground areas or as deposits in tailing ponds. As such, they pose potential long-term risks as environmental pollution. However, use of tailings is not only relevant to environmental conservation, but can also benefit the mining industry. These solid wastes contain compounds with potential pozzolanic properties and can decrease the amount of cement used to produce concrete, reducing simultaneously the ecological impact of the cement and mining industries. An additional benefit of mine tailings is that they are already finely ground. Most of the other SCMs require mechanical grinding, as a pre-treatment for use, to improve their reactivity. However, mine tailings already come with favourable physical and chemical properties as particle dimensions, crystal structures and even surface properties [96].

Therefore, depositories of tailings have lately gained global attention [95]. However, despite the potential of mine tailings as substitutes in the production of cementitious materials, there is a striking lack of studies regarding this topic in the cement and concrete literature.

Marble dust. Marble is a finely crystallized metamorphic rock originating from the low-intensity metamorphism of calcareous and dolomitic rocks. Calcium carbonate ($CaCO_3$) can form up to 99% of the total amount of this carbonated rock. Additional phases may also include SiO_2, MgO, $Fe2O_3$, Al_2O_3 and Na_2O and, in minor ratio, MnO, K_2O, P_2O_5, F, Cu, S, Pb and Zn [97]. For a long time, marble has been a significant building material. For instance, annual production of marble in Turkey accounts for 7 Mt and represents 40% of the global storage; Turkey has an annual total production of block marble of ca. 1,500,000 m^3, generating approximately 375,000 m^3 of marble powder [98]. Throughout the shaping, sawing and polishing operations, around 20%–25% of processed marble is converted into powder or lumps. As a result, dumps of marble dust have become an important environmental issue worldwide [99]. Marble powder (MP) has successfully been demonstrated as a viable SCM in self-compacting concrete (SCC). The research proved that marble powder used as mineral substitute of cement can enhance some properties of fresh concrete and/or hardened concrete [100]. In the cement-related literature, there are just a few research studies related to the application of marble powder in concrete or mortar production. Thus, more detailed studies are needed in order to define the properties of concretes or mortars with marble powder. The use of marble powder in ternary cementitious blends demands further caution to remove or reduce its adverse effects on the fresh properties of self-compacting concrete and/or mortar [98].

Construction and demolition debris (CDD) constitute one of the massive flows of solid waste generated from municipal and commercial activities of modern society [101]. Usually, CDD are in the shape of brick bats, mortars, aggregates, concrete, glass, ceramic tiles, metals and even plastics. They must be mechanically sorted according to size and quality level. They are then crushed down to desired size [102]. It is essential to study the life cycle of construction materials to develop a global understanding of sustainable building construction and the feasible use of CDD as SCMs for OPC substitution. The life cycle of some construction materials, such as concrete, has been analysed to evaluate their environmental consequences. While substantial effort has been applied to LCA-based sustainability assessment of construction materials and buildings, the specialty literature needs more detailed studies regarding the LCA of recycled construction materials, ones that take into account both process and supply chain-related outcomes as a whole [103]. The sheer mass and heterogeneity of CDD materials and absence of data classification across non-standardized tracking systems have led to new challenges [101]. In addition, the lack of knowledge about the possible savings and implications correlated with recycling of this kind of construction materials from the life cycle outlook still limits their use [103].

3. Life Cycle Assessment (LCA)

A life-cycle assessment (LCA) is a method for assessing all the potential environmental impacts of a product, process, or activity over its entire life cycle [104]. Several LCA studies have focused on the sustainability of concrete [105,106]. It is important to integrate recycled EOL products at the beginning of concrete's life cycle, and re-valorise it at the end of concrete's life cycle in another production process or even maybe, for the concrete production industry (see Figure 13) [107–110].

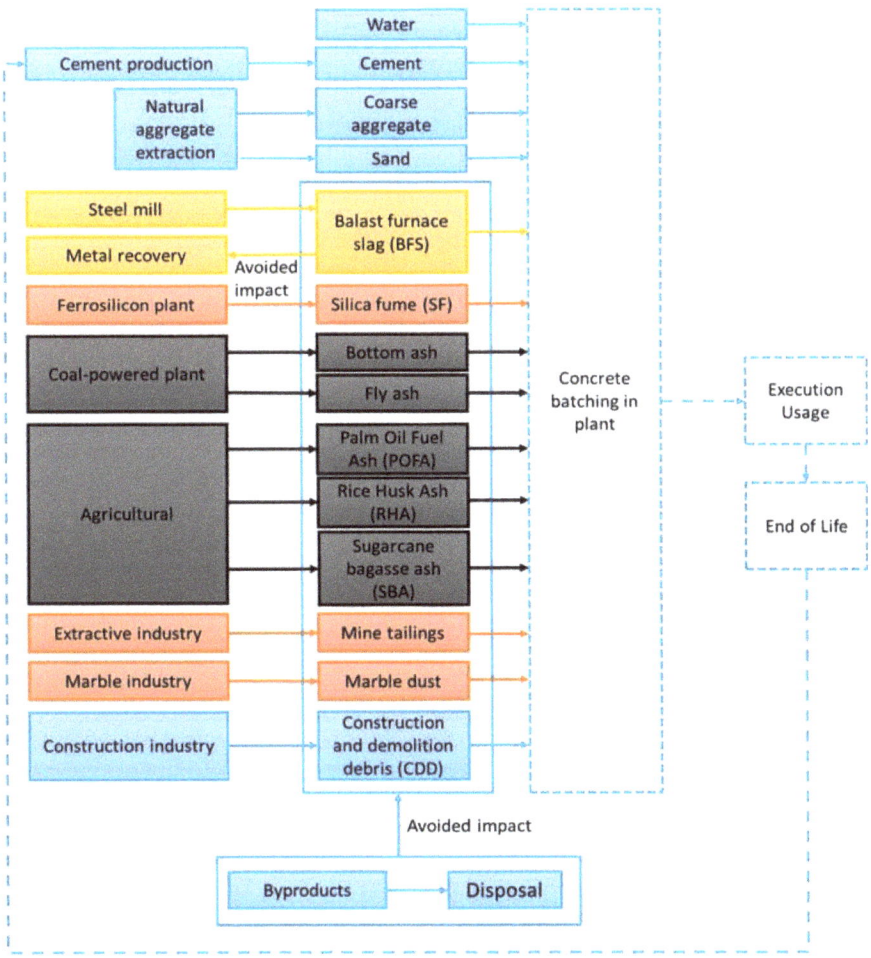

Figure 13. Cradle-to-gate life cycle assessment (LCA): studied system boundaries of the concretes (plain line—included processes; dashed line—non-included processes) [111].

Sustainability through re-use of accessible economic and social resources is a way to attain equilibrium with the environment, while ensuring long-term development and endurance [26]. A beneficial difference for the environment could result from substituting cement and sand with by-products or EOL products from intersectoral industrial activities. By this way, we may be able to reduce the adverse environmental impacts stemming from cement (74%–93%) and sand (0.3%–2%) consumption in the total LCA of EOL material-based concretes. The minimum contribution of sand to the entire environmental assessment of concrete makes this issue important to concrete design [111–113].

With regard to the LCA of concrete, 4 aspects should be considered: (1) design, (2) production/execution, (3) usage and (4) end-of-life disposal. Structural concrete has life expectancy differentiated by application, such as in pillars, beams and walls. While the durability over time for foundation or load-bearing structural elements is 50–300 years, the corresponding lifetime for cover walls is only 20–50 years [111]. Therefore, sufficient data is not presently available for the EOL stage of structural concrete and its disposal conditions [111,114–116].

The necessity for further LCA studies on the treatment and re-use of construction waste is clear. Instead of being released into the environment, it can be re-valorised in the life cycle of new designs of concrete. Use of waste like fly ash, blast furnace slag and other mineral admixtures as a binder for the production of concrete is becoming common in the construction industry. Replacing the principal factor responsible for the negative environmental effects of concrete is the key to generating an ecologically beneficial life cycle for concrete [107,117,118].

With regards to economic impact, regenerating alternative EOL materials for binder components in concrete would decrease the cost of construction without sacrificing performance. Other costs can also be considered, such as those concerning the source and transport of the alternative SCM materials, controlled combustion process and also savings as a result of diversion, such as disposal management. Consequently, the environmental advantages will reduce the enormous demand for Portland cement per unit volume of concrete, in addition to a concurrent and meaningful reduction in Greenhouse Gas (GHG) emissions [108,119,120].

4. Conclusions

Concrete represents one of the most widely used construction materials worldwide by volume. Portland cement production is highly energy intensive, and emits significant amounts of CO_2 through the calcination process, which contributes substantial adverse impact on global warming. Efforts are needed to design and develop more ecologically friendly concrete with improved performance in strength and durability.

SCMs are frequently applied in concrete mixtures as a substitute for clinker in cement or cement in concrete. This approach yields concrete with reduced cost, decreased environmental impact, higher long-term strength and better long-term durability. Presently the two most common SCMs are silica fume and fly ash. As a by-product of the silicon and ferrosilicon alloy fabrication, SF contains more than 90% SiO_2 and is present as spheres with average diameters about 100 times smaller than cement particles. The large specific surface of SF, ca. 20,000 m^2/kg, is 10 to 20 times greater than that of other pozzolanic materials, imbuing it with high pozzolanic activity and fluidization properties.

FA has also been widely used in concrete mix design, due to its growing availability and substantial environmental complications appeared by release of fly ash. Physical characteristics of FA can differ on the nature of coal, rank, mineral matter chemistry and mineralogy, furnace design, furnace operation and method of particulate control, while chemical characteristics are relatively insensitive those factors. FA has been used in the manufacturing of bricks, concrete and cement-composites like columns, beams, slabs, columns, sheets, pipes, wall panels, etc. Typically, about 25% of FA is used as a substitute for cement to achieve effective final products. FA increases durability, workability, density and workability of concrete, while simultaneously decreasing water demand, porosity and permeability of concrete.

In the present, the use of industrial EOL materials in concrete has been demonstrated. However, it is clear that more research is needed to assess the feasibility of long-term performance, develop more ecologically sound production, in addition to quality assessment of these materials. When ashes of high quality can be regularly obtained with reduced financial and most importantly environmental costs, their use in the engineering domain will become more widespread. Technical and economic performances of alternative SCM is evident, proving that along with the studying of the material's mechanical and physico-chemical properties, the review of its life cycle is as well important and will mention whether it will be environmentally feasible to apply the SCM admixture at the total scale of the life cycle of concrete.

Author Contributions: Conceptualization, A.I.N., B.S.V.; investigation, A.E.S.; data curation, B.S.V., A.E.S., A.I.N.; writing—review and editing, A.E.S., B.S.V., M.V., N.Š.R., S.S., C.O.-Y., M.A.G., Z.B.B., I.C.; All authors have read and agreed to the published version of the manuscript.

Funding: This research was possible with the support of European Commission, Horizon 2020, ERA-MIN2 Research & Innovation Programme on Raw Materials to Foster Circular Economy – "RECEMENT – Re-generating (raw) materials and-of-life products for re-use in Cement/Concrete" funded project.

Conflicts of Interest: The authors declare no conflict of interest.

References

1. Madani, H.; Norouzifar, M.N.; Rostami, J. The synergistic effect of pumice and silica fume on the durability and mechanical characteristics of eco-friendly concrete. *Constr. Build. Mater.* **2018**, *174*, 356–368. [CrossRef]
2. Alnahhal, M.F.; Alengaram, U.J.; Jumaat, M.Z.; Alqedra, M.A.; Mo, K.H.; Sumesh, M. Evaluation of Industrial By-Products as Sustainable Pozzolanic Materials in Recycled Aggregate Concrete. *Sustainability* **2017**, *9*, 767. [CrossRef]
3. Nili, M.; Sasanipour, H.; Aslani, F. The Effect of Fine and Coarse Recycled Aggregates on Fresh and Mechanical Properties of Self-Compacting Concrete. *Materials* **2019**, *12*, 1120. [CrossRef] [PubMed]
4. Sagoe-Crentsil, K.K.; Brown, T.; Taylor, A.H. Performance of concrete made with commercially produced coarse recycled concrete aggregate. *Cem. Concr. Res.* **2001**, *31*, 707–712. [CrossRef]
5. Limbachiya, M.C.; Leelawat, T.; Dhir, R.K. Use of recycled concrete aggregate in high-strength concrete. *Mater. Struct.* **2000**, *33*, 574. [CrossRef]
6. Gómez-Soberón, J.M.V. Porosity of recycled concrete with substitution of recycled concrete aggregate: An experimental study. *Cem. Concr. Res.* **2002**, *32*, 1301–1311. [CrossRef]
7. Berndt, M.L. Properties of sustainable concrete containing fly ash, slag and recycled concrete aggregate. *Constr. Build. Mater.* **2009**, *23*, 2606–2613. [CrossRef]
8. Yang, J.; Du, Q.; Bao, Y. Concrete with recycled concrete aggregate and crushed clay bricks. *Constr. Build. Mater.* **2011**, *25*, 1935–1945. [CrossRef]
9. Shi, C.; Li, Y.; Zhang, J.; Li, W.; Chong, L.; Xie, Z. Performance enhancement of recycled concrete aggregate—A review. *J. Clean. Prod.* **2016**, *112*, 466–472. [CrossRef]
10. Solak, A.M.; Tenza-Abril, A.J.; Saval, J.M.; García-Vera, V.E. Effects of Multiple Supplementary Cementitious Materials on Workability and Segregation Resistance of Lightweight Aggregate Concrete. *Sustainability* **2018**, *10*, 4304. [CrossRef]
11. Gomaa, E.; Gheni, A.A.; Kashosi, C.; ElGawady, M.A. Bond strength of eco-friendly class C fly ash-based thermally cured alkali-activated concrete to portland cement concrete. *J. Clean. Prod.* **2019**, *235*, 404–416. [CrossRef]
12. Yang, J.; Huang, J.; Su, Y.; He, X.; Tan, H.; Yang, W.; Strnadel, B. Eco-friendly treatment of low-calcium coal fly ash for high pozzolanic reactivity: A step towards waste utilization in sustainable building material. *J. Clean. Prod.* **2019**, *238*, 117962. [CrossRef]
13. Mironyuk, I.; Tatarchuk, T.; Paliychuk, N.; Heviuk, I.; Horpynko, A.; Yarema, O.; Mykytyn, I. Effect of surface-modified fly ash on compressive strength of cement mortar. *Mater. Today: Proc.* **2019**. [CrossRef]
14. Krishnaraj, L.; Ravichandran, P.T. Investigation on grinding impact of fly ash particles and its characterization analysis in cement mortar composites. *Ain Shams Eng. J.* **2019**, *10*, 267–274. [CrossRef]
15. Ávalos-Rendón, T.L.; Chelala, E.A.P.; Mendoza Escobedo, C.J.; Figueroa, I.A.; Lara, V.H.; Palacios-Romero, L.M. Synthesis of belite cements at low temperature from silica fume and natural commercial zeolite. *Mater. Sci. Eng. B* **2018**, *229*, 79–85. [CrossRef]

16. Sasanipour, H.; Aslani, F.; Taherinezhad, J. Effect of silica fume on durability of self-compacting concrete made with waste recycled concrete aggregates. *Constr. Build. Mater.* **2019**, *227*, 116598. [CrossRef]
17. Meddah, M.S.; Ismail, M.A.; El-Gamal, S.; Fitriani, H. Performances evaluation of binary concrete designed with silica fume and metakaolin. *Constr. Build. Mater.* **2018**, *166*, 400–412. [CrossRef]
18. Sekhar, C.D.; Nayak, S. Utilization of granulated blast furnace slag and cement in the manufacture of compressed stabilized earth blocks. *Constr. Build. Mater.* **2018**, *166*, 531–536. [CrossRef]
19. Jaya prithika, A.; Sekar, S.K. Mechanical and fracture characteristics of Eco-friendly concrete produced using coconut shell, ground granulated blast furnace slag and manufactured sand. *Constr. Build. Mater.* **2016**, *103*, 1–7. [CrossRef]
20. Du, S. Mechanical properties and reaction characteristics of asphalt emulsion mixture with activated ground granulated blast-furnace slag. *Constr. Build. Mater.* **2018**, *187*, 439–447. [CrossRef]
21. Hayles, M.; Sanchez, L.F.M.; Noël, M. Eco-efficient low cement recycled concrete aggregate mixtures for structural applications. *Constr. Build. Mater.* **2018**, *169*, 724–732. [CrossRef]
22. Valipour, M.; Pargar, F.; Shekarchi, M.; Khani, S. Comparing a natural pozzolan, zeolite, to metakaolin and silica fume in terms of their effect on the durability characteristics of concrete: A laboratory study. *Constr. Build. Mater.* **2013**, *41*, 879–888. [CrossRef]
23. Sánchez de Rojas Gómez, M.I.; Frías Rojas, M. 4 - Natural pozzolans in eco-efficient concrete. In *Eco-Efficient Concrete*; Pacheco-Torgal, F., Jalali, S., Labrincha, J., John, V.M., Eds.; Woodhead Publishing: Shaston, UK, 2013; pp. 83–104. [CrossRef]
24. Papadakis, V.G.; Tsimas, S. Supplementary cementing materials in concrete: Part I: Efficiency and design. *Cem. Concr. Res.* **2002**, *32*, 1525–1532. [CrossRef]
25. Franco de Carvalho, J.M.; Melo, T.V.D.; Fontes, W.C.; Batista, J.O.D.S.; Brigolini, G.J.; Peixoto, R.A.F. More eco-efficient concrete: An approach on optimization in the production and use of waste-based supplementary cementing materials. *Constr. Build. Mater.* **2019**, *206*, 397–409. [CrossRef]
26. Al-Mansour, A.; Chow, C.L.; Feo, L.; Penna, R.; Lau, D. Green Concrete: By-Products Utilization and Advanced Approaches. *Sustainability* **2019**, *11*, 5145. [CrossRef]
27. Rahla, K.M.; Mateus, R.; Bragança, L. Comparative sustainability assessment of binary blended concretes using Supplementary Cementitious Materials (SCMs) and Ordinary Portland Cement (OPC). *J. Clean. Prod.* **2019**, *220*, 445–459. [CrossRef]
28. Samad, S.; Shah, A. Role of binary cement including Supplementary Cementitious Material (SCM), in production of environmentally sustainable concrete: A critical review. *Int. J. Sustain. Built Environ.* **2017**, *6*, 663–674. [CrossRef]
29. Seifi, S.; Sebaibi, N.; Levacher, D.; Boutouil, M. Mechanical performance of a dry mortar without cement, based on paper fly ash and blast furnace slag. *J. Build. Eng.* **2019**, *22*, 113–121. [CrossRef]
30. Thomas, M. *Supplementary Cementing Materials in Concrete*; CRC Press: Boca Raton, FL, USA, 2013.
31. Juenger, M.C.G.; Siddique, R. Recent advances in understanding the role of supplementary cementitious materials in concrete. *Cem. Concr. Res.* **2015**, *78*, 71–80. [CrossRef]
32. Thomas, M.; Hooton, R.; Rogers, C. Prevention of Damage Due to Alkali-Aggregate Reaction (AAR) in Concrete Construction—Canadian Approach. *Cem. Concr. Aggreg.* **1997**, *19*, 26–30. [CrossRef]
33. Duchesne, J.; Bérubé, M.A. Effect of supplementary cementing materials on the composition of cement hydration products. *Adv. Cem. Based Mater.* **1995**, *2*, 43–52. [CrossRef]
34. Eymael, M.M.T.; Cornelissen, H.A.W. Processed pulverized fuel ash for high-performance concrete. *Waste Manag.* **1996**, *16*, 237–242. [CrossRef]
35. Reddy, M.V.S.; Ashalatha, K.; Surendra, K. Studies on eco-friendly concrete by partial replacement of cement with Alccofine and fine Fly Ash. *Micron* **2006**, *10*, d90.
36. Garg, C.; Jain, A. Green concrete: Efficient & eco-friendly construction materials. *Int. J. Res. Eng. Technol.* **2014**, *2*, 259–264.
37. Kavitha, O.; Shanthi, V.; Arulraj, G.P.; Sivakumar, V. Microstructural studies on eco-friendly and durable Self-compacting concrete blended with metakaolin. *Appl. Clay Sci.* **2016**, *124*, 143–149. [CrossRef]
38. Sathawane, S.H.; Vairagade, V.S.; Kene, K.S. Combine effect of rice husk ash and fly ash on concrete by 30% cement replacement. *Procedia Eng.* **2013**, *51*, 35–44. [CrossRef]
39. Zareei, S.A.; Ameri, F.; Bahrami, N.; Dorostkar, F. Experimental evaluation of eco-friendly light weight concrete with optimal level of rice husk ash replacement. *Civ. Eng. J.* **2017**, *3*, 972–986. [CrossRef]

40. Hwang, C.-L.; Huynh, T.-P. Investigation into the use of unground rice husk ash to produce eco-friendly construction bricks. *Constr. Build. Mater.* **2015**, *93*, 335–341. [CrossRef]
41. Zareei, S.A.; Ameri, F.; Bahrami, N. Microstructure, strength, and durability of eco-friendly concretes containing sugarcane bagasse ash. *Constr. Build. Mater.* **2018**, *184*, 258–268. [CrossRef]
42. Almeida, F.C.R.; Sales, A.; Moretti, J.P.; Mendes, P.C.D. Use of sugarcane bagasse ash sand (SBAS) as corrosion retardant for reinforced Portland slag cement concrete. *Constr. Build. Mater.* **2019**, *226*, 72–82. [CrossRef]
43. Embong, R.; Shafiq, N.; Kusbiantoro, A.; Nuruddin, M.F. Effectiveness of low-concentration acid and solar drying as pre-treatment features for producing pozzolanic sugarcane bagasse ash. *J. Clean. Prod.* **2016**, *112*, 953–962. [CrossRef]
44. Zhou, Y.-F.; Li, J.-S.; Lu, J.-X.; Cheeseman, C.; Poon, C.S. Recycling incinerated sewage sludge ash (ISSA) as a cementitious binder by lime activation. *J. Clean. Prod.* **2019**, *244*, 118856. [CrossRef]
45. Payá, J.; Monzó, J.; Borrachero, M.V.; Soriano, L. 5 - Sewage sludge ash. In *New Trends in Eco-Efficient and Recycled Concrete*; de Brito, J., Agrela, F., Eds.; Woodhead Publishing: Shaston, UK, 2019; pp. 121–152. [CrossRef]
46. Wi, K.; Lee, H.-S.; Lim, S.; Song, H.; Hussin, M.W.; Ismail, M.A. Use of an agricultural by-product, nano sized Palm Oil Fuel Ash as a supplementary cementitious material. *Constr. Build. Mater.* **2018**, *183*, 139–149. [CrossRef]
47. Alsubari, B.; Shafigh, P.; Ibrahim, Z.; Alnahhal, M.F.; Jumaat, M.Z. Properties of eco-friendly self-compacting concrete containing modified treated palm oil fuel ash. *Constr. Build. Mater.* **2018**, *158*, 742–754. [CrossRef]
48. Khankhaje, E.; Rafieizonooz, M.; Salim, M.R.; Khan, R.; Mirza, J.; Siong, H.C.; Salmiati. Sustainable clean pervious concrete pavement production incorporating palm oil fuel ash as cement replacement. *J. Clean. Prod.* **2018**, *172*, 1476–1485. [CrossRef]
49. Alsubari, B.; Shafigh, P.; Jumaat, M.Z. Utilization of high-volume treated palm oil fuel ash to produce sustainable self-compacting concrete. *J. Clean. Prod.* **2016**, *137*, 982–996. [CrossRef]
50. Pyo, S.; Tafesse, M.; Kim, B.-J.; Kim, H.-K. Effects of quartz-based mine tailings on characteristics and leaching behavior of ultra-high performance concrete. *Constr. Build. Mater.* **2018**, *166*, 110–117. [CrossRef]
51. Kiventerä, J.; Piekkari, K.; Isteri, V.; Ohenoja, K.; Tanskanen, P.; Illikainen, M. Solidification/stabilization of gold mine tailings using calcium sulfoaluminate-belite cement. *J. Clean. Prod.* **2019**, *239*, 118008. [CrossRef]
52. Lyu, X.; Yao, G.; Wang, Z.; Wang, Q.; Li, L. Hydration kinetics and properties of cement blended with mechanically activated gold mine tailings. *Thermochim. Acta* **2020**, *683*, 178457. [CrossRef]
53. Ince, C. Reusing gold-mine tailings in cement mortars: Mechanical properties and socio-economic developments for the Lefke-Xeros area of Cyprus. *J. Clean. Prod.* **2019**, *238*, 117871. [CrossRef]
54. Taji, I.; Ghorbani, S.; de Brito, J.; Tam, V.W.Y.; Sharifi, S.; Davoodi, A.; Tavakkolizadeh, M. Application of statistical analysis to evaluate the corrosion resistance of steel rebars embedded in concrete with marble and granite waste dust. *J. Clean. Prod.* **2019**, *210*, 837–846. [CrossRef]
55. Ghorbani, S.; Taji, I.; Tavakkolizadeh, M.; Davodi, A.; de Brito, J. Improving corrosion resistance of steel rebars in concrete with marble and granite waste dust as partial cement replacement. *Constr. Build. Mater.* **2018**, *185*, 110–119. [CrossRef]
56. Aliabdo, A.A.; Abd Elmoaty, A.E.M.; Auda, E.M. Re-use of waste marble dust in the production of cement and concrete. *Constr. Build. Mater.* **2014**, *50*, 28–41. [CrossRef]
57. Topçu, İ.B.; Bilir, T.; Uygunoğlu, T. Effect of waste marble dust content as filler on properties of self-compacting concrete. *Constr. Build. Mater.* **2009**, *23*, 1947–1953. [CrossRef]
58. Li, L.G.; Huang, Z.H.; Tan, Y.P.; Kwan, A.K.H.; Chen, H.Y. Recycling of marble dust as paste replacement for improving strength, microstructure and eco-friendliness of mortar. *J. Clean. Prod.* **2019**, *210*, 55–65. [CrossRef]
59. Cantero, B.; Sáez del Bosque, I.F.; Matías, A.; Sánchez de Rojas, M.I.; Medina, C. Inclusion of construction and demolition waste as a coarse aggregate and a cement addition in structural concrete design. *Arch. Civ. Mech. Eng.* **2019**, *19*, 1338–1352. [CrossRef]
60. Akhtar, A.; Sarmah, A.K. Construction and demolition waste generation and properties of recycled aggregate concrete: A global perspective. *J. Clean. Prod.* **2018**, *186*, 262–281. [CrossRef]
61. Silva, R.V.; de Brito, J.; Dhir, R.K. Use of recycled aggregates arising from construction and demolition waste in new construction applications. *J. Clean. Prod.* **2019**, *236*, 117629. [CrossRef]

62. Lothenbach, B.; Scrivener, K.; Hooton, R.D. Supplementary cementitious materials. *Cem. Concr. Res.* **2011**, *41*, 1244–1256. [CrossRef]
63. Pedro, D.; de Brito, J.; Evangelista, L. Evaluation of high-performance concrete with recycled aggregates: Use of densified silica fume as cement replacement. *Constr. Build. Mater.* **2017**, *147*, 803–814. [CrossRef]
64. Wu, Z.; Shi, C.; Khayat, K.H. Influence of silica fume content on microstructure development and bond to steel fiber in ultra-high strength cement-based materials (UHSC). *Cem. Concr. Compos.* **2016**, *71*, 97–109. [CrossRef]
65. Uzal, B.; Turanlı, L.; Yücel, H.; Göncüoğlu, M.C.; Çulfaz, A. Pozzolanic activity of clinoptilolite: A comparative study with silica fume, fly ash and a non-zeolitic natural pozzolan. *Cem. Concr. Res.* **2010**, *40*, 398–404. [CrossRef]
66. Kim, S.S.; Qudoos, A.; Jakhrani, S.H.; Lee, J.B.; Kim, H.G. Influence of Coarse Aggregates and Silica Fume on the Mechanical Properties, Durability, and Microstructure of Concrete. *Materials* **2019**, *12*, 3324. [CrossRef]
67. Zhao, S.; Zhang, Q. Effect of Silica Fume in Concrete on Mechanical Properties and Dynamic Behaviors under Impact Loading. *Materials* **2019**, *12*, 3263. [CrossRef] [PubMed]
68. Mansoor, J.; Shah, S.A.R.; Khan, M.M.; Sadiq, A.N.; Anwar, M.K.; Siddiq, M.U.; Ahmad, H. Analysis of Mechanical Properties of Self Compacted Concrete by Partial Replacement of Cement with Industrial Wastes under Elevated Temperature. *Appl. Sci.* **2018**, *8*, 364. [CrossRef]
69. Kwon, Y.-H.; Kang, S.-H.; Hong, S.-G.; Moon, J. Intensified Pozzolanic Reaction on Kaolinite Clay-Based Mortar. *Appl. Sci.* **2017**, *7*, 522. [CrossRef]
70. Gamage, N.; Liyanage, K.; Fragomeni, S. Overview of Different Type of Fly Ash and Their Use as a Building and Construction Material. In Proceedings of the Conference: International Conference of Structural Engineering, Construction and Management, Kandy, Sri Lanka, 16–18 December 2011.
71. Coppola, L.; Coffetti, D.; Crotti, E. Plain and Ultrafine Fly Ashes Mortars for Environmentally Friendly Construction Materials. *Sustainability* **2018**, *10*, 874. [CrossRef]
72. Jones, M.R.; McCarthy, A.; Booth, A.P.P.G. Characteristics of the ultrafine component of fly ash. *Fuel* **2006**, *85*, 2250–2259. [CrossRef]
73. Supit, S.W.M.; Shaikh, F.U.A.; Sarker, P.K. Effect of ultrafine fly ash on mechanical properties of high volume fly ash mortar. *Constr. Build. Mater.* **2014**, *51*, 278–286. [CrossRef]
74. Shaikh, F.U.A.; Supit, S.W.M. Compressive strength and durability properties of high volume fly ash (HVFA) concretes containing ultrafine fly ash (UFFA). *Constr. Build. Mater.* **2015**, *82*, 192–205. [CrossRef]
75. Saha, A.K. Effect of class F fly ash on the durability properties of concrete. *Sustain. Environ. Res.* **2018**, *28*, 25–31. [CrossRef]
76. Yan, B.; Kouame, K.J.A.; Lv, W.; Yang, P.; Cai, M. Modification and in-place mechanical characteristics research on cement mortar with fly ash and lime compound admixture in high chlorine environment. *J. Mater. Res. Technol.* **2019**, *8*, 1451–1460. [CrossRef]
77. Rakhimova, N.R.; Rakhimov, R.Z. Toward clean cement technologies: A review on alkali-activated fly-ash cements incorporated with supplementary materials. *J. Non Cryst. Sol.* **2019**, *509*, 31–41. [CrossRef]
78. El-Gamal, S.; Selim, F. Utilization of some industrial wastes for eco-friendly cement production. *Sustain. Mater. Technol.* **2017**, *12*, 9–17. [CrossRef]
79. Palankar, N.; Shankar, A.R.; Mithun, B. Studies on eco-friendly concrete incorporating industrial waste as aggregates. *Int. J. Sustain. Built Environ.* **2015**, *4*, 378–390. [CrossRef]
80. Talsania, S.; Pitroda, J.; Vyas, C.M. Effect of rice husk ash on properties of pervious concrete. *Int. J. Adv. Eng. Res. Studies/IV/II/Jan.-March* **2015**, *296*, 299.
81. Krishna, R.; Admixtures, P.K.C. Rice husk ash–an ideal admixture for concrete in aggressive environments. *Magnesium (as MgO)* **2012**, *1*, 0–35.
82. Chandra Paul, S.; Mbewe, P.B.K.; Kong, S.Y.; Šavija, B. Agricultural Solid Waste as Source of Supplementary Cementitious Materials in Developing Countries. *Materials* **2019**, *12*, 1112. [CrossRef] [PubMed]
83. Xu, W.; Lo, T.Y.; Wang, W.; Ouyang, D.; Wang, P.; Xing, F. Pozzolanic Reactivity of Silica Fume and Ground Rice Husk Ash as Reactive Silica in a Cementitious System: A Comparative Study. *Materials* **2016**, *9*, 146. [CrossRef] [PubMed]
84. Rorat, A.; Courtois, P.; Vandenbulcke, F.; Lemiere, S. 8 - Sanitary and environmental aspects of sewage sludge management. In *Industrial and Municipal Sludge*; Prasad, M.N.V., de Campos Favas, P.J., Vithanage, M., Mohan, S.V., Eds.; Butterworth-Heinemann: Oxford, UK, 2019; pp. 155–180. [CrossRef]

85. Lynn, C.J.; Dhir, R.K.; Ghataora, G.S.; West, R.P. Sewage sludge ash characteristics and potential for use in concrete. *Constr. Build. Mater.* **2015**, *98*, 767–779. [CrossRef]
86. Cyr, M.; Coutand, M.; Clastres, P. Technological and environmental behavior of sewage sludge ash (SSA) in cement-based materials. *Cem. Concr. Res.* **2007**, *37*, 1278–1289. [CrossRef]
87. Monzó, J.; Payá, J.; Borrachero, M.; Girbés, I. Reuse of sewage sludge ashes (SSA) in cement mixtures: The effect of SSA on the workability of cement mortars. *Waste Manag.* **2003**, *23*, 373–381. [CrossRef]
88. Kappel, A.; Ottosen, L.M.; Kirkelund, G.M. Colour, compressive strength and workability of mortars with an iron rich sewage sludge ash. *Constr. Build. Mater.* **2017**, *157*, 1199–1205. [CrossRef]
89. Mohammadhosseini, H.; Tahir, M.M.; Mohd Sam, A.R.; Abdul Shukor Lim, N.H.; Samadi, M. Enhanced performance for aggressive environments of green concrete composites reinforced with waste carpet fibers and palm oil fuel ash. *J. Clean. Prod.* **2018**, *185*, 252–265. [CrossRef]
90. Bamaga, S.; Hussin, M.; Ismail, M.A. Palm oil fuel ash: Promising supplementary cementing materials. *KSCE J. Civ. Eng.* **2013**, *17*, 1708–1713. [CrossRef]
91. Raut, A.N.; Gomez, C.P. Thermal and mechanical performance of oil palm fiber reinforced mortar utilizing palm oil fly ash as a complementary binder. *Constr. Build. Mater.* **2016**, *126*, 476–483. [CrossRef]
92. Awal, A.S.M.A.; Shehu, I.A.; Ismail, M. Effect of cooling regime on the residual performance of high-volume palm oil fuel ash concrete exposed to high temperatures. *Constr. Build. Mater.* **2015**, *98*, 875–883. [CrossRef]
93. Mohammadhosseini, H.; Tahir, M.M. Durability performance of concrete incorporating waste metalized plastic fibres and palm oil fuel ash. *Constr. Build. Mater.* **2018**, *180*, 92–102. [CrossRef]
94. Tangchirapat, W.; Saeting, T.; Jaturapitakkul, C.; Kiattikomol, K.; Siripanichgorn, A. Use of waste ash from palm oil industry in concrete. *Waste Manag.* **2007**, *27*, 81–88. [CrossRef]
95. Yao, G.; Liu, Q.; Wang, J.; Wu, P.; Lyu, X. Effect of mechanical grinding on pozzolanic activity and hydration properties of siliceous gold ore tailings. *J. Clean. Prod.* **2019**, *217*, 12–21. [CrossRef]
96. Kinnunen, P.; Ismailov, A.; Solismaa, S.; Sreenivasan, H.; Räisänen, M.-L.; Levänen, E.; Illikainen, M. Recycling mine tailings in chemically bonded ceramics – A review. *J. Clean. Prod.* **2018**, *174*, 634–649. [CrossRef]
97. Ruiz-Sánchez, A.; Sánchez-Polo, M.; Rozalen, M. Waste marble dust: An interesting residue to produce cement. *Constr. Build. Mater.* **2019**, *224*, 99–108. [CrossRef]
98. Güneyisi, E.; Gesoğlu, M.; Özbay, E. Effects of marble powder and slag on the properties of self compacting mortars. *Mater. Struct.* **2009**, *42*, 813–826. [CrossRef]
99. Aydin, E.; Arel, H.Ş. High-volume marble substitution in cement-paste: Towards a better sustainability. *J. Clean. Prod.* **2019**, *237*, 117801. [CrossRef]
100. Belaidi, A.S.E.; Azzouz, L.; Kadri, E.; Kenai, S. Effect of natural pozzolana and marble powder on the properties of self-compacting concrete. *Constr. Build. Mater.* **2012**, *31*, 251–257. [CrossRef]
101. Townsend, T.G.; Ingwersen, W.W.; Niblick, B.; Jain, P.; Wally, J. CDDPath: A method for quantifying the loss and recovery of construction and demolition debris in the United States. *Waste Manag.* **2019**, *84*, 302–309. [CrossRef]
102. Prabhu, K.R.; Yaragal, S.C.; Venkataramana, K. In Persuit of Alternative Ingredients to Cement Concrete Construction. *Int. J. Res. Eng. Technol.* **2013**, *02*, 404–410.
103. Kucukvar, M.; Egilmez, G.; Tatari, O. Life Cycle Assessment and Optimization-Based Decision Analysis of Construction Waste Recycling for a LEED-Certified University Building. *Sustainability* **2016**, *8*, 89. [CrossRef]
104. An, J.; Middleton, R.S.; Li, Y. Environmental Performance Analysis of Cement Production with CO2 Capture and Storage Technology in a Life-Cycle Perspective. *Sustainability* **2019**, *11*, 2626. [CrossRef]
105. Wang, X.-Y. Effect of Carbon Pricing on Optimal Mix Design of Sustainable High-Strength Concrete. *Sustainability* **2019**, *11*, 5827. [CrossRef]
106. Huntzinger, D.N.; Eatmon, T.D. A life-cycle assessment of Portland cement manufacturing: Comparing the traditional process with alternative technologies. *J. Clean. Prod.* **2009**, *17*, 668–675. [CrossRef]
107. Vieira, D.R.; Calmon, J.L.; Coelho, F.Z. Life cycle assessment (LCA) applied to the manufacturing of common and ecological concrete: A review. *Constr. Build. Mater.* **2016**, *124*, 656–666. [CrossRef]
108. Valderrama, C.; Granados, R.; Cortina, J.L.; Gasol, C.M.; Guillem, M.; Josa, A. Implementation of best available techniques in cement manufacturing: A life-cycle assessment study. *J. Clean. Prod.* **2012**, *25*, 60–67. [CrossRef]

109. Feiz, R.; Ammenberg, J.; Baas, L.; Eklund, M.; Helgstrand, A.; Marshall, R. Improving the CO2 performance of cement, part I: Utilizing life-cycle assessment and key performance indicators to assess development within the cement industry. *J. Clean. Prod.* **2015**, *98*, 272–281. [CrossRef]
110. Strazza, C.; Del Borghi, A.; Gallo, M.; Del Borghi, M. Resource productivity enhancement as means for promoting cleaner production: Analysis of co-incineration in cement plants through a life cycle approach. *J. Clean. Prod.* **2011**, *19*, 1615–1621. [CrossRef]
111. Pushkar, S. The Effect of Additional Byproducts on the Environmental Impact of the Production Stage of Concretes Containing Bottom Ash Instead of Sand. *Sustainability* **2019**, *11*, 5037. [CrossRef]
112. Salas, D.A.; Ramirez, A.D.; Rodríguez, C.R.; Petroche, D.M.; Boero, A.J.; Duque-Rivera, J. Environmental impacts, life cycle assessment and potential improvement measures for cement production: A literature review. *J. Clean. Prod.* **2016**, *113*, 114–122. [CrossRef]
113. García-Gusano, D.; Garraín, D.; Herrera, I.; Cabal, H.; Lechón, Y. Life Cycle Assessment of applying CO2 post-combustion capture to the Spanish cement production. *J. Clean. Prod.* **2015**, *104*, 328–338. [CrossRef]
114. Chen, C.; Habert, G.; Bouzidi, Y.; Jullien, A. Environmental impact of cement production: Detail of the different processes and cement plant variability evaluation. *J. Clean. Prod.* **2010**, *18*, 478–485. [CrossRef]
115. Gursel, A.P.; Masanet, E.; Horvath, A.; Stadel, A. Life-cycle inventory analysis of concrete production: A critical review. *Cem. Concr. Compos.* **2014**, *51*, 38–48. [CrossRef]
116. Josa, A.; Aguado, A.; Heino, A.; Byars, E.; Cardim, A. Comparative analysis of available life cycle inventories of cement in the EU. *Cem. Concr. Res.* **2004**, *34*, 1313–1320. [CrossRef]
117. Sui, X.; Zhang, Y.; Shao, S.; Zhang, S. Exergetic life cycle assessment of cement production process with waste heat power generation. *Energy Convers. Manag.* **2014**, *88*, 684–692. [CrossRef]
118. Song, D.; Yang, J.; Chen, B.; Hayat, T.; Alsaedi, A. Life-cycle environmental impact analysis of a typical cement production chain. *Appl. Energy* **2016**, *164*, 916–923. [CrossRef]
119. Aprianti S, E. A huge number of artificial waste material can be supplementary cementitious material (SCM) for concrete production—A review part II. *J. Clean. Prod.* **2017**, *142*, 4178–4194. [CrossRef]
120. Van den Heede, P.; De Belie, N. Environmental impact and life cycle assessment (LCA) of traditional and 'green'concretes: Literature review and theoretical calculations. *Cem. Concr. Compos.* **2012**, *34*, 431–442. [CrossRef]

© 2020 by the authors. Licensee MDPI, Basel, Switzerland. This article is an open access article distributed under the terms and conditions of the Creative Commons Attribution (CC BY) license (http://creativecommons.org/licenses/by/4.0/).

Article

Evaluation of Strength Development in Concrete with Ground Granulated Blast Furnace Slag Using Apparent Activation Energy

Hyun-Min Yang [1,2], Seung-Jun Kwon [3], Nosang Vincent Myung [2], Jitendra Kumar Singh [1], Han-Seung Lee [4,*] and Soumen Mandal [5]

1. Innovative Durable Building and Infrastructure Research Center, Department of Architectural Engineering, Hanyang University, 1271 Sa 3-dong, Sangnok-gu, Ansan 15588, Korea; yhm04@hanyang.ac.kr (H.-M.Y.); jk200386@hanyang.ac.kr (J.K.S.)
2. Department of Chemical and Environmental Engineering, University of California-Riverside, Riverside, CA 92521, USA; myung@engr.ucr.edu
3. Department of Civil and Environmental Engineering, Hannam University, Daejeon 34430, Korea; jjuni98@hannam.ac.kr
4. Department of Architectural Engineering, Hanyang University, 1271 Sa 3-dong, Sangnok-gu, Ansan 15588, Korea
5. Intelligent Construction Automation Center, Kyungpook National University, 80, Daehak-ro, Buk-gu, Daegu 41566, Korea; sou.chm@gmail.com
* Correspondence: ercleehs@hanyang.ac.kr; Tel.: +82-31-436-8159

Received: 19 November 2019; Accepted: 14 January 2020; Published: 17 January 2020

Abstract: Ground granulated blast furnace slag (GGBFS) conventionally has been incorporated with ordinary Portland cement (OPC) owing to reduce the environmental load and enhance the engineering performance. Concrete with GGBFS shows different strength development of normal concrete, but sensitive, to exterior condition. Thus, a precise strength evaluation technique based on a quantitative model like full maturity model is required. Many studies have been performed on strength development of the concrete using equivalent age which is based on the apparent activation energy. In this process, it considers the effect of time and temperature simultaneously. However, the previous models on the apparent activation energy of concrete with mineral admixtures have limitation, and they have not considered the effect of temperature on strength development. In this paper, the apparent activation energy with GGBFS replacement ratio was calculated through several experiments and used to predict the compressive strength of GGBFS concrete. Concrete and mortar specimens with 0.6 water/binder ratio, and 0 to 60% GGBFS replacement were prepared. The apparent activation energy (E_a) was experimentally derived considering three different curing temperatures. Thermodynamic reactivity of GGBFS mixed concrete at different curing temperature was applied to evaluate the compressive strength model, and the experimental results were in good agreement with the model. The results show that when GGBFS replacement ratio was increased, there was a delay in compressive strength.

Keywords: compressive strength; concrete; ground granulated blast furnace slag; apparent activation energy; equivalent age

1. Introduction

Ground granulated blast furnace slag (GGBFS) is widely used in various engineering applications to replace the ordinary Portland cement (OPC) [1–3]. However, GGBFS is sensitive to curing conditions and exhibit slow strength development [4–8]. In order to predict the strength behavior in the concrete with GGBFS, much research have been performed based on cement hydration phenomena [9]. Moreover, it is necessary to estimate the mechanical properties of GGBFS concrete, such as compressive strength, splitting tensile strength, elastic modulus, creep, shrinkage, etc. Among the mechanical properties used in the design, compressive strength is the most important. There are different models for strength development, but among them, the maturity model is the best model. This model is based on the age concept. It has been widely used to evaluate the compressive strength of the concrete on the assumption of linear relationship with temperature [10–12] or a nonlinear relationship with the chemical reaction rate of the cement [13]. By considering curing temperature range and the accuracy of the prediction result, the equivalent age model is widely used for the interpretation of strength development, which incorporates the chemical reaction rate of the cement [13]. In the chemical reaction rate model, it is considered that apparent activation energy (E_a) is a key parameter with the curing temperature on the hydration reaction [14–20]. The E_a is indirectly proportional to compressive strength, as suggested in ASTM C 1074-11 [21]. The application of E_a in the equivalent age model for normal concrete and the related strength prediction model is reasonably agreed with the test results. However, the concrete with GGBFS exhibits several differences compared to normal concrete in regards to the strength development attributed to a reduction of compressive strength in early age, caused by retardation of the setting. When the GGBFS comes into contact with concrete pore solution, the impermeable acid film surrounds the particles on the surface is destroyed; therefore, the chemical reaction in concrete would start. The delay in initial compressive strength of concrete with GGBFS is owing to insufficient alkalinity of cement paste [4–6,22,23]. In addition, Escalante et al. [4,5] studied the hydration of Portland cement with GGBFS under curing conditions. The hydration reaction was measured for six months at 10 °C, 30 °C and 50 °C curing temperature of cement pastes with 30%, 50% GGBFS replacement in 0.5 and 0.35 W/C ratio. The highest hydration reaction rate was found with 30% GGBFS at 50 °C in 0.5 W/C, while the lowest was shown by 50% GGBFS replacement at 10 °C in 0.35 W/C. Thus, a prediction of concrete compressive strength development with GGBFS by using a full maturity model by evaluating E_a is required. The present study is aimed to predict the compressive strength development of the concrete with GGBFS by calculating the E_a.

2. Prediction of the Compressive Strength Based on Maturity Theory

2.1. Maturity

Maturity is a function that quantitatively expresses the effect of curing temperature and time on strength development of the concrete. Therefore, the maturity theory by considering curing temperature and time function can be defined as:

$$M = \int_0^t H(T)dt, \tag{1}$$

where M and $h(T)$ is maturity and maturity function, respectively. T denotes curing temperature over the age of t.

The maturity function is expressed in linear Equation (2) considering temperature and the age [10], whereas, the equivalent age can be calculated by Arrhenius chemical reaction rate (Equation (3)) [12,13,24].

$$\sum_0^t M_s = \sum (T - T_0)\Delta t, \tag{2}$$

where M_s is maturity at age t. T is the average temperature (°C) of the concrete during the time interval, and T_0 is reference temperature (−10 °C) [11].

$$k_T = A \cdot \exp{\frac{E_a}{R \cdot T}}, \tag{3}$$

where k_T is the reaction rate constant, A is proportionality constant, E_a is apparent activation energy (kJ/mol), R is gas constant (8.314 J/mol·K), and T is the absolute temperature (Kelvin). Thus, the equivalent age (t_e) can be calculated by the following equation:

$$t_e = \frac{\int_0^t H(T) dt}{H(T_r)}, \tag{4}$$

where $H(T)$ is maturity function at different curing temperature (t), whereas, $H(T_r)$ represents the maturity function at fixed curing temperature, i.e., 20 °C (293 K). The equivalent age means curing time at standard temperature (20 °C). By considering E_a, t_e can be derived as:

$$te = \int_0^t \exp{\frac{E_a}{R} \cdot (\frac{1}{T_r} - \frac{1}{T})} dt, \tag{5}$$

where T is the average curing temperature of the concrete during a time interval and T_r is the absolute or fixed temperature, i.e., 20 °C (293 K).

The strength development analysis using maturity was carried out by considering curing temperature and duration. Other researchers have suggested a hyperbolic regression model (Equation (6)) which gives more accurate predictions compared to an exponential function [24–27]. In addition, the following model was implemented by introducing a third variable to explain the dormant period during the hydration process of cement.

$$S = \frac{S_u k_T (t_e - t_0)}{1 + k_T (t_e - t_0)}. \tag{6}$$

S is predicted compressive strength, whereas, S_u represents the obtained experimental compressive strength at 28 days of curing. k_T is the reaction rate constant at curing temperature (T). t_e and t_0 represent the equivalent age and the age when compressive development starts (final setting time), respectively.

2.2. Reaction Rate Constant (k_T) and Apparent Activation Energy (E_a)

The cement reacts with water at an early age resulting hydration reaction. Due to the hydration of cement paste in concrete, it develops strength. The hydration rate is governed by the reaction rate constant of the point when the cement paste and water react [24,28]. The hydration degree of Portland cement can be derived by the weight ratio of reaction products. The weight ratio of reaction products can be determined by non-hydrated cement using an electron microscope or X-ray diffraction analysis [29–32]. Also, the degree of hydration can be measured by the amount of water, as well as heat generation, and the compressive strength in an indirect way. The most common indirect hydration measurement method used in Portland cement is the micro-hydration method using conduction calorimeter. It measures the amount of hydration heat generated at the beginning when hydration of cement starts with time and represents the ratio of calorific value of final hydration per weight of cement [33]. This method accurately shows the degree of hydration at the beginning of cement hydration, but cannot change the degree of hydration, due to change in curing process [27]. The reaction rate constant is the indicator of the initial gradient for the degree of hydration. There are many factors which affect the reaction rate constant. However, it is difficult to quantitatively predict the effect of temperature on the reaction rate constant. Therefore, the reaction rate constant can be represented by the function of curing temperature if other conditions are identical. It is known that the reaction rate constant is influenced by the types of cement, curing temperature, W/C (%), admixture, and humidity conditions etc. [9,34–37]. Thus, activation energy (E_a) can help to calculate the reaction rate constant where minimum energy is required to occur the reaction. It has been reported that E_a can vary owing to the nature of Portland cement, which has different hydration reaction patterns with the setting process, curing period, and cementitious components [38]. Some researchers have reported that E_a is approximately 33.5 to 47.0 kJ/mol at early-age, and approximately 10 to 30 kJ/mol at long-term age [39–49]. Freisleben-Hansen and Pederson (FHP) [13] proposed an equation (Equation (7)) to estimate E_a of OPC concrete as a function of curing temperature.

$$E_a = 33.5 + 1.47(20 - T_a) \text{kJ/mol} (T_a < 20\,°C), \tag{7a}$$

$$E_a = 33.5 \text{kJ/mol} (T_a \geq 20\,°C), \tag{7b}$$

where E_a is apparent activation energy by Freisleben-Hansen and Pedersen with temperature parameter, T_a is the average curing temperature of concrete during a time interval.

However, Carino pointed out that E_a can be determined by the composition, powder level, type, amount, and admixture of the cement [25,26], and other researchers argue that the E_a is changed by W/C ratio.

The hydration reaction of Portland cement can be formulated with E_a. Therefore, ASTM C 1074 [21] has suggested the method to determine the E_a. The procedure for determining the E_a and the compressive strength prediction procedure, according to ASTM C 1074 is shown in Figure 1.

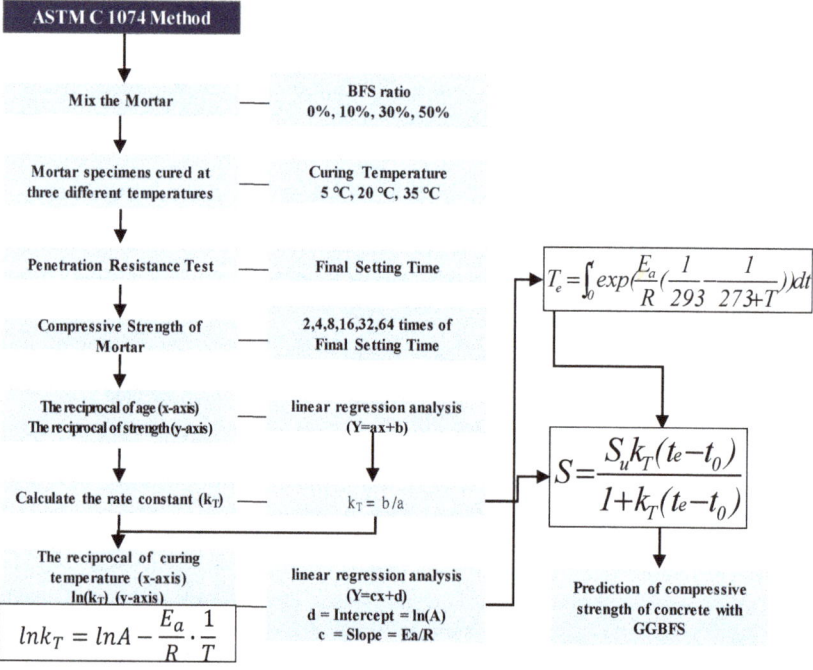

Figure 1. Calculation of E_a and estimation of compressive strength of concrete according to ASTM C 1074.

The final setting time of the mortar cured at three different temperatures is measured. The compressive strength of mortar is measured at 2, 4, 8, 16, 32 and 64 times as the final setting time. By plotting the reciprocal of the age(x-axis) versus the compressive strength (y-axis), the y-intercept of the linear regression line can be obtained from the regression analysis. E_a can be calculated by plotting the reciprocal of the curing temperature(x-axis) versus the reciprocal of lnk_T (y-axis) and dividing the gradient of the linear regression line obtained from the regression analysis. The equivalent age can be calculated by E_a of GGBFS concrete from Equation (5), and the compressive strength development is analyzed by the calculated equivalent age and k_T from Equation (6).

3. Experimental Program

3.1. Experimental Variables

Experimental variables were GGBFS replacement ratio and curing temperature at 0.60 water/binder (W/B) ratio. Details of these ratios are shown in Table 1.

Table 1. The experimental variables.

Experimental Level	Items
W/B	0.60
Replacement ratio of BFS (%)	0, 10, 30, 50
Curing temperature (°C)	5, 20, 35
Fresh mortar	setting time (initial and final)
Hardened mortar	compressive strength (2, 4, 8, 16, 32, 64 times of final setting time)
Hardened concrete	compressive strength (3, 7, 14, 28 days)

3.2. Materials

Cement and GGBFS

The cement used in the present study was ordinary Portland cement Type I containing 3.14 g/cm³ specific gravity. Blaine specific surface area of cement was 3230 cm²/g. GGBFS was used according to ASTM C 989 [50] with 2.84 g/cm³ specific gravity. Blaine specific surface area of cement was 4260 cm²/g. The chemical analysis of these materials are given in Table 2. The chemical composition OPC and GGBFS were carried out by X-ray fluorescence (XRF) instrument.

Table 2. Chemical analysis of materials. GGBFS, ground granulated blast furnace slag.

Component	Portland Cement %	GGBFS %
SiO_2	21.07	35.35
Al_2O_3	5.00	14
Fe_2O_3	2.92	0.36
CaO	62.40	41.91
MgO	2.07	7.74
SO_3	2.34	0.1
K_2O	0.59	–
Na_2O	0.26	–
LOI	1.19	0.31
Insoluble	0.41	0.21
Cl	0.05	0.02
Free Lime	1.70	–
Total (%)	100	100

3.3. Mixture Proportions

The concrete mixture proportion is presented in Table 3. Four types of specimens with different GGBFS replacement ratio were prepared. The mortar was prepared by removing less than 5 mm coarse particle as shown the mixture proportion in Table 3 using 600 Am sieve according to ASTM C33 [51] to maintain the quality of the concrete. Different amounts of superplasticizer were used to equalize the workability of GGBFS and OPC concrete. When we have added the high amount of superplasticizer in GGBFS, the workability was reduced (results are not shown). Therefore, it is required to maintain workability. Hence, we have chosen a different amount of superplasticizer. We have taken the high amount of superplasticizer in OPC (without GGBFS) to mix the concrete properly owing to delay the setting time whereas, in case of GGBFS replacement, specimens were mixed properly even at a low amount of superplasticizer.

Table 3. Concrete mixture proportions.

W/B	GGBFS (%)	S/a (%)	Mix Composition (kg/m^3)					
			W *	GGBFS	C *	S *	G *	SP *
0.60	0	46	217.2	0	362	798	912	1.156
	10			36.2	325.8	794	910	0.976
	30			108.6	253.4	792	908	0.659
	50			181	181	790	906	0.481

* W = water, C = cement, S = sand, G = coarse aggregates, and SP = Superplasticizer.

3.4. Prepared Specimens

Firstly, 0.1 m^3 of concrete mix was prepared for each batch. Cement, GGBFS, fine aggregate and coarse aggregate were mixed in a screw type mixer for 1 min and 30 s. Thereafter, water and superplasticizer were added, and the mixture was mixed for 3 min and 30 s. In order to obtain the mortar, the fresh concrete was passed through a 5 mm sieve, and cast into a mold (200 mm × 100 mm) for penetration resistance test according to ASTM C403 [52], whereas, compressive strength of a mold (50 mm × 50 mm × 50 mm) according to ASTM C109 [53]. All specimens were cured at 5 ± 2 °C, 20 ± 2 °C, 35 ± 2 °C and 60% (±2.5%) relative humidity. The curing condition was selected according to ASTM C39 [54].

3.5. Test Methods

3.5.1. Penetration Resistance Test for Determining Setting Time

In order to measure the mortar setting time, the penetration resistance of the mortar was measured using a standard needle at a regular time interval according to ASTM C 403 [52]. From the plot of penetration resistance versus elapsed time, the initial and final setting times were determined. For mortar with 50% GGBFS content, the setting time could not be measured under curing conditions of 5 ± 1.5 ° C. Therefore, the time when the compressive strength of the mortar reaches 4 MPa was considered as the final setting time by the alternative method proposed in ASTM C 1074 [21].

3.5.2. Compressive Strength Test of Mortar and Concrete

The compressive strength of mortar was measured by 30 t class universal testing machine, and the average value was used by measuring the compressive strength in a triplicate set of specimens at 2, 4, 8, 16, 32 and 64 times for the final setting time [21]. In addition, the compressive strength of the concrete was measured 200 t class universal testing machine to measure the compressive strength in a triplicate set of specimens at ages 3, 7, 14, and 28 days, and the average value was considered as a result.

4. Results and Discussion

4.1. Setting Time of Mortars

The initial and final setting time of mortar with GGBFS was measured by a penetration resistance test. Figure 2 shows the effect of GGBFS and temperature on the setting time of mortar. As expected, mortars of the same GGBFS replacement ratio show a decrease in setting time at higher curing temperatures. In addition, compared with OPC mortar, as the GGBFS replacement ratio increases, the setting time increases at the same curing temperature. In case of mortar with 50% of GGBFS compared to OPC mortar at the curing temperature of 5 °C, the final setting time was delayed by 15.7 h.

When mortar with 50% GGBFS was cured at 5° C and 35 °C, the final setting time was delayed by 16.9 h. The largest difference in the final setting time was 23.7 h between 50% GGBFS at 5 °C and OPC at 35 °C. It is considered that impermeable film has formed when GGBFS particle reacts with water. Due to the formation of the impermeable film results in the delay of hydration reaction, which restraining the penetration of water and ion [44–47]. Thus, this result suggests that the setting quickly occur owing to the fast reaction of hydration caused in early age at high temperature. Therefore, the setting time is affected by both curing temperature and GGBFS replacement ratio.

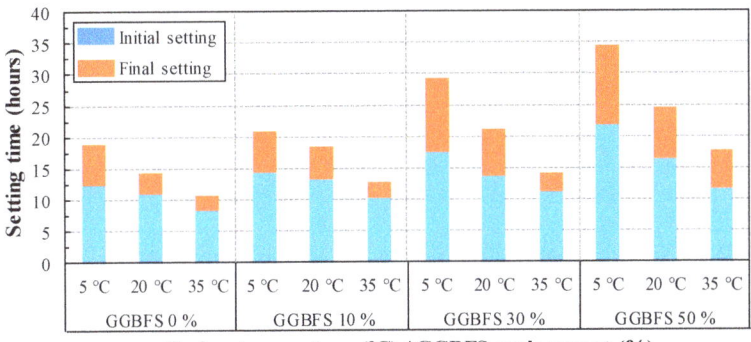

Figure 2. Setting time of mortar, according to curing temperature and replacement ratio of GGBFS.

4.2. Compressive Strength of Mortars

The compressive strength of mortar with GGBFS replacement ratio at each curing temperature is shown in Figure 3. The compressive strength of mortar at an early age is found to be lowest with increasing GGBFS replacement ratio. However, with an increase in curing temperature, the compressive strength is increased. It is reported by Karim et al. that supplementary cementitious materials (SCM) in cement significantly modify the hydration kinetics and gives better performance at higher curing temperature [55]. However, the long term strength of blended cement mortar depends on curing temperature. Moreover, SCM can recover the strength once it cured in high temperature at longer duration. The early strength of higher GGBFS replacement mortar is lower, but it can be improved when it would cure at elevated temperature attributed to the quick cement hydration reaction. At 5 °C curing temperature, the compressive strength of 50% GGBFS mortar exceeded the compressive strength compared to 30% GGBFS after 39 days (Figure 3a). In addition, at 20 °C (Figure 3b) and 35 °C (Figure 3c) curing temperature, the compressive strength of 50% GGBFS mortar exceeded compare to OPC after 28 and 22 days, respectively. As the GGBFS replacement ratio increases, the crossover effect of compressive strength is delayed compared to OPC. However, the crossover effect of compressive strength decreases as the curing temperature increases. Therefore, depending on the replacement ratio of GGBFS, the hydration and degree of hydration of mortar are very sensitive to temperature.

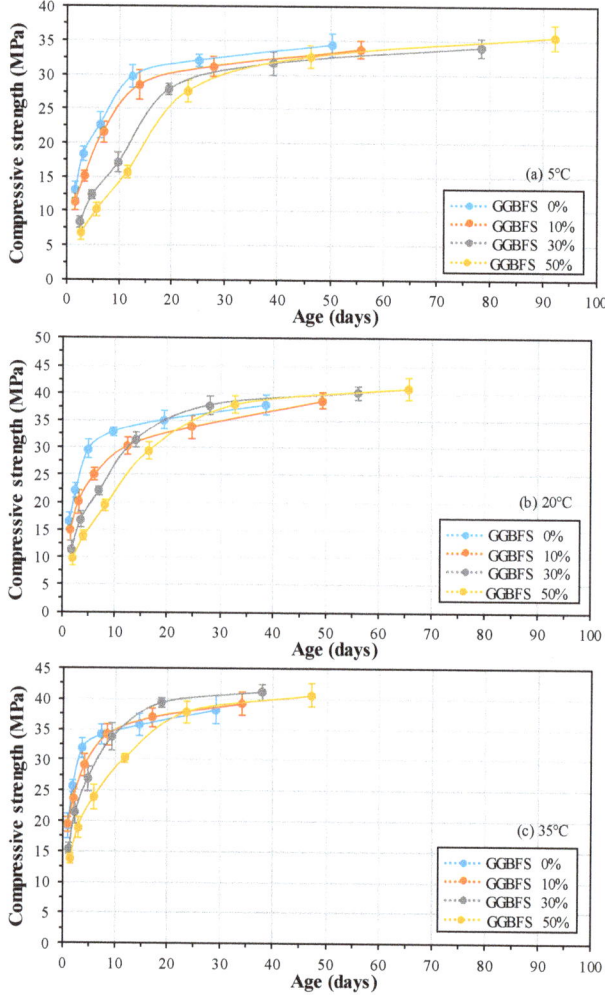

Figure 3. Compressive strength of mortar, according to curing temperature and replacement ratio of GGBFS (**a**) 5 °C, (**b**) 20 °C, (**c**) 35 °C.

4.3. Calculation of Reaction Rate Constant (k_T) and Apparent Activation Energy (E_a)

The relationship between age and compressive strength was analyzed, and k_T was obtained according to the GGBFS replacement ratio. Figure 4 shows the reciprocal of compressive strength versus reciprocal age. By dividing the y-intercept of the linear regression line by the slope, we can derive the k_T that takes the curing temperature and the replacement ratio as variables. Figure 5 shows k_T according to the GGBFS replacement ratio at each curing temperature. As the curing temperature increased, the k_T increased in the form of an exponential function and showed a high correlation between 0.87 and 0.99. In addition, the lower the GGBFS replacement ratio at the same curing temperature, the higher the k_T. The k_T of 50% GGBFS mortar is decreased by 75% and 70% at 5 °C and 35 °C curing temperature, respectively compared to OPC. In addition, the reaction rate constant of OPC and 50% GGBFS mortar is decreased by 76% and 81% at 5 °C compared to 35 °C curing temperature, respectively. Therefore, the curing temperature and GGBFS replacement ratio have a complex effect on the k_T.

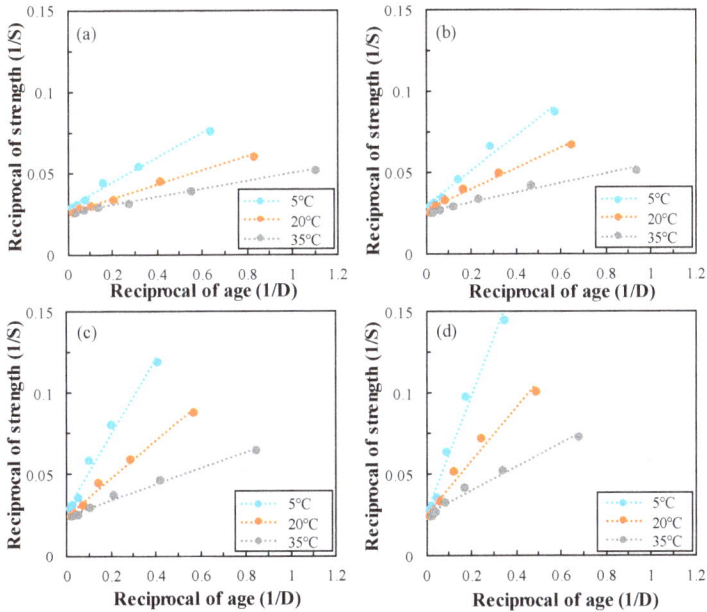

Figure 4. Regression analysis results for calculating k_T (**a**) 0%, (**b**) 10%, (**c**) 30%, (**d**) 50%.

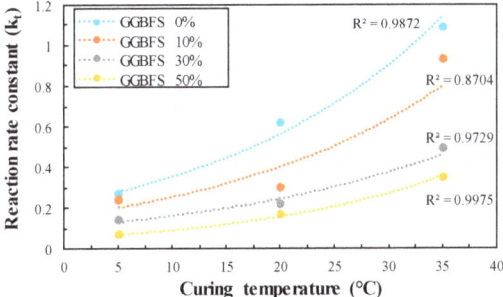

Figure 5. The effect of temperature on the rate constant with GGBFS replacement ratio.

By taking the natural logarithm of the calculated k_T and plotting the reciprocal of the curing temperature (K), we can represent the Arrhenius plot, as shown in Figure 6. The gradient of the linear regression line of each GGBFS replacement ratio represents E_a/R, and the y-intercept represents the value of $ln(A)$. As the GGBFS replacement rate increases, the gradients of the linear regression line decreases to a negative value.

Figure 7 shows the E_a results of GGBFS replacement ratio. As the GGBFS replacement ratio increases, the E_a increases linearly, which considered as the result of an increase in the minimum energy for the chemical reaction of cement, GGBFS and water. For OPC mortar, E_a is estimated to be 33.475 kJ/mol, which is very similar to the proposed value of Freisleben-Hansen and Pederson, i.e., 33.5 kJ/mol [13]. Wirkin et al. have found that superplasticizer has a little role on the hydration kinetics of cement and the difference in E_a, with or without the superplasticizer, is insignificant, i.e., 3 kJ/mol [44]. In the present study, the E_a value of 10%, 30% and 50% GGBFS is found to be 37.325 kJ/mol, 41.958 kJ/mol and 45.541 kJ/mol, respectively.

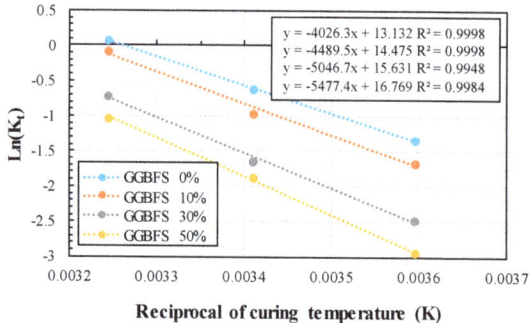

Figure 6. Arrhenius plot of ASTM C 1074 for calculating E_a.

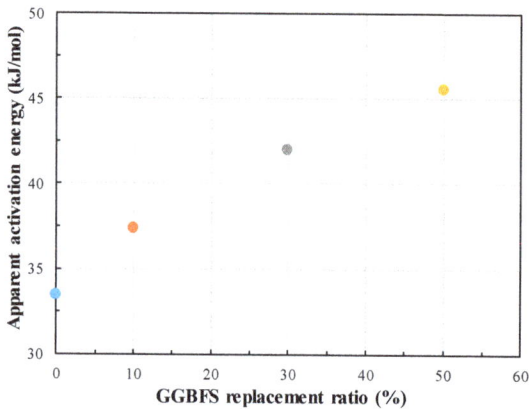

Figure 7. Apparent activation energy according to the GGBFS replacement ratio.

4.4. Prediction of Compressive Strength of Concrete with GGBFS

4.4.1. Compressive Strength of Concretes

This study measured the compressive strength of concrete with GGBFS at the age of 3, 7, 14, and 28 days and used the average of compressive strength of three specimens as a result. The results of compressive strength with varying curing temperature and GGBFS replacement ratio are shown in Figure 8. At higher curing temperatures, the compressive strength increased, and as the GGBFS replacement ratio increased, the compressive strength decreased. In addition, the difference in compressive strength with the change of curing temperature is the largest at three days of age. As the GGBFS replacement ratio increased, the difference in compressive strength, due to curing temperature increased. Especially, 28 days of age, the difference in compressive strength of OPC concrete at curing temperature of 5 °C and 35 °C was about 2.6 MPa, and the difference in compressive strength of 50% GGBFS is about 7.9 MPa. At curing temperatures of 5 °C, the compressive strengths of OPC, 10%, 30% and 50% GGBGS at three days were 12.1 MPa, 8.6 MPa, 4.3 MPa and 2.4 MPa, respectively. By increasing the GGBFS replacement ratio at low curing temperature causes delayed development of compressive strength in early age. Therefore, GGBFS concretes have different compressive strengths than OPC concrete. Thus, accurate prediction is necessary.

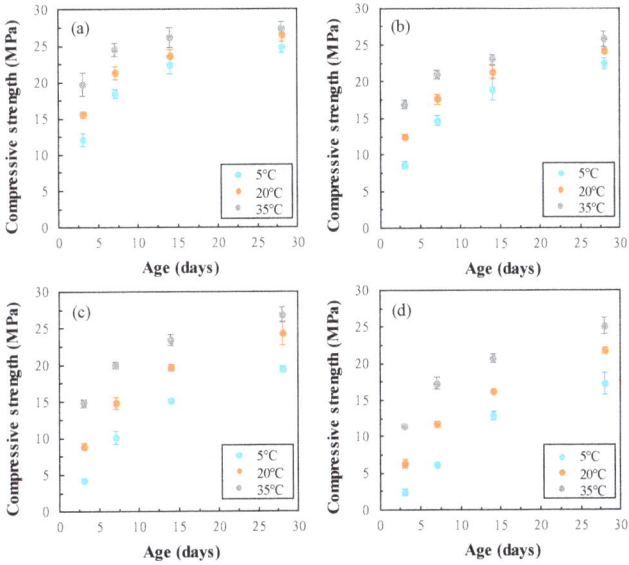

Figure 8. Compressive strength of concrete with curing temperature and GGBFS replacement ratio (**a**) 0%, (**b**) 10%, (**c**) 30%, (**d**) 50%.

4.4.2. Prediction of Compressive Strength of Concrete with GGBFS

The maturity model applied in this study is the equivalent age model, as shown in Equation (5), and the equivalent age was derived using calculated E_a. The compressive strength prediction, according to the equivalent age by GGBFS replacement ratio, is shown in Figure 9. From this figure, it is found that the compressive strength at an early age is delayed as GGBFS replacement ratio increased. The compressive strength development prediction curve approximately overestimates the compressive strength, but similarly simulates the delayed expression of initial compressive strength, due to the increase of GGBFS replacement ratio. Figure 10 compares the predicted compressive strength with the measured compressive strength. The compressive strength is predicted similarly in all compressive strength regions. Thus, very high correlation ($R^2 = 0.91$) is obtained. This proves that the compressive strength of concrete can be predicted in all strength at various curing temperatures and GGBFS replacement ratio.

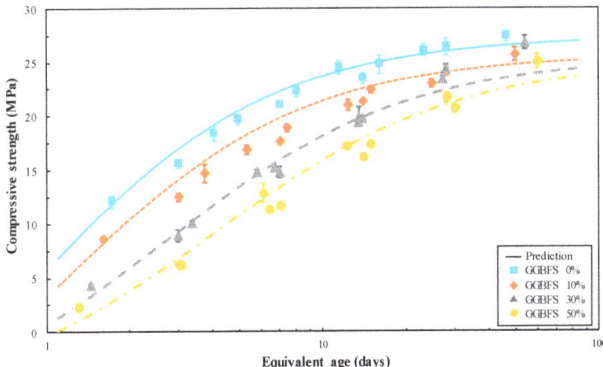

Figure 9. Prediction of compressive strength in concrete using GGBFS based on equivalent age.

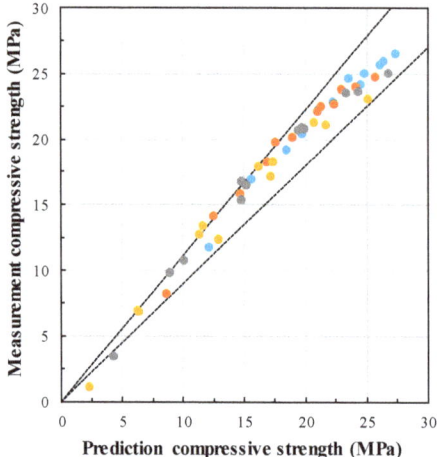

Figure 10. Comparison between predicted compressive strength and Measurement compressive strength.

5. Conclusions

From the above results and discussion, it is found that curing temperature and GGBFS replacement ratio have a significant effect on the development of concrete compressive strength. The addition of GGBFS also reduces the compressive strength of early age. As a result, the apparent activation energy and the compressive strength of concrete are affected by the GGBFS replacement ratio and curing temperature. Therefore, the existing compressive strength prediction model, i.e., Carino for OPC is not suitable for GGBFS concrete. The E_a of the proposed OPC, 10%, 30% and 50% of GGBFS are found to be 33.475 kJ/mol, 37.325 kJ/mol, 41.958 kJ/mol and 45.541 kJ/mol, respectively. Therefore, the equivalent age using E_a and a high level of accuracy ($R^2 = 0.91$) is obtained.

Author Contributions: Data curation, H.-M.Y. and N.V.M.; Formal analysis, H.-M.Y. and S.-J.K.; Funding acquisition, H.-S.L.; Investigation, H.-M.Y., J.K.S. and S.M.; Methodology, H.-M.Y., H.-S.L.; Supervision, S.-J.K., N.V.M. and H.-S.L.; Writing—original draft, H.-M.Y., S.-J.K., N.V.M., J.K.S., H.-S.L. and S.M.; Writing—review and editing, H.-M.Y., S.-J.K., N.V.M., J.K.S., H.-S.L. and S.M. All authors have read and agreed to the published version of the manuscript.

Funding: There is no external funding for this work.

Acknowledgments: This research was supported by a basic science research program through the National Research Foundation (NRF) of Korea funded by the Ministry of Science, ICT and Future Planning (No. 2015R1A5A1037548).

Conflicts of Interest: The authors declare no conflict of interest.

References

1. Lothenbach, B.; Scrivener, K.; Hooten, R.D. Supplementary cementitious materials. *Cem. Concr. Res.* **2011**, *41*, 1244–1256. [CrossRef]
2. ACI Committee 233. *Guide to the Use of Slag Cement in Concrete and Mortar*; ACI 223 American Concrete Institute: Farmington Hills, MI, USA, 2017.
3. Bijen, J. *Blast Furnace Slag Cement for Durable Marine Structures*; Stichting BetonPrisma: Den Bosch, The Netherlands, 1996; p. 62.
4. Escalante, J.I.; Sharp, J.H. The microstructure and mechanical properties of blended cements hydrated at various temperatures. *Cem. Concr. Res.* **2001**, *31*, 695–702. [CrossRef]
5. Escalante, J.I.; Gómez, L.Y.; Johal, K.K.; Mendoza, G.; Mancha, H.; Méndez, J. Reactivity of blast-furnace slag in portland cement blends hydrated under different conditions. *Cem. Concr. Res.* **2001**, *31*, 1403–1409. [CrossRef]
6. Roy, D.M.; Idorn, G.M. Hydration, structure and properties of blast furnace slag cements. mortars and concrete. *ACI J.* **1982**, *79*, 444–457.

7. Korde, C.; Cruickshank, M.; West, R.P.; Pellegrino, C. Activated slag as partial replacement of cement mortars: Effect of temperature and a novel admixture. *Constr. Build. Mater.* **2019**, *216*, 506–524. [CrossRef]
8. Soutsos, M.; Hatzitheodorou, A.; Kwasny, J.; Kanavaris, F. Effect of in situ temperature on the early age strength development of concretes with supplementary cementitious materials. *Constr. Build. Mater.* **2019**, *103*, 105–116. [CrossRef]
9. Mehta, P.K.; Monteiro, P. *Concrete: Microstructure, Properties, and Materials*, 4th ed.; McGraw-Hill Education: New York, NY, USA, 2014.
10. Saul, A.G.A. Principles Underlying the Steam Curing of Concrete at Atmospheric Pressure. *Mag. Concr. Res.* **1951**, *2*, 127–140. [CrossRef]
11. Nurse, R.W. Steam curing of concrete. *Mag. Concr. Res.* **1949**, *1*, 179–188. [CrossRef]
12. Rastrup, E. Heat of Hydration in Concrete. *Mag. Concr. Res.* **1954**, *6*, 79–92. [CrossRef]
13. Hansen, P.F.; Pedersen, E.J. Maturity Computer for Controlled Curing and Hardening of Concrete. *Nord Betong* **1977**, *1*, 21–25.
14. Neville, A.M. *Properties of Concrete*; Addison Wesley Longman Limited Edinburgh Gate: Harlow, UK, 1996.
15. Hansen, P.F.; Pedersen, E.J. Curing of concrete structure. *CEB Bull. d'Inf.* **1985**, *166*, 42.
16. Xiong, X.; Breugel, K.V. Isothermal calorimetry study of blended cements and its application in numerical simulations. *Heron. J.* **2001**, *46*, 151–159.
17. Schindler, A.K.; Folliard, K.J. Heat of hydration models for cementitious materials. *ACI Mater. J.* **2005**, *102*, 24–33.
18. Poole, J.L. Modeling temperature Sensitivity and Heat Evolution of Concrete. Ph.D. Thesis, University of Texas, Austin, TX, USA, 2007.
19. Mehdizadeh, H.; Kani, E.N. Rheology and apparent activation energy of alkali activated phosphorous slag. *Constr. Build. Mater.* **2018**, *171*, 197–204. [CrossRef]
20. Carette, J.; Staquet, S. Monitoring and modelling the early age and hardening behaviour of eco-concrete through continuous non-destructive measurements: Part I. Hydration and apparent activation energy. *Cem. Concr. Compos.* **2016**, *73*, 10–18. [CrossRef]
21. ASTM C1074-11. *Standard Practice for Estimating Concrete Strength by the Maturity Method*; ASTM International: West Conshohocken, PA, USA, 2011.
22. Kolani, B.; Buffo, L.L.; Sellier, A.; Escadeillas, G.; Boutillon, L.; Linger, L. Hydration of slag-blended cements. *Cem. Concr. Compos.* **2012**, *34*, 1009–1018. [CrossRef]
23. Sanjayan, J.G.; Sioulas, B. Strength of slag-cement concrete cured in place and in other conditions. *ACI Mater. J.* **2000**, *97*, 603–611.
24. Tank, R.C.; Carino, N.J. Rate constant functions for strength development of concrete. *ACI Mater. J.* **1991**, *88*, 74–83.
25. Carino, N.J. The maturity method: Theory and application. *ASTM J. Cem. Concr. Aggreg.* **1984**, *6*, 61–73.
26. Carino, N.J.; Tank, R.C. Maturity functions for concrete made with various cements and admixtures. *ACI Mater. J.* **1992**, *89*, 188–196.
27. Malhotra, V.M.; Carino, N.J. *CRC Handbook on Nondestructive Testing of Concrete*, 2nd ed.; CRC Press: Boca Raton, FL, USA, 2004.
28. Soutsos, M.N.; Barnett, S.J.; Bungey, J.H.; Millard, S.G. Fast track construction with high strength concrete mixes containing ground granulated blast furnace slag. In *Proceedings of ACI Seventh International Symposium on High Strength/High Performance Concrete, ACI SP-228*; Russell, H.G., Ed.; American Concrete Institute: Farmington Hills, MI, USA, 2005; Volume 1, pp. 255–270.
29. Wang, X.Y. Modeling of hydration, compressive strength, and carbonation of portland-limestone cement (PLC) concrete. *Materials* **2017**, *10*, 115. [CrossRef] [PubMed]
30. Abdelrazig, B.E.I.; Bonner, D.G. Estimation of the degree of hydration in modified ordinary Portland cement pastes by differential scanning calorimetry Thermochim. *Acta* **1989**, *145*, 203–217.
31. Lin, R.S.; Wang, X.Y.; Lee, H.S.; Cho, H.K. Hydration and microstructure of cement Pastes with Calcined Hwangtoh Clay. *Materials* **2019**, *12*, 458. [CrossRef] [PubMed]
32. Narmluk, M.; Nawa, M.T. Effect of fly ash on the kinetics of Portland cement hydration at different curing temperatures. *Cem. Concr. Res.* **2011**, *41*, 579–589. [CrossRef]
33. Pane, I.; Hansen, W. Investigation of blended cement hydration by isothermal calorimetry and thermal analysis. *Cem. Concr. Res.* **2005**, *35*, 1155–1164. [CrossRef]

34. Krstulović, R.; Dabić, P. A conceptual model of the cement hydration process. *Cem. Concr. Res.* **2000**, *30*, 693–698. [CrossRef]
35. Canut, M. Pore Structure in Blended Cement Pastes. Ph.D. Thesis, Technical University of Denmark, Kgs. Lyngby, Denmark, 2011.
36. Beaudoin, J.J.; Gu, P.; Marchand, J.; Tamtsia, B.; Myers, R.E.; Liu, Z. Solvent replacement studies of hydrated Portland cement systems: The role of calcium hydroxide, advanced. *Cem. Based Mater.* **1998**, *8*, 56–65. [CrossRef]
37. García, A.E.; Garcés, P.; Chinchón, S. General study of alkaline hydrolysis in calcium aluminate cement mortars under a broad range of experimental conditions. *Cem. Concr. Res.* **2000**, *30*, 1689–1699. [CrossRef]
38. Cura, G.; Garcés, P.; Alcocel, E. Petrographical analysis of calcium aluminate cement mortars. *Cem. Concr. Res.* **1999**, *29*, 1881–1885. [CrossRef]
39. Lachemi, M.; Hossain, K.M.A.; Anagnostopoulos, C.; Sabouni, A.R. Application of maturity method to slipforming operations: Performance validation. *Cem. Concr. Comp.* **2007**, *29*, 290–299. [CrossRef]
40. Turcry, P.; Loukili, A.; Barcelo, L.; Casabonne, J.M. Can the maturity concept be used to separate the autogenous shrinkage and thermal deformation of a cement paste at early age. *Cem. Concr. Res.* **2002**, *32*, 1443–1450. [CrossRef]
41. Garcia, A.; Castro-Fresno, D.; Polanco, J.A. Maturity approach applied to concrete by means of Vicat tests. *ACI Mater. J.* **2008**, *105*, 445–450.
42. Barnett, S.J.; Soutsos, M.N.; Millard, S.G.; Bungey, J.H. Strength development of mortars containing ground granulated blast-furnace slag: Effect of curing temperature and determination of apparent activation energies. *Cem. Concr. Res.* **2006**, *36*, 434–440. [CrossRef]
43. Voigt, T.; Sun, Z.; Shah, S.P. Comparison of ultrasonic wave reflection method and maturity method in evaluating early-age compressive strength of mortar. *Cem. Concr. Comp.* **2006**, *28*, 307–316. [CrossRef]
44. Wirkin, E.; Broda, M.; Duthoit, B. Determination of the apparent activation energy of one concrete by calorimetric and mechanical means: Influence of a superplasticizer. *Cem. Concr. Res.* **2002**, *32*, 1207–1213. [CrossRef]
45. Pinto, R.C.A.; Schindler, A.K. Unified modeling of setting and strength development. *Cem. Concr. Res.* **2010**, *40*, 58–65. [CrossRef]
46. Saadoon, T.; Gomes-Meijide, B.; Garcia, A. New predictive methodology for the apparent activation energy and strength of conventional and rapid hardening concretes. *Cem. Concr. Res.* **2019**, *115*, 264–273. [CrossRef]
47. Bie, Y.; Qiang, S.; Sun, X.; Song, J. A new formula to estimate final temperature rise of concrete considering ultimate hydration based on equivalent age. *Constr. Build. Mater.* **2017**, *142*, 514–520. [CrossRef]
48. Li, Q.B.; Guan, J.F.; Wu, Z.M.; Dong, W.; Zhou, S.W. Equivalent maturity for ambient temperature effect on fracture parameters of site-casting dam concrete. *Constr. Build. Mater.* **2016**, *120*, 293–308. [CrossRef]
49. Mi, Z.; Hu, Y.; Li, Q.; Gao, X.; Yin, T. Maturity model for fracture properties of concrete considering coupling effect of curing temperature and humidity. *Constr. Build. Mater.* **2019**, *196*, 1–13. [CrossRef]
50. *Standard Specification for Slag Cement for Use in Concrete and Mortars*; ASTM C 989-18a; ASTM International: West Conshohocken, PA, USA, 2018.
51. *Standard Specification for Concrete Aggregates*; ASTM C33; ASTM International: West Conshohocken, PA, USA, 2018.
52. *Standard Test Method for Time of Setting of Concrete Mixtures by Penetration Resistance*; ASTM C403; ASTM International: West Conshohocken, PA, USA, 2016.
53. *Standard Test Method for Compressive Strength of Hydraulic Cement Mortars (Using 2-in. or [50-mm] Cube Specimens)*; ASTM C109; ASTM International: West Conshohocken, PA, USA, 2016.
54. *Standard Test Method for Compressive Strength of Cylindrical Concrete Specimens*; ASTM C39; ASTM International: West Conshohocken, PA, USA, 2018.
55. Karim, E.; El, H.K.; Abdelkader, B.; Rachid, B. Analysis of Mortar Long-Term Strength with Supplementary Cementitious Materials Cured at Different Temperatures. *ACI Mater. J.* **2010**, *107*, 323–331.

© 2020 by the authors. Licensee MDPI, Basel, Switzerland. This article is an open access article distributed under the terms and conditions of the Creative Commons Attribution (CC BY) license (http://creativecommons.org/licenses/by/4.0/).

Article

The Properties of Composites with Recycled Cement Mortar Used as a Supplementary Cementitious Material

Katarzyna Kalinowska-Wichrowska, Marta Kosior-Kazberuk *and Edyta Pawluczuk

Faculty of Civil and Environmental Engineering, Bialystok University of Technology, 15-351 Bialystok, Poland; k.kalinowska@pb.edu.pl (K.K.-W.); e.pawluczuk@pb.edu.pl (E.P.)
* Correspondence: m.kosior@pb.edu.pl; Tel.: +48-797-995-935

Received: 21 November 2019; Accepted: 18 December 2019; Published: 21 December 2019

Abstract: The process of recycling concrete rubble is accompanied by the formation of a large amount of fine fraction, which cannot be reused as aggregate. The results of research on the possibility of using recycled cement mortar (RCM), obtained during concrete recycling, as a cementitious supplementary material, are presented. The experimental research was carried out on the basis of two variables determining the recycling process: X_1—temperature (range of variation 288–712 °C) and X_2—time (range of variation 30–90 min) of thermal treatment of concrete rubble. The experiment included 10 series of new composites made with RCMs subjected to different variants of thermal treatment, and two additional control series. The best treatment parameters were determined based on the assessment of selected physical and mechanical properties of the new cement composites, as well as the analysis of characteristics and microstructure of RCM. The test results showed that proper thermal treatment of concrete rubble makes it possible to obtain a high-quality fine fraction, which has the properties of an active addition and can be used as a partial replacement for cement in mortars and concretes.

Keywords: recycled cementitious supplementary material; comprehensive concrete recycling; recycled fine fraction; rehydration reactivity

1. Introduction

Recycling is currently one of the main ways of managing concrete rubble. The huge consumption of concrete in the world and the fact that its manufacturing consumes a large amount of non-renewable natural resources and other materials, e.g., aggregates (80% of concrete mass), Portland cement (10%), supplementary cementitious materials (3%) water (7%), and its production is responsible for 5% of anthropogenic worldwide CO_2 emissions, encourage a responsible approach to searching for methods and possibilities of its effective recycling [1]. The global aggregate production is currently estimated at 40 billion tons, which is leading to depletion of natural resources and high energy consumption and has negative impact on the environment [2].

The use of recycled aggregates (RA) from construction and demolition waste (CDW) in manufacturing concrete and mortar is a viable way to reduce the unsustainable level of consumption of natural aggregates worldwide and avoid landfilling CDW [3,4].

Research on the recycling of concrete is mainly devoted to finding the most effective way to obtain recycled aggregates of the best quality, which usually means removing the impurities from the surface of the natural aggregate grains. The recycled aggregate's quality is closely related to the adhered cement paste properties, since the bond between the natural aggregates and the cement paste is usually weak in the interfacial transition zone [5–8]. It is widely accepted that the presence of cement paste in recycling concrete aggregate causes its worse physical, mechanical and chemical properties compared to natural aggregates. During the process of crushing concrete, even 30–60% of its mass is a fine fraction (<4 mm) containing mainly cement mortar [9]. In order to improve the quality of coarse recycled aggregate, many refining methods (gravity concentration method, heating and rubbing method, mechanical grinding, etc.) have been developed for separating mortar from the surface of its grains [10]. As a result of these treatments, 35% of high-quality coarse aggregate and a total of 65% of the fine fraction containing mainly cement mortar are obtained [11]. An alternative method of treating recycling aggregates was proposed by Tam et al. [12]. They have studied three pre-soaking treatment methods—ReMortarHCl, ReMortarH$_2$SO$_4$ and ReMortarH$_3$PO$_4$—aiming at reducing the old cement mortar attached onto the RA. Experimental results show that the water absorption of the pre-treated RA has been significantly reduced with improvement of mechanical properties of the recycled aggregate concrete. A similar approach to construction waste recycling was presented by Robayo-Salazar et al. [13]. The results obtained in their investigation demonstrate the viability of reusing red clay brick waste, concrete waste and glass waste to produce alkali-activated cements that can be used to fabricate blocks, pavers and tiles. The alkaline activators used were solutions of either NaOH or NaOH and water glass.

According to other test results [14,15], the mortar content in recycled aggregates may be as high as 41% of the volume of the concrete rubble. Many studies indicate that this material cannot be used as a fine aggregate, as it significantly worsens the properties of concretes prepared with its use [16]. Therefore, other methods of using it in cement composites should be sought, and such attempts have been made by researchers for several years.

Gastaldi et al. [17] and Schoon et al. [15] used up to 30% of fine recycled material (grain diameter <63 μm) mainly for the production of Portland clinker, obtaining favorable results in terms of C$_2$S content in the clinker. Considering the significant amount of non-hydrated cement in the fine fraction of recycled concrete, Bordy et al. [18] presented the results of studies of composite in which part of the cement was replaced with finely ground (to diameter below 80 μm) powder made by crushing and milling of cement paste made in laboratory conditions. It was observed that there was about 24% of active clinker in the cement paste that could be rehydrated. Zhao et al. [16] observed the effect of water saturation of fine fraction from recycled aggregates, used as a partial cement replacement, on the properties of fresh mortar and mechanical properties of hardened new material and the microstructure of interfacial transition zone. Test results showed that mortars made of recycled dry fine aggregate were characterized by higher compressive strength than mortars with saturated aggregate due to the reduced interfacial transition zone.

Some researchers [19,20] studied the rehydration reactivity of fine fraction from recycled concrete after its heating at different temperatures. The results confirm that cement paste heated at a sufficiently high temperature is dehydrated. Particularly as a result of portlandite decomposition, reactive lime is formed. As a result of re-contact with water, it regains the ability to rehydrate. Ahmari et al. [21] proposed the production of a new geopolymeric binder from ground waste concrete powder mixed with fly ash, which can then be used with recycled concrete aggregates to produce new concrete. Tests carried out by some researchers [22,23] have proved that recycled mortar contains non-hydrated cement, calcium hydroxide (CH) and dicalcium silicate (C$_2$S), which are capable of hydration and creation of rehydration products. Some research has shown, however, that the mortar remaining on the surface of the recycled aggregates stored outside for a longer time does not show any rehydration reactivity [24]. Nevertheless, a study carried out by Vegas et al. [25] of the mix proportions and characteristics of mortars made with recycled concrete aggregate showed that up to 25% recycled aggregate can be used in cement-based masonry mortars with no significant decline in performance and no new admixtures or higher cement content required. Braga et al. [26] have analyzed the behavior of cement mortars using fine recycled fractions as a substitute for natural sand. In this study, 15% of the required natural sand was replaced by recycled aggregate. An increase in compressive strength with a simultaneous decrease in modulus of elasticity and an increase in water absorbability in comparison to traditional cement mortars were observed.

Considering the need to manage whole concrete rubble, the authors have developed a method for comprehensive recycling of concrete. The method allows obtaining high-quality secondary aggregate and a fine fraction that can be used as a partial cement replacement. Research on the effect of recycled aggregate on new concrete properties has been described in [8].

The aim of the now presented research work was to determine the effect of thermal and mechanical treatment of concrete rubble on the properties of the fine fraction and to assess the possibilities of using waste cement mortar as cementitious supplementary material. For this purpose, two variables have been analyzed: calcination temperature and time of thermal treatment of concrete rubble. The objects of the study were the composites in which part of the Portland cement had been replaced with recycled cement mortar (RCM) after thermal treatment. The RCM applicability as a reactive supplementary cementitious material was assessed based on such composite properties as compressive strength, flexural strength and water absorbability. X-ray diffractometry (XRD), differential thermal analysis (DTA), thermogravimetry analysis (TG) and scanning electron microscopy were used to characterize the microstructure of RCM and also to explain its rehydration reactivity. The conducted research partially resulted in issuance of a patent (PAT 229887 [27]).

2. Technology of Recycled Cement Mortar (RCM) Production

2.1. The Initial Concrete Rubble

The recycled cement mortar (RCM) was obtained from 2-year-old laboratory cube specimens 100 × 100 × 100 mm due to the recycling process according to PAT 229887 [28]. The composition of the initial concrete mix is presented in Table 1. The concrete was classified as C30/37 strength class.

Table 1. The composition of the initial concrete mix on 1 m^3.

Component	Content
Cement CEM I 42.5 R, kg	360
Sand 0–2 mm, kg	641
Gravel 2–16 mm, kg	1170
Plasticizer, dm^3	3.2
Water, dm^3	144

2.2. The Recycling Process of Concrete (Thermo-Mechanical Treatment According to PAT.229887)

The initial crushing of previously disintegrated concrete specimens (100 × 100 × 100 mm) was carried out in a jaw crusher. The concrete was crushed to the grain size $d \leq 40$ mm. In the next step, the concrete rubble was calcined in a ceramic laboratory furnace. Calcination of the recycled aggregate is necessary in order to dehydrate the cement paste. Dehydration of the cement paste reduces adhesion of the hardened mortar to the aggregate grains and eventually makes it easier to remove it from the surface of the aggregate. After thermal treatment and cooling, the recycled material was mechanically processed in a Los Angeles machine (Merazet, Poland). The grinding time was 15 min. This allowed for the final separation of the recycled cement mortar (RCM) from the surface of coarse aggregate grains. In the last step, the recycled material (fine and coarse fraction) was sieved in order to separate recycled cement mortar (<4 mm) from coarse aggregate (>4 mm). The recycled coarse aggregate and recycled cement mortar after the full process of recycling are shown in Figures 1 and 2. The coarse aggregate (>4 mm) obtained in recycling process can be used in new concrete as a substitute of natural aggregate. The test results of properties of concrete with recycled aggregate were presented in [8].

Figure 1. Recycled coarse aggregate obtained after thermal and mechanical treatment.

Figure 2. Recycled cement mortar (fine fraction) obtained after thermal and mechanical treatment, used to tests.

3. Materials and Methods

3.1. Materials

Portland cement CEM I 42.5 R, meeting the requirements of the EN 197-1:2011 [28] standard, was used for manufacturing cement composites. The natural aggregate sand with a maximum size of 2 mm was used. The procedure of obtaining RCM was described in Section 2.2 and in [23]. For further studies, only the fraction <250 µm of RCM was used as 25% cement replacement. RCM sieve analysis was performed in accordance with EN 933-1:2012 [29], and the grading curve of RCM is presented in Figure 3. As shown in Figure 3, after the recycling process, approximately 20% of the fraction classified as fine aggregate (<4000 µm) according to EN 206:2016 [30] is the dust fraction (<63 µm). For further studies, the fraction <250 µm was used, which was about 50% of the whole material. In future applications, an additional grinding of RCM is recommended immediately after the recycling process to increase the proportion of the finest particles.

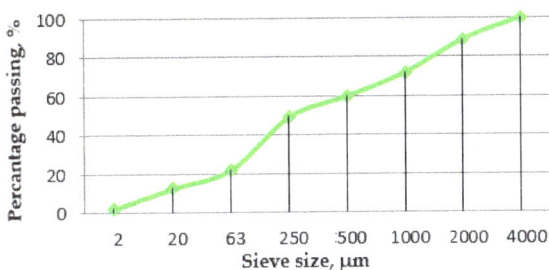

Figure 3. Grading curve of RCM after recycling process.

3.2. The Cement Composites with RCM

The mix compositions of composites with RCM used as a partial cement substitute are shown in Table 2. The compositions were designed as standard cement mortar according to EN 196-1:2016 [31]. In the series 1–10, 25% of the mass of Portland cement was replaced by RCM after thermal and mechanical treatment (RCM calcination temperature and time of treatment were based on the assumptions of the experimental plan, which is described in detail in Section 4). Series 12 had an analogous composition, but the RCM used was not subjected to calcination. Series 11 was made as a standard cement mortar: it did not contain RCM.

Table 2. The composition of cement composites.

Component	Unit	Cement Composites	
		Series 1–10; 12	Series 11
Cement CEM I 42.5 R	g	337.5	450
Water	mL	225	225
Sand 0–2 mm	g	1350	1350
Recycled Cement Mortar	g	112.5	-

3.3. Methods

3.3.1. Physical and Mechanical Properties Test Methods

Specimens of composites (40 × 40 × 160 mm) were prepared in accordance with EN 196-1:2016 [31]. After 28 days of curing, the compressive strength and flexural strength tests were carried out [31]. The water absorbability test was performed by determining the percentage increase in the weight of the specimens saturated with water in relation to the weight of the specimen in the dry state. The consistency measurement of fresh cement composite mixtures was made by the flow table method according to EN 1015-3:1999 [32]. The strength activity index (SAI) of the RCM was determined according to EN 450-1:2012 [33]. The heat of hydration of RCM was tested using a semi-adiabatic method based on the standard EN 196-9:2010 [34].

3.3.2. Analysis of RCM Properties and Microstructure

In order to determine the effect of the thermal and mechanical treatment on the phase composition of the RCM, X-ray diffraction analysis was conducted using a D8 Discover A25 instrument (Bruker) with CuKα radiation. All diffraction patterns were obtained by scanning the goniometer from 10 to 70 (2θ) at the rate of 0.05 min^{-1}.

The differential thermal analysis and thermogravimetric analysis were carried out using a model STA 409 PG analyzer (Netzsch, Selb, Germany) under a nitrogen atmosphere. The specimens were heated at rate of 10 °C/min to the temperature 1100 °C. The content of RCM components was calculated using DTA/TG DTG curves based on the instructions [35,36].

The morphology of RCM and cement composites with RCM was investigated using a Tescan high-resolution scanning microscope (Aztek Automated, Oxford Instruments, UK) equipped with an X-ray microanalysis system based on the method of X-ray spectrometry with energy dissipation (EDS) and a high-resolution microscope (Quanta 250 FEG, FEI, ThermoFisher Scientific, USA), digitally controlled and equipped with an electron gun with thermal field emission (the Schottky emitter). The shapes of fine particles were classified according to EN ISO 3252:2002 [37].

4. Design of the Experiment

Selection of Variables and Development of the Experimental Plan

For better understanding the relations among the factors determining the characteristics of RCM as a partial substitute for cement, an experiment was performed based on two variables: X_1—temperature of concrete rubble calcination, X_2—time of thermal treatment. The range of variation and the levels of analyzed factors are shown in detail in Table 3.

Table 3. Variables in the plan of experiment.

X_1	Temperature of concrete rubble treatment, °C	288	350	500	650	712
X_2	Thermal treatment time, min	30	40	60	80	90

The calcination temperatures at which the effects of phase changes can be expected were selected. At temperatures up to 350 °C, dehydration of C–S–H silicates, hydrated aluminates and aluminum calcium sulphates occurs along with gypsum decomposition. However, up to 650 °C, portlandite breaks down into CaO and H_2O. The temperature 500 °C is the center of the plan, and the other extreme like 288 °C and 712 °C are the star points and result from the construction of the adopted rotational plan.

Statistical analysis was carried out in accordance with the rotatable central composite design with a double repetition of the experiment at a central point. The design of the experiment (DoE) enables to check repeatability of results, to find which input factors and their interactions can influence the output properties significantly, to calculate regression equation and to check its adequacy with the test results. The following output properties were selected for analysis: compressive strength, flexural strength and water absorbability of composites with RCM.

On the basis of the above-mentioned variables, the experimental plan including 10 test series and 2 additional control series was established. Table 4 shows the detailed experimental plan with the real and normalized values of the variables.

Table 4. The rotatable central composite design of experiment.

Series	Real Values		Normalized Values	
	X_1, °C	X_2, min	x_1	x_2
1	350	40	−1	−1
2	350	80	−1	1
3	650	40	1	−1
4	650	80	1	1
5	288	60	−1.414	0
6	712	60	1.414	0
7	500	30	0	−1.414
8	500	90	0	1.414
9	500	60	0	0
10	500	60	0	0

Apart from the series described in Table 4, additional series were also tested for comparison 11 and 12. For the first control series 11, the cement composite included only cement as a binder, while the other control series 12 was made with RCM without thermal treatment.

Combinations of real values of the examined factors X_1 and X_2 were established on the basis of the assumptions of the design of experiment [38]. The dimensionless normalized values—x_1 and x_2—related to them were used to develop the functions describing the influence of the analyzed factors on the resulting quantities.

The test results were statistically analyzed in order to determine an approximating function describing the influence of the tested variables on the selected properties of the composites with RCM. The analyses included analysis of variance, calculation of regression coefficients and assessment of the regression coefficients' significance. The function describing the changes in the physical and mechanical properties of cement composites adopted the form of a second-degree polynomial (1):

$$y = b_0 + b_1 x_1 + b_2 x_2 + b_3 x_1 x_2 + b_4 x_1^2 + b_5 x_2^2 \quad (1)$$

where: y—dependent variable, explained; x_1, x_2—independent variables; b_i—coefficients; b_0—free term in expression. Calculations were performed according to [38] using software Statistica Version 13 (StatSoft, Poland).

5. Test Results and Discussion

5.1. Characteristics of Recycled Cement Mortar (RCM)

5.1.1. Heat of Hydration

In order to assess the rehydration reactivity of RCM, calorimetric tests were performed, and the results are presented in Figure 4. Two types of RCM were analyzed: calcined at the temperature 650 °C for 60 min and non-calcined material (without thermal treatment). CEM I 42.5 R was used as a standard for comparison. The specific density of RCM equal to 2.66 g/cm^3 was slightly lower than cement CEM I 42.5 R density, which was 3.05 g/cm^3.

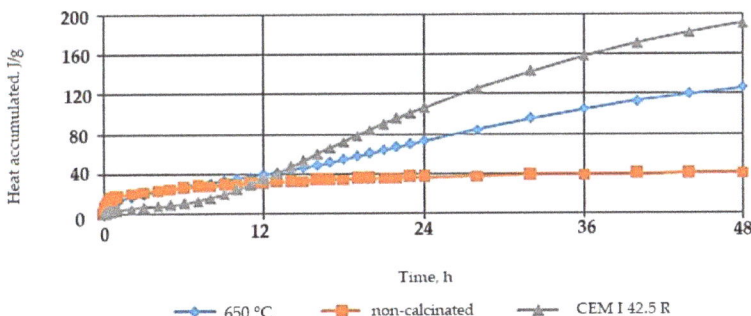

Figure 4. Changes in the amount of heat accumulated of tested materials.

Based on the analysis of data in Figure 4, it can be concluded that the thermal treatment of RCM has a significant impact on its rehydration reactivity properties and applicability as an active supplementary material. Non-calcined RCM showed no change in the value of heat released over a 48-h measurement period, and the maximum recorded value of heat accumulated was 40 J/g. The RCM calcined at the temperature 650 °C showed the best rehydration reactivity, the amount of accumulated heat increased successively during the test, and after a 48-h measurement period it reached the level of 125 J/g. In relation to the value of accumulated heat obtained for CEM I 42.5 R, it was a decrease of 35%, while compared to non-calcined RCM, an increase of almost 70% was observed.

5.1.2. Strength Activity Index (SAI)

The RCM's ability to function as an active addition was assessed on the basis of its pozzolanic properties. The SAI index was calculated according EN 450-1:2012 [33] as mentioned in Section 3.3.1. The SAI of the tested material should be ≥0.75 after 28 days of maturation. Figure 5 presents the strength activity index (SAI) test results for composites with 25% addition of RCM as a cement supplement (compositions according to Table 2) after 28 days of curing for different calcination temperatures and the same time of calcination, which was 60 min. The results were compared to control mortar made of Portland cement (series 11, Table 2).

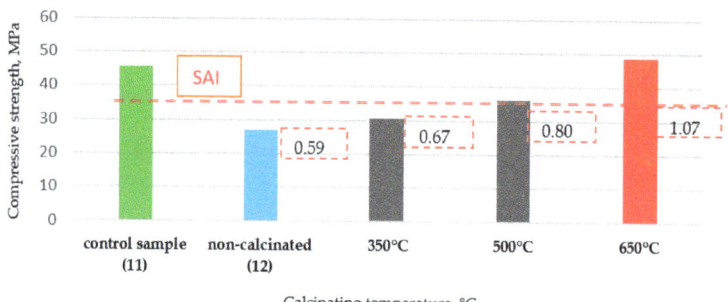

Figure 5. Compressive strength of the composites with and without RCM after 28 days of curing with the strength activity index (SAI).

As seen in Figure 5 the highest compressive strength was obtained for the composite containing 25% of the addition in the form of the RCM calcined at the temperature 650 °C for 60 min (SAI reached the highest level, SAI = 1.07). The increase in SAI value for these samples compared to the non-calcined RCM (series 12, SAI = 0.59) was 79%, and it was 66% compared to RCM calcined in 350 °C (SAI = 0.67). The composites with RCM calcined at temperature 500 °C also confirmed the requirements of standards [34], (SAI = 0.80), but the compressive strength reached 25 and 20% lower than the strength of the composites calcined at 650 °C and the control series, respectively.

5.1.3. X-ray Diffractometry

In order to determine the effect of thermal treatment on the phase composition of RCM, the X-ray diffractometry test was performed. The X-ray patterns for non-calcined RCM and thermally treated specimen are shown in Figures 6 and 7, respectively.

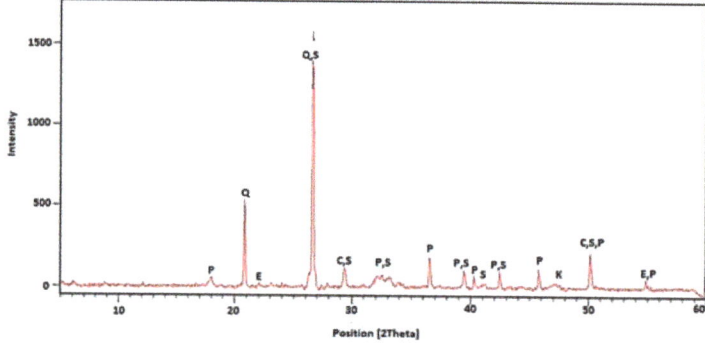

Figure 6. XRD pattern for RCM without thermo-mechanical treatment (Q—quartz, P—portlandite, C—calcium silicate hydrate, S—belite, K—calcite, E—ettringite).

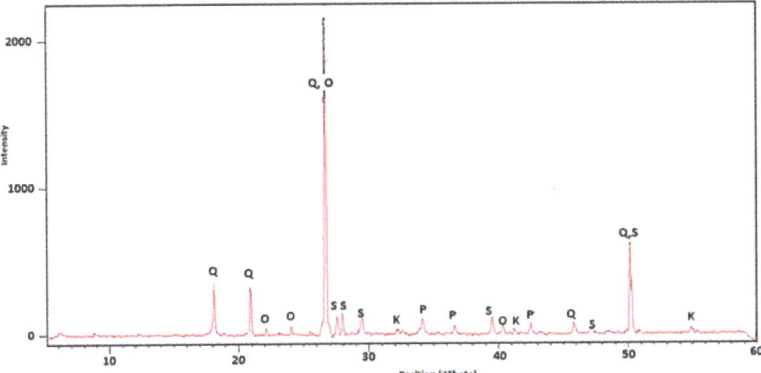

Figure 7. XRD pattern for RCM after thermo-mechanical treatment at temperature 650 °C (Q—quartz, P—portlandite, S—belite, K—calcite, O—calcium oxide).

The main crystalline phases found in RCM without thermo-mechanical treatment (Figure 6) are C–S–H gel, belite (C_2S), portlandite (CH), ettringite, and calcite ($CaCO_3$). The XRD pattern of the RCM after thermo-mechanical treatment indicates that some diffraction peaks of the preheated samples gradually disappear, such as ettringite and C–S–H gel. When the heating temperature is raised to 650 °C, it is mainly composed of belite (C_2S), CaO, partial CH and non-crystalline dehydrated phases.

As expected, in case of material not subjected to the calcination process (Figure 6) the peaks associated with the presence of portlandite were quite intensive and frequent. In case of RCM after thermal treatment, only a single peak was observed. This indicated a properly selected treatment temperature, allowing for almost complete $Ca(OH)_2$ decomposition. The small peaks from portlandite in the specimen subjected to calcination (Figure 7) can be explained by high hygroscopicity of disintegrated RCM. The finely ground and calcined RCM contains active CaO, which may react with moisture contained in the air. This phenomenon could not be avoided during sample preparation for XRD, hence the presence of a secondary portlandite in phase composition. Moreover, the calcined sample revealed higher and more frequent peaks, indicating the presence of Portland clinker components (belite), which are responsible for reactivity with water and for the hydration process. This could explain the applicability of calcined RCM as a pozzolanic additive and active filler [35,36]. The above-mentioned components of RCM facilitate further hydration and reaction with new cement paste, which cause the improvement of physical properties of cement composites. This is confirmed by the observations of other authors, who have noticed the content of non-hydrated cement, calcium hydroxide (CH), and dicalcium silicate (C_2S) in RCM, which are capable of hydration and creation of rehydration products [19,20,22].

5.1.4. Thermogravimetry (TG) and Differential Thermal Analysis (DTA)

In order to determine the effect of thermal treatment on the content of calcium hydroxide and calcium carbonate in RCM, the samples were subjected to thermogravimetry and differential thermal analysis. In Figures 8 and 9, the weight losses of material when heated to 1100 °C are presented for non-calcined and calcined RCM (at a temperature of 650 °C), respectively.

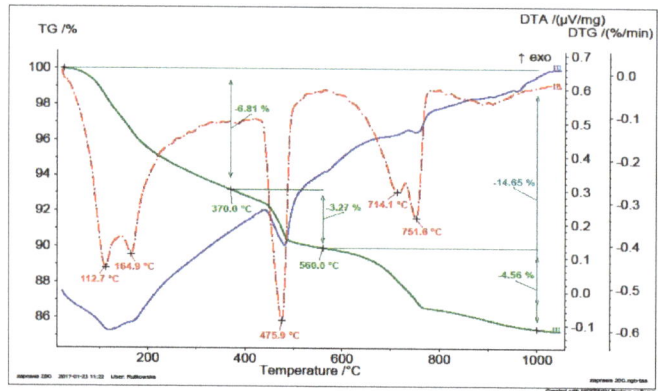

Figure 8. Thermal changes of RCM without thermal and mechanical treatment.

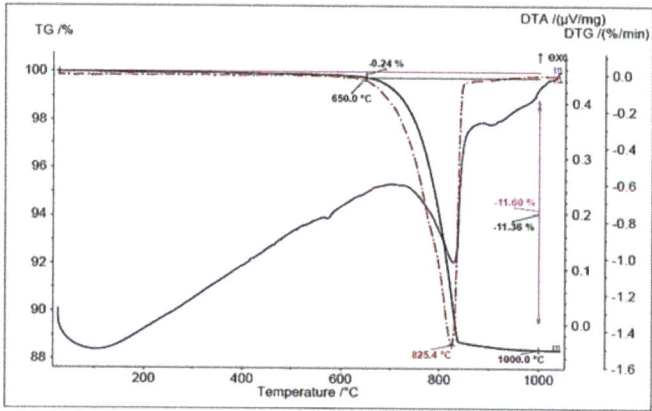

Figure 9. Thermal changes of RCM after thermal and mechanical treatment at temperature 650 °C.

Table 5 presents the content of bound water, portlandite and calcite in RCM specimens calculated on the basis of plots in Figures 8 and 9 and according to [35,36].

Table 5. Content of selected components of RCM.

Type of RCM	Components of the RCM, % of Mass				
	Bound Water			Ca(OH)$_2$	CaCO$_3$
	H$_I$	H$_{CH}$	Σ		
Non-calcined	6.81	3.27	10.08	13.44	10.44
Calcined	-	0.24	0.24	0	25.79

The quite high content of calcium carbonates in both tested specimens is noteworthy. It results from the applied heating temperature equal to 650 °C, which does not cause the decomposition of CaCO$_3$. This phenomenon occurs at a temperature above 750 °C, as indicated by peaks associated with mass losses in Figures 8 and 9. The higher CaCO$_3$ content in calcined RCM can be explained by the presence of aggregate in the tested specimen. In the RCM specimen after heat treatment, however, there was no peak of portlandite, which confirmed a sufficiently well-selected RCM treatment temperature (Figure 9). Earlier heating of concrete rubble resulted in the disintegration of calcium hydroxide into calcium oxide and water, as evidenced by its lack in the tested RCM specimen in comparison with the untreated specimen. In the presence of water, free calcium oxide has the ability to undergo rehydration, as evidenced by the test results obtained.

5.1.5. SEM Images Analysis

Micrographs of RCM are shown in Figures 10–14. The shape of RCM particles (Figures 10 and 11) was defined as irregular polyhedral [37], and they were very similar to cement grains. On the surface of the RCM particles, crystalline inclusions have also been observed, which may probably be the remains of hydrates from the primary hardening process, e.g., parts of non-hydrated cement, free calcium, as well as calcium silicates, as indicated by the results of the phase composition test presented in Section 5.1.3.

Figure 10. SEM micrograph of RCM particles, mag. 2000×.

Figure 11. SEM micrograph of RCM particles, mag. 10,000×.

Micrographs of the microstructure of cement composites with RCM are presented in Figures 12 and 13. Observations of the microstructure of cement composites (Figures 12 and 13) confirmed that RCM can be perfectly embedded in the cement matrix, forming a compact microstructure. The compact C-S-H phase can be observed in Figures 12 and 13. The compact microstructure may result from strong bonding of high-quality recycling mortar containing C-S-H phases to the components of new paste, as well as the developed surface of the recycled fraction due to applied thermal and mechanical treatment [20].

As it has been noted, the rough surface of the aggregate (in this case also RCM particles) deteriorates the portlandite orientation [39]. The arrangement of portlandite plates (CH) and the C-S-H phase surrounding them are presented in Figure 14.

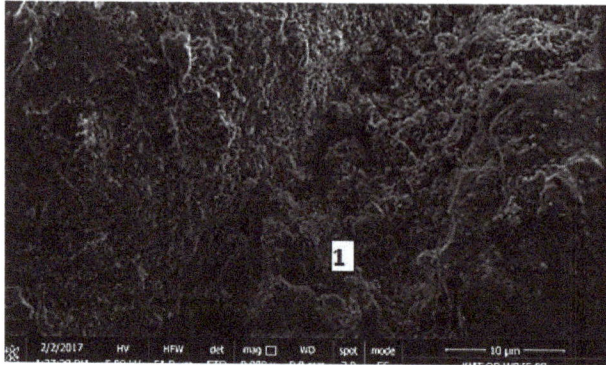

Figure 12. SEM micrograph of cement composite with RCM addition, 1—C-S-H gel (mag.10,000×).

Figure 13. SEM micrograph of cement composite with RCM, 1—C-S-H gel (mag. 16,000×).

Figure 14. SEM micrograph of microstructure of composite with RCM after thermal and mechanical treatment, 1—portlandite, 2—C-S-H gel (mag. 8000×).

The orientation of portlandite crystals is additionally disturbed by the presence of $Ca(OH)_2$, which, being the nuclei of crystallization of this phase, causes the growth of calcium hydroxide crystals in various directions, improving mortar strength properties [39].

5.2. Properties of Fresh Mixtures

Selected test results of the consistency of cement composite mixtures measured according to [28] are presented in Table 6.

Table 6. The flow diameter of cement composites mixes.

Type of Mixture	Flow Diameter, mm
Standard cement mortar (series 11)	170
Cement composite with RCM (series 4)	150

The fresh mortar with RCM obtained as a result of the thermal and mechanical treatment (650 °C, 60 min) of concrete rubble was characterized by slightly limited flow diameter in comparison to control mortar. The RCM, due to the developed specific surface (about 3000 cm^2/g), was also characterized by high water demand. Despite this, it was not necessary to increase the amount of water in the recipe in order to thoroughly mix the ingredients and achieve the required workability.

5.3. Properties of Hardened Cement Composites

5.3.1. Compressive Strength

The average compressive strength results obtained for the test series in comparison to the control samples (series 11) and (series 12) are given in Figure 15.

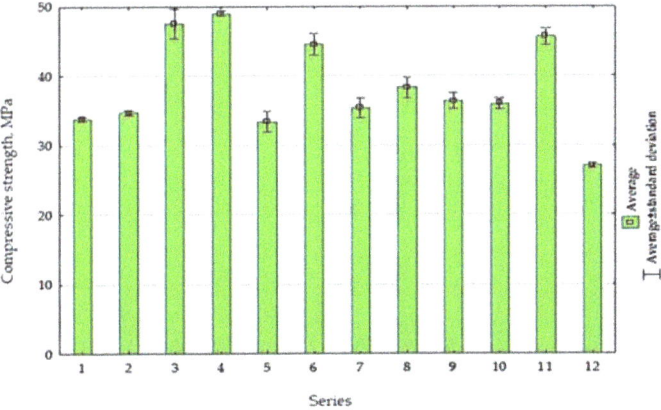

Figure 15. The compressive strength of composites after 28 days.

The changes in cement composite compressive strength depending on the calcination temperature of rubble (x_1) and treatment time (x_2) are presented in Figure 16. The function describing the dependence of the compressive strength on the tested variables for composites with RCM is expressed in the following equation:

$$f_{cm,28} = 36.16 + 5.48x_1 + 0.82x_2 + 2.24x_1^2 + 1.18x_2^2 \quad R^2 = 0.83 \tag{2}$$

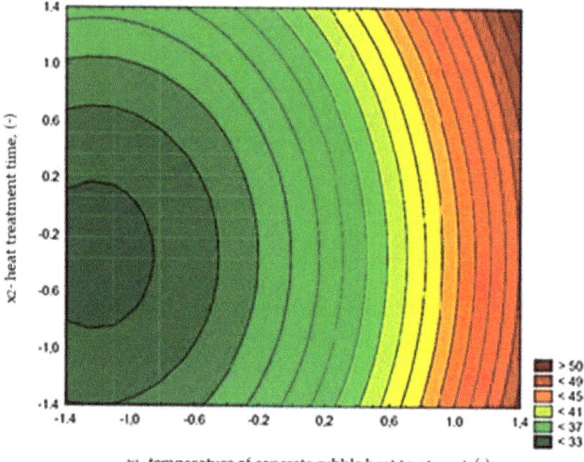

Figure 16. The changes in compressive strength of cement composites (MPa), depending on x_1 and x_2.

The obtained compressive strength test results indicate that the increase in calcination temperature from 288 to 712 °C caused the increase in compressive strength of about 32%. The highest strength values, exceeding those obtained for the control series, were recorded for composites with RCM calcined at temperature 650 °C (series 3 and 4, Figure 16). The compressive strength for these series exceeded the results obtained for the control series 11 by 4 and 7%, respectively. However, in case of using the RCM calcined at lower temperatures (up to 350 °C), as well as in the presence of non-calcined addition, the compressive strength results of the composite did not exceed 35 and 30 MPa, respectively. Thus, only a sufficiently high temperature of rubble processing allows to obtain RCM that causes a favorable effect on the compressive strength of cement composites.

5.3.2. Flexural Strength

The average flexural strength results obtained for the test series in comparison to the control samples (series 11) and (series 12) are given in Figure 17.

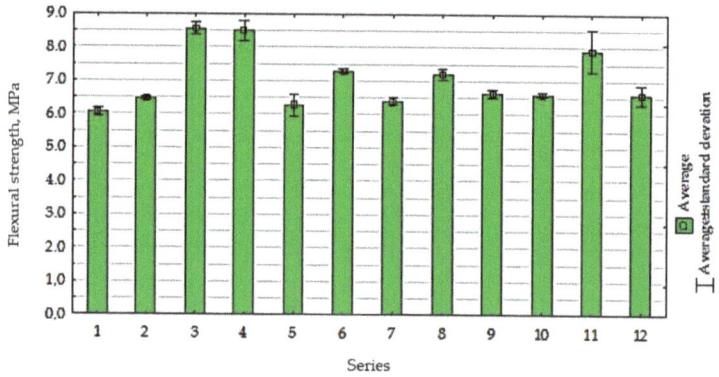

Figure 17. The flexural strength of composites with RCM.

The changes in flexural strength depending on the calcination temperature of rubble (x_1) and treatment time (x_2) are presented in Figure 18. The function describing the dependence of the flexural strength on the tested variables for composites with RCM is expressed in the following equation:

$$f_{fm,28} = 6.76 + 0.75x_1 + 0.18x_2 + 0.24x_1^2 + 0.24x_2^2 \quad R^2 = 0.78 \qquad (3)$$

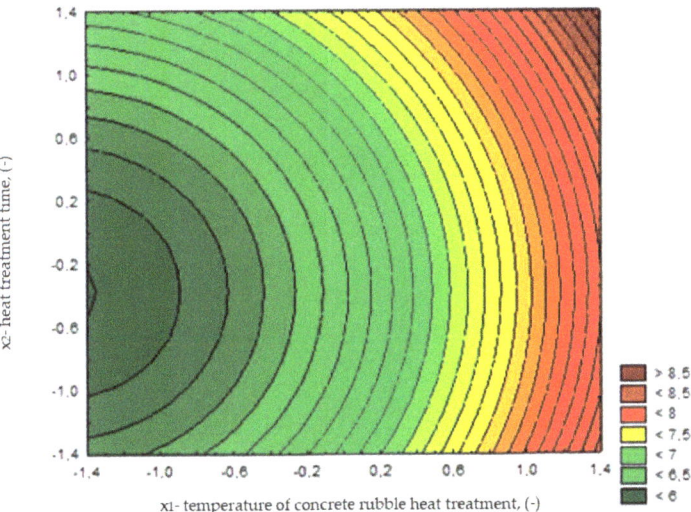

Figure 18. The changes in flexural strength of cement composites (MPa), depending on x_1 and x_2.

The analysis of test results showed that the most favorable results of flexural strength of cement composites containing RCM were obtained in cases of using the recycled material after calcination at temperature higher than 650 °C (series 3, 4, 6), and the results were similar to those obtained in the control series. The use of RCM from rubble after thermal treatment at lower temperatures (<650 °C) had no significant effect on flexural strength compared to the composite with non-calcined RCM.

5.3.3. Water Absorbability

The average results of water absorbability test were given in Figure 19.

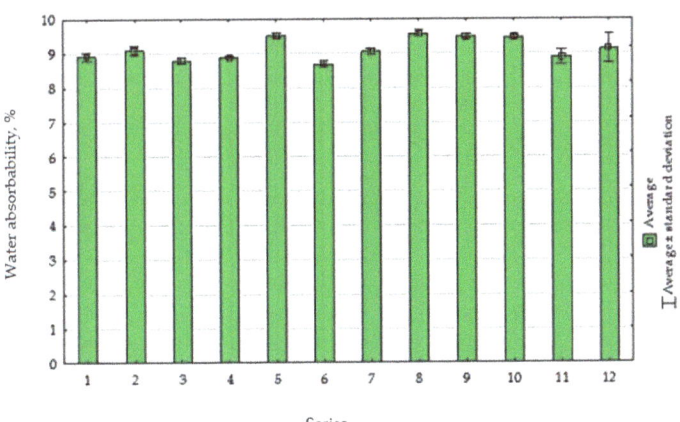

Figure 19. The water absorbability of composites with RCM.

The changes in cement composite water absorbability depending on the calcination temperature of rubble (x_1) and thermal treatment time (x_2) are presented in Figure 20. The function describing the dependence of water absorbability on the tested variables for composites with RCM is expressed in the following equation:

$$WA = 9.49 - 0.18x_1 + 0.13x_2 - 0.2x_1^2 - 0.16x_2^2 \quad R^2 = 0.68 \tag{4}$$

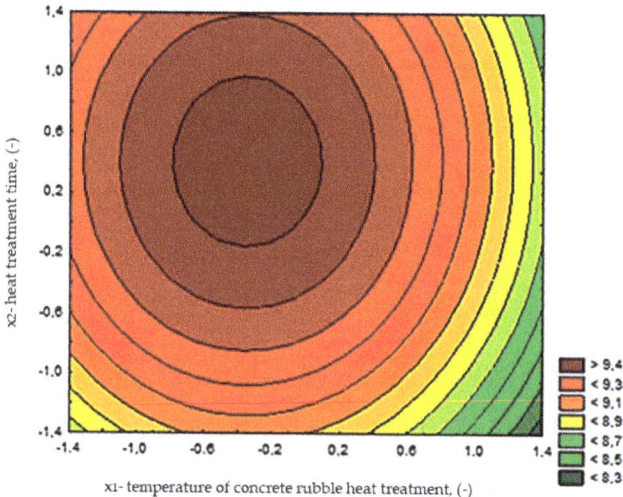

Figure 20. The water absorbability (%) changes of cement composites depending on x_1 and x_2.

Based on the test results, it can be concluded that a decrease in the absorbability of the cement composite occurs with the increase in RCM calcination temperature. In case of using RCM calcined at a temperature 650 °C or higher, water absorbability comparable to the absorbability of the control mortar was achieved. However, extended calcination time had a relatively negative influence on this property. It should be noted that the observed changes in water absorbability were relatively small and ranged from 8.7 to 9.6%. This is probably due to the fact that the RCM has a developed specific surface area similar to the cement used. The lowest water absorbability was recorded from the composite with RCM calcined at temperature 712 °C (series 6, Figure 20), which is also related to the test results of other parameters such as compressive strength or flexural strength.

6. Conclusions

Cementitious supplementary material used in cement composites was obtained as a result of thermal and mechanical treatment of concrete rubble as part of comprehensive recycling of reinforced concrete structures.

The statistical analysis of test results of compressive strength, flexural strength and water absorbability of mortars with RCM made it possible to determine the optimal conditions for production of cementitious supplementary material. It was found that the calcination temperature of concrete rubble had the most significant effect on the analyzed parameters of cement composites. The effect of calcination time was statistically less significant. The regression equations can be useful for estimation of the physical properties of composites with RCM considering the conditions of thermal treatment of concrete rubble.

The calcination of concrete rubble at a temperature of about 650 °C caused partial dehydration of cement hydration products, mainly the disintegration of portlandite ($Ca(OH)_2$) into CaO and H_2O.

This treatment partially removed the hydration reactivity of old cement mortar, which resulted in improved physical properties of cement composites with RCM.

The results of extensive microstructural analysis, including X-ray diffractometry (XRD), differential thermal analysis (DTA), thermogravimetry analysis (TG) and scanning electron microscopy, confirmed the presence of non-hydrated cement, calcium hydroxide (CH), calcium oxide (CaO) and dicalcium silicate (C_2S) in RCM, which are capable of hydration and creation of rehydration products. The influence of RCM treatment temperature on its rehydration reactivity properties was assessed based on the analysis of heat of hydration.

The proposed highly ecological solution for the management of waste generated in the concrete recycling process supports the idea of sustainable development by limiting the consumption of natural resources and reducing CO_2 emissions generated during the cement production process. The test results showed that appropriate treatment of concrete rubble allows to obtain high-quality fine fraction which may be successfully used as a cement substitute or as pozzolanic additive for cement composites.

Author Contributions: Conceptualization, K.K.-W., M.K.-K. and E.P.; methodology, E.P.; software, M.K.-K.; validation, M.K.K, E.P. and K.K.-W.; investigation K.K.-W., M.K.-K., E.P. resources K.K.-W., E.P.; data verification, K.K.-W.; writing—original draft preparation, K.K.-W. and M.K.-K.; writing—review and editing, M.K.-K., K.K.-W., E.P.; visualization, K.K.-W., M.K.-K., E.P. supervision, M.K.-K.; project administration, M.K.-K., K.K.-W., E.P. All authors have read and agreed to the published version of the manuscript.

Funding: This research received no external funding.

Acknowledgments: The study was performed under the research project number S/WBiIŚ/2/2017 funded by the Polish Ministry of Science and Higher Education.

Conflicts of Interest: The authors declare no conflict of interest.

References

1. Hasanbeigi, A.; Price, L.; Lin, E. Emerging Energy-Efficiency and CO_2 Emission-Reduction Technologies for Cement and Concrete Production: A Technical Review. *Renew. Sustain. Energy Rev.* **2012**, *16*, 6220–6238. [CrossRef]
2. Rodríguez-Robles, D.; Van den Heede, P.; De Belie, N. Life Cycle Assessment Applied to Recycled Aggregate Concrete. In *New Trends in Eco-Efficient and Recycled Concrete*, 1st ed.; de Brito, J., Agrela, F., Eds.; Woodhead Publishing: Cambridge, UK, 2019; pp. 207–256. [CrossRef]
3. Fernández-Ledesma, E.; Jiménez, J.R.; Ayuso, J.; Corinaldesi, V.; Iglesias-Godino, F.J. A Proposal for the Maximum Use of Recycled Concrete Sand in Masonry Mortar Design. *Mater. Constr.* **2016**, *66*, 75. [CrossRef]
4. Tam, V.W.; Soomro, M.; Evangelista, A.C.J. A Review of Recycled Aggregate in Concrete Applications (2000–2017). *Constr. Build. Mater.* **2018**, *172*, 272–292. [CrossRef]
5. Evangelista, L.; Guedes, M. Microstructural Studies on Recycled Aggregate Concrete. In *New Trends in Eco-Efficient and Recycled Concrete*, 1st ed.; de Brito, J., Agrela, F., Eds.; Woodhead Publishing: Cambridge, UK, 2019; pp. 425–451. [CrossRef]
6. Evangelista, L.; Guedes, M.; de Brito, J.; Ferro, A.C.; Pereira, M.F. Physical, Chemical and Mineralogical Properties of Fine Recycled Aggregates Made from Concrete Waste. *Constr. Build. Mater.* **2015**, *86*, 178–188. [CrossRef]
7. Esmaeeli, H.S.; Shishehbor, M.; Weiss, W.J.; Zavattieri, P.Z. A Two-Step Multiscale Model to Predict Early Age Strength Development of Cementitious Composites Considering Competing Fracture Mechanisms. *Constr. Build. Mater.* **2019**, *208*, 577–600. [CrossRef]
8. Pawluczuk, E.; Kalinowska-Wichrowska, K.; Bołtryk, M.; Jiménez, J.R.; Fernández, J.M. The Influence of Heat and Mechanical Treatment of Concrete Rubble on the Properties of Recycled Aggregate Concrete. *Materials* **2019**, *12*, 367. [CrossRef]
9. Ismail, S.; Ramli, M. Engineering Properties of Treated Recycled Concrete Aggregate for Structural Applications. *Constr. Build. Mater.* **2013**, *44*, 464–476. [CrossRef]
10. Shaban, W.M.; Yang, J.; Su, H.; Mo, K.H.; Li, L.; Xie, J. Quality Improvement Techniques for Recycled Concrete Aggregate: A Review. *J. Adv. Concr. Technol.* **2019**, *17*, 151–167. [CrossRef]

11. Dosho, Y. Development of a Sustainable Concrete Waste Recycling System-Application of Recycled Aggregate Concrete Produced by Aggregate Replacing Method. *J. Adv. Concr. Technol.* **2007**, *5*, 27–42. [CrossRef]
12. Tam, V.W.Y.; Tam, C.M.; Lea, K.N. Removal of Cement Mortar Remains from Recycled Aggregate Using Pre-Soaking Approaches. *Resour. Conserv. Recycl.* **2007**, *50*, 82–101. [CrossRef]
13. Robayo-Salazar, R.A.; Rivera, J.F.; de Gutiérrez, R.M. Alkali-Activated Building Materials Made with Recycled Construction and Demolition Wastes. *Constr. Build. Mater.* **2017**, *149*, 130–138. [CrossRef]
14. Kim, Y.C.; Choi, Y.W. Utilization of Waste Concrete Powder as a Substitution Material for Cement. *Constr. Build. Mater.* **2012**, *30*, 500–504. [CrossRef]
15. Schoon, J.; De Buysser, K.; Van Driessche, I.; De Belie, N. Fines Extracted from Recycled Concrete as Alternative Raw Material for Portland Cement Clinker Production. *Cem. Concr. Compos.* **2015**, *58*, 70–80. [CrossRef]
16. Zhao, Z.; Remond, S.; Damidot, D.; Xu, W. Influence of Fine Recycled Concrete Aggregates on the Properties of Mortars. *Constr. Build. Mater.* **2015**, *81*, 179–186. [CrossRef]
17. Gastaldi, D.; Canonico, F.; Capelli, L.; Buzzi, L.; Boccaleri, L.; Irico, S. An Investigation on the Recycling of Hydrated Cement from Concrete Demolition Waste. *Cem. Concr. Compos.* **2015**, *61*, 29–35. [CrossRef]
18. Bordy, A.; Younsi, A.; Aggoun, S.; Fiorio, B. Cement Substitution by a Recycled Cement Paste Fine: Role of the Residual Anhydrous Clinker. *Constr. Build. Mater.* **2017**, *132*, 1–8. [CrossRef]
19. Shui, Z.; Xuan, D.; Wan, H.; Cao, B. Rehydration Reactivity of Recycled Mortar from Concrete Waste Experienced to Thermal Treatment. *Constr. Build. Mater.* **2008**, *22*, 1723–1729. [CrossRef]
20. Xuan, D.X.; Shui, Z.H. Rehydration Activity of Hydrated Cement Paste Exposed to High Temperature. *Fire Mater.* **2011**, *35*, 481–490. [CrossRef]
21. Ahmari, S.; Ren, X.; Toufigh, V.; Zhang, L. Production of Geopolymeric Binder from Blended Waste Concrete Powder and Fly Ash. *Constr. Build. Mater.* **2012**, *35*, 718–729. [CrossRef]
22. Chai, L.; Monismith, C.L.; Harvey, J. *Re-Cementation of Crushed Material in Pavement Bases*; UCPRC-TM-2009-04; Pavement Research Center, University of California: Berkeley, CA, USA, 2009.
23. Paige-Green, P. A Preliminary Evaluation of the Reuse of Cementitious Materials. In Proceedings of the 29th Annual Southern African Transport Conference, Pretoria, South Africa, 16–19 August 2010; pp. 520–529.
24. Kim, J.; Nam, B.H.; Behring, Z.; Al Muhit, B. Evaluation of Recementation Reactivity of Recycled Concrete Aggregate Fines. *Transp. Res. Rec.* **2014**, *2401*, 44–51. [CrossRef]
25. Vegas, I.; Azkarate, I.; Juarrero, A.; Frias, M. Diseño y prestaciones de morteros de albañilería elaborados con áridos reciclados procedentes de escombro de hormigón. *Mater. Constr.* **2009**, *59*, 5–18. [CrossRef]
26. Braga, M.; de Brito, J.; Veiga, R. Incorporation of Fine Concrete Aggregates in Mortars. *Constr. Build. Mater.* **2012**, *36*, 960–968. [CrossRef]
27. Bołtryk, M.; Kalinowska-Wichrowska, K.; Pawluczuk, E. Method for Separation of Set Cement Mortar from Coarse Aggregate and for Crushing that Mortar, and the Device for the Application of This Method, PAT.229887. Available online: http://regserv.uprp.pl/register/application?number=P.417362 (accessed on 1 December 2018).
28. *EN 197-1:2011 Cement. Composition, Specifications and Conformity Criteria for Common Cements*; European Committee for Standardization: Brussels, Belgium, 2011.
29. *EN 933-1:2012 Test for Geometrical Properties of Aggregates. Determination of Particle Size Distribution. Sieving Method*; European Committee for Standardization: Brussels, Belgium, 2012.
30. *EN 206:2013+A1:2016 Concrete. Specification, Performance Production and Conformity*; European Committee for Standardization: Brussels, Belgium, 2012.
31. *EN 196-1:2016 Methods of Testing Cement. Determination of Strength*; European Committee for Standardization: Brussels, Belgium, 2016.
32. *EN 1015-3:1999 Methods of Test for Mortar for Masonry. Determination of Consistence of Fresh Mortar (by Flow Table)*; European Committee for Standardization: Brussels, Belgium, 1999.
33. *EN 450-1:2012 Fly Ash for Concrete. Definition, Specifications and Conformity Criteria*; European Committee for Standardization: Brussels, Belgium, 2012.
34. *EN 196-9:2010 Methods of Testing Cement. Heat of Hydration. SEMI-Adiabatic Method*; European Committee for Standardization: Brussels, Belgium, 2010.
35. Ramachandran, V.S.; Paroli, R.M.; Beaudoin, J.J.; Delgado, A.H. *Handbook of Thermal Analysis of Construction Materials*; Noyes Publications/William Andrew Publishing: New York, NY, USA, 2003.

36. Handoo, S.K.; Agarwal, S.; Agarwal, S.K. Physicochemical, Mineralogical, and Morphological Characteristics of Concrete Exposed to Elevated Temperatures. *Cem. Concr. Res.* **2002**, *32*, 1009–1018. [CrossRef]
37. *EN ISO 3252:2002 Powder Metallurgy. Vocabulary*; European Committee for Standardization: Brussels, Belgium, 2002.
38. Brandt, S. *Data Analysis. Statistical and Computational Methods for Scientists and Engineers*, 3rd ed.; Springer: New York, NY, USA, 1999.
39. Kurdowski, W. *Cement and Concrete Chemistry*; Springer: Dordrecht, The Netherlands, 2014; ISBN 978-94-007-7944-0.

 © 2019 by the authors. Licensee MDPI, Basel, Switzerland. This article is an open access article distributed under the terms and conditions of the Creative Commons Attribution (CC BY) license (http://creativecommons.org/licenses/by/4.0/).

Article

Tailoring Confining Jacket for Concrete Column Using Ultra High Performance-Fiber Reinforced Cementitious Composites (UHP-FRCC) with High Volume Fly Ash (HVFA)

Alessandro P. Fantilli [1,*], Lucia Paternesi Meloni [1,2], Tomoya Nishiwaki [2] and Go Igarashi [2,3]

1. Department of Structural, Geotechnical and Building Engineering, Politecnico di Torino, 10129 Torino, Italy; s238354@studenti.polito.it
2. Department of Architecture and Building Science, Tohoku University, Sendai 980-8579, Japan; tomoya.nishiwaki.e8@tohoku.ac.jp (T.N.); go.igarashi@concrete.t.u-tokyo.ac.jp (G.I.)
3. Department of Civil Engineering, The University of Tokyo, Tokyo 113-8656, Japan
* Correspondence: alessandro.fantilli@polito.it; Tel.: +39-011-090-4900

Received: 4 October 2019; Accepted: 26 November 2019; Published: 3 December 2019

Abstract: Ultra-High Performance Fibre-Reinforced Cementitious Composites (UHP-FRCC) show excellent mechanical performances in terms of strength, ductility, and durability. Therefore, these cementitious materials have been successfully used for repairing, strengthening, and seismic retrofitting of old structures. However, UHP-FRCCs are not always environmental friendly products, especially in terms of the initial cost, due to the large quantity of cement that is contained in the mixture. Different rates of fly ash substitute herein part of the cement, and the new UHP-FRCCs are used to retrofit concrete columns to overcome this problem. To simulate the mechanical response of these columns, cylindrical specimens, which are made of normal concrete and reinforced with different UHP-FRCC jackets, are tested in uniaxial compression. Relationships between the size of the jacket, the percentage of cement replaced by fly ash, and the strength of the columns are measured and analyzed by means of the eco-mechanical approach. As a result, a replacement of approximately 50% of cement with fly ash, and a suitable thickness of the UHP-FRCC jacket, might ensure the lowest environmental impact without compromising the mechanical performances.

Keywords: high volume fly ash (HVFA); steel reinforcing fiber; jacketing; carbon footprint; substitution strategy; environmental impact

1. Introduction

In the last decades, Ultra-High Performance Fiber-Reinforced Cementitious Composites (UHP-FRCC) have been developed to meet the requests of the construction industry [1–3]. UHP-FRCCs can enhance the resistance of buildings and infrastructures due to the ultra-high strength, high ductility, durability, and energy absorption capacity, when compared with normal strength concrete or traditional FRCC. In fact, UHP-FRCCs show a compressive strength that was larger than 150 MPa, combined with high tensile and flexural strengths. Such performances are achieved with a low water/binder ratio, high content of cementitious materials, and by incorporating a copious amount of fibers (steel, polymeric, glass, etc.) [4–7].

One of the most well-known and relevant applications of UHP-FRCC is the retrofitting of existing structures, especially the jacketing of concrete columns and beams, due to these mechanical properties. The aim is to harden those parts of the existing structures that are exposed to high environmental and mechanical actions, especially in the most highly stressed cross-sections and in the structural joints [8–10].

However, this high-performance material has, in parallel, high environmental impact, because of the high content of cement in the mixture [3,5,6,11,12]. Indeed, it is unanimously accepted that the compressive strength of concrete is in direct proportion with the power 2 of the cement content and, consequently, this strength is proportional to the power 2 of the CO_2 emission per cubic meter of concrete. Therefore, the "material substitution strategy" that Habert and Roussel introduced [11] is generally adopted. As the name suggests, it consists of a partial replacement of cement with supplementary cementitious materials, in order to reduce the environmental impact. Specifically, different amounts of cement are substituted by fly ash, a waste by-product that is derived from coal burning. When the mass of fly ash is higher than 50% of the total cementitious materials, the concrete system takes the name of High Volume Fly Ash (HVFA) [13]. It must be remarked that the substitution of cement with fly ash might not always be beneficial. HVFA significantly reduces the environmental impact, but it also leads to a decrease of concrete strength, especially in the UHP-FRCC [14,15].

Some researches, regarding the high strength concrete and UHP-FRC using high quantity of admixtures from by-products, were performed in the last years [16–18]. On the other hand, most of these studies only focused on the development of the cement-based material, without focusing on the structural applications. Accordingly, a new experimental campaign has been carried out on normal-strength concrete cylinders that were reinforced with UHP-FRCC jackets, with the aim of simulating the confinement effect in columns. Different jackets are tailored to investigate the relationship between the thickness of the reinforcing layer and the mechanical performances of the reinforced column. In some of them, different percentages of fly ash substitute the cement for achieving the best mechanical and ecological performances [19]. A design procedure is also proposed to select the best solution by defining the optimal replacement rate of cement with fly ash.

2. Experimental Investigation: Materials and Methods

Figure 1 illustrates the cylindrical sample subjected to uniaxial compression. It consists of a normal-strength concrete core, with a radius r_0 = 50 mm and a height H = 200 mm, confined by a UHP-FRCC jacket of H_1 = 178 mm and different thickness (t_i = 25, 37.5, 50, and 75 mm). As no standard exists to test the confinement effects, the minimum thickness of 25 mm was determined due to the limitation of the casting procedures of the UHP-FRCC layer into a narrow gap.

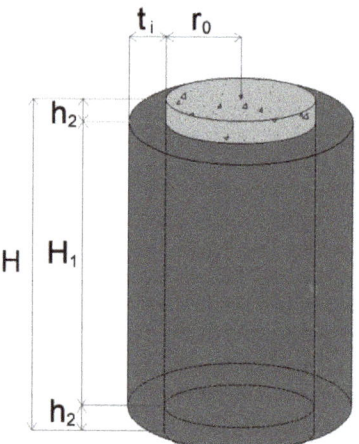

Figure 1. Geometrical properties of the specimens.

According to the mix proportion that is shown in Table 1, the concrete cores have been made with High Early Strength Portland Cement (HESP; Density: 3.14 g/cm^3, Specific surface area: 4490 cm^2/g Ignition loss: 1.08%), the combination of land sand and crushed sand as fine aggregates (S), crushed stone as coarse aggregates (G), tap water (W), and superplasticizer (SP$_1$; Polycarboxylate-based, Density: 1.03 g/cm^3).

Table 1. Concrete mixture used to cast the cores.

HESP (kg/m^3)	S (kg/m^3)	G (kg/m^3)	W (kg/m^3)	SP$_1$ (kg/m^3)
300.3	836	900.3	171.7	1.7

The UHP-FRCCs used herein are made by the following materials [3,13,14]:

- Low Heat Portland Cement (LHC; Density: 3.24 g/cm^3, Specific surface area: 3640 cm^2/g, C$_2$S: 57%, C$_3$A: 3%, MgO: 0.6%, SO$_3$: 2.78%, Ignition loss: 0.72%)
- Undensified Silica Fume (SF; Density: 2.20 g/cm^3, Bulk density: 0.20–0.35 g/cm^3, Coarse particles >45 μm: less than 1.5, SiO$_2$: more than 90%, Ignition loss: less than 3.0%)
- Fly Ash (FA; Density: 2.31 g/cm^3, Specific surface area: 4050 cm^2/g, Coarse particles >45 μm: 5%, SiO$_2$: 54.8%, Ignition loss: 1.2%)
- Silica sand (Ss; Density: 2.60 g/cm^3, Average particle size: 0.212 mm, SiO$_2$: 98.49%, Al$_2$O$_3$: 0.49%)
- Wollastonite mineral fibers (Wo; CaSiO$_3$, Density: 2.60 g/cm^3, Length: 50–2000 μm, Aspect ratio: 3–20, SiO$_2$: 49.71%, CaO: 45.87%, Ignition loss: 1.94%)
- Specific superplasticizer (SP$_2$; Polycarboxylate-based, Density: 1.05 g/cm^3, Solid part: 30%)
- De-foaming Agent (DA; Density: 1.05 g/cm^3)
- Steel micro-fibers (OL—1% in volume), with a length of 6 mm (see Figure 2a)
- Steel macro-fibers (HDR—1.5% in volume), with a length of 30 mm (see Figure 2b)

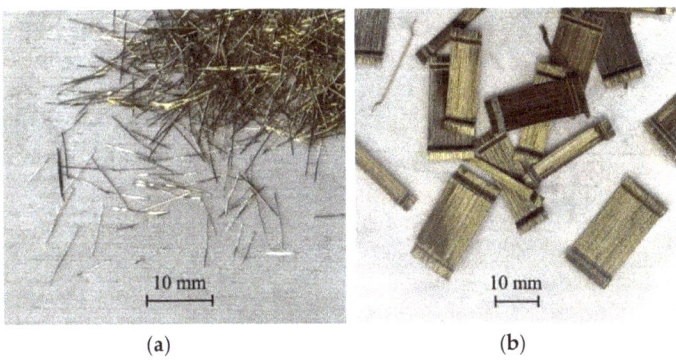

Figure 2. Steel Micro-fibers OL (a) and steel Macro-Fibers HDR (b) used to reinforce the Ultra-High Performance Fibre-Reinforced Cementitious Composites (UHP-FRCC) layers.

Following the material substitution strategy, four mixtures have been tailored for the UHP-FRCC jackets, modifying only the percentage of cement and fly ash. Table 2 shows the mix proportion of the four series of UHP-FRCC (FA0, FA20, FA50, and FA70), with, respectively, 0%, 20%, 50%, and 70% of cement replaced by fly ash. In this Table, the components of each series are reported as a percentage with respect to the weight of the binder, whereas Table 3 shows the mix proportion of UHP-FRCC jackets in kg/m^3. In each series, the water-binder ratio is constant and equal to 0.16, because it guarantees a good balance between the flowability properties and the strength of the hardened concrete, according to the authors' previous study [3]. Note that SP$_2$ contains 30% of the solid part, which is taken into account to calculate the water-binder ratio. The flow table tests were performed in accordance with the Japanese Standard JIS R 5201 [20], which complies with ASTM C 1437 [21].

In particular, for FA0, FA20, FA50, and FA70, a diameter of 190 mm, 260 mm, 290 mm, and 250 mm, was, respectively, measured in the tests. Having these consistencies, all of the UHP-FRCs were cast and compacted in the jacket mould without any segregation.

Ten days after casting the concrete cores (which simulate herein an existing column), the UHP-FRCC jackets are cast around using paper cylinders with a height of 178 mm as a disposable formwork, as the proposed jacket has to be applied to existing structures.

Table 2. Mix proportion of the UHP-FRCC (in weight %) referred to the binder (B).

Series	Binder (B)			Ss/B	Wo/B	W/B	SP$_2$/B	DA/B
	LH C/B	FA/B	SF/B					
FA0	82		18	35	13	14.5	2.2	0.02
FA20	65.6	16.4	18	35	13	14.3	2.6	0.02
FA50	41	41	18	35	13	14.3	2.6	0.02
FA70	24.6	57.4	18	35	13	14.3	2.6	0.02

Table 3. Mix proportion of UHP-FRCC jackets in kg/m^3.

Series	LHC (kg/m^3)	FA (kg/m^3)	SF (kg/m^3)	Ss (kg/m^3)	Wo (kg/m^3)	W (kg/m^3)	SP$_2$ (kg/m^3)	DA (kg/m^3)
FA0	1217.52		267.26	519.67	193.02	214.70	32.67	0.30
FA20	939.24	234.81	257.72	501.12	186.13	204.74	37.23	0.29
FA50	558.33	558.33	245.12	476.63	177.03	194.74	35.41	0.27
FA70	324.43	757.00	237.39	461.59	171.45	188.59	34.29	0.26

Subsequently, the paper formworks were removed two days later and the specimens were subjected to steam curing for the following 48 h. In this way, the hydration process of the binder was accelerated, especially in the UHP-FRCC jackets containing large amounts of fly ash. As depicted in Figure 3, the temperature was slowly increased at a rate of 15 °C per hour up to 90 °C to avoid any crack generation produced by the thermal gradient. Uniaxial compression tests are performed after 28 days from the concrete core casting. A universal testing machine (UTM) with a maximum capacity of 1000 kN is used to apply the compressive load by moving the displacement of the stroke (of the loading cell) at a constant velocity of less than 0.3 mm/min. external strain gauges and embedded strain gauges were used to measure the strain during the tests.

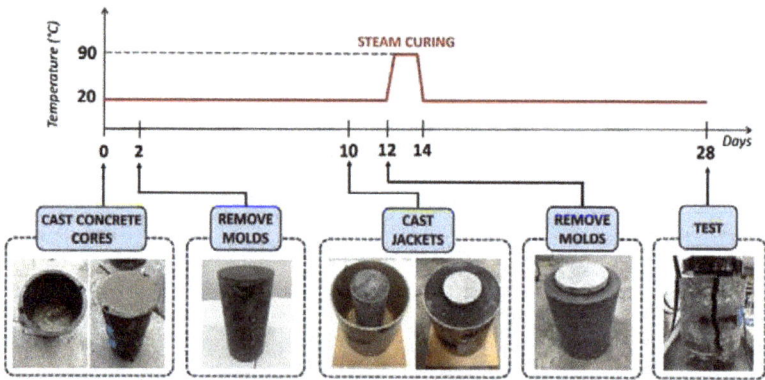

Figure 3. The preparation of the specimens.

Four strain gauges are applied on the jacket's surface: two in the vertical direction and two in the horizontal direction. The vertical gauges measure the strain along the longitudinal axis, while those horizontally oriented gauges measure the swelling of the specimen. Embedded strain gauges were installed inside in the middle of the moulds before casting the concrete cores to record the vertical strain of the concrete core during the compression test.

As shown in Table 4, for some of the thickness of the jacket, three samples are tested for each mixture. In addition, for each series, one concrete core with no reinforcement is tested to estimate the mechanical properties without UHP-FRCC jackets.

Table 4. List of specimens tested in uniaxial compression.

Series	Thickness of the UHP-FRCC Jacket			
	25 (mm)	37.5 (mm)	50 (mm)	75 (mm)
FA0	3 specimens		3 specimens	
FA20	3 specimens	3 specimens	3 specimens	3 specimens
FA50	3 specimens		3 specimens	
FA70	3 specimens	3 specimens	3 specimens	

3. Results

3.1. Mechanical Performances

Table 5 shows the mechanical properties of the UHP-FRCC used in the jackets. The compressive strength and the elastic modulus both do not change with age because of the acceleration of hydration due to the steam curing. Table 5 summarizes the results of the uniaxial compression tests on the jacketed specimens. In this table, the average values of the maximum load (P_{max})—and of the compressive strength (σ_{max}) as well, the Young Modulus (E_{cm}) and the Poisson's Ratio (ν) of the composite specimens are reported. E_{cm} and ν were both calculated at one-third of the maximum load, according to the Japanese Standard JIS A 1149 [22], which complies with ISO 6784 [23]. In Table 6, the strength is also normalized with respect to the strength of the unconfined concrete cylinders (i.e., f_{c_CORE}). E_{cm} increased with the thickness of the jacket due to the confined effect. In the same manner, ν decreased as the jacket thickness increased. The tests on the specimen that is shown in Figure 1 were performed until the failure, which is generally produced by the formation of a large tensile crack in the UHP-FRCC jacket (see the right edge photo of Figure 3). During the first stage of loading, the jacked was uncracked and a linear relationship between the stress (calculated by dividing the load by the cross-sectional area of the core concrete) and the strain can be observed (see Figure 4). When the first crack appeared in the jacket, the slope of this relationship drastically reduces in all of the specimens. This is due to the fact that multiple fine cracks, having a width lower than 0.1 mm, formed in the jacket. Despite the growing number of cracks, the stress continuously increased up to peak, where the tensile strains localised in a single crack of the jacket and the failure occurred.

Table 5. The compressive strength and Young's modulus of UHP-FRCC used in the jackets.

Series	Age	Compressive Strength (MPa)	Young' Modulus (GPa)
FA0	1 week	193.8	46.34
	4 weeks	197.93	45.90
FA20	1 week	193.31	43.39
	4 weeks	179.65	43.38
FA50	1 week	146.97	36.20
	4 weeks	154.55	37.66
FA70	1 week	121.50	33.40
	4 weeks	121.98	34.12

Table 6. The average values of the parameters measured in the uniaxial compression tests.

Series	Jacket (mm)	P_{max} (kN)	σ_{max} (MPa)	σ/f_{c_CORE}	E_{cm} (MPa)	ν
FA0	25	481.47	61.30	1.32	37.90	0.120
	50	595.93	75.88	1.63	40.40	0.104
FA20	25	463.86	59.06	1.27	37.48	0.179
	37.5	499.80	63.64	1.37	38.30	0.157
	50	515.27	65.61	1.41	38.67	0.125
	75	663.53	84.48	1.82	41.70	0.121
FA50	25	415.40	52.89	1.14	36.25	0.195
	50	541.00	68.88	1.48	39.23	0.162
FA70	25	396.07	50.43	1.08	35.74	0.216
	37.5	418.40	53.27	1.15	36.66	0.168
	50	516.60	65.78	1.41	38.70	0.123
Unconfined cylinders		356.37	46.52	1		

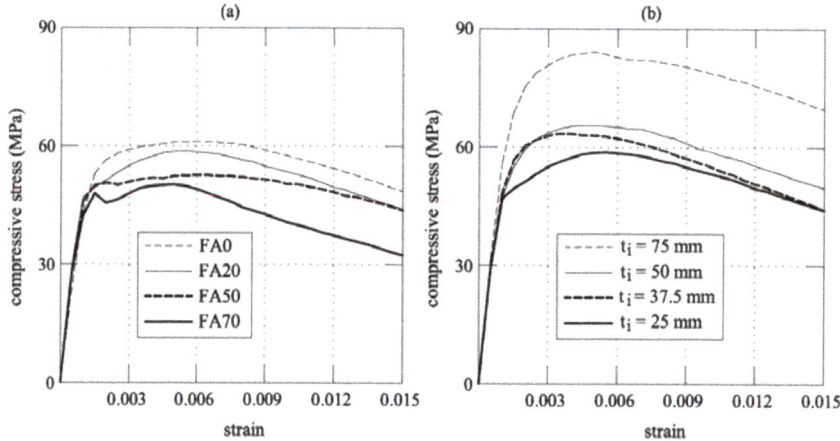

Figure 4. Stress-strain curves of the jacketed cylinders: (a) behaviour of different UHP-FRCC jackets having a constant of thickness $t_j = 25$ mm; (b) behaviour of the same UHP-FRCC jacket (FA20) having different thickness.

The composition of the UHP-FRCC binder considerably influences the mechanical performance of the column. In particular, the effect of the substitution strategy on the mechanical properties is evident in Figure 4a, which reports the stress-strain curves of the concrete cylinders that were confined with a jacket of 25 mm, but with 0%, 20%, 50%, and 70% of cement being replaced by fly ash. The results that were obtained in the case of the FA20 mixture (σ_{max} = 59.06 MPa) are close to those that were achieved from the FA0 mixture (σ_{max} = 61.30 MPa). This is the same tendency of the results of UHP-FRCC material itself, as shown in Table 5. In other words, the replacement of 20% in weight of cement content is paid with a loss of 4% of the maximum compressive strength. When considering the cylinder confined with a jacked of FA0 and $t_j = 25$ mm as a reference, the decrement of the compressive strength is about 14% and 18% when 50% and 70% of cement is replaced by fly ash, respectively.

On the other hand, Figure 4b reports the average stress-strain curves of the cylinders that were reinforced with the same type of jacket (FA20), but of different thickness (i.e., 25, 50, 37.5, and 70 mm). In this case, the strength increases with the thickness of the reinforcing UHP-FRCC layer.

For each UHP-FRCC series, Figure 5a shows the relationship between the thickness of the jacket and the strength of the reinforced cylinders. In all of the cases, the following linear relationship can be used to predict the compressive strength:

$$\sigma_{max} = s \cdot t_i + f_{c\_CORE} \tag{1}$$

where s = slope of the linear relationship. The coefficient s can be separately computed for each series, as shown in Figure 5b, and the values can be plotted as a function of the replacement rate of cement with fly ash (see Figure 6). The slope gradually reduces as the percentage of substitution of cement increases (C_{sub}). Thus, the following linear correlation can be introduced:

$$s = -0.004 \cdot C_{sub} + 0.584 \tag{2}$$

where s is measured in MPa/mm.

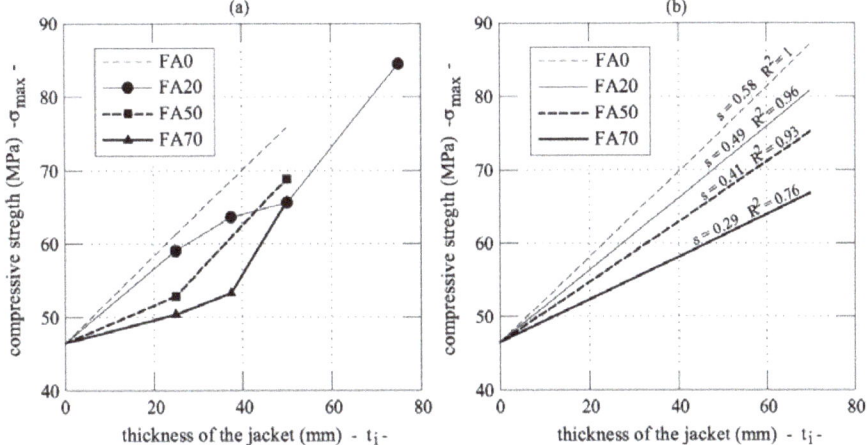

Figure 5. Compressive strength vs. thickness of jacket in the four ultra high performance fiber reinforced concrete (UHP-FRC) series investigated herein: (**a**) results from the tests; and, (**b**) the trend lines of the experimental data.

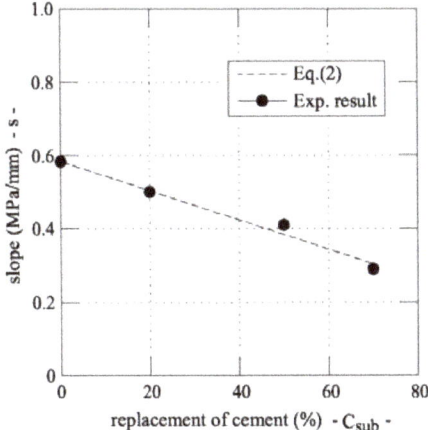

Figure 6. Formula for predicting the slope of the linear approximation of Equation (1).

3.2. Ecological Performances

The parameter considered evaluating the ecological performances is the amount of CO_2 related to the production of 1 m^3 of UHP-FRCC. The amount of CO_2 emitted per unit volume of each mixture is calculated in accordance with the inventory analysis [24] while using the values that were provided by the Japan Concrete Institute (JCI) [25], which listed the main materials used in cementitious composites and the relative carbon footprint. Such values are reported in Table 7 in terms of kg of CO_2 released in the atmosphere for the production of one-ton of material.

Table 7. CO_2 emissions of UHP-FRCC components [25].

Components	kg of CO_2/t
LHC	769
FA	29
Sand	4.9
Water	34.8
SP	150
Fibers	1320

Figure 7 compares the environmental impacts of the various jackets, which were obtained by multiplying the values reported in Table 7 and the mass of materials used to cast the four series of UHP-FRCC. For the sake of completeness, Figure 7 also shows the values of some specimens that has not been tested. In this Figure, when the amount of cement replaced with fly ash is quite high (>50%), the environmental impact is considerably reduced with respect to FA0 (it is halved for FA70). Moreover, it can be noticed that a reduced environmental impact is attained by decreasing the thickness of the jacket and, in parallel, by increasing the percentage of fly ash in the mixture. Here, the CO_2 emission due to the steam curing is not taken into account, because all of the series were subjected to the same procedures and, subsequently, a comparative analysis among all of the specimens is performed.

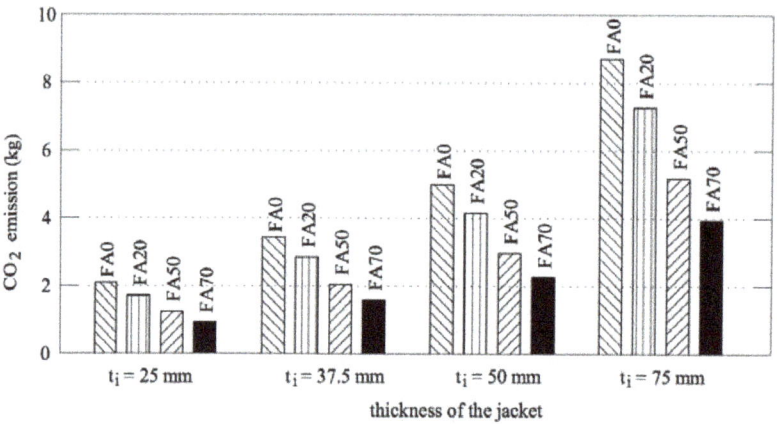

Figure 7. The ecological performances of the UHP-FRCC jackets used to reinforce concrete cores.

3.3. Eco-Mechanical Analysis and Design Procedure

By means of the approach that was proposed by Fantilli and Chiaia [19], ecological and mechanical analyses can be combined to define the best material. Compressive strength (σ_{max}) is considered as the functional unit, herein called the mechanical index (MI), whereas the ecological impact is evaluated through the carbon footprint (ecological index—EI). The reference values MI_{inf} (i.e., the minimum mechanical performance) and EI_{sup} (i.e., the maximum impact) are those of the concrete cylinder reinforced with the UHP-FRCC jacket without the substitution of cement with fly ash (i.e., FA0) and with a thickness t_i = 25 mm.

Figure 8 shows the non-dimensional chart that was used to perform the comparative analysis among the concrete specimens reinforced with the UHP-FRCC layer. In this diagram, the ratios MI/MI$_{inf}$ and EI$_{sup}$/EI are the abscissa and the ordinate, respectively. In other words, the following formula are used to define the non-dimensional axes in Figure 8:

$$\frac{MI}{MI_{inf}} = \frac{\sigma_{max}}{\sigma_{max} \; of \; FA0 \; with \; t_i = 25 \; mm} \qquad (3)$$

$$\frac{EI_{sup}}{EI} = \frac{kg \; CO_2 \; of \; FA0 \; with \; t_i = 25 \; mm}{kg \; CO_2} \qquad (4)$$

Most of the experimental results fall in Zone 2, where the mechanical performances are increased at the expense of an environmental impact higher than that of FA0 with t_i = 25 mm (for which MI/MI$_{inf}$ = EI$_{sup}$/EI = 1).

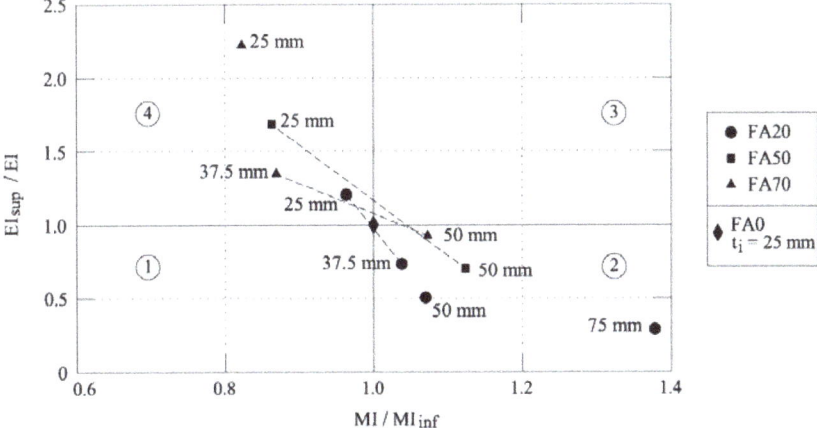

Figure 8. Eco-Mechanical analysis of the UHP-FRCC jackets [19].

However, a group of experimental values falls within Zone 4, which shows ecological performances greater than those shown by the reference specimen, but lower mechanical performances. Although none of the test results fall within Zone 3, where the ecological and mechanical performances are both improved, an area of the possible best solutions can be defined in Figure 8. More precisely, the UHP-FRCC jackets with a thickness between 37.5–50 mm and made with FA70 series, and those of the FA50 series with thickness being included in the range 25–50 mm (see the dashed lines), might perform better than the reference cylinders (i.e., part of the dashed lines falls within Zone 3). The same is also valid for some thickness (within the range 25~37.5 mm) of the series FA20, as evidenced by the dashed line that is reported in Figure 8.

Nevertheless, Equations (1) and (2) can be used to relate the compressive stress, the thickness of the jacket, and the type of fiber-reinforced concrete to design the exact UHP-FRCC layer. In particular, the design procedure that is shown in Figure 9 can be introduced with the aim of increasing the strength of the UHP-FRCC jacketing system, and reducing the CO_2 emissions as well.

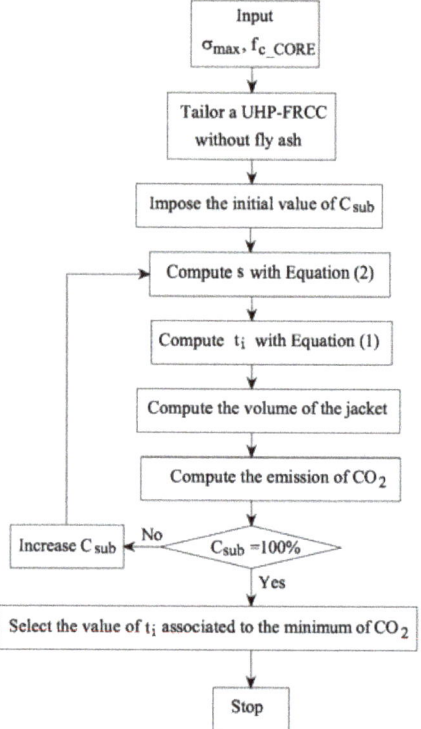

Figure 9. The design procedure used to optimize the UHP-FRCC jacket of the concrete columns.

As a result, Figure 10 shows the curves that relate the CO_2 emissions with the percentage of cement replaced by fly ash for three values of the compressive strength σ_{max} (i.e., 55, 65, and 75 MPa). For each load carrying capacity, four thicknesses of the jackets have been obtained referring to the mixtures FA0, FA20, FA50, and FA70.

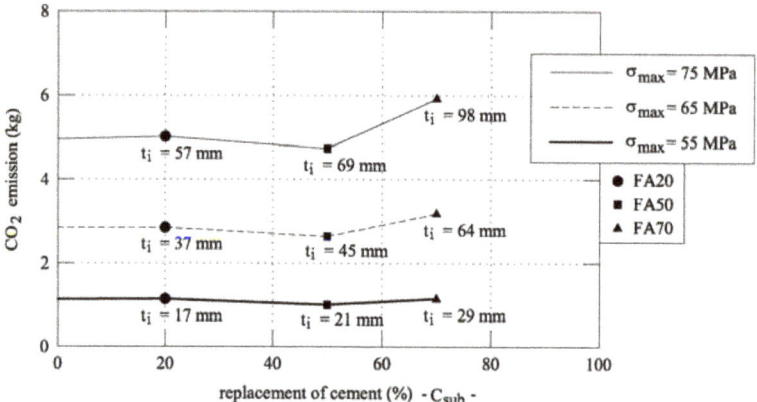

Figure 10. The ecological impact of UHP-FRCC jackets made with different mixtures (having different σ_{max}).

If $\sigma_{max} \leq 55$ MPa is required, the CO_2 emission does not change with the thickness or the percentage of cement replacement with fly ash. Conversely, as the required strength increases, the corresponding CO_2 emission has a minimum in correspondence of a specific thickness. For instance, the optimal replacement of cement with fly ash to achieve a compressive stress of 75 MPa is 50% and the corresponding thickness is 67 mm. In the three curves that are shown in Figure 10, the best substation rate is at 50%, as obtained by Fantilli et al. [26] in the reinforced concrete beam. Indeed, for lower substitution rates, thickness reduces, but the CO_2 emissions are higher.

When the substitution strategy is forced to higher percentages, the same mechanical performances can only be obtained through solutions with a high environmental impact, as the thickness of the jacket increases. In other words, the UHP-FRCC jacket with a low fly ash content shows mechanical performances that are not compensated by the environmental impact. In the same way, the jacketing system is so thick to produce a large environmental impact for large substitutions of cement.

4. Conclusions

According to the results of the tests previously described, the following conclusions are drawn:

- A linear relationship between the thickness of the jacket and the compressive strength of concrete columns has been found. By means of this relationship, the thickness of the UHP-FRCC jacket might be adjusted to achieve the desired mechanical performances.
- The strength of the confined column linearly decreases as the substitution of cement with fly ash increased in the UHP-FRCC jacket.
- Through the eco-mechanical analysis, it is possible to demonstrate that the partial replacement of cement with fly ash, combined with a suitable thickness of the jacket, might simultaneously guarantee the best mechanical and ecological performances.
- A design procedure has been introduced to optimise the UHP-FRCC jacketing system and reduce the environmental impact as well. Particularly, for a given strength of the column, the impact of the UHP-FRCC jacket has a minimum in correspondence of a substitution rate of cement with fly ash that is close to 50%.

Author Contributions: Conceptualization, T.N. and A.P.F.; Methodology, T.N., G.I. and A.P.F.; Validation, T.N. and A.P.F.; Investigation, L.P.M., T.N. and A.P.F.; Resources, L.P.M. and T.N.; Writing—Original Draft Preparation, L.P.M. and A.P.F.; Writing—Review & Editing, T.N. and A.P.F.; Visualization, L.P.M. and A.P.F.; Project Administration, T.N. and A.P.F.

Funding: This work was partially supported by JSPS KAKENHI Grant Number 17H03337. The authors would like to express their deep gratitude.

Conflicts of Interest: The authors declare no conflict of interest.

References

1. Richard, P.; Cheyrezy, M. Composition of reactive powder concretes. *Cem. Concr. Res.* **1995**, *25*, 1501–1511. [CrossRef]
2. Yu, R.; Spiesz, P.; Brouwers, H.J.H. Mix design and properties assessment of Ultra-High Performance Fibre Reinforced Concrete (UHPFRC). *Cem. Concr. Res.* **2014**, *56*, 29–39. [CrossRef]
3. Kwon, S.; Nishiwaki, T.; Kikuta, T.; Mihashi, H. Development of ultra-high-performance hybrid fiber-reinforced cement-based composites. *ACI Mater. J.* **2014**, *111*, 309–318. [CrossRef]
4. Resplendino, J. Introduction: What is UHPFRC? In *Designing and Building with Ultra-High Performance Fibre-Reinforced Concrete (UHPFRC): State of the Art and Development*; Toutlemonde, F., Resplendino, J., Eds.; John Wiley & Sons, Inc.: London, UK, 2011; pp. 3–12.
5. Park, S.H.; Kim, D.J.; Ryu, G.S.; Koh, K.T. Tensile behaviour of Ultra-High Performance Hybrid Fibre Reinforced Concrete. *Cem. Concr. Compos.* **2012**, *34*, 172–184. [CrossRef]
6. Rossi, P. Influence of fibre geometry and matrix maturity on the mechanical performance of ultra-high-performance cement-based composites. *Cem. Concr. Compos.* **2013**, *37*, 246–248. [CrossRef]

7. Wille, K.; El-Tawil, S.; Naaman, A.E. Properties of strain hardening ultra high performance fiber reinforced concrete (UHP-FRC) under direct tensile loading. *Cem. Concr. Compos.* **2014**, *48*, 53–66. [CrossRef]
8. Brühwiler, E. "Structural UHPFRC": Welcome to the post-concrete era! In Proceedings of the First International Interactive Symposium on UHPC, Des Moines, IA, USA, 18–20 July 2016.
9. Kosaka, Y.; Imai, T.; Kunieda, M.; Mitamura, H.; Matsui, S. Development of High Performance Fibre Reinforced Concrete for Rehabilitation of Bridge Deck Slab. In Proceedings of the International Conference on the Regeneration and Conservation of Concrete Structures RCCS, Nagasaki, Japan, 1–3 June 2015.
10. Noshiravani, T.; Brühwiler, E. Rotation capacity and stress redistribution ability of R-UHPFRC–RC composite continuous beams: An experimental investigation. *Mater. Struct.* **2013**, *46*, 2013–2028. [CrossRef]
11. Habert, G.; Roussel, N. Study of two concrete mix-design strategies to reach carbon mitigation objectives. *Cem. Concr. Compos.* **2009**, *31*, 397–402. [CrossRef]
12. Turner, L.; Collins, F. Carbon dioxide equivalent (CO2-e) emissions: A comparison between geopolymer and OPC cement concrete. *Constr. Build. Mater.* **2013**, *43*, 125–130. [CrossRef]
13. Mehta, P.K.; Monteiro, P.J.M. *Concrete-Microstructure, Properties and Materials*; McGraw-Hill: New York, NY, USA, 2014.
14. Nishiwaki, T.; Suzuki, K.; Kwon, S.; Igarashi, G.; Fantilli, A.P. Ecological and Mechanical Properties of Ultra High Performance—Fiber Reinforced Cementitious Composites containing High Volume Fly Ash. In Proceedings of the 4th International RILEM Conference on Strain-Hardening Cement-Based Composites (SHCC4), Dresda, Germany, 18–20 September 2017.
15. Shaikh, F.U.; Nishiwaki, T.; Kwon, S. Effect of fly ash on tensile properties of ultra-high performance fiber reinforced cementitious composites (UHP-FRCC). *J. Sustain. Cem. Based Mater.* **2018**, *7*, 1–15. [CrossRef]
16. Yu, R.; Spiesz, P.; Brouwers, H.J.H. Development of an eco-friendly Ultra-High Performance Concrete (UHPC) with efficient cement and mineral admixtures uses. *Cem. Concr. Compos.* **2015**, *55*, 383–394. [CrossRef]
17. Yalçınkaya, Ç.; Yazıcı, H. Effects of ambient temperature and relative humidity on early-age shrinkage of UHPC with high-volume mineral admixtures. *Constr. Build. Mater.* **2017**, *144*, 252–259. [CrossRef]
18. Aghdasi, P.; Ostertag, C.P. Green ultra-high performance fiber-reinforced concrete (G-UHP-FRC). *Constr. Build. Mater.* **2018**, *190*, 246–254. [CrossRef]
19. Fantilli, A.; Chiaia, B. The work of fracture in the eco-mechanical performances of structural concrete. *J. Adv. Concr. Technol.* **2013**, *11*, 282–290. [CrossRef]
20. Japan Industrial Standards. *JIS R 5201:2015 Physical Testing Methods for Cement*; Japanese Standards Association: Tokyo, Japan, 2015.
21. ASTM International. *ASTM C1437-15 Standard Test Method for Flow of Hydraulic Cement Mortar*; ASTM International: West Conshohocken, PA, USA, 2015.
22. Japan Industrial Standards. *JIS A 1149:2010 Method of Test for Static Modulus of Elasticity of Concrete*; Japanese Standards Association: Tokyo, Japan, 2010.
23. International Organization for Standardization. *ISO 6784:1982 Concrete—Determination of Static Modulus of Elasticity in Compression*; American National Standards Institute (ANSI): New York, NY, USA, 1982.
24. Itsubo, N.; Sakagami, M.; Washida, T.; Kokubu, K.; Inaba, A. Weighting across safeguard subjects for LCIA through the application of conjoint analysis. *Int.J. Life Cycle Assess.* **2004**, *9*, 196–205. [CrossRef]
25. Japan Concrete Institute. *Textbook for Environmental Aspects of Concrete (Tentative)*; Concrete Institute: Tokyo, Japan, 2013. (In Japanese)
26. Fantilli, A.P.; Tondolo, F.; Chiaia, B.; Habert, G. Designing Reinforced Concrete Beams Containing Supplementary Cementitious Materials. *Materials* **2019**, *12*, 1248. [CrossRef] [PubMed]

© 2019 by the authors. Licensee MDPI, Basel, Switzerland. This article is an open access article distributed under the terms and conditions of the Creative Commons Attribution (CC BY) license (http://creativecommons.org/licenses/by/4.0/).

Article

Study of Strain-Hardening Behaviour of Fibre-Reinforced Alkali-Activated Fly Ash Cement

Hyuk Lee [1,*,†], Vanissorn Vimonsatit [1,†], Priyan Mendis [2,†] and Ayman Nassif [3]

1. School of Civil and Mechanical Engineering, Curtin University, Perth, WA 6102, Australia; V.Vimonsatit@curtin.edu.au
2. Department of Infrastructure Engineering, University of Melbourne, Parkville, VIC 3010, Australia; pamendis@unimelb.edu.au
3. School of Civil Engineering and Surveying, The University of Portsmouth, Portsmouth, Hampshire PO1 2UP, UK; ayman.nassif@port.ac.uk
* Correspondence: lee.lee@curtin.edu.au; Tel.: +61-892-664-160
† These authors contributed equally to this work.

Received: 25 October 2019; Accepted: 29 November 2019; Published: 3 December 2019

Abstract: This paper presents a study of parameters affecting the fibre pull out capacity and strain-hardening behaviour of fibre-reinforced alkali-activated cement composite (AAC). Fly ash is a common aluminosilicate source in AAC and was used in this study to create fly ash based AAC. Based on a numerical study using Taguchi's design of experiment (DOE) approach, the effect of parameters on the fibre pull out capacity was identified. The fibre pull out force between the AAC matrix and the fibre depends greatly on the fibre diameter and embedded length. The fibre pull out test was conducted on alkali-activated cement with a capacity in a range of 0.8 to 1.0 MPa. The strain-hardening behaviour of alkali-activated cement was determined based on its compressive and flexural strengths. While achieving the strain-hardening behaviour of the AAC composite, the compressive strength decreases, and fine materials in the composite contribute to decreasing in the flexural strength and strain capacity. The composite critical energy release rate in AAC matrix was determined to be approximately 0.01 kJ/m^2 based on a nanoindentation approach. The results of the flexural performance indicate that the critical energy release rate of alkali-activated cement matrix should be less than 0.01 kJ/m^2 to achieve the strain-hardening behaviour.

Keywords: fibre reinforced; alkali-activated; strain hardening

---

1. Introduction

Alkali-activated cement (AAC) is a potential cementitious system to be introduced as an alternative cement [1,2]. AAC-based concrete exhibits a variety of advantageous properties and characteristics, such as high strength, low shrinkage, fast setting time, good acid and fire resistance, and low thermal conductivity. A highly concentrated alkali hydroxide solution or silicate solution that reacts with solid aluminosilicate produces synthetic alkali aluminosilicate materials [2]. These materials are classified as polymers because their structures are large molecules formed by number of group of smaller molecule [3]. The form of one such polymer is the product of the reaction of an alkali solution and source materials, such as fly ash—which is rich in aluminosilicate and includes organic minerals, such as kaolinite and inorganic material [4].

Cementitious materials, such as mortar and concrete, generally show brittle behaviour. Historically, traditional reinforcement in concrete was in the form of continuous reinforcing bars, which should be in an appropriate location to resist the imposed tensile and shear stresses. In a fibre reinforced cementitious composite, fibres are discontinuous and are randomly distributed throughout the cementitious matrix. They tend to be more closely located than conventional reinforcing bars, and are therefore better at controlling cracking. High performance fibre reinforced cementitious composite (HPFRCC) is a type of material that exhibits a pseudo strain-hardening characteristic under uniaxial tensile stress in fibre reinforced cementitious composites. The "high performance" refers to the quality a fibre reinforced cementitious composite based on the shape of its stress–strain curve in fibre orientations [5]. HPFRCC can be generally classified by composite mechanics, energy, and numerical approaches. One way to define the condition to accomplish strain hardening behaviour is that post-cracking strength of the composite is higher than its cracking strength. It is, therefore, necessary to understand some important parameters which are related to the shape of the stress–strain relationship of HPFRCC [6]. Several research works [7–10] reported strain-hardening behaviour of cementitious materials; however, the performance of fibre reinforced AAC composite is still an enigma. Fly ash is a common aluminosilicate source in AAC; therefore, in this research, an investigation was carried out on the affects of fibre contents in alkali-activated fly ash cement (AAFA) composites. The experimental works were to determine fibre interfacial strength in AAFA matrices, and the numerical analysis approach using Taguchi's DOE method was to determine the effect of the parameters on AAFA matrix. Furthermore, the compressive strength development and the strain-hardening behaviour of AAFA composites were studied to examine the structural performance under compression and flexure.

2. Materials and Methods

Class F (low calcium) fly ash available locally in Australia was used to prepare AAFA matrices. The summary of chemical compositions of fly ash is presented in Table 1. The specimens were cast in 25 mm cubic moulds for the compressive strength test, which was modified based on ASTM C109, and in prismatic specimens of 160 × 40 × 40 mm for a 3-point flexural performance test according to ASTM C78 as shown in Figure 1. The monofilament polyvinyl alcohol (PVA) fibre was used in this research; its diameter and length are 38 μm and 8 mm, respectively. PVA fibre has high chemical bond strength due to the hydrophilic nature and highly alkali resistant characteristic. The tensile strength and elastic modulus of PVA fibre were reported as 1600 MPa and 40 GPa, respectively.

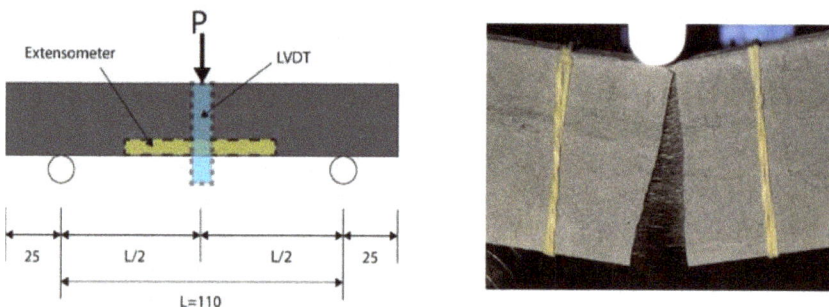

Figure 1. Configuration of flexural performance.

The specimens were cured for 24 h at 60 °C which is a common curing temperature for AAC [3,11,12]. After that, the specimens were placed in a curing room at 23 °C ± 3 until testing. The compressive strength test was conducted at 7, 14, and 28 days of curing age while the flexural test was conducted on day 28 of curing. Each test was repeated on six samples. The selected mixing proportion is the process of choosing suitable fibre volume fraction of AAFA mixtures, as shown in Table 2; there were two main groups, with and without silica fume, and with varying fibre volume fraction in AAFA mixtures. The liquid to solid ratio and the content of superplasticiser were 0.5 and 0.02%, respectively.

Table 1. Chemical composition of low calcium fly ash (wt. %).

SiO_2	Al_2O_3	CaO	Fe_2O_3	K_2O
65.9	24.0	1.59	2.87	1.44

Table 2. Mix proportion.

Group	Index	Fly Ash	Silica Fume	PVA Fiber *
A	F1	1	-	-
	F2	1	-	0.5%
	F3	1	-	1.0%
	F4	1	-	2.0%
B	FS1	1	0.2%	-
	FS2	1	0.2%	0.5%
	FS3	1	0.2%	1.0%
	FS4	1	0.2%	2.0%

* by volume fraction.

It is important to note that interface's properties between fibre and matrix significantly influence the performance of a composite. The interfacial properties are also important in the fracture mechanism and the fracture toughness of the composite. The failure process in a composite material when a crack propagates is complex and involves matrix cracking. The bonding strength between fibre and matrix is to be considered as a source of energy dissipation of HPFRCC. The single fibre pull out test is the most common method to understanding the interfacial strength. Generally, the fibre pull out has three stages during debonding [13–15], as shown in Figure 2. Each stage of a single fibre pull out test can be expressed by:

- The first stage, S_0: the fibre and matrix is bonded until reaching the maximum interfacial bond strength τ_{max};
- The second stage, $S_0 - S_1$: a crack propagation could occur along the interface between the fibre and matrix which leads to complete debonding;
- The third stage, $S_1 - S_{ref}$: fibre is pulled out from the matrix and starts to slip;
- Thus, the maximum pull out force is the most important parameter of HPFRCC, which can present maximum interfacial bond strength.

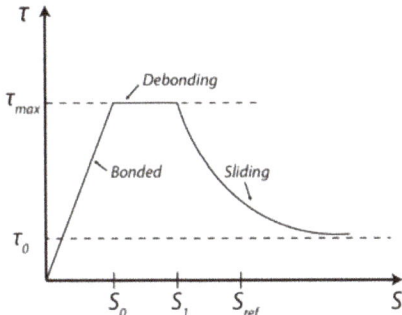

Figure 2. Idealised interface law in three stages of single fibre pull out (adopted after [16]).

A numerical study for the behaviour of single fibre pull out was carried using commercial finite element (FE) software package ANSYS [17]. A 2-D axisymmetric model was employed for the simulation of the single fibre pull out process. In the developed model, a PVA fibre with a radius R_f was embedded at the centre of the cylindrical matrix, and L_d was the total embedded length of the fibre. The bottom of the model was constrained in both radial and axial directions. The interfacial properties were modelled using the bilinear cohesive zone model (CZM) in mode II, which was established by fracture mechanic models, such as the interface traction and separation. The relationship between normal critical energy G_{cn} and tangential critical energy G_{ct} can be expressed by the maximum normal contact stress σ_{max}, the maximum tangential contact stress τ_{max}, the complete normal displacement δ_n, and the complete tangential displacement δ_t [17]. Figure 3 presents the model of the FE single fibre pull out test with the fibre and matrix model which were meshed with 122,406 six node quadrilateral elements. The model was analysed using a non-linear geometrical method with convergent displacement control. To confirm the validity of the FE analysis of the single fibre pull out, an analytical fibre pull out test was conducted. An interfacial friction law for the slip mechanism between the fibre and the matrix has been investigated by several authors [16,18,19]. For an analytical fibre pull out, a proposed model by Zhan et al. [16], which was based on the interfacial law that could capture the major mechanism involved in various situations, was used to obtain the fibre pull out force. The results of the analytical and the FE analyses of the single fibre pull out model were overall in good agreement, with around 2% difference, as illustrated in Figure 4. Thus, the FE simulation can be used for investigating the interfacial behaviour between the fibre and the AAFA matrix.

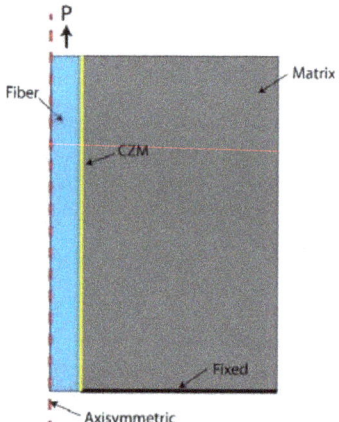

Figure 3. Configuration of single fibre pull out simulation without an inclined angle.

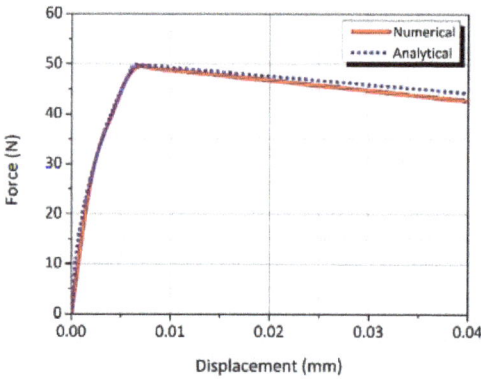

Figure 4. Validation of finite element (FE) model with the analytical model by Zhan et al. [16].

Taguchi's DOE approach with eight parameters and three levels of test variables were selected in accordance to the literature [16,18,20–23], as shown in Table 3. The standard L_{27} (3^{13}) orthogonal array was used in accordance to these parameters, and the detail of L_{27} orthogonal array is shown in Table 4.

Table 3. Variation parameters and levels.

Parameter	Level 1	Level 2	Level 3
Elastic modulus of matrix, E_m (GPa)	20	25	30
Diameter of matrix, d_m (mm)	5	10	15
Poisson's ratio of matrix ν_m	0.2	0.22	0.25
Elastic modulus of fibre E_f (GPa)	40	120	210
Diameter of fibre, d_f (mm)	0.038	0.5	1
Fibre embedded length, L_d (mm)	4	10	12
Maximum tangential traction, τ_t^{max} (MPa)	0.5	1	1.5
Complete tangential displacement, δ_t (mm)	0.1	0.25	0.4

Table 4. Standard L_{27} orthogonal array.

No.	E_m	d_m	ν_m	E_f	d_f	L_d	τ_t^{max}	δ_{max}
1	1	1	1	1	1	1	1	1
2	1	1	1	1	2	2	2	2
3	1	1	1	1	3	3	3	3
4	1	2	2	2	1	1	1	2
5	1	2	2	2	2	2	2	3
6	1	2	2	2	3	3	3	1
7	1	3	3	3	1	1	1	3
8	1	3	3	3	2	2	2	1
9	1	3	3	3	3	3	3	2
10	2	1	2	3	1	2	3	1
11	2	1	2	3	2	3	1	2
12	2	1	2	3	3	1	2	3
13	2	2	2	3	1	1	3	2
14	2	2	3	1	2	3	1	3
15	2	2	3	1	3	1	2	1
16	2	3	1	2	1	2	3	3
17	2	3	1	2	1	2	3	3
18	2	3	1	2	3	1	2	2
19	3	1	3	2	1	3	2	1
20	3	1	3	2	1	3	3	2
21	3	1	3	2	3	2	1	3
22	3	2	1	3	1	3	2	2
23	3	2	1	3	2	1	3	3
24	3	2	1	3	3	2	1	1
25	3	3	2	1	1	3	2	3
26	3	3	2	1	2	1	3	1
27	3	3	2	1	3	2	1	2

3. Results and Discussion

3.1. Single Fibre Pull Out

Based on the Taguchi's DOE approach, a statistical signal to noise (S/N) ratio analysis was performed to determine the effect of these parameters on the maximum fibre pull out force P_{max}, as illustrated in Table 5 and Figure 5. The S/N ratio shows that the diameter of the fibre has the most effect on the fibre pull out force. The elastic modulus of the fibre and the matrix has a minor effect on the pull out force. A further analysis of the single fibre pull out behaviour was done using analysis of variance (ANOVA) and the results indicate that the contribution of the fibre diameter on pull out force is 44.69% of the total contribution factors. The overall results are presented in Table 6. It can be observed that increasing the elastic modulus of the matrix, the diameter of the fibre, the tangential traction, and the embedded length of the fibre results in increasing the pull out force. The contributions of the elastic modulus of the matrix, the tangential traction and the embedded length of the fibre on the pull out force are 14.48%, 8.92%, and 9.47%, respectively. At the same time, increasing the Poisson's ratio results in decreasing in the pull out force but the contribution is minor. The contributions of the diameter of the fibre, Poison's ratio, the elastic modulus of matrix, and complete tangential displacement on the pull out force are fairly similar at about 2.5%.

Table 5. Numerical studies of single fibre pull out with Taguchi's DOE.

No.	P_{max} (N)	No.	P_{max} (N)	No.	P_{max} (N)
1	0.23	10	1.27	19	1.43
2	15.51	11	9.42	20	9.41
3	56.09	12	15.71	21	12.57
4	0.24	13	1.09	22	1.67
5	15.67	14	9.41	23	9.41
6	55.95	15	15.59	24	12.56
7	0.24	16	1.36	25	1.51
8	15.61	17	9.40	26	9.19
9	14.61	18	15.69	27	12.56

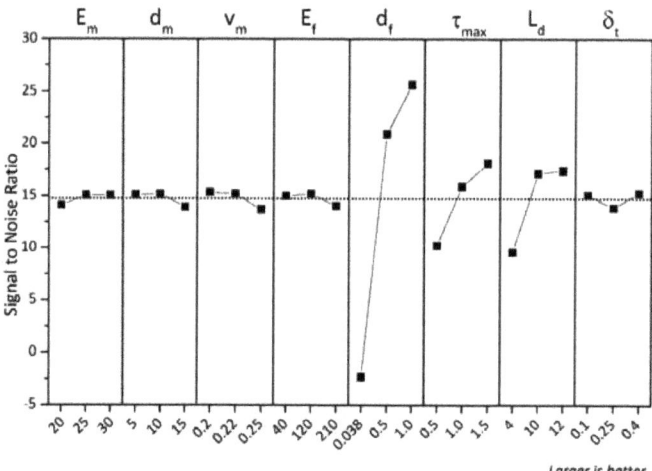

Figure 5. Signal to noise ratio of single fibre pull out.

Table 6. Analysis of Variance of fibre pull out force.

Source	DF [a]	SS [b]	MS [c]	Contribution %
E_m	2	737.6	368.8	14.48
d_m	2	127.2	63.6	2.50
v_m	2	129.3	64.7	2.54
E_f	2	124.0	61.9	2.43
d_f	2	2276.5	1138.3	44.69
L_d	2	454.4	227.2	8.92
τ_t^{max}	2	482.5	241.2	9.47
δ_{max}	2	126.7	63.4	2.49
Error	10	653.8	63.6	12.48

[a] degree of freedom; [b] sum of squares; [c] mean of squares.

The failure process in a composite material when a crack propagates is complex and involves matrix cracking. The bonding strength between fibre and matrix is to be considered as a source of energy dissipation. Thus, a single fibre pull out test of PVA fibre conducted with AAFA paste matrix (l/s = 0.6) was also conducted with OPC paste matrix (w/c = 0.3) to compare the interfacial bonding strength between AAFA and OPC matrices. The embedded length (L_d) of the fibre was around 4 mm which is half of the total length of the fibre, and the diameter of fibre (d_f) was 38 μm. Assuming uniform bonding, the maximum interfacial bonding stre. The results of the single fibre pull out test of AAFA and OPC matrices show that the pull out force is similar. A comparison of the maximum pull out force between the numerical and experimental results are presented in Table 7. The input parameters such as elastic modulus and Poisson's ratio of the FE model were adopted from the authors' previous work [24]. The comparison of the maximum pull out force between the numerical and experimental results are in good agreement, thus, validating the numerical analysis of the single fibre pull out with Taguchi's DOE.

Table 7. Maximum pull out force in the finite element and the experimental results.

Matrix	FEM	Experimental	Ratio
OPC	0.482	0.480	1.001
AAFA	0.530	0.530	1.000

3.2. Compressive Strength

Figure 6 shows the average compressive strength development from 7 to 28 days of curing age in each composite. It can be seen that the compressive strength of AAFA composite generally decreases with increasing fibre volume fraction ratio. Also, it was observed that the compressive strength development was not significantly increased by the fibre volume fraction ratio in F2 mixture, which exhibited a high rate of compressive strength development between 7 to 14 days of curing ages. The test results indicate that the compressive strength development is not significantly affected by the fibre volume fraction ratio.

The behaviour and the ultimate compressive failure mode of AAFA composites are shown in Figure 7. It is known that PVA fibre matrix can exhibit ductile behaviour after reaching its compressive strength because of the transverse confinement effect of the PVA fibre, while normal AAFA mixtures without PVA fibre (F1 and FS1 mixtures) present a significant decrease in stress after reaching their ultimate compressive strength. However, OPC composites (w/c = 0.4) have more ductile behaviour after reaching their ultimate compressive strength than that of AAFA composites, as shown in Figure 7. It can also be seen that the post-peak behaviour depends on the fibre content; those mixes with the same fibre content show similar post-peak behaviour. It can be seen that the compressive strain is not significantly affected by the fibre volume fraction ratio. Further, the compressive strain corresponding to the compressive strength is not meaningfully affected. However, the compressive strength generally decreases with increasing fibre volume fraction ratio and the content of the added silica fume in AAFA composites led to lower compressive strength. In a previous research study [24], it was observed that silica fume in AAFA matrix contributed to a significant decrease in the compressive strength due to a decrease in the cohesion of the reaction products.

3.3. Flexural Performance

The flexural behaviour of composites will exhibit deflection-hardening, or softening behaviour after, first, cracking. The first cracking point of the composite is defined as limit of proportionality (LOP), and the maximum equivalent flexural strength point of the composite is defined as modulus of rupture (MOR) [25]. The flexural behaviours of AAFA composites are shown in Figures 8 and 9. The flexural performance of F1 mixture shows a typical form of deflection-softening behaviour; F2 mixture shows quasi-deflection-softening behaviour; and F4 mixture shows deflection-hardening behaviour. However, for F3 mixture, some of the specimens had complex behaviours, which were deflection-hardening and quasi-deflection-softening behaviours. The maximum loading capacity of F4 mixture was observed to be about 74% greater than that of other mixtures, and the deflection capacity of F4 mixtures was also observed to be greater than that of F1, F2, and F3 mixtures. Similarly, the flexural performance of FS1, FS2, and FS3 mixtures showed typical deflection-softening behaviours, while F4 mixture presented deflection-hardening behaviour. The maximum loading and deflection capacity of F4 mixture were found to be around 65% and 85% greater than the maximum loading and the deflection capacities, respectively, of other mixtures. As the volume fraction ratio of fibre in the AAFA composite increased from 0% to 2.0%, the effects of the fibre volume fraction ratio on the deflection capacities of different mixtures of AAFA composites were plotted in Figure 10. There results show an increasing trend of the deflection capacity at LOP as the linear relationship, and an increasing trend of the deflection capacity at MOR, as the exponential relationship. The improvement of deflection at MOR in Group A was observed to have much higher deflection capacity than that of Group B. Li et al. [6,22,26] reported that adding fine aggregates in OPC composite could improve the pseudo-strain hardening behaviour. In AAFA composite, however, adding fine aggregates (SF) in this case does not improve the flexural deflection and strength capacity. The flexural behaviour of AAFA composite with SF as added fine aggregates shows a decrease in the flexural strength and no improvement in the flexural deflection and strain capacity.

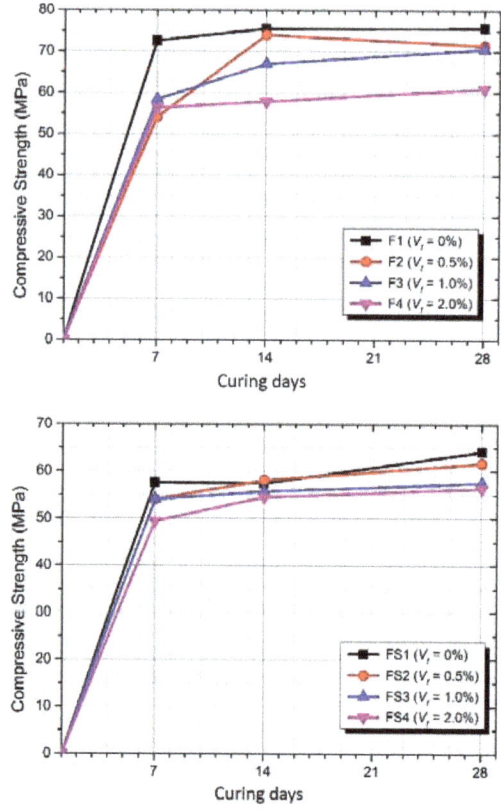

Figure 6. Compressive strength developments of Group A (**top**) and Group B (**bottom**).

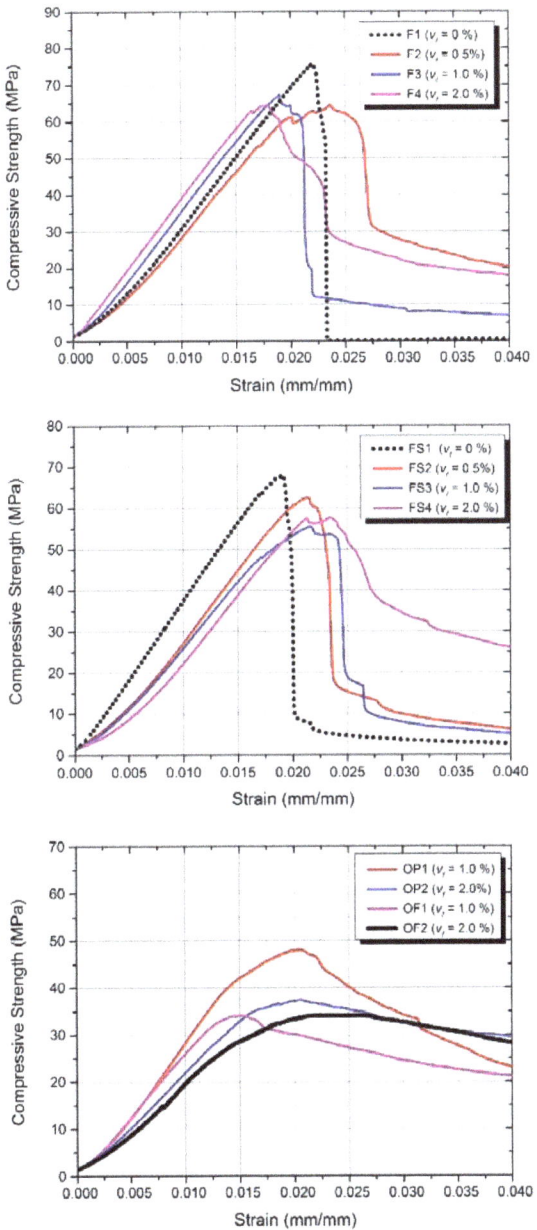

Figure 7. Compressive stress–strain curves of Group A (**top**) and Group B (**bottom**).

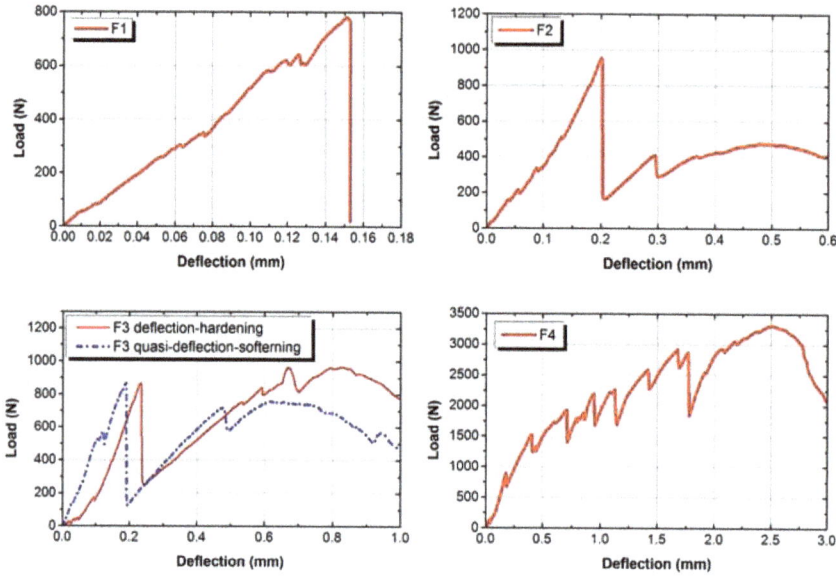

Figure 8. Flexural behaviour of Group A.

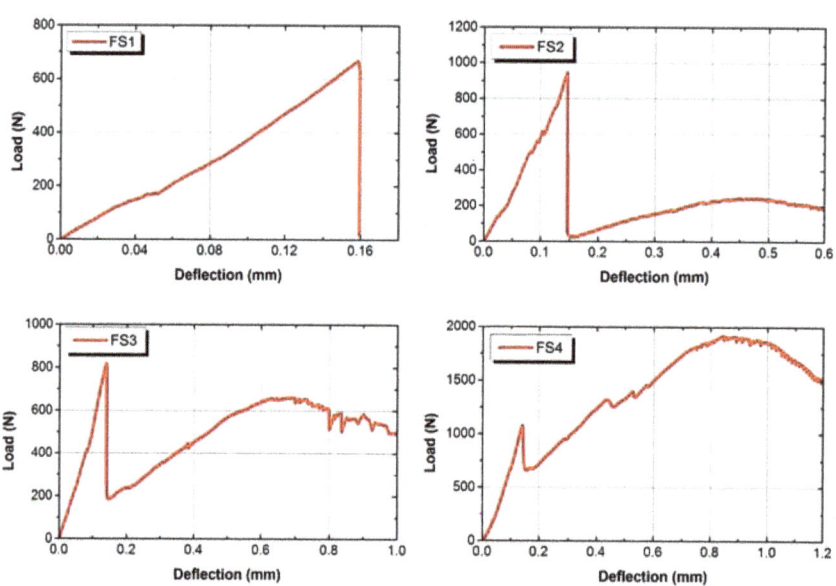

Figure 9. Flexural behaviour of Group B.

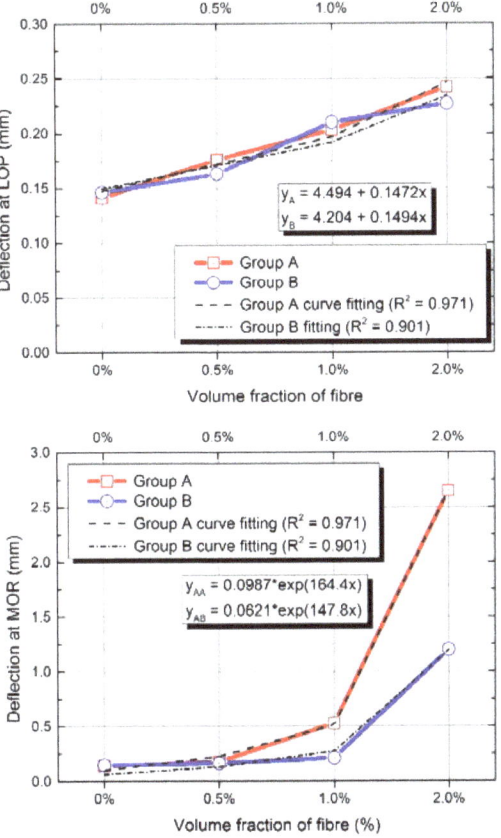

Figure 10. Effect of fibre volume fraction on deflection at limit of proportionality (LOP) (**top**) and modulus of rupture (MOR) (**bottom**).

According to the literature [5,27,28], the tensile and compressive behaviour of a composite material strongly influence the flexural performance. Also, the strain-hardening behaviour in tension leads to a deflection-hardening behaviour when the flexural behaviour of the composites is strongly associated with its tensile characteristic [29]. Thus, the results of the flexural performance obtained in this research could be related with the tensile behaviour of AAFA composites. Based on the theoretical discussions by several researchers [5–7,9,22,26,30], the critical energy release rate G_c and the interfacial bond strength τ of the composites are important parameters to be considered in a design of the composite's matrix to achieve the strain-hardening behaviour of the composite. In addition, the matrix properties, such as elastic modulus and fracture toughness, which are linked to the composite's critical energy release rate G_c, are affected by several parameters [7,22]. Using nanoindentation data, the composite critical energy release rate G_c of AAFA matrix was found to be 0.010 kJ/m^2. Based on the fracture toughness and the elastic modulus of AAFA matrix [24], the interfacial bond strength of AAFA composite was plotted against the critical fibre volume fraction ratio and the corresponding strain-hardening behaviour with the snubbing coefficient f, which is in term of the inclining angle between fibre and matrix, as illustrated in Figure 11. It can be seen that with 2.0% of the fibre volume fraction ratio in the AAFA composites, F4 and FS4 mixtures are in the region of strain-hardening, whereas, with less than 0.5% of the fibre volume fraction ratio, F2 and FS2 are not in the region of strain-hardening. It can also be noticed that with 1.0% of the fibre volume fraction ratio, F3 and FS3 mixtures are partly in the region of strain-hardening with other parts falling in the region of strain-hardening. This is consistent with F3 mixture, which shows a combination of strain-hardening and quasi-strain-hardening behaviour.

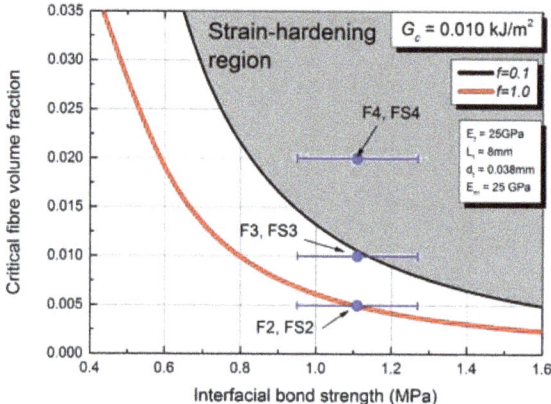

Figure 11. Critical volume fraction against interfacial bond strength.

4. Conclusions

The experimental and theoretical studies of parameters affecting the fibre pull out capacity and the strain-hardening behaviour of AAFA composites have been presented. Based on the results obtained in this research, the following conclusion can be drawn:

- The interfacial bond strength between the fibre and the AAFA matrix was determined to be in a range of 0.8 to 1.0 MPa. A numerical analysis coupled with a statistical analysis tool shows that an increase in the fibre diameter and embedded length would increase the interfacial bond strength.
- The strain corresponding to the compressive strength is not significantly affected by the fibre volume fraction ratio. However, while achieving the strain-hardening behaviour of the AAFA composites, the compressive strength decreased. In addition, using silica fume as a fine material in AAFA composite is not suitable as it decreases the flexural strength and strain capacity of the composite.
- The critical energy release rate G_c of AAFA matrix determined from the indentation fracture toughness was approximately 0.01 kJ/m^2. The results of the flexural behaviour showed the relationship between the strain-hardening behaviour of AAFA composite and the indentation G_c.
- For a mix design of AAFA matrix, it is recommended that G_c should be less than 0.01 kJ/m^2. It is theoretically impossible to achieve the strain-hardening behaviour when G_c is more than 0.015 kJ/m^2.

Author Contributions: H.L. and V.V. conceived of the presented idea. H.L. and V.V. developed the theory and performed the approach. V.V., P.M., and A.N. supervised H.L. in the interpretation of the findings of this work. All authors discussed the results and contributed to the final manuscript.

Funding: This research received no external funding.

Conflicts of Interest: The authors declare no conflict of interest.

References

1. Shi, C.; Roy, D.; Krivenko, P. *Alkali-Activated Cements and Concretes*; CRC Press/Taylor and Francis: London, UK; New York, NY, USA, 2006.

2. Shi, C.; Jiménez, A.F.; Palomo, A. New cements for the 21st century: The pursuit of an alternative to Portland cement. *Cem. Concr. Res.* **2011**, *41*, 750–763. [CrossRef]
3. Davidovits, J. *Geopolymer Chemistry and Applications*; Geopolymer Institute: Saint-Quentin, France, 2008.
4. Kumar, S.; Kumar, R.; Alex, T.; Bandopadhyay, A.; Mehrotra, S. Effect of mechanically activated fly ash on the properties of geopolymer cement. In Proceedings of the 4th World Congress on Geopolymer, Saint-Quentin, France, 29 June–1 July 2005; pp. 113–116.
5. Naaman, A.; Reinhardt, H. High performance fiber reinforced cement composites HPFRCC-4: International RILEM Workshop. *Mater. Struct.* **2003**, *36*, 710–712. [CrossRef]
6. Li, V.C. Engineered cementitious composites-tailored composites through micromechanical modeling. *J. Adv. Concr. Technol.* **1998**. [CrossRef]
7. Li, V.C.; Leung, C.K. Steady-state and multiple cracking of short random fiber composites. *J. Eng. Mech.* **1992**, *118*, 2246–2264. [CrossRef]
8. Berry, E.; Hemmings, R.; Cornelius, B. Mechanisms of hydration reactions in high volume fly ash pastes and mortars. *Cem. Concr. Compos.* **1990**, *12*, 253–261. [CrossRef]
9. Lee, B.Y.; Kim, J.K.; Kim, Y.Y. Prediction of ECC tensile stress-strain curves based on modified fiber bridging relations considering fiber distribution characteristics. *Comput. Concr.* **2010**, *7*, 455–468. [CrossRef]
10. Rigaud, S.; Chanvillard, G.; Chen, J. Characterization of bending and tensile behavior of ultra-high performance concrete containing glass fibers. In *High Performance Fiber Reinforced Cement Composites 6*; Springer: Dordrecht, The Netherlands, 2012; pp. 373–380.
11. Hardjito, D.; Rangan, B.V. *Development and Properties of Low-Calcium Fly Ash-Based Geopolymer Concrete*; Curtin Research Publications: Perth, Australia, 2005.
12. Rangan, B.V. *Fly Ash-Based Geopolymer Concrete*; Curtin Research Publications: Perth, Australia, 2008.
13. Chen, X.; Beyerlein, I.J.; Brinson, L.C. Curved-fiber pull-out model for nanocomposites. Part 1: Bonded stage formulation. *Mech. Mater.* **2009**, *41*, 279–292. [CrossRef]
14. Chen, X.; Beyerlein, I.J.; Brinson, L.C. Curved-fiber pull-out model for nanocomposites. Part 2: Interfacial debonding and sliding. *Mech. Mater.* **2009**, *41*, 293–307. [CrossRef]
15. Herrera-Franco, P.; Drzal, L. Comparison of methods for the measurement of fibre/matrix adhesion in composites. *Composites* **1992**, *23*, 2–27. [CrossRef]
16. Zhan, Y.; Meschke, G. Analytical model for the pullout behavior of straight and hooked-end steel fibers. *J. Eng. Mech.* **2014**, *140*, 04014091. [CrossRef]
17. *ANSYS Academic Research Mechanical Release 19.2*; ANSYS: Canonsburg, PA, USA, 2018.
18. Kullaa, J. Dimensional analysis of bond modulus in fiber pullout. *J. Struct. Eng.* **1996**, *122*, 783–787. [CrossRef]
19. Naaman, A.E.; Namur, G.G.; Alwan, J.M.; Najm, H.S. Fiber pullout and bond slip. I: Analytical study. *J. Struct. Eng.* **1991**, *117*, 2769–2790. [CrossRef]
20. Budiansky, B.; Cui, Y.L. On the tensile strength of a fiberreinforced ceramic composite containing a crack-like flaw. *J. Mech. Phys. Solids* **1994**, *42*, 1–19. [CrossRef]
21. Kanda, T.; Lin, Z.; Li, V.C. Tensile stress-strain modeling of pseudostrain hardening cementitious composites. *J. Mater. Civ. Eng.* **2000**, *12*, 147–156. [CrossRef]
22. Li, V.C.; Mishra, D.K.; Wu, H.C. Matrix design for pseudo-strain-hardening fibre reinforced cementitious composites. *Mater. Struct.* **1995**, *28*, 586–595. [CrossRef]
23. Li, V.C.; Wang, Y.; Backer, S. Effect of inclining angle, bundling and surface treatment on synthetic fibre pull-out from a cement matrix. *Composites* **1990**, *21*, 132–140. [CrossRef]
24. Lee, H.; Vimonsatit, V.; Chindaprasirt, P. Mechanical and micromechanical properties of alkali activated fly-ash cement based on nano-indentation. *Constr. Build. Mater.* **2016**, *107*, 95–102. [CrossRef]
25. Active Standard. Standard test method for flexural performance of fiber-reinforced concrete (using beam with third-point loading). *ASTM-C1609* **2012**._C1609M-19. [CrossRef]
26. Li, V.C.; Wu, H.C.; Maalej, M.; Mishra, D.K.; Hashida, T. Tensile behavior of cement-based composites with random discontinuous steel fibers. *J. Am. Ceram. Soc.* **1996**, *79*, 74–78. [CrossRef]
27. Soranakom, C.; Mobasher, B. Correlation of tensile and flexural responses of strain softening and strain hardening cement composites. *Cem. Concr. Compos.* **2008**, *30*, 465–477. [CrossRef]
28. Ward, R.; Li, V.C. Dependence of flexural behaviour of fibre reinforced mortar on material fracture resistance and beam size. *Constr. Build. Mater.* **1991**, *5*, 151–161. [CrossRef]

29. Kim, D.J.; Park, S.H.; Ryu, G.S.; Koh, K.T. Comparative flexural behavior of hybrid ultra high performance fiber reinforced concrete with different macro fibers. *Constr. Build. Mater.* **2011**, *25*, 4144–4155. [CrossRef]
30. Lee, B.Y.; Kim, J.K.; Kim, J.S.; Kim, Y.Y. Quantitative evaluation technique of Polyvinyl Alcohol (PVA) fiber dispersion in engineered cementitious composites. *Cem. Concr. Compos.* **2009**, *31*, 408–417. [CrossRef]

© 2019 by the authors. Licensee MDPI, Basel, Switzerland. This article is an open access article distributed under the terms and conditions of the Creative Commons Attribution (CC BY) license (http://creativecommons.org/licenses/by/4.0/).

Article

Performance of Fly Ash Geopolymer Concrete Incorporating Bamboo Ash at Elevated Temperature

Shafiq Ishak [1], Han-Seung Lee [1], Jitendra Kumar Singh [2,*], Mohd Azreen Mohd Ariffin [3], Nor Hasanah Abdul Shukor Lim [4] and Hyun-Min Yang [2]

1. Department of Architectural Engineering, Hanyang University, 1271 Sa 3-dong, Sangnok-gu, Ansan 15588, Korea; shafiq94@hanyang.ac.kr (S.I.); ercleehs@hanyang.ac.kr (H.-S.L.)
2. Innovative Durable Building and Infrastructure Research Center, Department of Architectural Engineering, Hanyang University, 1271 Sa 3-dong, Sangnok-gu, Ansan 15588, Korea; yhm04@hanyang.ac.kr
3. Forensic of Engineering Centre, School of Civil Engineering, Universiti Teknologi Malaysia, Johor Bahru 81310, Johor, Malaysia; mohdazreen@utm.my
4. Department of Structure and Materials, School of Civil Engineering, Universiti Teknologi Malaysia, Johor Bahru 81310, Johor, Malaysia; norhasanah@utm.my
* Correspondence: jk200386@hanyang.ac.kr; Tel.: +82-3-1436-8159

Received: 23 August 2019; Accepted: 14 October 2019; Published: 17 October 2019

Abstract: This paper presents the experimental results on the behavior of fly ash geopolymer concrete incorporating bamboo ash on the desired temperature (200 °C to 800 °C). Different amounts of bamboo ash were investigated and fly ash geopolymer concrete was considered as the control sample. The geopolymer was synthesized with sodium hydroxide and sodium silicate solutions. Ultrasonic pulse velocity, weight loss, and residual compressive strength were determined, and all samples were tested with two different cooling approaches i.e., an air-cooling (AC) and water-cooling (WC) regime. Results from these tests show that with the addition of 5% bamboo ash in fly ash, geopolymer exhibited a 5 MPa (53%) and 5.65 MPa (66%) improvement in residual strength, as well as 940 m/s (76%) and 727 m/s (53%) greater ultrasonic pulse velocity in AC and WC, respectively, at 800 °C when compared with control samples. Thus, bamboo ash can be one of the alternatives to geopolymer concrete when it faces exposure to high temperatures.

Keywords: fly ash; bamboo ash; supplementary materials; geopolymer concrete; elevated temperature

1. Introduction

Production of Portland cement consumes energy and releases a massive volume of carbon dioxide (CO_2) into the atmosphere but is still considered as a conventional binder owing to its excellent performance in most civil engineering applications [1]. In addition, in some instances, the production of the concrete with Portland cement is less durable in an aggressive environment and at high temperature conditions [2]. However, it was observed that the geopolymer has become a problem solver to all these issues [3]. High demand for conventional concrete, which is known as environmentally friendly concrete, can also solve landfill problems by leading to recycling and reusing waste material. These problems can be eliminated by utilizing the industrial waste products in construction purposes. Using waste products as a cementitious material in geopolymer concrete would also maximize its recycling potential throughout the industrial sector.

Geopolymer is an inorganic composite which is produced by synthesizing pozzolanic materials under highly alkaline hydroxide and/or alkaline silicate [4]. Apart from that, the geopolymer concrete has superb resistance to chemical attack and exhibits great ability against aggressive environments with a high amount of CO_2, high content of sulfate, and acid resistance [5]. A previous study by Wallah and Rangan [6] concluded that geopolymer concrete revealed small changes in the length and few increases in mass after one-year of exposure to sulfate solution. Bakharev [7] has studied the properties of concrete in different concentrations of sulfate solution with different type of activators and found that the properties of the concrete depended on the quality of materials and activators.

Geopolymer concrete can be used in various kinds of applications including as a fire resistant, sealants, concretes, ceramics, etc. It was reported that geopolymer concrete can withstand high temperature exposure [8,9]. Therefore, geopolymer concrete may possess a superior fire resistance compared to conventional concrete i.e., Ordinary Portland Cement (OPC).

Interest in using fly ash as a sustainable material in geopolymer concrete has increased since 2000 [10]. Hardjito and Rangan [11] have investigated the effects of alkaline parameters, water content, and curing conditions in their research. In Malaysia, some researchers focused on geopolymer concrete as well [12,13]. As a result, geopolymers have become prominent among researchers due to its environmentally friendly and high performance characteristics.

Agriculture waste is a serious environmental problem in many countries. This waste is being mainly produced from gardens and rice fields. The majority of previous research involving agriculture waste involved Palm Oil Fuel Ash (POFA) [14–16] and Rice Husk Ash (RHA) [17–19] as binders. Navid Ranjbar et al., [20] have conducted an experiment regarding the performance of POFA and fly ash (FA) based geopolymer mortar exposed to elevated temperatures. It was concluded by them that all FA/POFA based geopolymers gained strength when exposed to temperatures up to 500 °C. However, by increasing FA content in samples, they produced higher compressive strength at 300 °C, while on the other hand increasing POFA content delays attaining maximum strength. They have also suggested that when the temperature was increased above 500 °C, all samples lost their strength. Besides this, a study was conducted which focused on the effect of pretreatment of FA and POFA on mechanical properties after the geopolymerization process [21]. It was shown that when FA is heated up to 800 °C, sintering of the particles was observed which led to a deformation and reactivation, thus leading to a reduction in the setting time and increased early compressive strength.

Interest in bamboo for construction has grown continuously as the focus shifts towards reducing the environment impact and embodied energy of the built environment. For developing countries, bamboo is considered as an ideal crop for rural development. Bamboo production and utilization are considered relevant to many in the UN for sustainable development goals. Naturally, the bamboo is found in cylindrical pole or culm. Bamboo is also part of the grass family. There are over 1200 species of bamboo all over the world, with structural species varying by locations. The different species can be categorized into three types of root systems: sympodial (clumping), monopodial (running), and amphipodial (clumping and running). According to the Food and Agriculture Organization of the United Nations (UNFAO), a total of 72% of land area in Malaysia is filled with forests. Bamboo is an easy plant to grow. Tropical rainforest areas found in Malaysia provide ideal growing conditions for the bamboo plant. The production of bamboo charcoal has increased and its applications especially in healthcare, cooking, water purification, and gardening have grown significantly [22]. Consequently, bamboo ash is the waste from the production of bamboo charcoal. Although rich in silica, the poor performance of bamboo ash is owing to the presence of silicate material. Due to the absence of alumina, it is attractive in combination with other materials which are rich in alumina i.e., fly ash. Commonly found in Malaysia and Indonesia, bamboo plant has been used as a fire protection material [23,24]. Bamboo is fire resistant even at higher temperatures, thus, it can be used in the construction industry.

There is lack of research on the use of bamboo ash in the construction industry. Therefore, in present study, we have experimentally investigated the properties and performance of fly ash geopolymer concrete incorporating with bamboo ash under elevated temperature. In addition, there is no previous study that has used bamboo ash as a construction material. This investigation includes the effect of physical appearances, compressive strength, weight loss, and ultrasonic pulse velocity loss after the desired exposure.

2. Experimentation

2.1. Materials

2.1.1. Binder

In this study, fly ash (FA) as low calcium fly ash (class F) and bamboo ash was used. The chemical composition and particle size analysis for both materials will be discussed in the results and discussion section.

Fly ash was obtained from Tanjung Bin, Johor, Malaysia. The bamboo ash (BA) was obtained from Lanchang, Pahang, Malaysia. First, collected bamboo ash was dried in the oven for 24 hours at 110 °C (± 5 °C) to ensure that there was no moisture available. Then, the bamboo ash was ground in an abrasion test machine for 6 hours to improve the fineness of the ash. Then, the bamboo ash was sieved through a 45 μm sieve in order to remove bigger size of ash particle and impurities. Only the fine bamboo ash passing through the sieve were collected following the standard size of Portland cement used in the mixing. The specific gravity of FA and BA was 2.20 and 2.05, respectively.

2.1.2. Aggregates

The standards used to determine the properties of aggregates are ASTM C127 [25], ASTM C128-15 [26], ASTM C29 [27], and BS EN 933-1:2012 [28]. ASTM C127 was used to determine the specific gravity and water absorption of coarse aggregate. ASTM C128 was used to obtain specific gravity and water absorption of fine aggregate. Apart from that, ASTM C29 was conducted to acquire the bulk density of both aggregates and BS EN 933-1:2012 [28] was used to check the grading requirement of the aggregates.

In this research, crushed granite with nominal size of 10 mm was used as a coarse aggregate. The specific gravity, water absorption, and bulk density of the coarse aggregate was 2.7, 0.5%, and 1551 kg/m^3, respectively. River sand was used in fine aggregates and obtained from a local source in Johor, Malaysia. Specific gravity, water absorption, and bulk density of the fine aggregate was 2.6, 0.7%, and 1650 kg/m^3, respectively. To ensure the aggregates did not absorb alkaline solution during the mixing process, saturated surface dry (SSD) conditions for both aggregates were conducted. For this purpose, both coarse and fine aggregates were soaked separately with clean tap water. Then, the aggregates were placed on a plastic sheet until the surface became dry.

2.1.3. Alkaline Solution

Alkali Sodium based activator purchased from QReC^{TM}, Auckland, New Zealand was used in this research. The alkaline solution was prepared by mixing 10 M sodium hydroxide (NaOH) with sodium silicate (Na_2SiO_3). The activator to binder ratio was different starting from 0.40, 0.45, and 0.5. Optimization of the activator to binder ratio was carried on the basis of workability results.

2.1.4. Superplasticizer

In this research, Master Glenium ACE 8589 (Master®Builders Solutions, Kuala Lumpur, Malaysia) was used as a superplasticizer (SP). SP is a chemical admixture that was added to the concrete during the mixing process. It is also known as a water reducer. SP provides exceptionally good early strength development and maintains flowability for a considerable period of time. Behzad and Jay have studied the effect of different SPs on the workability and strength of fly ash based geopolymer and they have found that SP is effective in improving the properties of concrete, which are directly dependant on the type of activator and the SP [29].

2.2. Mix Proportions

The preliminary mix design was optimized based on several factors. In this study, the mass ratio of binder to aggregates and coarse to fine aggregates was set to be 1. The mass ratio for coarse to fine aggregates and sodium silicate to sodium hydroxide was set to be 1 and 2.5, respectively, along with 3% superplasticizer. Huseien et al. [30], have found that this ratio for binder to aggregates provided optimum results of flowability, compressive strength, and bending stress. Besides, increased binder content led to reducing the workability, strength of samples, and reduced initial and final setting times of the geopolymer.

Table 1 shows the mix proportion used in this research. The mixing phase was an important part in the production of geopolymer concrete. The mixing process and curing was carried out at 25 ± 2 °C. This proportion is generally being used for ultra-high performance fiber reinforced concrete [31–35]. Huseien et al. have considered 1100 kg/m³ binder content in alkali activated mortar for durability properties [36] and they found that this composition has increased the durability performance of mortar. In present study, this proposed mix proportion was not acceptable, but we have tried to consume the maximum waste utilized in the construction industry. Therefore, we have followed the Huseien et al. mortar proportion for the present study [36]. During mixing of the proposed proportion, it was very hard to mix the geopolymer concrete properly however, when using 3% SP it was easy to mix.

2.3. Flow Table Test

The determination for the workability of the geopolymer mortar was conducted in accordance with ASTM C1437 [37]. The standard conical frustum with a 100 mm diameter was used. Table 2 shows the workability criteria.

Table 1. Mixed proportion of fly ash geopolymer concrete incorporated with bamboo ash.

No.	Fly Ash (kg/m³)	Bamboo Ash (kg/m³)	NaOH (kg/m³)	Na₂SiO₃ (kg/m³)	Activator/Binder	Water (kg/m³)
1.	1215	–	138.86	347.14		63.18
2.	1154	61(5%)	138.60	347.14		63.18
3.	1093	122(10%)	138.60	347.14		63.18
4.	1033	182(15%)	138.60	347.14		63.18
5.	972	243(20%)	138.86	347.14	0.40	63.18
6.	729	486(40%)	138.86	347.14		63.18
7.	486	729(60%)	138.86	347.14		63.18
8.	243	972(80%)	138.86	347.14		63.18
9.	–	1215(100%)	138.86	347.14		63.18

Table 2. Workability criteria for geopolymer mortar.

No	Flow Diameter	Workability
1	Above 250 mm	Very High
2	180 mm to 250 mm	High
3	150 mm to 180 mm	Moderate
4	150 mm to 120 mm	Stiff
5	Below 120 mm	Very Stiff

2.4. Compressive Strength Test

The compressive strength test measurement was carried out on a 100cm × 100cm × 100cm concrete sample. This test was conducted according to ASTM C109-16 standard [38].

2.5. Testing Procedures

2.5.1. X-ray Fluorescence (XRF)

The chemical compositions of bamboo ash and fly ash were acquired by XRF test (Rigaku NEX CG, Tokyo, Japan). 100 grams of both fly ash and bamboo ash was sealed safely in a plastic bag before being tested.

2.5.2. X-ray Diffraction (XRD)

X-ray Diffraction (XRD: Rigaku Smartlab Diffractometer) of fly ash, bamboo ash, 100% fly ash and 5% bamboo ash + 95% fly ash geopolymer concrete after 7 days of curing were carried out from $2\theta = 5$–$85°$ using Cu radiation ($\lambda = 1.54182$ Å) at 25 kV.

2.5.3. Particle Size Analysis (PSA)

Particle size analysis (PSA) was conducted to investigate the size distribution of the binder. This analysis was performed by laser scattering technique (Mastersizer 3000). This test was carried out using the wetting method where the particles were dispersed using distilled water to avoid an agglomerated condition.

2.5.4. Scanning Electron Microscopy (SEM)

The surface morphology of both materials (bamboo ash and fly ash) as well as 100% fly ash and 5% bamboo ash+95% fly ash geopolymer concrete after 7 days of curing were conducted by scanning electron microscopy (SEM, Jeol, Tokyo, Japan) operated at 15 kV.

2.5.5. Ultrasonic Pulse Velocity (UPV) Test

The UPV value of concrete samples was determined according to BS EN 12504-4:2004 [39] after exposure to high temperature.

2.5.6. Fire Endurance Test

The fire endurance test was carried out in accordance with ASTM E119-12a [40] using automatic electrical furnace (Figure 1) with 100 mm × 100 mm × 100 mm concrete. All samples were cured for 28 days at 25 ± 2 °C before being subjected to high temperature. An electrically-heated furnace designed for a maximum 1200 °C was used. All samples were heated for the duration of 1 hour at 200 °C, 400 °C, 600 °C and 800 °C as targeted temperature. Two different cooling approaches were tested which were air cooling (AC) and water cooling (WC) regimes with curing conditions at 25 °C and 60% relative humidity. Different cooling regimes for normal cement composite had a significant influence on the mechanical properties of the concrete after exposure [41,42]. Before the test were conducted, all samples were weighed to determine their initial density and initial ultrasonic pulse velocity (UPV) value. After acquiring the temperature level, a further experiment was then carried out to determine the UPV loss, weight loss, physical appearances, and residual compressive strength.

Figure 1. Overview of an automatic electrical furnace.

3. Results and Discussion

3.1. Characterization of the Binder

3.1.1. X-ray Fluorescence (XRF)

The chemical composition of bamboo ash and fly ash were determined using XRF. Based on the results shown in Table 3, the main oxides composition of bamboo ash is silica, potassium oxide, calcium oxide, and sulfur trioxide containing 35.2%, 33.1%, 13.5%, and 8.3%, respectively. For fly ash, most of the compounds are silica and alumina. Furthermore, silica/alumina ratio for bamboo ash is unidentified while fly ash is around 2.0. Silica and alumina are very important for geopolymer synthesis. A nil percentage of alumina in bamboo ash is too unrealistic for geopolymerization to occur. The presence of high amount of calcium oxide in bamboo ash reduces the setting time of fly ash containing geopolymer concrete. The fly ash used in the present study was considered as Class F, revealing that the summation of silica, alumina and iron oxide is more than 88%.

Table 3. Chemical composition (%) of BA and FA.

Chemical Compounds	BA	FA
Silica (SiO_2)	35.2	55.92
Alumina (Al_2O_3)	–	28.8
Calcium oxide (CaO)	13.5	5.16
Potassium oxide (K_2O)	33.1	0.94
Sulfur trioxide (SO_3)	8.3	–
Iron oxide (Fe_2O_3)	1.0	3.67
Manganese oxide (MnO)	3.1	–
Phosphorus pentaoxide (P_2O_5)	–	0.69
Titanium oxide (TiO_2)	–	2.04
Magnesium oxide (MgO)	–	1.48
Loss of Ignition (LOI)	5.8	1.3

3.1.2. X-ray Diffraction (XRD)

XRD of both binder materials is shown in Figure 2. Fly ash shows the presence of quartz (SiO_2, JCPDS = 88-2487) and mullite ($3Al_2O_3 \cdot 2SiO_2$, JCPDS = 06-0259) in the XRD pattern (Figure 2a), while bamboo ash diffraction patterns show the presence of quartz (JCPDS = 88-2487), potassium oxide (K_2O, JCPDS = 26-1327), rosenhahnite i.e., calcium hydroxide silicate ($Ca_3Si_3O_8(OH)_2$, JCPDS = 83-1242), and sulfur trioxide (SO_3, JCPDS = 76-0760) (Figure 2b). Potassium oxide, rosenhahnite, and sulfur trioxide are found only in bamboo ash. This result corroborates with XRF data (Table 3). A sharp peak at 28.28° indicates a mainly crystalline structure consisting of quartz, as well as potassium oxide and rosenhahnite in bamboo ash.

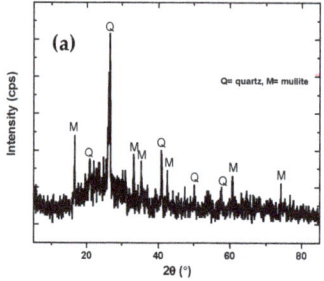 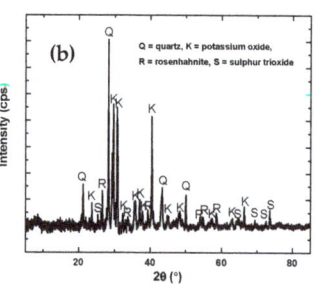

Figure 2. X-ray Diffraction (XRD) of (**a**) fly ash and (**b**) bamboo ash.

3.1.3. Particle Size Analysis

The particle size distribution of fly ash and bamboo ash was characterized by their D50 values as shown in Figure 3. The particle of bamboo ash is in micro size where the biggest size is determined at 163 µm with a median size of 14.5 µm. Also, the biggest size of particles in fly ash is revealed to be 66.9 µm with a median size of 15 µm. It is revealed that 83% of the bamboo ash particles are smaller than 45 µm, while almost 96% of fly ash particles have a size less than 45 µm. The specific area is found to be 510.4 and 658.4 m^2/kg for bamboo ash and fly ash particles, respectively.

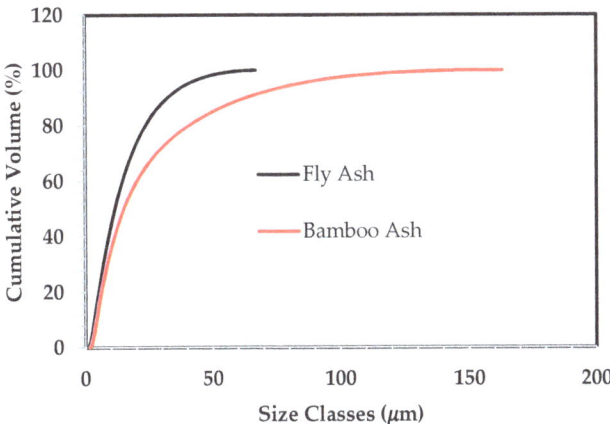

Figure 3. Particle size analysis of fly ash and bamboo ash.

3.1.4. Scanning Electron Microscopy (SEM) Analysis

Figure 4 shows the surface and shape morphology of the fly ash as well as the bamboo ash used in this study. Most of the fly ash particles have a glassy and spherical structure (Figure 4a), and are also known as cenospheres [43]. This ash contains a series of spherical vitreous particles of different sizes. Meanwhile, Figure 4b shows the SEM micrographs of bamboo ash, which depicted that bamboo ash has a rectangular structure and different particle sizes. It also has a porous structure, as can be seen from Figure 4b.

Figure 4. Scanning Electron Microscopy (SEM) micrographs of (a) Fly ash (b) Bamboo ash.

3.2. Appropriate Mix Proportion

3.2.1. Workability

Table 4 represents the workability result of geopolymer concrete using different mix proportions. The standard deviation of the different mix proportions is found to be 2–8 mm. It is depicted that the slump value increases with the increase of the activator to binder ratio. The mix was very stiff and no flow was observed when the activator to binder ratio was 0.40 for both types of sample. Based on the visual observation, the mix was not homogenous and difficult to mix and compact. Apparently, the bamboo ash absorbs more water during the mixing, thus reducing the workability of the mix.

Once the activator to binder ratio was reached at 0.45, the slump value increases compared to 0.40 for 100% FA (Csample) and 5% BA+ 95% FA (Asample). Based on the observations, the mix was quite homogenous and the mixing process was easier compared to for the 0.40 ratio. The workable mix with the highest flow value was obtained when the activator to binder ratio was 0.50 for both samples.

In conclusion, the small percentage i.e., 5% addition of bamboo ash at a 0.45 ratio, gave a much smaller slump loss in Asample compared to Csample. However, with 0.40 and 0.50 ratios, the slump loss is 10 mm for Asample. Thus, we decided to perform another experiment in a 0.45 activator to binder ratio.

Table 4. Slump test result.

Sample	Activator to Binder Ratio	Slump Value (mm)			Average (mm)	Standard Deviation (mm)	Slump Loss (mm)
		Sample 1	Sample 2	Sample 3			
100% FA (Csample)	0.40	140	142	138	140	2.00	0
	0.45	202	187	190	193	7.94	0
	0.50	225	220	215	220	5.00	0
5% BA+ 95% FA (Asample)	0.40	139	126	125	130	7.81	10
	0.45	185	183	187	185	2.00	8
	0.50	219	203	208	210	8.19	10

3.2.2. Compressive Strength

Table 5 shows the compressive strength as well as the standard deviation of samples with curing duration. This table shows the effect of bamboo ash to fly ash ratio (%) on the compressive strength of geopolymer concrete at a 0.45 activator to binder ratio. The results show that the compressive strength of geopolymer concrete is increased with the increasing of curing time. This is due to the geopolymerization process between alumina and silica from the binder with the alkaline solution. But our studies failed to get desired ultra-high performance fiber reinforced concrete compressive strength [31–35]. However, as the percentage of bamboo ash increased, the compressive strength of the geopolymer concrete was decreased. This was probably due to a lack of the amount of alumina in bamboo ash, which reduced the geopolymerization product and contributed to the strength of concrete [44]. For 7 days of curing, replacement with 5% bamboo ash shows higher compressive strength at 18.94 MPa compared to 100% of fly ash i.e., 16.15 MPa. The increased early age compressive strength of BA with FA is due to the hydrolysis which imposed to form crystalline phases and reduced the porosity of the samples. Besides, it was also concluded that surface area of the mixture was increased due to the addition of bamboo ash, since the specific gravity of BA (2.05) is less than FA (2.20).

The standard deviation in the compressive strength for different mix proportions was found to be 0.11 to 1.31 MPa. A very stable proportion was needed in order to obtain a very good concrete mix for construction purposes. The addition of 5% bamboo ash was shown to be the most suitable mixture to get a better performance in regards to early compressive strength compared to other mixtures.

Table 5. Effect of curing period on compressive strength of the bamboo ash (BA) to fly ash (FA) ratio (%) in the 0.45 activator to binder ratio.

Sample ID	Curing Time (Days)	Compressive Strength (MPa)			Average (MPa)	Standard Deviation (MPa)
		Sample 1	Sample 2	Sample 3		
100FA	7	16.31	16.19	15.96	16.15	0.18
	14	32.48	33.76	33.93	33.39	0.79
	28	42.81	42.53	41.06	42.13	0.94
5BA:95FA	7	18.67	19.1	19.05	18.94	0.24
	14	31.01	30.34	32.14	31.16	0.91
	28	42.56	41.03	39.96	41.18	1.31
10BA:90FA	7	15.87	16.03	16.21	16.04	0.17
	14	24.52	25.1	24.86	24.83	0.29
	28	39.44	40.12	39.68	39.75	0.34
15BA:85FA	7	13.1	14.05	13.85	13.67	0.50
	14	22.66	23.81	23.06	23.18	0.58
	28	32.28	33.14	33.56	32.99	0.65
20BA:80FA	7	11.01	12.56	12.14	11.90	0.80
	14	21.41	22.35	21.03	21.60	0.68
	28	34.8	35.24	34.78	34.94	0.26
40BA:60FA	7	7.5	6.96	8.5	7.65	0.78
	14	6.86	8.96	7.48	7.77	1.08
	28	9.21	10.04	9.86	9.70	0.44
60BA:40FA	7	2.35	2.37	2.5	2.41	0.08
	14	5.12	5.74	6.01	5.62	0.46
	28	7.03	7.56	8.05	7.55	0.51
80BA:20FA	7	0	0	0	0	0.00
	14	1.06	1.24	1.33	1.21	0.14
	28	2.44	2.22	2.29	2.32	0.11
100BA	7	0	0	0	0	0.00
	14	0	0	0	0	0.00
	28	0	0	0	0	0.00

To understand the formation of phases after 7 days of curing, XRD of Csample (100% FA) and Asample (5% BA + 95% FA) was performed and results are shown in Figure 5. Quartz and mullite were obtained in both samples. There is a possibility that bamboo ash in Asample was completely mixed and dissolved in the matrix of fly ash attributed to the high dissolution rate of potassium oxide, rosenhahnite ,and sulfur trioxide in high alkaline condition, as explained by other researchers [45], during 7 days of curing. Thus, only quartz and mullite are found as present in 100% fly ash. This result confirms that owing to the dissolution of oxides present in bamboo ash control, the crack formed and filled out the porosity of geopolymer concrete (SEM results will be shown in Figure 6), leading to higher compressive strength being observed for 5%BA + 95%FA (Table 5) after 7 days of curing. The presence of quartz and mullite in Csample and Asample explained that SiO_2 and Al_2O_3 are not fully utilized for geopolymer formation. Bamboo ash and activator are the materials that are involved in the synthesis of fly ash based geopolymer concrete. The pure geopolymer network actually consists mainly of Si, Al, and O with alkali Na^+ or K^+. In this reaction, all minerals did not participate in the geopolymerization process.

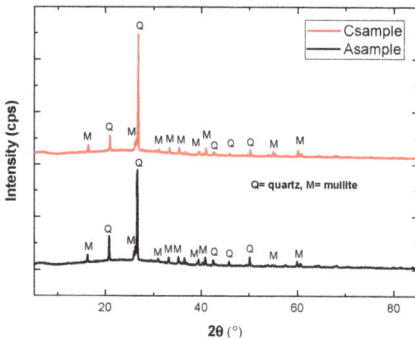

Figure 5. XRD analysis for Csample (100% FA) and Asample (5% BA + 95% FA) after 7 days of curing.

Figure 6 shows the SEM results of 100% fly (Csample) and 5% bamboo ash + 95% fly ash (Asample) after 7 days of curing at 0.45 activator to binder ratio. 100% FA (Csample) containing sample shows macro and micro cracks in Figure 6a, however, 5% bamboo ash + 95% fly ash (Asample) does not show any defect in Figure 6b which indicated that the potassium oxide, sulfur trioxide, and rosenhahnite has dissolved and filled out the cracks/pores in the concrete. Thus, after 7 days, the 5% bamboo ash + 95% fly ash sample shows higher compressive strength (Table 5). The bamboo ash has the property to hold the paste together, thus controlling the cracks within the concrete owing to the reactive compositions which contain potassium oxide and other oxides. Generally, previous researchers [46] agreed with the addition of fiber into concrete, which increases the strength of the concrete structure. Moreover, Csample exhibits cracks, pores, and defects, which decreases the positive properties of concrete.

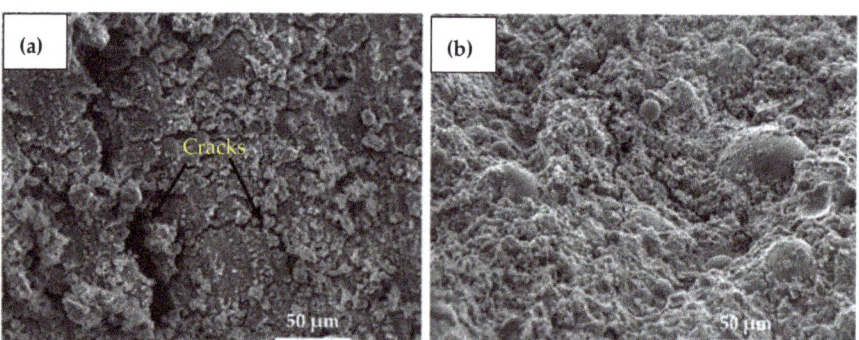

Figure 6. SEM micrographs of (**a**) 100% fly ash (Csample) and (**b**) 5% bamboo ash + 95% fly ash (Asample) after 7 days of curing.

3.3. Fire Endurance Test

3.3.1. Cooling Effect on the Physical Appearance of the Concrete

High temperature i.e., heat exposure is one of the most important parameters which affects the surface characteristics, surface outlook, shape, and color of concrete. Although it does not give significant information regarding the distortion suffered by the concrete, it will give an immediate impression of the failure tendency of the concrete. Tables 6 and 7 represent a detailed picture of the concrete after exposure to different temperature and cooling regimes for Csample and Asample, respectively.

It can be observed from the entire cooling regime at 200 °C that the color of the sample was grey for both Csample and Asample with a smooth and sharp edges. These characteristics are maintained up to 400 °C. At 600 °C, the light grey color is observed for air cooling (AC) with rough surface while water cooling (WC) exhibits yellowish grey with cracks appeared in Csample whereas in Asample there is no cracks. At 800 °C, the Csample develops heavy cracks throughout the surface in AC and WC, whereas Asample shows lesser cracks compared to Csample in both AC and WC.

The change in the matrix structure is a consequence of the dehydration in the phases of the binder during exposure to different temperatures. It is manifested as changes in porosity and color, as well as physical defects such as the presence of cracks. It is matched with the results presented in Tables 6 and 7 for Csample and Asample, respectively, which show the physical appearance of the concrete after exposure to high temperature. In general, the color changes experienced in both samples as a result of heating may be linked to the chemical transformations taking place in the heated samples.

Table 6. Cooling type, color, and texture of Csample.

Parameters	Temperature (°C)							
	200		400		600		800	
Type of cooling	AC	WC	AC	WC	AC	WC	AC	WC
Color	Grey	Grey	Grey	Grey	Light grey	Yellowish grey	Light grey	Grey
Texture	Smooth	Smooth	Smooth	Smooth	Rough	Crack	Fragmentation	Fragmentation

Table 7. Cooling type, appearance, color, and texture of Asample.

Parameters	Temperature (°C)							
	200		400		600		800	
Type of cooling	AC	WC	AC	WC	AC	WC	AC	WC
Color	Grey	Grey	Grey	Grey	Light grey	Light grey	Grey	Grey
Texture	Smooth	Smooth	Smooth	Smooth	Rough	Rough	Rough	Fragmentation

3.3.2. Residual Compressive Strength

The residual compressive strength of Csample and Asample at different temperature and cooling regimes is shown in Table 8. For both cooling types, the residual compressive strength increases at 200 °C and achieved a maximum strength for Csample and Asample. At 400 °C, the Csample has the least strength loss of 41% and 43% for AC and WC, respectively. For 600 °C, the loss trend continues to increase for both types of samples and achieved a maximum loss of compressive strength at 800 °C. 78% (AC) and 80% (WC) of strength loss gained by Csample which is the highest loss of compressive strength, compared with Asample with 65% (A/C) and 66% (W/C) loss.

Table 8. Residual compressive strength of samples exposed at different temperature.

Temperature (°C)	Residual Compressive Strength (MPa)			
	Csample (100% FA)		Asample (95% FA + 5% BA)	
	AC	WC	AC	WC
27	42.6	42.8	41.1	41.5
200	57.93	56.10	57.28	55.82
400	25.14	24.43	24.69	22.48
600	15.40	14.70	15.12	14.89
800	9.47	8.54	14.51	14.19

According to a previous researcher [40], geopolymer concrete strength was enhanced in heat conditions. Changes in chemical structure and the dehydration of free and chemically-bound water was caused from the exposure to high temperature. As the temperature increases, the moisture particles inside the samples tend to escape to the surface.

In this research, it was proved that fly ash geopolymer concrete (Asample and Csample) does not complete a geopolymerization process until 28 days have passed. Within 200 °C exposure, both type of samples gain an improvement in terms of strength and also the matrix structure. These are associated with the reported changes in the value of compressive strength after exposure.

It is seen that the water molecule is expelled from the geopolymer concrete during the presence of heat, which improves the strength of the concrete and also leads to the discontinuous nano-pores of the matrix. It is possible that not all water molecules were expelled due to high temperature, especially in a larger sample. Higher surface tensions in larger samples will dissipate moisture at slower rates compared to in smaller samples.

Starting from 400 °C, 600 °C and 800 °C levels of heat exposure, the the compressive strength decreased for both type of samples. But in the presence of bamboo ash, the residual compressive strength of Asample tends to show a positive result at 800 °C where the strength is higher compared to Csample.

This clearly shows that at 800 °C, the presence of bamboo ash contains potassium oxide, thus producing a significant compressive strength compared to Csample. If alkali metal oxides content is above a certain limit, it exhibits a high coefficient of thermal expansion. With the presence of bamboo ash at high temperature, the tendency of the oxides lead to change its shape to have a higher area and volume. Thus, more solid and packed molecules are formed and provide better structural components for the geopolymer concrete.

3.3.3. Ultrasonic Pulse Velocity (UPV) Value after Exposure

The change in UPV value due to exposure to high temperature is depicted in Table 9. The UPV values at the initial temperature (27 °C) were 3854 m/s (AC) and 3850 m/s (WC) for Csample, and 3810 m/s (AC) and 3802 m/s (WC) for Asample. UPV value increases at 200 °C yielded values of 4451 m/s (AC) and 4417 m/s (W/C) for Csample, and 4438 m/s (AC) and 4394 m/s (WC) for Asample, which is the highest value among all studied temperature. At 400 °C, the trends changed, after which the Csample experienced UPV losses of 13.3% (AC) and 14.0% (WC) compared to Asample, which experienced 12.9% (AC) and 16.1% (WC) UPV losses. At 600 °C, the loss trends continued to increase and achieved a maximum loss at 800 °C for both types of samples. 67.8% (AC) and 64.7% (WC) losses was achieved by Csample, while 42.8% (WC) and 45.1% (AC) were achieved for Asample. The Asample exhibited a smaller loss of UPV compared to Csample. Generally, the reduction of the velocity in the concrete was due to the deformation of the microstructure in the geopolymer concrete. A rise in the temperature increases the amount of air voids in the concrete samples. Thus, the transmission speed of sound waves decreased with the increase in the traveling time of the ultrasonic pulse transmission.

Table 9. UPV value after exposure to different temperatures.

Temperature (°C)	UPV (m/s)			
	Csample		Asample	
	AC	WC	AC	WC
27	3854	3850	3810	3802
200	4451	4417	4438	4394
400	3340	3314	3320	3191
600	2241	2210	2235	2140
800	1240	1360	2180	2087

The quality of the concrete can be classified based on Table 10 [47]. Based on Table 10 values, good quality concrete can be produced at 200 °C. It was proven that the highest compressive strength corresponded with the highest UPV value. Apart from that, at 800 °C, Asample gained a higher UPV value compared to Csample, which corresponded with the residual compressive strength results.

Table 10. Quality of concrete based on UPV.

UPV (m/s)	Quality of Concrete
>4500	Excellent
3500–4500	Good
3000–3500	Doubtful
2000–3000	Poor
<2000	Very Poor

3.3.4. Weight Loss of Concrete

The effect of elevated temperature on the weight loss of geopolymer concrete for all curing regimes is depicted in Table 11. The initial weight of the samples is expressed as the density of the sample exposed to a 25 °C temperature. The weight loss of Csample and Asample increased as the temperature increased. The weight loss increased significantly from 200 °C to 400 °C where the mass loss observed was a result of moisture movement out of the geopolymer matrix. The reduction in weight loss gradually decreased from 11.96% to 10.83% for AC to WC, respectively, in Csample at 600 °C. However, a greater loss is to be found at 800 °C with ranges of 12.3% (AC) and 10.79% (WC). Apart from that, the weight loss for Asample was slightly higher compared to Csample at 800 °C, having reached 12.62% (AC) and 11.28% (WC). The exposure to elevated temperature can lead to changes in the stiffness and mechanical properties of geopolymer concrete. It was proved that high temperature can affect the stiffness and mechanical properties of geopolymer concrete [48].

In addition, an increase in temperature and weight reduction lead to deterioration of the structural integrity of geopolymer concrete. Asample yielded a higher weight loss compared to Csample. Furthermore, weight loss was observed to be lower in WC compared to AC regime. This may be due to water being absorbed during the application of water-cooling to make sure the temperature of the concrete was under control.

Table 11. Weight loss of concrete samples.

Sample	Temperature (%)	Type of Cooling	Weight Loss (%)
Csample	200	AC	2.41
		WC	1.19
	400	AC	8.9
		WC	6.38
	600	AC	11.96
		WC	10.83
	800	AC	12.3
		WC	10.79
Asample	200	AC	2.89
		WC	1.62
	400	AC	9.1
		WC	6.65
	600	AC	12.09
		WC	11.21
	800	AC	12.62
		WC	11.28

The experimental results indicated that Csample had a lower confidence level i.e., $R^2 = 0.8949$ (Figure 7a) that lay outside of the recommendation range (0.90–1.00) whereas Asample (Figure 7b) showed a strong relationship between these two parameters with a high confidence level i.e. 0.9939.

From this result, it is illustrated that the experimental results show fitted well for Asample but not for Csample. Thus, it can be concluded that the presence of 5% bamboo ash in 95% fly ash gives a better confidence level compared to the control sample.

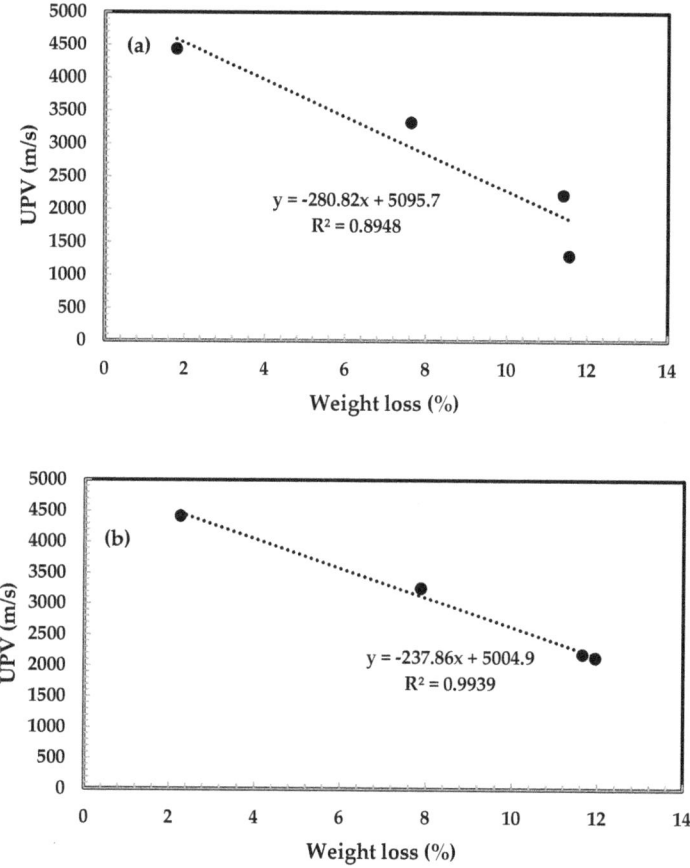

Figure 7. Ultrasonic pulse velocity (UPV) value of (**a**) Csample and (**b**) Asample as a function of weight loss.

4. Conclusions

Based on the experimental results and discussion, the following conclusions can be drawn:

1. SEM analysis shows a rectangular shape of bamboo ash and it is believed that it can hold the paste together and help to provide resistance at high temperatures.
2. A 5% addition of bamboo ash in 95% fly ash provided better compressive strength after a short period of time i.e., 7 days of curing, compared to the control sample.
3. A light grey color was observed for air cooling (AC) at 600 °C with a rough surface and cracks that appeared in Csample during water cooling (WC), whereas there were no cracks in Asample.
4. Csample and Asample gained approximately 36% and 39% residual compressive strength at 200 °C of exposure, respectively, whereas losses of approximately 62% and 41% from 400 °C to 800 °C, respectively, were found after using the air cooling regime.
5. Asample exhibited the highest residual compressive strength at 800 °C of exposure compared to Csample for the AC and WC regimes.

6. UPV values for both types of samples increased tremendously at 200 °C of exposure, which was concluded to be good in terms of concrete quality. The UPV values tended to decrease from 400 °C to 800 °C of concrete exposure. A greater loss was found at 800 °C of exposure and it led to a change in the stiffness and mechanical properties of geopolymer concrete.
7. Asample shows a high confidence level and the best fitted value compared to Csample in UPV vs. weight loss results, which reveals that the velocity of heat passing through the concrete is slower.

Author Contributions: Data curation, S.I.; Formal analysis, S.I., J.K.S., M.A.M.A., N.H.A.S.L. and H.-M.Y.; Funding acquisition, H.-S.L. and M.A.M.A.; Investigation, H.-S.L. and J.K.S.; Methodology, S.I., N.H.A.S.L. and H.-M.Y.; Supervision, M.A.M.A. and N.H.A.S.L.; Writing—original draft, S.I., H.-S.L., J.K.S., M.A.M.A., N.H.A.S.L. and H.-M.Y.; Writing—review & editing, S.I., H.-S.L., J.K.S., M.A.M.A., N.H.A.S.L. and H.-M.Y.

Acknowledgments: This research was supported by a basic science research program through Research University Grant (RUG) (No. Q.J130000.2522.17H81), with the funding from the Ministry of High Education (MOHE) and Universiti Teknologi Malaysia, as well as the basic science research program through the National Research Foundation (NRF) of Korea funded by the Ministry of Science, ICT and Future Planning (No. 2015R1A5A1037548). The authors wish to express their utmost gratitude to these funding entities, which have allowed the research endeavour to be carried out.

Conflicts of Interest: The authors declare no conflict of interest.

References

1. Rajamane, N.P.; Nataraja, M.C.; Lakshmanan, N.; Dattatreya, J.K.; Sabitha, D. Sulphuric acid resistant ecofriendly concrete from geopolymerisation of blast furnace slag. *Indian J. Eng. Mater. Sci.* **2012**, *19*, 357–367.
2. Venkata, R.V.; Mahindrakar, A.B. Impact of aggressive environment on concrete—A review. *Int. J. Civ. Eng. Technol.* **2017**, *8*, 777–788.
3. Duxson, P.; Provis, J.L.; Lukey, G.C.; van Deventer, J.S.J. The role of inorganic polymer technology in the development of green concrete. *Cem. Concr. Res.* **2007**, *37*, 1590–1597. [CrossRef]
4. Davidovits, J. *Geopolymer Chemistry and Applicatons*, 3rd ed.; Institut Géopolymère: Saint-Quentin, France, 2011; Volume 9, p. 552.
5. Rangan, B.V. Upcycling Fly Ash into Geopolymer Concrete Products. *Int. J. Adv. Str. Geotech Eng.* **2014**, *3*, 349–353.
6. Wallah, S.E.; Rangan, B.V. Low Calcium Fly Ash Based Geopolymer Concrete: Long Term Properties. Ph.D. Thesis, Curtin University, Perth, Australia, 2006.
7. Bakharev, T. Geopolymeric materials prepared using Class F fly ash and elevated temperature curing. *Cem. Concr. Res.* **2005**, *35*, 1224–1232. [CrossRef]
8. Ul Haq, E.; Kunjalukkal Padmanabhan, S.; Licciulli, A. Synthesis and characteristics of fly ash and bottom ash based geopolymers—A comparative study. *Ceram. Int.* **2014**, *40*, 2965–2971. [CrossRef]
9. Cheng, T.W.; Chiu, J.P. Fire-resistant geopolymer produce by granulated blast furnace slag. *Min. Eng.* **2003**, *16*, 205–210. [CrossRef]
10. Muhammad, N. Effect of Heat Curing Temperatures on Fly Ash-Based Geopolymer Concrete. *Int. J. Eng. Technol.* **2019**, *8*, 15–19.
11. Hardjito, D.; Rangan, B.V. *Development and Properties of Low Calcium Fly Ash Based Geopolymer Concrete*; Curtin Research Publications: Perth, Australia, 2005; Volume 8, pp. 10–103.
12. Nuruddin, M.F.; Demie, S.; Shafiq, N. Effect of mix composition on workability and compressive strength of self-compacting geopolymer concrete. *Can. J. Civ. Eng.* **2011**, *38*, 1196–1203. [CrossRef]
13. Azreen, M.A.; Hussin, M.W.; Nor Hasanah, A.S.L.; Mostafa, S. Effect of ceramic aggregate on high strength multi blended ash geopolymer mortar. *J. Teknol.* **2015**, *16*, 33–36.
14. Abdul Awal, A.S.M.; Abubakar, S.I. Properties of Concrete Containing High Volume Palm Oil. *Malays. J. Civ. Eng.* **2011**, *2*, 54–66.
15. Yong, M.; Liu, J.; Chua, C.P.; Alengaram, U.J.; Jumaat, M.Z. Utilization of Palm Oil Fuel Ash as Binder in Lightweight Oil Palm Shell Geopolymer Concrete. *Adv. Mater. Sci. Technol.* **2014**, *2014*, 610274.
16. Ahmad, M.H.; Omar, R.C.; Malek, M.A.; Md Noor, N.; Thiruselvem, S. Compressive Strength of Palm Oil Fuel Ash Concrete. In Proceedings of the International Conference on Construction and Building Technology, Kuala Lumpur, Malaysia, 16–20 June 2008; Volume 27, pp. 297–306.

17. Habeeb, G.A.; Mahmud, H.B. Study on Properties of Rice Husk Ash and Its Use as Cement Replacement. *Mater. Res.* **2010**, *13*, 185–190. [CrossRef]
18. Kartini, K. Rice Husk Ash-Pozzolanic Material for Sustainability. *Int. J. Appl. Sci. Technol.* **2011**, *1*, 169–178.
19. Kawabata, C.Y.; Junior, H.S.; Sousa-coutinho, J. Rice Husk Ash derived Waste Materials as Partial Cement Replacement in Lighweight Concrete. *Agric. Eng.* **2012**, *36*, 26–31.
20. Ranjbar, N.; Mehrali, M.; Alengaram, U.J.; Metselaar, H.S.C.; Jumaat, M.Z. Compressive strength and microstructural analysis of fly ash/palm oil fuel ash based geopolymer mortar under elevated temperatures. *Constr. Build. Mater.* **2014**, *65*, 114–121. [CrossRef]
21. Ranjbar, N.; Kuenzel, C. Influence of preheating of fly ash precursors to produce geopolymers. *J. Am. Ceram. Soc.* **2017**, *100*, 3165–3174. [CrossRef]
22. Guan, M. *Manual for Bamboo Charcoal Production and Utilization*; Bamboo Engineering Research Centre (BERC): Nanjing, China, 2004; Volume 4, pp. 1–24.
23. Nordahlia, A.S.; Anwar, U.M.K.; Hamdan, H.; Latif, M.A.; Mahanim, S.M.A. Anatomical, physical and strength properties of Shizostachyum brachycladum (Buluh lemang). *J. Bamboo Ratt.* **2010**, *10*, 111–122.
24. Araminta, B.; Octavia, F.A.; Hadipraja, M.; Isnaeniah, S.; Viriani, V. Lemang (Rice Bamboo) as representative of typical malay food in Indonesia. *J. Ethic Foods* **2017**, *4*, 3–7.
25. ASTM C127-15. Standard Test Method for Density. In *Relative Density (Specific Gravity), and Absorption of Coarse Aggregates*; ASTM International: West Conshohocken, PA, USA, 2015.
26. ASTM C128-15. *Standard Test Method for Relative Density (Specific Gravity) and Absorption of Fine Aggregate*; ASTM International: West Conshohocken, PA, USA, 2015.
27. ASTM C29/C29M-17a. *Standard Test Method for Bulk Density ("Unit Weight") and Voids in Aggregate*; ASTM International: West Conshohocken, PA, USA, 2017.
28. BS EN 933-1. *Tests for Geometrical Properties of Aggregates. Determination of Particle Size Distribution*; Sieving method; British Stanards Institution: London, UK, 2012.
29. Behzad, N.; Jay, S. Effect of different superplasticizers and activator combinationson workability and strength of fly ash based geopolymer. *Mater. Des.* **2014**, *57*, 667–672.
30. Huseien, G.F.; Ismail, M.; Tahir, M.; Mirza, J.; Hussein, A.; Khalid, N.H.; Sarbini, N.N. Effect of Binder to Fine Aggregate Content on Performance of Sustainable Alkali Activated Mortars Incorporating Solid Waste Materials. *Chem. Eng. Trans.* **2018**, *63*, 667–672. [CrossRef]
31. Shen, P.; Lu, L.; He, Y.; Wang, F.; Lu, J.; Zheng, H.; Hu, S. Investigation on expansion effect of the expansive agents in ultra-high performance concrete. *Cem. Concr. Compos.* **2020**, *105*, 103425. [CrossRef]
32. Zhang, X.; Zhao, S.; Liu, Z.; Wang, F. Utilization of steel slag in ultra-high performance concrete with enhanced eco-friendliness. *Constr. Build. Mater.* **2019**, *214*, 28–36. [CrossRef]
33. Liang, X.; Wu, C.; Yang, Y.; Wu, C.; Li, Z. Coupled effect of temperature and impact loading on tensile strength of ultra-high performance fibre reinforced concrete. *Compos. Struct.* **2019**, *229*, 111432. [CrossRef]
34. Kim, S.; Yoo, D.Y.; Kim, M.-J.; Banthia, N. Self-healing capability of ultra-high-performance fiber-reinforced concrete after exposure to cryogenic temperature. *Cem. Concr. Compos.* **2019**, *104*, 103335. [CrossRef]
35. Turker, K.; Hasgul, U.; Birol, T.; Yavas, A.; Yazici, H. Hybrid fiber use on flexural behavior of ultra high performance fiber reinforced concrete beams. *Compos. Struct.* **2019**, *229*, 111400. [CrossRef]
36. Huseien, G.F.; Sam, A.R.M.; Shah, K.W.; Mirza, J.; Tahir, M.M. Evaluation of alkali-activated mortars containing high volume waste ceramic powder and fly ash replacing GBFS. *Constr. Build. Mater.* **2019**, *210*, 78–92. [CrossRef]
37. ASTM C1437-15. *Standard Test Method for Flow of Hydraulic Cement Mortar*; ASTM International: West Conshohocken, PA, USA, 2015.
38. ASTM C 109-16. *Standard Test Method for Compressive Strength of Hydraulic Cement Mortars*; ASTM International: West Conshohocken, PA, USA, 2016.
39. BS EN 12504-4. *Testing Concrete. Determination of Ultrasonic Pulse Velocity*; British Stanards Institution: London, UK, 2004.
40. ASTM E119-12a. *Standard Test Methods for Fire Tests of Building Construction and Materials*; ASTM International: West Conshohocken, PA, USA, 2012.
41. Luo, X.; Sun, W.; Chan, S.Y.N. Effect of heating and cooling regimes on residual strength and microstructure of normal strength and high-performance concrete. *Cem. Concr. Res.* **2000**, *30*, 379–383. [CrossRef]

42. Peng, G.; Bian, S.; Guo, Z.; Zhao, J.; Peng, X.; Jiang, Y. Effect of thermal shock due to rapid cooling on residual mechanical properties of fiber concrete exposed to high temperatures. *Constr. Build. Mater.* **2008**, *22*, 948–955. [CrossRef]
43. Kuenzel, C.; Ranjbar, N. Dissolution mechanism of fly ash to quantify the reactive aluminosilicates in geopolymerisation. *Resour. Conserv. Recycl.* **2019**, *150*, 104421. [CrossRef]
44. Davidovits, J. *Geopolymer Cement a Review*; Technical Paper #21 for Geopolymer Science and Technics; Geopolymer Institute Library: Saint-Quentin, France, 2013.
45. Sukumar, A.; John, E. Fiber addition and its effect on concrete strength. *Int. J. Innov. Res. Adv. Eng. (IJIRAE)* **2014**, *1*, 144–149.
46. Kong, D.L.Y.; Sanjayan, J.G. Effect of elevated temperatures on geopolymer paste, mortar and concrete. *Cem. Concr. Res.* **2010**, *40*, 334–339. [CrossRef]
47. Neville, A.M. *Properties of Concrete*, 5th ed.; Pearson Education Limited: Edinburgh Gate, UK, 2011.
48. Farhad, A.; Zohaib, A. Properties of Ambient-Cured Normal and Heavyweight Geopolymer Concrete Exposed to High Temperatures. *Materials* **2019**, *5*, 740.

© 2019 by the authors. Licensee MDPI, Basel, Switzerland. This article is an open access article distributed under the terms and conditions of the Creative Commons Attribution (CC BY) license (http://creativecommons.org/licenses/by/4.0/).

Article

Experimental Tests on Fiber-Reinforced Alkali-Activated Concrete Beams Under Flexure: Some Considerations on the Behavior at Ultimate and Serviceability Conditions

Linda Monfardini, Luca Facconi and Fausto Minelli *

DICATAM—Department of Civil, Environmental, Architectural Engineering and Mathematics, University of Brescia, 25123 Brescia, Italy; l.monfardini001@unibs.it (L.M.); luca.facconi@unibs.it (L.F.)
* Correspondence: fausto.minelli@unibs.it

Received: 1 September 2019; Accepted: 9 October 2019; Published: 15 October 2019

Abstract: Alkali-activated concrete (AAC) is an alternative concrete typology whose innovative feature, compared to ordinary concrete, is represented by the use of fly ash as a total replacement of Portland cement. Fly ash combined with an alkaline solution and cured at high temperature reacts to form a geopolymeric binder. The growing interest in using AACs for structural applications comes from the need of reducing the global demand of Portland cement, whose production is responsible for about 9% of global anthropogenic CO_2 emissions. Some research studies carried out in the last few years have proved the ability of AAC to replace ordinary Portland cement concrete in different structural applications including the construction of beams and panels. On the contrary, few experimental results concerning the structural effectiveness of fiber-reinforced AAC are currently available. The present paper presents the results of an experimental program carried out to investigate the flexural behavior of full-scale AAC beams reinforced with conventional steel rebars, in combination with fibers uniformly spread within the concrete matrix. The experimental study included two beams containing 25 kg/m^3 (0.3% in volume) of high-strength steel fibers and two beams reinforced with 3 kg/m^3 (0.3% in volume) of synthetic fibers. A reference beam not containing fibers was also tested. The discussion of the experimental results focuses on some aspects significant for the structural behavior at ultimate limit states (ULS) and serviceability limit states (SLS). The discussion includes considerations on the flexural capacity and ductility of the test specimens. About the behavior at the SLS, the influence of fiber addition on the tension stiffening mechanism is discussed, together with the evolution of post-cracking stiffness and of the mean crack spacing. The latter is compared with the analytical predictions provided by different formulations developed over the past 40 years and adopted by European standards.

Keywords: alkali-activated concrete; fly ash; geopolymer concrete; flexure; beams; fiber-reinforced concrete; crack spacing; tension stiffening

1. Introduction

Alkali-activated concrete (AAC) has been studied over the past years as a "green" alternative to ordinary Portland cement (OPC), whose production is energy intensive and responsible for about 8–9% of CO_2 emissions worldwide [1]. Alkali-activated binders can be generated from different types of aluminosilicate precursors, with differing availability, reactivity, cost, and value worldwide [2]. Because of the need for careful control of formulation, practical difficulties in application and supply chain limitations, geopolymers are still far from a total replacement of OPC across its full range of applications. However, alkali-activated binders may become sustainable and cost-effective construction materials, especially in the case that they are produced using locally-available raw materials [3]. The Australian experience in the field of geopolymers shows that the use of these binders may lead to a potential reduction of 40–60% in greenhouse gas emission, while the financial costs can be even 7% lower compared with OPC [2].

Currently, most literature regarding fly ash alkali-activated concrete focuses mainly on the study of the material properties, whereas limited attention has been paid to the structural behavior of AAC structures. The latter were first investigated by Hardjito et al. (2004) [4], Sumajouw et al. (2005) [5], and Sumajouw & Rangan (2006) [6], who performed a series of flexural tests on reinforced AAC beams and conventional reinforced concrete (RC) beams with different reinforcement ratios (0.64–2.69%). Test results showed a similar behavior of ACC and RC beams in terms of capacity and ductility evaluated for the same reinforcement ratio. A few years later, Dattereya et al. (2011) [7] compared the flexural behavior of reinforced geopolymer and conventional RC beams with reinforcement ratios ranging from 1.82 to 3.33%. According to their results, the normalized ultimate bending moments of all the test beams were quite similar, whereas the normalized bending moment at first cracking of the ACC specimens was generally lower (15–30%) than that observed for the RC beams.

Other studies have been devoted to investigating the ability of existing analytical models, which were originally developed for RC elements made with OPC, to predict the behavior of AAC elements. Yost et al. (2013) [8,9] carried out a series of tests on under-reinforced beams and then applied the models reported by the ACI 318-08 [10] to predict the behavior of the samples at the service and ultimate limit state. The authors concluded that the equations usually used to predict both the elastic behavior and the flexural/shear strength of RC beams can also be applied to get a reasonable estimation of AAC beam responses. Based on different experimental results reported by the literature, Prachasaree et al. (2014) [11] proved the inadequacy of the rectangular stress-block parameters typically used for OPC and proposed new stress-block design equations suitable for ash-based geopolymer concrete. The latter provided a rather good prediction of the flexural response of a series of AAC beams found in the literature.

It is well known that ordinary concrete exhibits brittle behavior because of its low uniaxial tensile strength and mode-I fracture energy. The addition of fibers randomly spread within the concrete matrix is a well acknowledged methodology to improve the tensile strength and toughness of concrete [12–15]. Experimental and numerical studies recently carried out by Mastali et al. [16] and Kheradmand et al. [17] proved that short hybrid polymeric fibers can be successfully employed to improve the flexural performance of geopolymeric mortar and concrete. As an additional benefit, fibers allow to better control the effects of shrinkage [18], thermal gradients, and any factor determining volumetric instability of the composite material.

The improvement of the tensile behavior of concrete due to the use of fibers can be exploited to enhance the ultimate and the serviceability performance of different kinds of structures [19]. Several research studies have proven the ability of fibers to partially or even totally replace conventional steel reinforcement in structures characterized by a significantly high degree of redundancy, such as slab-on-grade and elevated slabs [20]. Other authors have shown the possibility of partially replacing either the flexural or the shear reinforcement in concrete beams [21].

The results of an experimental research performed on full-scale reinforced AAC beams subjected to flexure are herein presented. This study aimed at evaluating the effectiveness of ACC as a structural material in view of its use for the construction of typical pre-cast elements such as ducts, manholes, beams, columns, and roof elements. The experimental program included ACC beams not containing fibers and beams made with ACC reinforced either with steel (rigid) or polymeric (deformed) fibers. The paper will describe and discuss the test results by referring to both the ultimate (ULS) and the serviceability (SLS) loading conditions. Particular attention will be devoted to the analysis of the tension stiffening effect, as well as to the prediction of crack spacing and width according to the analytical models available from different European structural codes and pre-standards.

2. Experimental Program

2.1. Properties of the Test Beams

A total of five beams were cast and tested in the laboratory of structural engineering of the University of Brescia. As shown in Figure 1, the beams had an overall length of 4500 mm (L), a span length of 4400 mm, and a cross-section of 200 mm (B) × 500 mm (h). Specimens were longitudinally reinforced with two 16 mm diameter (2 Ø16 − $A_{s,long}$ = 402 mm^2) bottom deformed rebars, resulting in a longitudinal steel ratio (ρ_l) of 0.43% (concrete clear cover = 26 mm). In the middle portion of the element (in the flexural span between the two point loads), no stirrups and reinforcement bars in compression were provided to promote a flexural collapse governed by concrete crushing. In the remaining portions of the beam, two Ø12 mm bars were provided in the compression zone and closed stirrups (Ø6 mm@75 mm) were placed to prevent shear failure. The main properties of the beams are summarized in Table 1. As one may note, except for the specimen AAC that was cast without fiber reinforcement, the beams SFRAAC-1 and SFRAAC-2 were reinforced with steel (rigid) fibers, whereas the specimens PFRAAC-1 and PFRAAC-2 contained macro-synthetic (deformed) fibers.

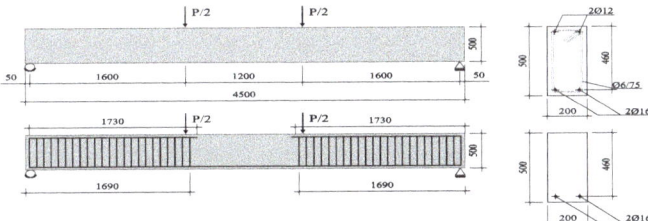

Figure 1. Sample details for full-scale beams, dimensions in millimeters.

Table 1. Main properties of the test beams.

Specimen Designation	B	h	L	ρ_l	Type of Fibers	Fiber Content	Fiber Volume Fraction (V_f)
	[mm]	[mm]	[mm]	[%]	[-]	[kg/m^3]	[%]
AAC	200	460	4500	0.44	No fibers	-	-
SFRAAC-1	200	460	4500	0.44	Hooked-end steel	25	0.3
SFRAAC-2							
PFRAAC-1					Synthetic Macrofiber—Embossed	3	0.3
PFRAAC-2							

After casting, beams and companion samples for material characterization were left in the wooden molders for 2 days (rest period). At the end of the rest stage, all the specimens were demolded, wrapped with a polyethylene sheet to prevent moisture loss and, finally, they were placed in a climate chamber to undertake curing. During the curing process, the ambient temperature was increased up to 60 °C and then kept constant for 24 h. More details about the adopted curing method are described in [22].

2.2. Materials

Steel fiber-reinforced alkali-activated concrete (SFRAAC) and polymeric fiber-reinforced alkali-activated concrete (PFRAAC) were used to cast the specimens SFRAAC-1 and 2 and PFRAAC-1 and 2, respectively. A single control beam made with AAC was cast.

High-strength hook-ended steel fibers, with a length (L_f) of 30 mm and a diameter (d_f) of 0.35 mm (aspect ratio L_f/d_f = 85.7), and synthetic embossed macro-fibers with a length of 54 mm and a diameter of 0.80 mm (aspect ratio L_f/d_f = 67.5), were added to the mixture in two different contents (25 and 3 kg/m^3, respectively, for steel and synthetic fibers) corresponding to a volume fraction (V_f) of 0.3%. The fiber tensile strength was 2200 MPa (minimum value according to the producer) and 585 MPa, respectively, for steel and synthetic fibers.

The mixture used for casting the AAC beam consisted of coarse aggregate 4–10 mm (1141 kg/m^3), sand 0–4 mm (615 kg/m^3), and class F fly ash (472 kg/m^3) mixed with an alkaline solution composed of 8 M sodium hydroxide (48 kg/m^3) and sodium silicate (119 kg/m^3). The chemical composition of the fly ash is reported in Table 2. The silica modulus (i.e., the SiO$_2$ to Na$_2$O ratio) that characterizes the activator solution was found to be 1.99. Extra water (35 kg/m^3) was added at the end of mixing just to promote suitable workability. The resulting volumetric mass density (2430 kg/m^3) of the hardened material was comparable to that of ordinary concrete.

Table 2. Chemical composition of fly ash.

Al$_2$O$_3$	SiO$_2$ *	CaO	Fe$_2$O$_3$	MgO	K$_2$O	Na$_2$O	TiO$_2$	SO$_3$
(%)	(%)	(%)	(%)	(%)	(%)	(%)	(%)	(%)
28	56	2	5.5	0.2 ÷ 3	0.2 ÷ 2	0.1 ÷ 0.6	0.1 ÷ 1.7	0.2 ÷ 2

* 40% is composed by reactive silica, representative of pozzolanic potential of fly ash.

The same mixture adopted for the AAC beam was used to cast the SFRAAC and the PFRAAC beams. In both cases, fibers were added to the mixture at two different times: Half of the total amount of fibers was mixed together with the dry components (i.e., aggregates and fly ash), whereas the remaining part was mixed after the addition of the liquid components (i.e., alkaline solution and extra water). This procedure led to an improved material workability as it allowed an enhanced fiber distribution and prevention of any fiber-balling phenomenon. The volumetric mass density of the hardened material resulted 2455 kg/m^3 and 2433 kg/m^3 for the SFRAAC and PFRAAC beams, respectively.

The mixtures were all characterized by a liquid to fly ash ratio of 0.43, an alkaline solution to fly ash ratio of 0.35, and by a sodium silicate to sodium hydroxide ratio equal to 2.5. The alkaline solution (mixing of sodium silicate and sodium hydroxide) had the same composition and chemical properties of that used in a previous work [23]. It has to be highlighted that extra water had the only aim of improving the workability of the fresh material. This fact explains why water was added some minutes later than the alkaline solution.

A series of tests were carried out to characterize the mechanical properties of the three materials used in the present research investigation.

To determine the average cube compressive strength (R_{cm}) (see Table 3), uniaxial compression tests were performed on cubes (side = 100 mm) according to EN 12390-3 [24]. Before being tested after 28 days from casting and about 25 days from the end of the curing period, the cubic samples were stored at room temperature of 15–25 °C and relative humidity of 45–60%. As also observed by other authors [2,10,25], the results of the present experimental program [18] show that the compressive strength of the adopted geopolymers stabilizes at the maximum value right after the end of the curing phase, namely 4–5 days after casting. Therefore, the tests performed at 28 days certainly provided the maximum strength of the materials.

Uniaxial compressive tests on cylinders (height = 200 mm; diameter = 100 mm) were carried out under displacement control in order to determine the Young's modulus and the compressive stress–strain constitutive law of the materials. Table 3 reports the mean elastic modulus (E_m), determined according to EN 12390-13 [26], as well as the mean values of the cylindrical compressive strength (f_{cm}) and of the corresponding strain (ε_{cm}). All the cylinders were tested after 28 days from the casting date. The results show that all the elastic moduli were lower than that (e.g., ~30 GPa) typically exhibited by a traditional OPC-based concrete. As observed by others [2,6,21,27], such a difference can be explained by considering that the C–S–H gel produced by the hydration process of OPC has a higher elastic modulus compared to N–A–S–H gel resulting from the alkaline activation process [28,29]. It also has to be noted that the addition of fibers decreases the workability and introduces air in the matrix, with a consequent possible further reduction of the modulus of elasticity. Conversely, the compressive strain at peak strength (ε_{cm}) obtained from the tests on cylinders were about 40–70% higher than the corresponding values reported by the Eurocode 2 [30] (clause 3.1) for OPC concrete with the same mean compressive strengths of the materials tested herein.

Table 3. Mechanical properties of materials.

Property	Unit	Test Beams				
		AAC	SFRAAC-1	SFRAAC-2	PFRAAC-1	PFRAAC-2
E_m	[GPa]	23.5 (4.0%)	18.2 (3.0%)	18.7 (1.0%)	18.1 (1.0%)	18.6 (2.0%)
ε_{cm}	[‰]	3.1 (1.8%)	5.0 (6.0%)	3.4 (12.8%)	2.8 *	2.8 *
R_{cm}	[MPa]	37.0 (4.1%)	45.0 (2.2%)	34.0 (8.5%)	40.0 (2.0%)	36.0 (3.3%)
f_{cm}	[MPa]	37.0 (1.0%)	27.0 (0.4%)	24.0 (2.9%)	24.0 (11.0%)	24.0 (11.0%)
R_{cm}/f_{cm}	[-]	1.0	1.7	1.4	1.7	1.5
f_{ck}	[MPa]	34.5	25.5	20.9	22.8	22.7
f_{ctm}	[MPa]	3.2	2.6	2.3	2.4	2.4
f_L	[MPa]	3.6 *	3.6 (0.6%)	3.9 (5.7%)	3.8 (15.7%)	3.6 (6.0%)
f_{R1}	[MPa]	0.37 *	3.3 (31.5%)	3.3 (2.4%)	0.7 (4.0%)	0.9 (4.7%)
f_{R2}	[MPa]	-	3.2 (39.8%)	3.3 (2.0%)	0.7 (5.0%)	0.9 (3.5%)
f_{R3}	[MPa]	-	2.9 (43.2%)	3.0 (0.6%)	0.8 (4.2%)	1.0 (8.1%)
f_{R4}	[MPa]	-	2.6 (46.3%)	2.7 (3.3%)	0.8 (1.9%)	1.0 (8.3%)

Coefficient of variation in round brackets; $f_{ctm} = 0.3 \cdot (f_{ck})^{2/3}$ according Eurocode 2 (2005); f_{ck} = characteristic value of the cylindrical compressive strength; * only one sample available for the test beam.

Figure 2 represents the flexural tensile stress vs. CMOD (crack mouth opening displacement) curves of the notched beams (150 × 150 × 500 mm^3) tested to characterize the tensile post-cracking behavior of the materials. Tests were carried out according to EN 14651-5 [31], which requires the evaluation of the limit of proportionality f_L and the residual flexural tensile strengths $f_{R,1}$, $f_{R,2}$, $f_{R,3}$, $f_{R,4}$, corresponding, respectively, to CMOD values of 0.5, 1.5, 2.5, and 3.5 mm. As usually observed for OPC concretes, the response of the AAC material not containing fibers was characterized by a significant reduction of the tensile resistance after cracking (i.e., after the achievement of the peak strength) due to the low mode-I fracture energy. Unlike AAC, the material containing steel fibers exhibited a significantly higher post-cracking strength in correspondence of the CMOD values 0.5 mm and 2.5 mm, which are representative of the serviceability and ultimate conditions according fib Model 2010 (MC2010) [32]. Because of the lower tensile performance of the polymeric fibers, the PFRAAC specimens experienced a brittle response, very similar to that observed for the AAC beams except for high values of CMOD, corresponding to residual strengths characterizing ultimate conditions.

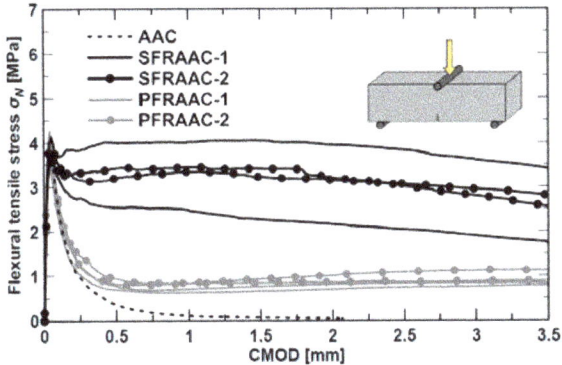

Figure 2. Flexural tensile stress–CMOD curves.

Table 3 summarizes the results from material characterization tests carried out for each of the full-scale test beams. This exhaustive characterization allows a better interpretation of the experimental results described in Section 3.

The longitudinal deformed rebars (B450 C according to Eurocode 2 [29]) were mechanically characterized by testing samples according EN 15630-1 [33]. The characterization tests provided a yielding strength (f_y) of 535 MPa, an ultimate strength (f_{tu}) of 646 MPa, and an ultimate tensile strain of about 13%.

2.3. Test Set-Up and Instrumentation

Figure 3a shows the loading set-up adopted to perform the flexural tests on the full-scale beams. The specimen was supported by two steel rollers located at the two ends of the beam (Figure 3b). A couple of steel rollers were also located on the top side of the specimen in order to support the longitudinal spreader steel beam used to apply the two point loads (P/2) acting at a distance of 600 mm from the middle of the beam. To prevent load concentration and possible local failure, each roller was laid on a steel plate positioned on a 25-mm-thick neoprene sheet. The total vertical load (P) was applied to the spreader beam by an electromechanical jack that allowed to perform the test under displacement control.

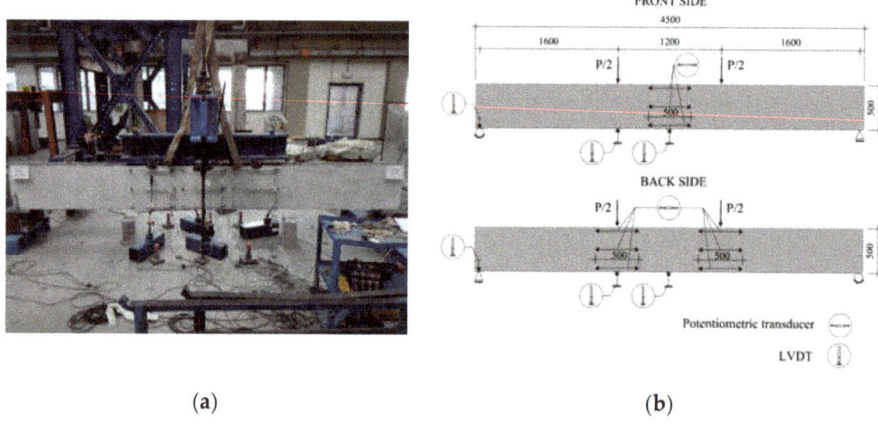

Figure 3. View of the typical 4-point bending test set-up (**a**). Schematic of the instrumentation (dimensions in millimeters) (**b**).

Figure 3b shows also the typical instrumentation set-up adopted to monitor the specimens. Six linear variable differential transformers (LVDTs) were used to measure deflections at midspan (front and back side) and at supports (front and back side). A total of nine potentiometric transducers were adopted for measuring the horizontal deformations along the height of the cross-sections located at midspan (front side), as well as under the two loading points (back side). In more detail, each cross-section was instrumented by three potentiometers installed, respectively, at 40, 250, and 460 mm from the top of the beam. A load cell was used to monitor the total load applied by the thrust jack. Tests were carried out by monotonically increasing the vertical displacement up to failure. The screw rate of the thrust jack was set at 0.75 mm/min in the initial stage, and then reduced to 0.5 mm/min after rebar yielding.

3. Experimental Results and Discussion

3.1. Behavior at Ultimate Limit State (ULS)

The total load–midspan deflection (δ) curves of the five test beams are depicted in Figure 4. Note that the load reported in the diagram includes both the weight of the loading system and the self-weight of the specimens.

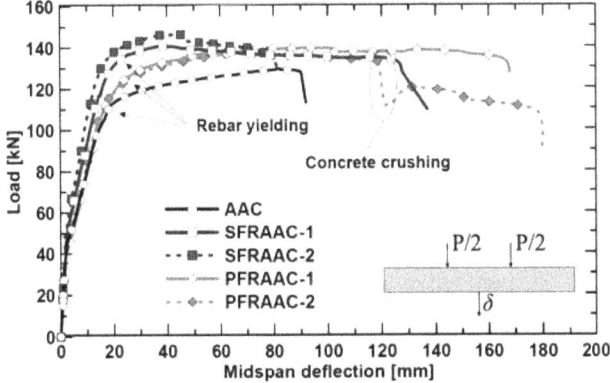

Figure 4. Total load–midspan deflection (δ) curves.

All beams exhibited the same initial elastic stiffness and similar first cracking loads that preceded the onset of the cracked stage. The kink point at the end of the second branch marks off the limit between the cracked stage and the plastic stage. In the plastic stage (i.e., the third branch), the applied load tended to slowly increase up to the maximum value.

As expected, the AAC element achieved the lowest maximum capacity (P_{max} = 130 kN) slightly before the final collapse. The latter occurred in the middle portion of the beam through concrete crushing (Figure 5).

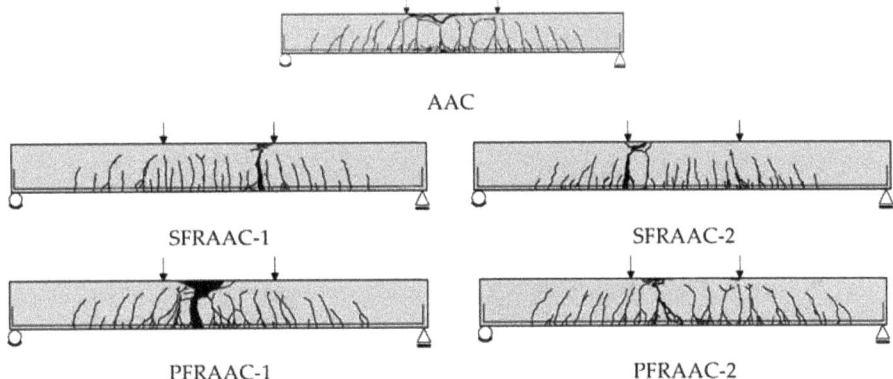

Figure 5. Crack patterns at failure.

The maximum capacity (P_{max}) of the beams reinforced with steel and synthetic fibers (Table 4) were, respectively, 8–12% and 8% higher than that reached by the AAC specimen. The low longitudinal reinforcement ratio (ρ_l = 0.43%) made the contribution of fibers more significant in governing the structural response of the beams. Because of their ability of enhancing concrete toughness in compression, fibers were able to avoid a sudden and brittle crushing of concrete. Moreover, compared to the reference beam, all fiber-reinforced beams experienced a more ductile behavior after yielding, and the final collapse was due to the tensile rupture of the longitudinal reinforcement. In fact, the improved steel-to-AAC interfacial bond [34] prevents or delays the development of plastic deformations in the yielded rebar. This phenomenon leads to a strain localization with a reduction of the portion of the bar under yielding which, with increasing fiber effect, determines an early rebar collapse.

Table 4. Behavior at ultimate limit states (ULS): Main test results and resisting moment obtained from the simplified cross-sectional model.

Specimen	P_{max}	$M_{R,max,exp}$	x	$M_{R,max,anl}$	$\varnothing_u$	$\varnothing_y$	$\mu_\varnothing$	δ_u	δ_y	μ_δ
	[kN]	[kNm]	[mm]	[kNm]	[1/km]	[1/km]	[-]	[mm]	[mm]	[-]
AAC	130	104	36.1	95	80	12	6.67	92	20	4.65
SFRAAC-1	141	113	59.4	116	246	18	13.67	132	19	6.96
SFRAAC-2	146	117	59.4	116	139	16	8.69	93	17	5.49
PFRAAC-1	140	112	43.8	102	175	20	8.75	194	18	10.8
PFRAAC-2	140	112	43.8	102	174	26	6.69	173	19	9.1

Beams PFRAAC-1 and PFRAAC-2 presented a slight decrease of the load due to the progression of a rather well controlled concrete crushing mechanism. The development of crushing particularly affected the response of the specimen PFRAAC-1, which exhibited a 15% decrease of the bearing capacity at a midspan deflection of about 120 mm (Figure 4). Figure 5 clearly shows that the significant damage caused by crushing occurred in the central portion of the specimen PFRAAC-1. In spite of concrete crushing, both PFRAAC beams continued to keep a significant load carrying capacity until final rupture of the longitudinal reinforcement occurred. The SFRAAC beams experienced the same failure mechanism, but the sharp reduction of resistance due to crushing was generally not observed. In conclusion, both the SFRAAC and the PFRAAC specimens presented a flexural failure mode characterized by the tensile rupture of longitudinal reinforcement.

The moment–curvature responses detected at different cross-sections, i.e., Section 1 (mid-span) and Sections 2 and 3 (under the point loads), are potted in Figure 6. Only the curves related to the beams AAC, SFRAAC-1, and PFRAAC-1 are reported, as representative of the typical behavior of each mixture. Based on the classical Navier's hypotheses, the curvature was calculated by the horizontal deformations measured by the potentiometers installed on the beams.

The observed responses appeared very similar to those reported by other studies [35] for OPC concrete beams with or without fiber reinforcement.

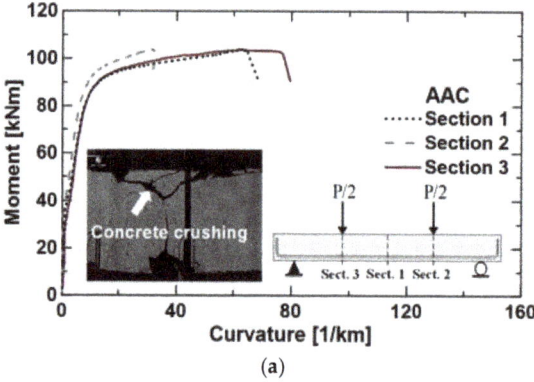

(a)

Figure 6. *Cont.*

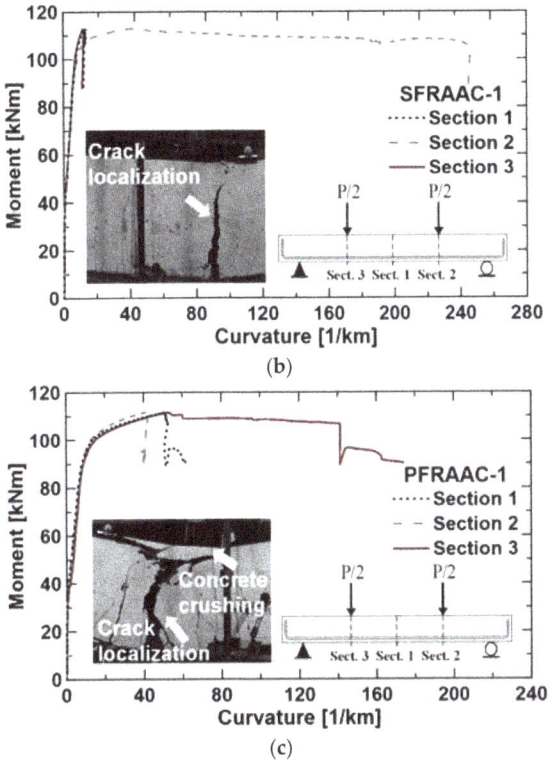

Figure 6. Experimental moment–curvatures curves and pictures of collapse for beams AAC (**a**), SFRAAC-1 (**b**), and PFRAAC-1 (**c**).

Moreover, the diagrams allow appreciating the strain localization (i.e., higher deformation) that generally involved one of the three monitored sections in which the ultimate mechanism took place (with significantly high values of local curvature, up to 240 km^{-1}, in specimen SFRAAC-1). The strain localization appeared to be more pronounced in the beams containing fibers, where it was promoted by the higher post-cracking tensile strength of the material (Figure 2), which allowed the compression chord to strongly delay its crushing with, conversely, a steady progressive material degradation in compression. Once the strain localization occurred, the beams started to behave like two rigid blocks, able to rotate about both the beam supports and the section subjected to the crack localization (a sort of plastic hinge).

The small picture reported in Figure 6b shows the flexural collapse mode of the SFRAAC-1 beam, which was not affected by significant damage of the top side of the cross-section. On the contrary, the picture of Figure 6c illustrates a high level of strain both at the bottom (steel strain) and at the top chord (concrete crushing). They both concurred in determining the collapse of the specimen PFRAAC-1, which experienced rebar collapse in the end.

In order to predict the flexural resistance of the test beams, the simplified rectangular stress-block model schematized in Figure 7 can be adopted. It is seen that for fiber-reinforced beams, the tensile resistance of the material is considered by means of the rigid-plastic model reported by the MC2010 (clause 5.6.4) [33], which considers a constant residual tensile strength (f_{Ftu}) over the depth (h-x) of the cross-section. Based on the previous assumptions, by also supposing that tensile reinforcement is yielded (reasonable assumption considering the low reinforcement ratio selected), the resisting moment ($M_{R,max,anl}$) can be calculated as follows:

$$M_{R,max,anl} = f_y \cdot A_s \cdot (d - 0.4 \cdot x) + f_{Ftu} \cdot B \cdot (h - x) \cdot (0.5 \cdot h + 0.1 \cdot x) \tag{1}$$

where $f_{Ftu} = f_{R3}/3$ is the tensile strength of fiber-reinforced concrete; f_{R3k} is the residual flexural strength at a crack mouth opening displacement (CMOD) of 2.5 mm (see Table 3); h = 500 mm is the section height; B = 200 mm is the section width; A_s = 400 mm² is the total area of bottom longitudinal rebars; f_y = 535 MPa is the yielding strength of rebars; and d = 460 mm is the effective depth. From the equilibrium of horizontal forces, the neutral axis depth (x) is obtained:

$$x = \frac{f_y \cdot A_s + f_{Ftu} \cdot B \cdot h}{B(0.8 \cdot f_{cm} + f_{Ftu})} \tag{2}$$

in which f_{cm} is the cylindrical compressive strength of concrete, as reported in Table 3. The results reported in Table 4 show that the resisting moment obtained from Equation (1) slightly underestimates (−9%) the maximum experimental bending moment ($M_{R,max,exp}$). On the contrary, the predicted bending moments related to beams containing steel fibers were basically equal or slightly higher (+2%) than the experimental ones. The analytical model confirmed the minor contribution provided by fibers to the flexural resistance of the beams. However, the simplified cross-sectional model proved to be able to well predict the resistance of all the AAC beams tested in this research study, in contrast to what is stated in [11].

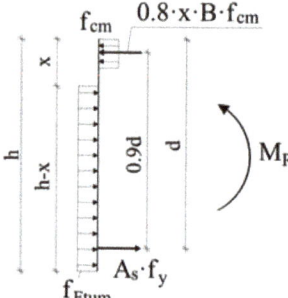

Figure 7. Simplified stress-block model for flexural strength calculation according MC2010 [33].

Table 4 also reports the ductility indexes evaluated as the ratio between the ultimate ($\emptyset_u$) and the yielding ($\emptyset_y$) curvature ($\mu_\emptyset = \emptyset_u/\emptyset_y$), as well as the ratio between the ultimate (δ_u) and the yielding (δ_y) midspan deflections ($\mu_\delta = \delta_u/\delta_y$). Irrespective of the ductility index considered, FRAAC beams resulted to be more ductile than the reference specimen. When considering the ductility in terms of curvature, the average ductility of the SFRAAC beams appears to be higher than that observed for the specimen containing synthetic fibers. Contrariwise, whether the ductility is assessed by the index, which better represents the overall response of the specimen instead of the behavior of the single local cross-section, the ductility of the beams containing synthetic fibers results to be the highest. In fact, once strain localization occurred, the lower post-cracking resistance of concrete reinforced with synthetic fibers allowed to delay the rebar failure, thus promoting a higher ductility (+60%) than that exhibited by the SFRAAC specimens.

It seems that, in flexural elements with low reinforcement ratios, synthetic deformed fibers are able to postpone (compared to steel fibers) the strain localization at the rebar level; in addition, their influence on the compression chord, even though less pronounced compared to steel fibers, allows anyway a progressive and controlled decay of the compression chord resistance. The combination of these two effects is beneficial in terms of overall structural ductility (i.e., ductility on terms of displacement). However, this trend should be cautiously evaluated by testing different longitudinal reinforcement ratios.

3.2. Behavior at Serviceability Limit State (SLS)

The change of stiffness occurring between the first and the second branch of the curves, shown in Figure 4, defines the onset of the cracking stage. All of the beams cracked at a load of about 45 kN, which corresponded to a midspan deflection approximately equal to $\delta_{cr} = 1.7$ mm. Considering that the materials used in the present investigation had basically the same peak tensile flexural strength (f_L) (see Table 3), similar first cracking loads were expected.

Crack control is one of the main features that must be considered in designing concrete members under service loading conditions. To estimate the mean crack width (w) in a concrete member, the following general equation can be used:

$$w = S_{rm} \cdot (\varepsilon_{sm} - \varepsilon_{cm}) \quad (3)$$

where ε_{sm} is the mean tensile strain in the reinforcement; ε_{cm} is the mean tensile strain in concrete between cracks; and S_{rm} is the mean crack spacing. A reasonable prediction of crack width necessarily requires the estimation of the crack spacing parameter.

The results of the tests performed herein allowed to calculate the crack spacing as the mean distance between cracks, namely the distance between the point loads (i.e., 1200 mm) divided by the number of cracks detected in the constant bending moment region. Figure 8a reports the evolution of the mean crack spacing observed in the pre-yielding stage against the deflection at midspan. The evolution of the number of cracks detected during the execution of each bending test is represented in Figure 8b.

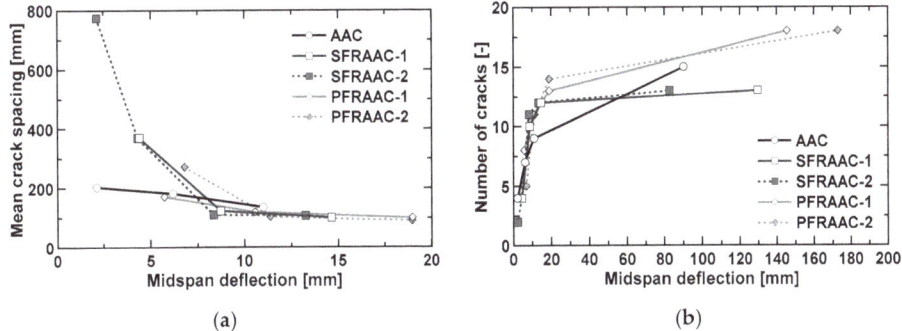

Figure 8. Mean crack spacing in the pre-yielding phase (**a**) and number of cracks (**b**) vs. midspan deflection.

As usually observed in OPC concrete members, the mean crack spacing tended to decrease with increasing vertical deflection and then stabilized at a minimum value (stabilizing cracking stage). The crack spacing became almost constant at a deflection of about 8 mm, which corresponded to a vertical load equal to 60% ($p \approx 80$ kN) of the maximum capacity. The attainment of a constant crack spacing represents the onset of the so-called stabilized cracking stage. The crack spacing detected in the stabilized cracking stage for the fiber-reinforced beams was similar or slightly lower (−10%) than that presented by the reference beam not containing fibers. On the contrary, for vertical deflections ranging from first cracking ($\delta_{cr} = 1.7$ mm) to 7.5 mm, the crack spacing of the reference beam resulted to be the lowest, together with that presented by beam PFRAAC-1. This result appears to be in contrast to what is reported by other authors [36,37], especially with regard to the specimens made with steel fiber-reinforced concrete. Because of its higher post-cracking strength, fibers are usually able to reduce the crack spacing even if provided in low (<0.5%) volume fractions. Further investigation needs to be undertaken to better understand these experimental observations.

As the crack spacing decreased in the pre-yielding stage, the number of cracks (Figure 8b) increased at a very high rate. After yielding, the formation of new cracks still took place, but at a slower rate compared to the pre-yielding stage. It is worth noting that the beams characterized by the lowest number of cracks at failure were those containing steel fibers (Figure 8b). This fact can be explained by considering the ability of steel fibers to localize the deformation in a single section, as already shown in the paragraph related to the behavior at ultimate limit states.

To estimate the mean crack spacing for RC members, European structural codes have proposed different relationships over the past years. The CEB-FIP Model Code 1978 [38] proposed the following equation:

$$S_{rm} = 2 \cdot \left(c + \frac{s}{10}\right) + k_1 \cdot k_2 \cdot \frac{\emptyset}{\rho_{eff}} \qquad (4)$$

where:
- c = concrete clear cover;
- s = distance between longitudinal reinforcement bars;
- k_1 = coefficient regarding bond between bars and concrete (=0.4 for deformed bars);
- k_2 = coefficient regarding stress distribution in the cross-section (=0.125 for flexure);
- $\emptyset$ = longitudinal rebar diameter; and
- ρ_{eff} = effective reinforcement ratio.

In 1990, the updated version of the same European code (CEB-FIP Model Code 1990 [39]) proposed a simplified formulation depending only on the effective reinforcement ratio and the bar diameter, as follows:

$$S_{rm} = \frac{2}{3} \cdot \frac{\emptyset}{3.6 \cdot \rho_{eff}} \qquad (5)$$

The first version of the Eurocode 2 published in 1991 [40] reported a modified version of the CEB-FIP Model Code 1978's [39] model:

$$S_{rm} = 50 + 0.25 \cdot k_1 \cdot k_2 \cdot \frac{\emptyset}{\rho_{eff}} \qquad (6)$$

in which k_1 and k_2 can be assumed equal to 0.8 (deformed bars) and 0.5 (members in flexure), respectively. The Eurocode 2 released in 2003 [41] re-introduced the clear cover as a parameter affecting the crack spacing:

$$S_{rm} = 3.4c + 0.425 \cdot k_1 \cdot k_2 \cdot \frac{\emptyset}{\rho_{eff}} \qquad (7)$$

Finally, the MC2010 [33] defined the crack spacing as a function of the transition length ($l_{s,max}$) according to the following relation:

$$S_{rm} = 1.17 \cdot l_{s,max} = k \cdot c + 0.25 \cdot \frac{f_{ctm}}{\tau_{bm}} \cdot \frac{\emptyset}{\rho_{eff}} \qquad (8)$$

where f_{ctm} is the mean tensile strength of concrete (Table 3) and $\tau_{bm} = 1.8 f_{ctm}$ is the mean value of bond stress between concrete and rebars.

Regarding steel fiber-reinforced concrete elements, one of the earliest models for predicting the crack spacing was proposed by the RILEM committee TC 162-TDF (2003) [42], which modified the Eurocode 2 (1991)'s relation [41] for plain concrete as follows:

$$S_{rm} = \left(50 + 0.25 \cdot k_1 \cdot k_2 \cdot \frac{\emptyset}{\rho_{eff}}\right) \cdot \frac{50}{L_f/d_f} \qquad (9)$$

Note that in the previous equation, the effect of fibers was considered by the factor $50/(L_f/d_f) \leq 1$, which includes the fiber aspect ratio L_f/d_f. Despite the fact that this formulation was initially meant only for elements reinforced with steel fibers, it will be here applied also to the specimens containing synthetic fibers.

The latest model for predicting the crack spacing in fiber-reinforced concrete elements was reported by the MC2010 (2013) [33]. The latter adjusted the formulation originally developed for RC concrete as follows:

$$S_{rm} = 1.17 \cdot l_{s,max} = k \cdot c + 0.25 \cdot \frac{(f_{ctm} - f_{Ftsm})}{\tau_{bm}} \cdot \frac{\varnothing}{\rho_{eff}} \qquad (10)$$

It is seen that the ability of fibers to reduce the crack spacing is considered by the term $f_{Ftsm} = 0.45 f_{R1m}$, which is related to the residual strength of concrete (f_{R1m}) corresponding to a crack width (CMOD) of 0.5 mm significant for serviceability loading conditions [33].

The diagrams of Figure 9 compare the experimental mean crack spacings, detected at different loading levels (i.e., 60 kN, 85 kN, 115 kN, failure load), with the corresponding values predicted by the models described above. From simple calculations, the tensile stress expected in the longitudinal reinforcement at the minimum loading level (i.e., 60 kN) is about 260 MPa. The latter is generally considered as the maximum value for the adopted reinforcing steel able to limit deflections and ensure good crack control at SLS conditions. It is worth noting that 60 kN and 85 kN correspond, respectively, to 40–46% and 60–65% of the maximum load attained by the test beams.

Regarding the AAC beam (Figure 9a), a good agreement between experimental and analytical data was observed except for the Eurocode 2 (2003) [42], whose predictions overestimated the experimental results both for low and high values of the applied load.

About the beams reinforced with steel fibers, both RILEM TC 162-TDF (2003) [42] and MC2010 [33] provided crack spacings 30–60% lower than those detected in the stabilized cracking stage (i.e., $p \geq 85$ kN). When considering the specimens containing synthetic fibers, the prediction of RILEM TC 162-TDF (2003) (Figure 9c) appeared to be very close to the experimental results. On the contrary, the MC2010 (Figure 9b) tended to overestimate by about 20% the stabilized value of crack spacing.

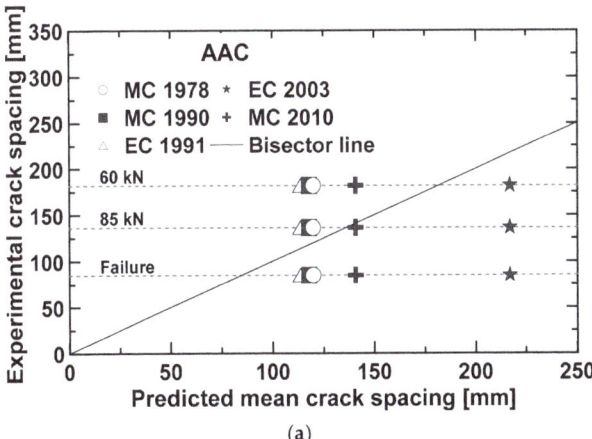

(a)

Figure 9. Cont.

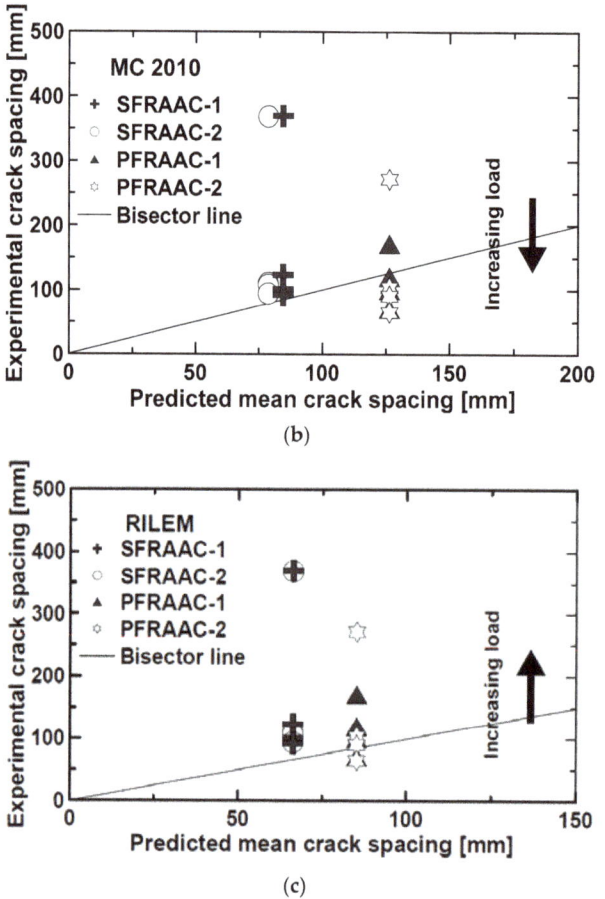

Figure 9. Comparison between predicted and experimental crack spacing. AAC beam (**a**). Predictions according MC2010 [33] (**b**) and RILEM [42] (**c**) for the beams reinforced with fibers.

The effect of fibers on tension stiffening can be highlighted by comparing midspan deflection of the fiber reinforced concrete beams (see Figure 4) with those exhibited by the reference beam (AAC) at a certain loading level. Here, the comparison was carried out by considering three different loading levels, i.e., 40 kN, 58 kN, and 90 kN, which corresponded, respectively, to 30%, 45%, and 70% of the maximum capacity (P_{max} = 130 kN) of the reference beam. The resulting relative variation of midspan deflections will be here referred to as $\Delta_{0.30}$ (p = 40 kN), $\Delta_{0.45}$ (p = 58 kN), and $\Delta_{0.70}$ (p = 90 kN) (Table 5). As shown in Table 5, both SFRAAC beams were characterized by a remarkable increment of the flexural post-cracking stiffness that, in turn, led to a reduction of the maximum deflection ranging from 16% to 40%. Compared to steel fibers, the synthetic fibers resulted to be less effective as the deflection reduction was approximately equal to 7% for the beam PFRAAC-2, and 11–17% for the specimen PFRAAC-1.

This enhanced post-cracking stiffness, related to low level of strains in the rebars (for the same load levels), is a key factor determining a decrease of the crack widths in flexure, confirming several findings in the literature [37].

Table 5. Behavior at serviceability limit states (SLS): Relative variation of midspan deflection referred to the beam AAC.

Specimen	$\Delta_{0.30}$	$\Delta_{0.45}$	$\Delta_{0.70}$
	[mm]	[mm]	[mm]
AAC	-	-	-
SFRAAC-1	−0.32 (−16%)	−1.84 (−35%)	−3.14 (−26%)
SFRAAC-2	−0.36 (−17%)	−2.11 (−40%)	−3.12 (−26%)
PFRAAC-1	−0.34 (−17%)	−0.67 (−13%)	−1.64 (−11%)
PFRAAC-2	−0.03 (−1.2%)	−0.37 (−7%)	−0.80 (−7%)

4. Concluding Remarks

Experimental results on full-scale beams made of AAC under flexure were presented and discussed in this paper, focusing on the structural response and fiber influence on the global and local behavior.

It was observed that the post-cracking response of FRAACs is the most influencing parameter for the structural behavior of beams herein discussed. For this particular geometry and longitudinal steel ratio, fibers promoted enhancements both at SLS and ULS conditions.

Based on the results herein presented and discussed, the main following conclusions can be drawn:

- For the studied element geometry and longitudinal reinforcement ratio (ρ_l = 0.44%), the presence of fibers affected the structural failure mode by promoting a collapse due to rebar rupture; on the contrary, the reference element without fibers (AAC) experienced a classical failure due to concrete crushing.
- The post-cracking strength of fibers, at ULS, developed a strain localization with a reduction of the portion of the bar under yielding. This determined an early rebar collapse, especially for steel fibers, which are more rigid and tougher compared to synthetic fibers. This peculiarity promoted a quite different local behavior well captured by the local moment–curvature diagrams observed: The section where the collapse occurred experienced a greater curvature, resulting in a corresponding larger local ductility, especially for FRC elements with higher post-cracking strengths (steel fibers). Conversely, the lower post-cracking strength observed in polymeric fibers tended to delay, in terms of midspan deformation, the final rebar collapse, resulting in a greater overall ductility (in terms of displacement), also possible thanks to the fiber ability to effectively promote a progressive decay of the concrete in compression.
- Elements reinforced with steel fibers developed the lowest number of cracks at ULS, due to the strain localization, which might have caused a limited (not full) development of crack spacing.
- The crack stabilizing stage for elements reinforced with fibers took place at a higher load level compared to non-fibrous elements; at the crack stabilizing stage, a good agreement between analytical and experimental crack spacing values was observed.
- At SLS conditions, despite AAC beams experiencing a greater elastic modulus, fibers promoted an enhancement of the post-cracking stiffness with a consequent reduction of the midspan displacement and crack widths; a tension stiffening effect was seen to be more noticeable in elements reinforced with steel fibers where post-cracking strengths were greater.

Author Contributions: L.M. and F.M. conceived and designed the experiments; L.M. performed the experiments; L.M. and F.M. analyzed the data; L.M., F.M. and L.F. wrote the paper. F.M. and L.F. revised the paper.

Funding: This research received no external funding.

Acknowledgments: The authors would like to give their appreciation to *General Admixtures Spa*, and *RAM. Italia Srl* for supplying most of the materials herein utilized. The authors would also like to thank Engs. Elena Bonvini, Claudio Romano, and Stefano Zubani and the technicians from Material Laboratory of the University of Brescia for their precious contribution in the experiments. Giovanni A. Plizzari is finally gratefully acknowledged for the fruitful discussion regarding this research project.

Conflicts of Interest: The authors declare no conflicts of interest.

References

1. Malhotra, V.M. Introduction: Sustainable development and concrete technology. *Concr. Int.* **2002**, *24*, 1147–1165.
2. McLellan, B.C.; Williams, R.P.; Lay, J.; Van Riessen, A.; Corder, G.D. Costs and carbon emissions for geopolymer pastes in comparison to ordinary portland cement. *J. Clean. Prod.* **2011**, *19*, 1080–1090. [CrossRef]
3. Attanasio, A.; Pascali, L.; Tarantino, V.; Arena, W.; Largo, A. Alkali-activated mortars for sustainable building solutions: Effect of binder composition on technical performance. *Environments* **2018**, *5*, 35. [CrossRef]
4. Hardjito, D.; Wallah, S.E.; Sumajouw, D.M.; Rangan, B.V. On the development of fly ash-based geopolymer concrete. *Mater. J.* **2004**, *101*, 467–472.
5. Sumajouw, D.M.J.; Hardjito, D.; Wallah, S.E.; Rangan, B.V. Behaviour and Strength of Reinforced Fly Ash-Based Geopolymer Concrete Beams. In Proceedings of the Australian Structural Engineering Conference, Newcastle, Australia, 11–14 September 2005.
6. Sumajouw, M.D.J.; Rangan, B.V. Low-calcium fly ash-based geopolymer concrete: Reinforced beams and columns. In *Research Report GC 3*; Faculty of Engineering, Curtin University of Technology: Perth, Australia, 2006.
7. Dattatreya, J.K.; Rajamane, N.P.; Sabitha, D.; Ambily, P.S.; Nataraja, M.C. Flexural Behaviour of Reinforced Geopolymer Concrete Beams. *Int. J. Civ. Struct. Eng.* **2011**, *2*, 138–159.
8. Yost, J.R.; Radlińska, A.; Ernst, S.; Salera, M. Structural behavior of alkali activated fly ash concrete. Part 1: Mixture design, material properties and sample fabrication. *Mater. Struct.* **2013**, *46*, 435–447. [CrossRef]
9. Yost, J.R.; Radlińska, A.; Ernst, S.; Salera, M.; Martignetti, N.J. Structural behavior of alkali activated fly ash concrete. Part 2: Structural testing and experimental findings. *Mater. Struct.* **2013**, *46*, 449–462. [CrossRef]
10. American Concrete Institute (ACI) Committee 318. *Building Code Requirements for Structural Concrete (ACI 318-08) and Commentary*; American Concrete Institute: Farmington Hills, MI, USA, 2008.
11. Prachasaree, W.; Limkatanyu, S.; Hawa, A.; Samakrattakit, A. Development of Equivalent Stress Block Parameters for Fly-Ash-Based Geopolymer Concrete. *Arab. J. Sci. Eng.* **2014**, *39*, 8549–8558. [CrossRef]
12. Balaguru, P.N.; Shah, S.P. *Fiber-Reinforced Cement Composites*; McGraw-Hill Inc.: New York, NY, USA, 1992.
13. Rossi, P.; Acker, P.; Malier, Y. Effect of steel fibres at two different stages: The material and the structure. *Mater. Struct.* **1987**, *20*, 436–439. [CrossRef]
14. Minelli, F.; Plizzari, G. Derivation of a simplified stress–crack width law for Fiber Reinforced Concrete through a revised round panel test. *Cem. Concr. Compos.* **2015**, *58*, 95–104. [CrossRef]
15. Conforti, A.; Minelli, F.; Plizzari, G.A.; Tiberti, G. Comparing test methods for the mechanical characterization of fiber reinforced concrete. *Struct. Concr.* **2018**, *19*, 656–669. [CrossRef]
16. Mastali, M.; Abdollahnejad, Z.; Dalvand, A. Increasing the flexural capacity of geopolymer concrete beams using partial deflection hardening cement-based layers: A numerical study. *Sci. Iranica. Trans. ACiv. Eng.* **2017**, *24*, 2832–2844.
17. Kheradmand, M.; Mastali, M.; Abdollahnejad, Z.; Pacheco-Torgal, F. Experimental and numerical investigations on the flexural performance of geopolymers reinforced with short hybrid polymeric fibres. *Compos. Part B Eng.* **2017**, *126*, 108–118. [CrossRef]
18. Chilwesa, M.; Facconi, L.; Minelli, F.; Reggia, A.; Plizzari, G. Shrinkage induced edge curling and debonding in slab elements reinforced with bonded overlays: Influence of fibers and SRA. *Cem. Concr. Compos.* **2019**, *102*, 105–115. [CrossRef]
19. Facconi, L.; Minelli, F. Verification of structural elements made of FRC only: A critical discussion and proposal of a novel analytical method. *Eng. Struct.* **2017**, *131*, 530–541. [CrossRef]
20. Facconi, L.; Plizzari, G.; Minelli, F. Elevated slabs made of hybrid reinforced concrete: Proposal of a new design approach in flexure. *Struct. Concr.* **2019**, *20*, 52–67. [CrossRef]
21. Germano, F.; Tiberti, G.; Plizzari, G. Post-peak fatigue performance of steel fiber reinforced concrete under flexure. *Mater. Struct.* **2016**, *49*, 4229–4245. [CrossRef]
22. Monfardini, L. Alkali Activated Materials for Sustainable Structural Applications. Ph.D. Thesis, Department of Civil, Environmental, Architectural Engineering and Mathematics, University of Brescia, Brescia, Italy, 2017.
23. Monfardini, L.; Minelli, F. Experimental Study on Full-Scale Beams Made by Reinforced Alkali Activated Concrete Undergoing Flexure. *Materials* **2016**, *9*, 739. [CrossRef]

24. EN 12390-3. *Testing Hardened Concrete—Part 3: Compressive Strength of Test Specimens*; CEN-European Committee for Standardization: Brussels, Belgium, 2009.
25. Fernandez-Jimenez, A.M.; Palomo, A.; Lopez-Hombrados, C. Engineering properties of alkali-activated fly ash concrete. *ACI Mater. J.* **2006**, *103*, 106–112.
26. UNI EN 12390-13. *Testing Hardened Concrete–Part 13: Determination of Secant Modulus of Elasticity in Compression*; British Standards Institution: London, UK, 2013.
27. Sofi, M.; Van Deventer, J.S.J.; Mendis, P.A.; Lukey, G.C. Engineering properties of inorganic polymer concretes (IPCs). *Cem. Concr. Res.* **2007**, *37*, 251–257. [CrossRef]
28. Puertas, F.; Palacios, M.; Manzano, H.; Dolado, J.S.; Rico, A.; Rodríguez, J. A model for the C-A-S-H gel formed in alkali-activated slag cements. *J. Eur. Ceram. Soc.* **2011**, *31*, 2043–2056. [CrossRef]
29. Provis, J.L.; van Deventer, J.S.J. *Alkali-Activated Materials: State-Of-The-Art Report, RILEM TC 224-AAM*; Springer/RILEM: Dordrecht, The Netherlands, 2014.
30. EN 1992-1-1: Eurocode 2. Design of Concrete Structures. In *Part 1-1: General Rules and Rules for Buildings*; CEN-European Committee for Standardization: Brussels, Belgium, 2004.
31. EN 14651-5. *Precast Concrete Products—Test Method for Metallic Fibre Concrete-Measuring the Flexural Tensile Strength*; CEN-European Committee for Standardization: Brussels, Belgium, 2005.
32. *Fib: Model Code for Concrete Structures 2010*; Ernst & Son: Berlin, Germany, October 2013; 434p, ISBN 978-3-433-03061-5.
33. EN 15630-1. *Steel for the Reinforcement and Prestressing of Concrete—Part 1: Test Methods*; CEN-European Committee for Standardization: Brussels, Belgium, 2004.
34. Sarker, P.K. Bond strength of reinforcing steel embedded in fly ash-based geopolymer concrete. *Mater. Struct.* **2011**, *44*, 1021–1030. [CrossRef]
35. Meda, A.; Minelli, F.; Plizzari, G.A. Flexural behaviour of RC beams in fibre reinforced concrete. *Compos. Part B Eng.* **2012**, *43*, 2930–2937. [CrossRef]
36. Jansson, A.; Flansbjer, M.; Löfgren, I.; Lundgren, K.; Gylltoft, K. Experimental investigation of surface crack initiation, propagation and tension stiffening in self-compacting steel–fibre-reinforced concrete. *Mater. Struct.* **2012**, *45*, 1127–1143. [CrossRef]
37. Minelli, F.; Tiberti, G.; Plizzari, G. Crack Control in RC Elements with Fiber Reinforcement. ACI Special Publication, ACI SP-280. Available online: http://www.cias-italia.it/PDF/PLIZZARI%20-%20Crack%20control%20in%20RC%20elements%20with%20Fiber%20Reinforcement.pdf (accessed on 12 October 2019).
38. Euro-International du Béton and Fédération Internationale de la Précontrainte. In *CEB–FIP Model Code for Concrete Structures*, 3rd ed.; CEB–FIP: Paris, France, 1978.
39. Comité Euro-International du Béton and Fédération Internationale de la Précontrainte. In *CEB-FIP Model Code*; CEB–FIP: Paris, France, 1990.
40. EN 1992-1-1: Eurocode 2 (1991): *Design of Concrete Structures—Part 1-1: General Rules and Rules for Buildings*; CEN-European Committee for Standardization: Brussels, Belgium, 1991.
41. EN 1992-1-1: Eurocode 2 (2003): *Design of Concrete Structures—Part 1-1: General Rules and Rules for Buildings*; CEN-European Committee for Standardization: Brussels, Belgium, 2003.
42. RILEM T C162-T. Test and design methods for steel fiber reinforced concrete. *Mater. Struct.* **2003**, *36*, 560–567. [CrossRef]

 © 2019 by the authors. Licensee MDPI, Basel, Switzerland. This article is an open access article distributed under the terms and conditions of the Creative Commons Attribution (CC BY) license (http://creativecommons.org/licenses/by/4.0/).

Article

The Alternatives to Traditional Materials for Subsoil Stabilization and Embankments

Mirjana Vukićević, Miloš Marjanović *, Veljko Pujević and Sanja Jocković

Faculty of Civil Engineering, University of Belgrade, Bulevar kralja Aleksandra 73, 11000 Belgrade, Serbia; mirav@grf.bg.ac.rs (M.V.); vpujevic@grf.bg.ac.rs (V.P.); borovina@grf.bg.ac.rs (S.J.)
* Correspondence: mimarjanovic@grf.bg.ac.rs

Received: 26 July 2019; Accepted: 14 September 2019; Published: 18 September 2019

Abstract: Major infrastructure projects require significant amount of natural materials, often followed by the soft soil stabilization using hydraulic binders. This paper presents the results of a laboratory study of alternative waste materials (fly ash and slag) that can be used for earthworks. Results of high plasticity clay stabilization using fly ash from Serbian power plants are presented in the first part. In the second part of the paper, engineering properties of ash and ash-slag mixtures are discussed with the emphasis on the application in road subgrade and embankment construction. Physical and mechanical properties were determined via following laboratory tests: Specific gravity, grain size distribution, the moisture–density relationship (Proctor compaction test), unconfined compressive strength (UCS), oedometer and swell tests, direct shear and the California bearing ratio (CBR). The results indicate the positive effects of the clay stabilization using fly ash, in terms of increasing strength and stiffness and reducing expansivity. Fly ashes and ash-slag mixtures have also comparable mechanical properties with sands, which in combination with multiple other benefits (lower energy consumption and CO_2 emission, saving of natural materials and smaller waste landfill areas), make them suitable fill materials for embankments, especially considering the necessity for sustainable development.

Keywords: fly ash; slag; soil stabilization; embankment; cement; lime

1. Introduction

The modern world is facing the consequences of the technological development followed by a huge environmental impact. This has stimulated recent scientific research in the field of identifying pollutants and the possibility of reducing the harmful effects of pollution. One of the major pollutants is fossil fuel, which produce huge amounts of CO_2 in the combustion process. According to the World Bank data for 2015 [1], the total share of fossil fuels in energy production in the world is 65.2%. According to the same data in Serbia, this share is 73.1%, since the thermal power plants are the main producers of electricity. There are six thermal power plants within the Electric Power Industry of Serbia, which use lignite as the main fuel.

Thermal power plants have multiple harmful effects on the environment: They pollute air by emitting CO_2, SO_2, N_2O and fly ash; landfills of ash and slag occupy large areas of mainly agriculture land; deposited ash can potentially pollute land and water due to the presence of microelements and radionuclides. The harmful effects of flue gases can be reduced by gas desulphurization, installation of efficient electro filters and application of methods for reducing the concentration of nitrogen oxides. The amount of deposited ash and slag can be significantly reduced by use in the construction industry.

According to the ECOBA (European Coal Combustion Products Association) data for 2016 [2], annual production of ash in the European Union (EU 15) was about 40 million tons, of which 64% was fly ash. About 50% of the produced amount was used in the construction industry, 41.5% was used for land reclamation–restoration and only 6.7% was deposited. In the construction industry, it was mostly used for the production of cement and concrete (about 25%), while much less (about 6% today vs. 25% ten years ago) was used in the road construction. The data show that ash in developed EU countries is successfully used as raw material in the industry. In Serbia, the situation is completely different. About 7 million tons of ash is produced annually. A very small part of the ash is deposited in silos, while most of the total produced amount is deposited with the slag at the landfills. The landfills occupy an area of about 1600 ha, with about 300 million tons of ash and slag [3,4]. So far, only 3% of ash has been used for the production of cement [4].

In Serbia, the first major research related to the possibility of using ash in road construction began in the first decade of this century, with the aim of reducing the large amount of deposited ash. Since then, four extensive studies have been done [5–8]. Based on the results of these studies, in 2015 the Serbian Government has passed a regulation [9] on the use of ash from thermal power plants in Serbia, thus creating a legal framework for the use of ash.

During the research [7,8], about 1000 laboratory tests of physical and mechanical properties of fly ash, ash and slag, mixtures of ash and soil with or without hydraulic binders (cement and lime) were done in the Laboratory for Soil Mechanics of the Faculty of Civil Engineering in Belgrade. Additionally, chemical composition of fly ash was investigated. The main results and conclusions from these studies are presented in this paper.

The aforementioned studies included a very important ecological aspect of the use of ash, bearing in mind that ash contains harmful substances that can be a potential source of pollution of soil and water. Ash can be disposed of as waste material if the content of artificial and natural radionuclides is less than the values prescribed in the Rulebook on the Limits of Radioactive Contamination of the Environment [10]. Ash and slag from the Serbian thermal power plants meet the prescribed requirements [11]. Ash also contains trace elements that are hazardous to the environment, such as As, B, Cr, Mo and Se [12]. These elements could contaminate the soil, water and marine ecosystems in case of their leaching. The main factor in the control of leaching is the control of the mobility of the trace elements. There are appropriate procedures that can reduce or eliminate the leaching of toxic trace elements such as As, B, Cr, Sb and Se [11,12]. If it is proven that there is no risk of leaching, the use of ash for embankments provides economic and environmental benefits.

In the first part of the paper, the results of high plasticity clay stabilization using fly ash from Serbian power plants (with and without binders) were presented. In addition, the effects of ash application as a soil stabilizer were compared with the effects of chemical additives for the same purpose. In the second part, engineering properties of ash and ash-slag mixtures as embankment material in road construction were investigated. The mechanical properties important for fulfillment of the technical requirements for road subgrade were tested. The influence of common binders (activators) was also investigated.

2. Literature Review

2.1. Ash Utilization for Soil Stabilization

The factors on which soil stabilization with ash depend on are: The type of ash and its characteristics, characteristics of the soil to be stabilized, the percentage of fly ash, the time period between wetting of the mixture and compaction and soil water content at the time of compaction.

According to ASTM C618 [13], fly ash types are class C and class F. This classification mainly depends on the content of SiO_2, Al_2O_3 and Fe_2O_3—minimum percentage of SiO_2 + Al_2O_3 + Fe_2O_3 for class F fly ash is 70%, and for class C is 50%. The percentage of sulfur trioxide (SO_3) is max 5% for both ash classes (fly ash with a sulfate content greater than 10% may cause soils to expand more than desired [14]). According to EN 14227-4 [15], fly ash is classified into calcareous (type W, equivalent to ASTM class C), and siliceous (type V, equivalent to ASTM class F).

Class C fly ash is mainly produced from lignite or subbituminous coal. This coal has a higher content of calcium carbonate, so class C fly ash is rich in calcium (more than 20% CaO), resulting in the self-cementing characteristics. Studies concerning fly ash utilization for soil stabilization indicate that class C fly ash is an effective and economical stabilizer for broad engineering applications [14,16–23].

Class F fly ash is mainly produced from burning anthracite or bituminous coal. This class of fly ash has pozzolanic properties, but has no self-cementing characteristics due to its lower CaO content (less than 10%). According to [24], class F fly ash should be used in soil stabilization with the addition of cementitious agent (lime, lime kiln dust, cement and cement kiln dust). However, there are researches indicating that this fly ash can effectively improve some engineering properties of soil (unconfined compressive strength (UCS), California bearing ratio (CBR) and swell potential) without activators [25–29].

According to [14,17,30], the optimal fly ash content for soil stabilization is in the range from 10% to 30%, depending on soil and ash type. Recent studies have shown that compaction properties and the strength of the mixture decreases with the increase in compaction delay time, which is a consequence of the loss of established cement bonds between particles and lower density [14,16]. According to the same research, it is proposed to carry out compaction within two hours after mixing. Water content of the soil during compaction has a major impact on density and strength of the mixture. According to [14,16], the water content for achieving the maximum strength is typically the optimal water content or up to 8% lower than the optimal.

The effects of applying fly ash for soil stabilization are the reduction in the plasticity and soil swell potential and increasing the soil strength and CBR values. The size of fly ash particles is commonly larger than the clay particles, thus the addition of fly ash changes the grain size distribution of the clay and reduces the liquid limit. The chemical composition of ash and treated soil also affects the Atterberg limits. Reduction of plasticity of the clay soil leads to a decrease in swell potential. Çoçka [17] as well as Nalbantoglu and Gucbilmez [30] have found that plasticity and swelling potential decrease with the increase in the content of class C fly ash. Ramadas et al. [26] have analyzed the characteristics of three expansive soils with the addition of class F fly ash of 0–50%, which resulted in significant decrease in liquid limit, swelling pressure and potential.

The increase in strength is the main reason for the use of fly ash for soil stabilization [24]. The California bearing ratio value is the primary parameter in the evaluation of suitability of fly ash stabilized soil utilization in road construction [14,16,20,22,31]. Clays generally have low CBR, and that makes them inappropriate for the use in base layers of pavements. Zia and Fox [32] have found that CBR values of loess increased five times by the addition of 10% class C fly ash. By adding 20% of self-cementing fly ash to fine-grained soil, White et al. [22] obtained CBR values that correspond to well-compacted gravel (~75%). Acosta et al. [18] investigated different soil types with very low CBR values (0–5%) and by the addition of 18% class C fly ash, achieved a significant increase in CBR values (20–56%). Vukićević et al. [29] analyzed the effect of class F fly ash on the strength of expansive clay. The highest increase in strength was obtained with the addition of 15% fly ash. The CBR value increased almost three times.

Increase of fly ash stabilized soil strength is a time-dependent process. The study of White et al. [22] on self-cementing fly ash showed a rapid increase in strength during the first 7 to 28 days, after which the slow down trend was registered due to prolonged pozzolanic reactions.

2.2. Ash Utilization as a Material for Embankment

Ash has been used for many years in construction as fill material in road construction, embankment construction and land reclamation [33]. Low compacted unit weight of fly ash makes it very suitable material in embankment construction.

Class F fly ash is more often applied as the material for embankments and backfills, in comparison with the class C fly ash [34], because of self-cementing characteristics of C class fly ash, which hardens within 2–4 h after the addition of water [35].

The important engineering properties of ash for its utilization in roads construction are: The moisture–density relationship, shear strength and compressibility.

Fly ash has a lower compacted density compared to traditional materials, which leads to smaller applied load and settlement of the subsoil. DiGioia et al. [36] have investigated the maximum dry density and the optimum water content for Western Pennsylvania class F fly ash and Western USA class C fly ash. Values of the maximum dry density varied from 11.9 to 18.7 kN/m^3 and values of the optimum water content from 13% to 32%. They concluded that the large variations were due to different physical and chemical characteristics of the ashes, which in the turn depend on the source of coal and the condition of coal combustion.

Shear strength tests on compacted ash specimens indicate that ash strength is mostly generated by internal friction [37]. Class F fly ash has a friction angle usually in the range of 26° to 42° [38]. Kim et al. [37] conducted tests on a mixture of fly ash and bottom ash and obtained friction angles in the range of 28–48°, which is in the rank of the shear strength of dense sandy soil.

There is not much published data for the California bearing ratio of ash. According to [39], CBR for class F fly ash ranges between 6.8% and 13.5% in the soaked conditions, and between 10.8% and 15.4% in the unsoaked conditions. For natural soils, CBR values normally range between 3% and 15% (fine-grained soils), 10–40% (sand and sandy soils) and 20–80% for gravels and gravelly soils [40].

Generally, technical standards prescribe that embankment must have small compressibility to reduce roadway settlements. The compressibility can be expressed through compression index Cc and recompression index Cr or through compressibility (constrained) modulus Mv (Cm). Kaniraj and Gayathri [41] carried out consolidation tests on the specimens of class F fly ash from the Dadri thermal power plant (New Delhi, India). They found that the compression indices Cc of the specimens were 0.041 or 0.084, depending on level of effective stress. The average recompression index Cr was 0.008. For the fly ash from the Rajghat thermal power plant (New Delhi), reported Cc and Cr were 0.072 and 0.017, respectively [42]. For the fly ashes from USA and Canada, McLaren and DiGioia [43] found the mean value of 0.13 for Cc. Kim and Prezzi [44] determined the tangent constrained modulus Cm at vertical stresses, from zero to 200 kPa, which is the range of stresses expected in highway embankments. The fly ash used in the study was class F, from three power plants in Indiana (USA). The obtained values were compared with the tangent constrained modulus available in the literature for compacted sand at different densities (at relative compaction of 99% and 85%). Specifically, the values of constrained moduli for the tested fly ashes (in the range of 10 MPa to 30 MPa at stress level 100–200 kPa) correspond to the lower end of the sand moduli range. This indicates that for the same compaction levels, the fly ashes are slightly more compressible than sand.

3. Materials

3.1. Soil

Soil was sampled from the location Radljevo, municipality Ub, Serbia. Based on the modified Unified Soil Classification System, the tested soil was high plasticity (CH) clay. Nevertheless, due to demonstrated shortcomings of the Casagrande chart [45], as an alternative, the authors used a new classification approach proposed in [46]. Moreno-Maroto and Alonso-Azcárate [46] classified clay by PI/LL (Plasticity Index vs. Liquid Limit) ratio. The PI/LL ratio for the tested clay was 0.38, which characterized the used material as moderately or slightly clayey soil. The maximum toughness, T_{max}, parameter that best represents plasticity [47], estimated by the Moreno-Maroto and Alonso-Azcárate [46] equation was 5.54, which indicates a low influence of clay minerals. Basic physical properties of CH clay are given in Table 1. The tested soil had a low to moderate swelling potential, with a swell deformation of 2.2% [48], which makes it generally unusable for most engineering purposes.

Table 1. Physical properties of the high plasticity (CH) clay.

G_s	Grain Size Distribution					Atterberg Limits			Swell %
	Clay <0.002 mm	Silt 0.002–0.06 mm	Sand 0.06–2 mm	Gravel 2–60 mm	Fines <0.075 mm	LL %	PL %	PI %	
2.67	22	72	6	-	96	51.0	31.5	19.5	2.2

Note: Testing methods are described in Section 4.

3.2. Ash and Slag

In the scope of this paper, the following waste materials from Serbian power plants were used:

(1) KOL FA—fly ash from electrostatic precipitators in thermal power plant "Kolubara";

(2) KOS FA—fly ash from electrostatic precipitators in thermal power plants "Kostolac A" and "Kostolac B";

(3) KOS AB—ash and slag mixture from the landfills of thermal power plants "Kostolac A" and "Kostolac B";

(4) TENT A—ash and slag mixture from the landfill of thermal power plant "Nikola Tesla A";

(5) TENT B—fly ash from the silos in thermal power plant "Nikola Tesla B".

Basic physical properties of tested waste materials are given in Table 2.

Table 2. Physical properties of tested waste materials.

Material	G_s	Grain Size Distribution (%)				
		Clay <0.002 mm	Silt 0.002–0.06 mm	Sand 0.06–2.0 mm	Gravel 2–60 mm	Fines <0.075 mm
KOL FA	2.13	0–2	60–65	35–38	-	67–72
KOS FA	2.22	-	75	25	-	80
KOS AB	2.41	2	10–22	77–89	-	14–27
TENT A	2.39	0–1	40–41	57–58	-	49–50
TENT B	2.26	2	14–31	65–82	2	22–40

Note: Testing methods are described in Section 4.

According to the standard ASTM C618 [13], the used materials belonged to class F. Chemical composition of all waste materials within this paper is given in Table 3.

Table 3. Chemical composition of the used waste materials.

Material	Chemical Composition (%)									
	SiO_2	Al_2O_3	Fe_2O_3	CaO	MgO	K_2O	Na_2O	TiO_2	SO_3	P_2O_5
KOL FA [7]	50.21	23.83	9.89	4.79	3.12	0.44	0.35	0.54	5.24	0.060
KOS FA [7]	56.38	17.57	10.39	7.46	2.13	0.57	0.38	0.52	0.95	0.025
KOS AB [6]	53.61	17.72	8.05	7.44	1.78	1.22	0.86	0.51	0.12	0.068
TENT A [6]	56.14	15.93	5.77	7.54	1.48	1.23	0.86	0.52	0.12	0.058
TENT B [6]	59.73	20.97	5.99	5.83	2.21	1.18	0.41	0.57	0.48	0.023

Note: Presented values may not entirely represent the tested material, since the chemical composition of the coal used in the power plants can change over time.

3.3. Binders (Activators)

The influence of common hydraulic binders on soil stabilization (cement and hydrated lime) was investigated in this paper. Specifically, Portland cement PC 20M (S-L) 42.5R "Beočin Profi" with the mixed addition of granular slag and limestone from the manufacturer "Lafarge"(Beočin, Serbia) was used. Important technical specifications for used cement are given in Table 4. On the other hand, in the case of lime, hydrated lime from the "NEXE" (Jelen Do, Serbia) manufacturer was employed. Besides binders, liquid additive Polybond™ from "Superroads Technologies" (Lausanne, Switzerland) was also used. Polybond™ is based on sulfuric acid and surfactant, which improves soil strength and soil resistance to moisture infiltration and frost. The Polybond™ stabilization effect is based on its ability to perform ionic water substitution on the soil particles' surface using stabilizing molecules. The main feature of the stabilizing molecules attached to the soil particles' surface is to repel moisture, thereby reducing the ability of clay particles to attract water [49]. Addition of Polybond™ increases the ultimate compression strength of specimens after 28 days by 1.5–2 times compared to reference specimens [49].

Table 4. Technical properties for cement PC 20M (S-L) 42.5R.

Consistency (%)	Setting Time (min)	Compressive Strength After 2 Days (MPa)	Compressive Strength After 28 Day (MPa)
27–29	160–250	26–28	49.5–54.5

4. Testing Methods and Laboratory Program

4.1. Testing Methods

Physical and mechanical properties were determined via the following laboratory tests: Specific gravity, grain size distribution, the moisture–density relationship (Proctor compaction test), unconfined compressive strength (UCS), direct shear, consolidation, CBR and swell tests. Tests were performed in accordance with Serbian (SRPS/EN) standards (see References 50–58). Additional details of laboratory tests are as follows:

- Specific gravity was determined in accordance with [50].
- Grain size distribution was determined using the hydrometer method, in accordance with [51].
- Atterberg limits were determined using a motorized Casagrande liquid limit device (Controls, Milan, Italy), in accordance with [52].
- The Proctor compaction test was done in accordance with [53]. Optimum moisture content (OMC) and maximum dry density $\gamma_{d,max}$ were determined using a compaction energy of 600 kJ/m^3.
- Unconfined compression (UCS) tests were done using a controlled strain rate machine (Controls, Milan, Italy), on the cylindrical specimens with a diameter of 38 mm and height of 76 mm. The tests were done in accordance with SRPS U.B1.029:1996 [54]. The rate of vertical displacement was 0.5 mm/min.
- Direct shear tests were performed in drained conditions, using machines with a constant strain rate and square shear box (60 mm × 60 mm × 30 mm), in accordance with [55]. Specimens were initially saturated in a separate consolidation device (Controls, Milan, Italy) during 24 hours. After saturation, specimens were consolidated with vertical loading of 100, 200 and 400 kPa and then sheared with the constant velocity of 5–15 μ/min (CH clay stabilization) and of 20–40 μ/min (fly ash and ash-slag mixtures).
- One-dimensional consolidation tests were done in accordance with [56], on cylindrical specimens with diameter of 70 mm and height of 20 mm. The specimens were soaked for 24 hours prior to compression. After soaking, the vertical load was applied step by step to achieve the maximum vertical stress of 400 kPa, according to the following scheme: 25/50/100/200/400/200/100/50/25 kPa
- California bearing ratio (CBR) tests were done on fully soaked samples, in accordance with [57].
- Frost resistance tests were done in accordance with [58]. After 15 cycles of freezing and thawing, the UCS reduction was measured. One cycle consisted of 16 hours freezing on temperature −10 °C and 8 hours thawing on temperature +25 °C.
- Free swell tests were performed in the oedometer apparatus (Controls, Milan, Italy) on remolded samples compacted at standard Proctor's maximum dry density and optimum moisture content and without any vertical surcharge load [48]. Upon completion of the swelling process, in order to capture the swelling pressure, the vertical load was gradually applied until swelling deformation was eliminated.

4.2. Laboratory Testing Program

The laboratory testing program within this study consisted of two parts. In the first part, the high plasticity clay stabilization using fly ash was investigated. In the second part, engineering properties of ash and ash-slag mixtures as an embankment material in road construction were studied. The influence of common binders (activators) was also investigated. The laboratory testing program is outlined in flowcharts in Figures 1 and 2.

A total of 24 combinations (mixtures) of soil, waste material and binders (activators) were tested. Untreated materials (without binders) were tested first, in order to determine initial physical (Tables 1 and 2) and mechanical properties (Tables 5 and 6), which were used later for comparison with treated materials. For all physical and mechanical tests two specimens were used for the determination of engineering properties, except for UCS where five specimens were tested.

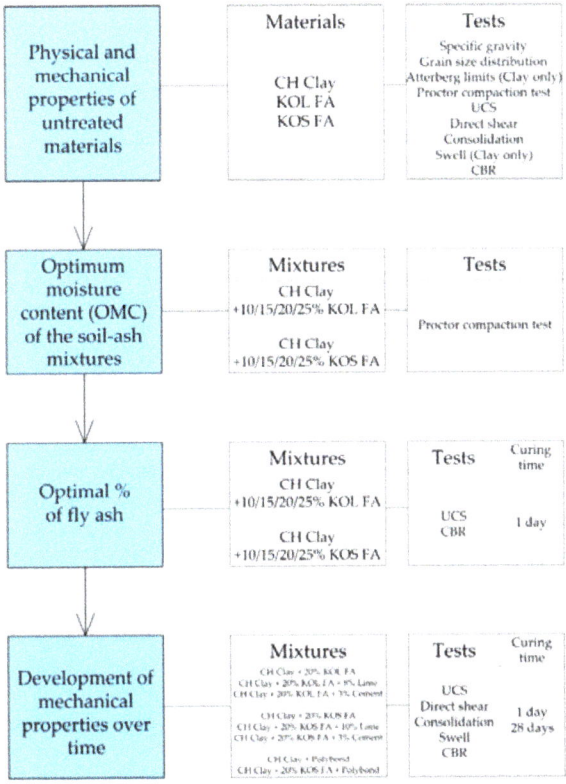

Figure 1. Flowchart of the laboratory testing program—high plasticity clay stabilization.

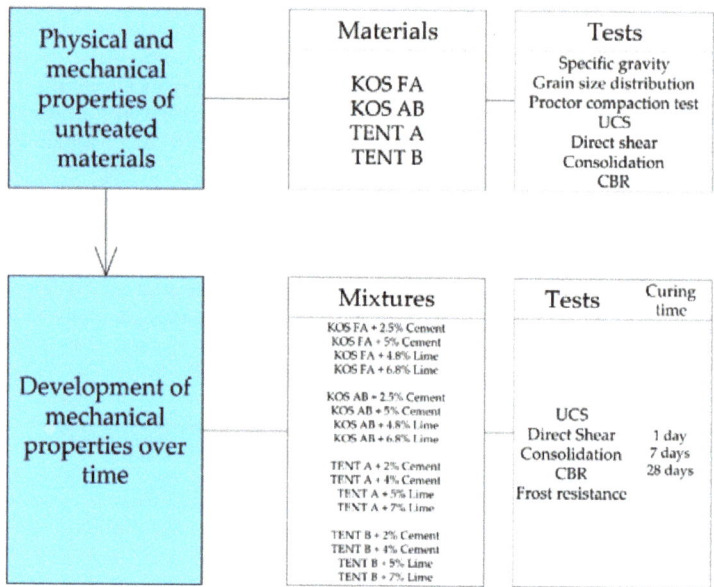

Figure 2. Flowchart of the laboratory testing program—ash and ash-slag mixtures.

Table 5. Mechanical properties of tested materials without binders.

Material	Compaction Proctor Test (600 kJ/m³)		Compressibility M_v (kPa)			Strength			
						Direct shear		UCS	
	OMC (%)	$\gamma_{d,max}$ (kN/m³)	50–100 kPa	100–200 kPa	200–400 kPa	φ' (°)	c' (kN/m²)	q_u (kN/m²)	CBR (%)
CH clay	19.1	16.6	14300	10400	10800	25.5	26.0	231	4.5
KOL FA	49.8–55.0	8.0	17700	24900	31400	29.5	36.5	83	13
KOS FA	37.5–43.9	9.0–9.8	25800	39200	42900	31.0	28.5	87	57

Table 6. Mechanical properties of tested materials without binders.

Material	Compaction Proctor Test (600 kJ/m³)		Compressibility M_v (kPa)			Strength			
						Direct Shear		UCS	
	OMC (%)	$\gamma_{d,max}$ (kN/m³)	50–100 kPa	100–200 kPa	200–400 kPa	φ' (°)	c' (kN/m²)	q_u (kN/m²)	CBR (%)
KOS FA	37.5–43.9	9.0–9.8	25800	39200	42900	31.0	28.5	87	57
KOS AB	48.1	9.1	10000	12600	22700	35.0	18.5	37	24
TENT A	48.5	8.9	14100	23300	34200	34.5	7.0	49	7
TENT B	33.7	10.4	19300	20700	26200	33.5	20.0	87	12

4.3. Specimen Preparation and Curing

In order to compare the results of different test mixtures, specimens for mechanical tests (UCS, direct shear, CBR, consolidation and swell) were prepared by compaction under the same conditions. First, premeasured amount of dried components (ash, slag, soil and binder) were mixed intensively to create a homogeneous dry mixture. After that the water was added and, after mixing, compaction was done immediately. Late compaction can reduce the effects of stabilization—during the hydration process, fly ash cements particles in the mixture, and more compaction effort is required. The smaller strength gain, and sometimes strength reduction after late compaction, is explained by the loss of hydration products, and by the loss of connections between the cemented particles [14,59]. According to [14,60], it is recommended that the amount of added water should be about 80–110% of OMC. In this study, the 100% of OMC was adopted for specimen preparation. After compaction, the specimens were extruded from compaction molds.

Specimens without binders and specimens with cement were kept (cured) in a plastic wrap, hermetically sealed, at laboratory temperature of 20 °C. Specimens with lime and Polybond^{TM} were not hermetically closed before testing. Specimens were cured in moist chamber at relative humidity RH > 95% and laboratory temperature of 20 °C.

4.4. Optimal % of Fly Ash (Only for Soil Stabilization)

For the successful soil stabilization it is important to use the optimal fly ash content, in order to create the conditions necessary for all chemical reactions and for changing the soil microstructure.

In order to determine the optimal content of fly ash, UCS tests were done on the specimens with different fly ash–soil ratios (10%, 15%, 20% and 25%), one day after compaction. Due to the fact that the increase of UCS was not significant (may be in the domain of scattering of the results), it was not possible to select the optimal % of fly ash. Therefore, additional CBR tests were performed on the specimens with the same fly ash–soil ratios. The highest CBR value was achieved for the mixture with 20% of fly ash, which was adopted as the optimal amount.

4.5. Used % of Binders

Used amount of cement was determined preserving the homogeneity of the mixture with minimum cement consumption.

Used amount of lime was set from the condition that the pH value of mixture shall be 12.4, securing the optimal conditions for the hydration process [61].

Test specimens with Polybond^{TM} were prepared with the minimum recommended Polybond^{TM} content according to organization standard [62] and the manufacturers' recommendations [49]. First the Polybond^{TM} solution with water was formed, which is then added into the dry material before compaction. During preparation of the soil–fly ash mixtures, the previously determined optimal ash percentage (20%) was used. Used amounts of binders are given in Figures 1 and 2.

5. Results and Discussion

5.1. Stabilization of High Plasticity (CH) Clay

The results of the CH clay stabilization using fly ash and binders are given in following subsections. Engineering properties of stabilized mixtures were compared with the properties of untreated soil. Mechanical properties of tested materials without binders are shown in Table 5.

For mixtures with lime, a significant increase in shear strength, compressibility parameters (constrained modulus M_v) and CBR was obtained after one day and therefore no further testing was performed after 28 days.

5.1.1. Unconfined Compressive Strength (UCS)

Increased soil strength is the main indicator of successful soil stabilization [14,16,18,23,31]. The results of UCS tests are presented in Figure 3. For mixtures with fly ash without a binder, the effects of stabilization were negligible because of low UCS of used fly ashes. With the addition of lime or cement, the pozzolanic reaction started and there was significant strength gain over time. Strength gain was more pronounced with the addition of lime.

The addition of Polybond^{TM} in the minimum recommended amount led to an increase of UCS after one day, but results indicate that there was insensitivity of UCS to the elapsed time. Since the Polybond^{TM} stabilization mechanism is primarily based on the reduction of bound water, the observed trend was expected.

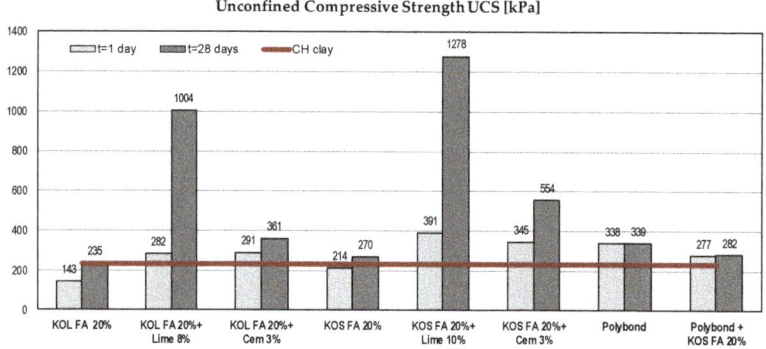

Figure 3. Unconfined compressive strength (UCS) of different mixtures with CH clay.

5.1.2. Shear Strength Parameters in Terms of Effective Stresses

Shear strength parameters affect the safety of any geotechnical structure. They are essential for earth structures design, calculation of bearing capacity and earth pressures, stability analysis of natural slopes, cuts and fills [63,64]. Effective shear strength parameters were determined using a direct shear test and they are presented in Figure 4.

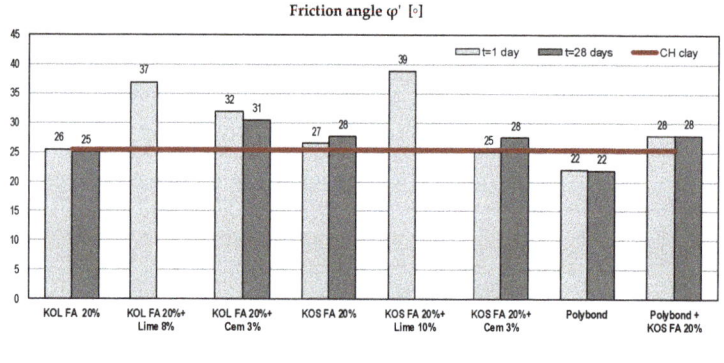

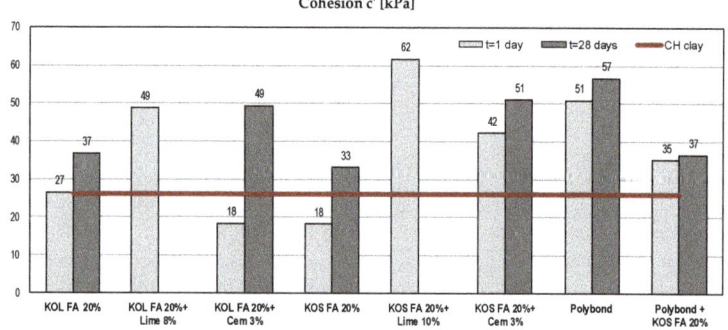

Figure 4. Shear strength parameters of different mixtures with CH clay.

Obtained results show that the friction angle does not substantially change with the addition of fly ash and Polybond[TM]. With the addition of cement, there was a mild increase of the friction angle, but a significant increase was noted with the addition of lime after only one day. The cohesion significantly increased with time for all tested mixtures. For mixtures with fly ash without binders, a slow pozzolanic reaction occurred due to the presence of reactive CaO. After the addition of cement or lime, a more pronounced pozzolanic reaction occurred as well as the creation of cement joints. The effects of treating CH clay with Polybond^{TM} were particularly expressed in terms of soil cohesion—the increase of cohesion was evident.

5.1.3. Compressibility Parameters

In order to calculate the consolidation settlement of soils, compressibility parameters are required. The constrained modulus from a one-dimensional compression (oedometer) test is a commonly used parameter to determine the settlement of a tested material. The vertical effective stress level 100–200 kPa was selected to display the results. A similar trend was observed for other stress levels. Constrained moduli were increased for all mixtures (Figure 5). Stabilization effects were greater with the addition of cement or lime.

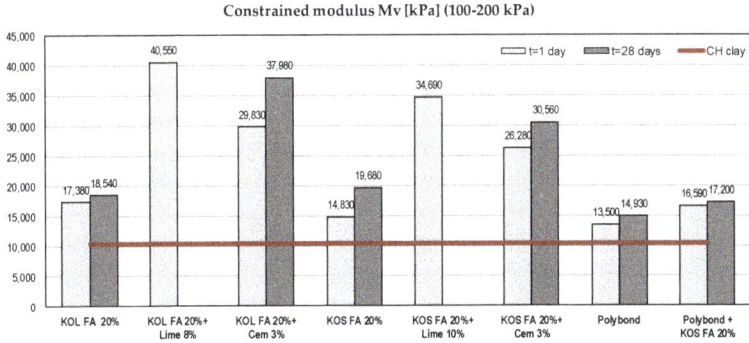

Figure 5. Compressibility parameters of different mixtures with CH clay.

5.1.4. California Bearing Ratio (CBR)

The California bearing ratio (CBR) is a parameter that describes the strength of roads subgrade. It is used for the determination of pavement thickness and its component layers [65,66]. Clays generally have low CBR values (<5), which make them inappropriate for road subgrade construction. Obtained results showed significant CBR gain. In the case of mixtures with fly ash and Polybond^{TM}, there was a mild, but important increase of strength, which made the tested soil usable for road construction. Test results (Figure 6) were in line with [18,19,22,23,31]. It is obvious that used binders had a significant influence on CBR gain.

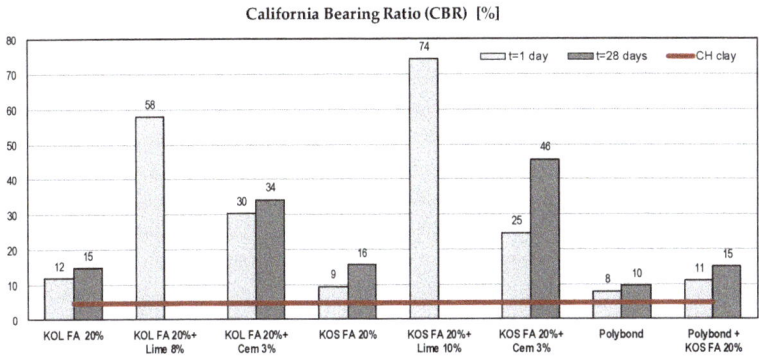

Figure 6. The California bearing ratio of different mixtures with CH clay.

5.1.5. Swell Potential

The volumetric change of soil causes movements in structures and imposes additional loads to structures [67,68]. According to [16], fly ash replaces some of the volume held by clay particles and acts as a mechanical stabilizer.

By addition of fly ash, the swell potential of all tested mixtures was entirely eliminated, which was somewhat expected considering the medium degree of expansivity of tested CH clay [48]. On the other hand, the addition of Polybond^{TM}, reduced the swelling deformation to about 1%.

5.2. Fly Ash and Ash-Slag Mixtures as a Material for Embankment

The engineering properties of ash and ash-slag mixture were discussed below. As in the case of high plasticity clay, tests were performed on the samples with and without binders and the results were compared. For all tested materials similar trends were observed and therefore test results for fly ashes and ash-slag mixtures would be considered together. Mechanical properties of tested materials without binders are given in Table 6.

5.2.1. Unconfined Compressive Strength (UCS)

The results of UCS tests are presented in Figure 7. Samples without a binder had very low UCS (Table 6). The pozzolanic reaction started with the addition of binders and water and constant strength gain over time could be observed. The substantial increase was recorded for fly ash samples. The strength gain was more pronounced with lime addition.

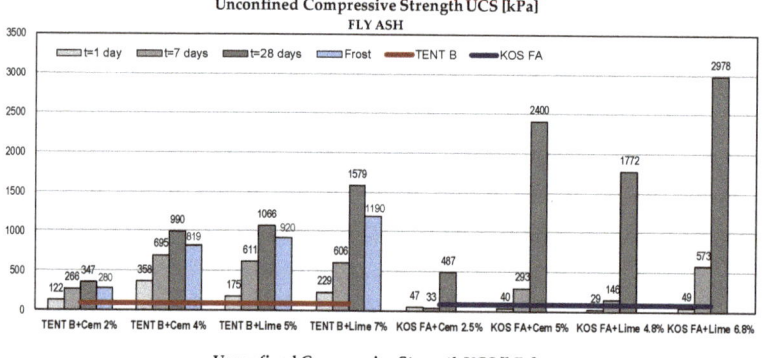

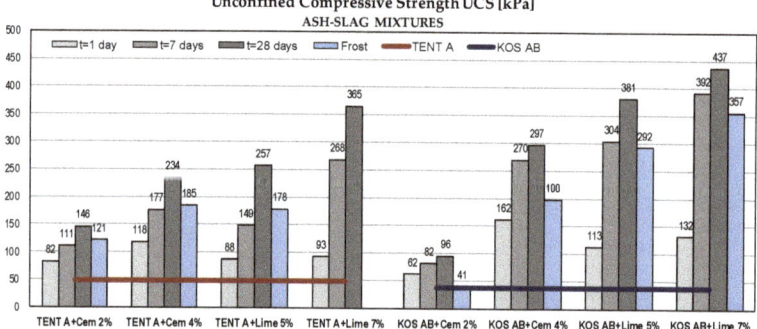

Figure 7. UCS of waste materials with binders.

5.2.2. Effect of Frost

The frost resistance of fly ash and ash-slag mixtures treated with binders was tested by measuring the UCS reduction. Samples aged 28 days were exposed to repeated freezing and thawing (15 cycles) and the UCS was determined. The frost resistance index is represented by the relation between the UCS of the sample after 15 freezing/thawing cycles and UCS of the reference sample (28 days old). According to standard SRPS U.B1.050:1970 [58], the mixture is frost resistant if the index is greater than 80%.

Results for ash-slag mixture TENT A with 7% of lime were missing because the samples were damaged during testing. Tests were not performed for fly ash KOS FA. The results are shown in Figure 7. The obtained frost resistance indices were within: 75–86% for TENT B, 69–83% for TENT A and 43–82% for KOS AB. Absolute values of UCS after freezing/thawing cycles classify mixtures as stiff to hard [64] and despite some lower indices values, the mixtures could be considered as frost resistant. The low index value for KOS AB with 2% of cement might be due to damage of samples during testing.

5.2.3. Shear Strength Parameters in Terms of Effective Stresses

Considering fly ash and ash-slag mixtures as a fill material for embankments, the strength of compacted material is of major importance to geotechnical engineers. High shear strength ensures higher bearing capacity and slope stability. Shear strength parameters are given in Figures 8 and 9.

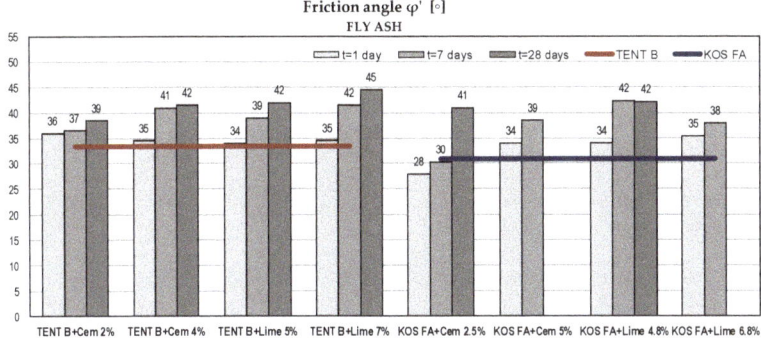

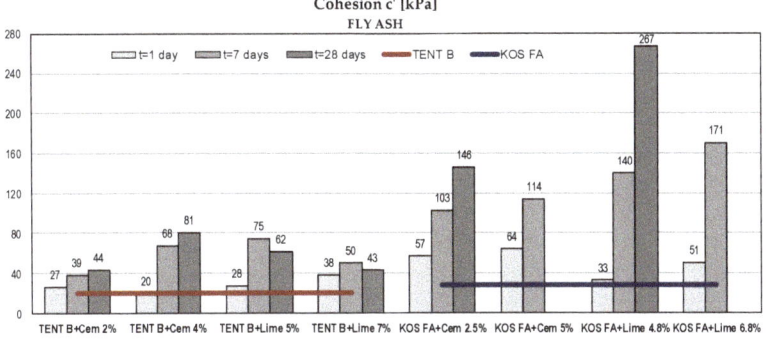

Figure 8. Shear strength parameters of fly ash samples with binders.

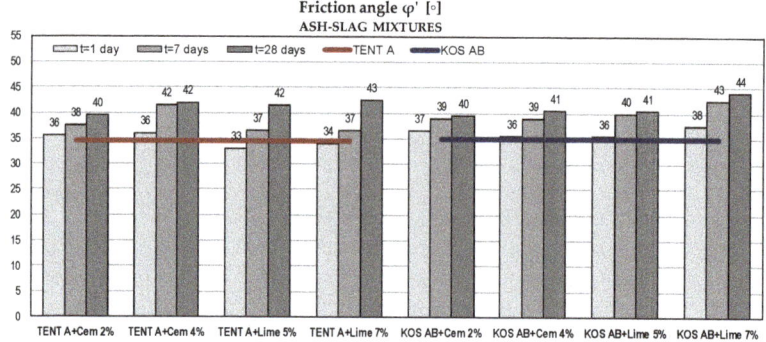

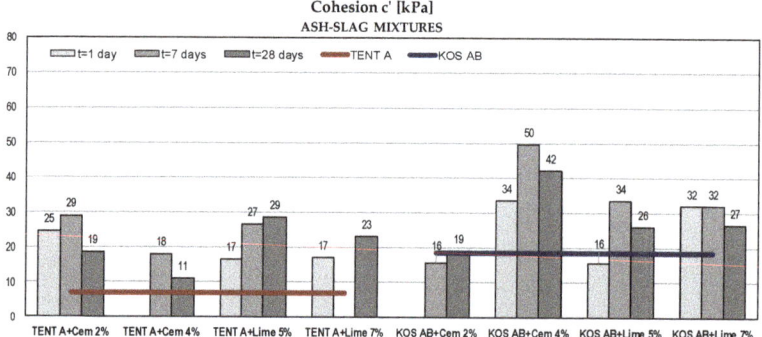

Figure 9. Shear strength parameters of ash-slag samples with binders (some test results were omitted because obtained values were too high).

Compared to the strength of the compacted sand (as traditional fill material) [37], all tested mixtures without binders had high values of friction angle ($\varphi' = 31$–$35°$). According to USA Navy [69] the friction angle for compacted sandy soils typically range from 31° to 45°. Test results show an increase of shearing resistance for all samples over time and with the increase of binder amount. Similar results were obtained for both binders. Obtained friction angles after 28 days were within the range 39° to 45°, which made tested materials generally comparable to the traditional compacted sandy soils.

Considering cohesion as an apparent shear strength parameter that captures the effects of intermolecular forces, soil tension or cementation, class F fly ash and ash-slag mixtures exhibit no cohesive characteristic in the saturated state [44]. In this case, cohesion is a consequence of the approximation of the non-linear failure envelope with a linear one. The failure envelope obtained from the strength test is a curved line for mostly granular materials, but solving the majority of soil mechanics problems, it is sufficient to approximate the shear stresses as a linear function of the normal stresses. The magnitude of cohesion is thus defined by the intersect segment on the shear stress axis. After adding binders and with the addition of water, the pozzolanic reaction occurred as well as the formation of bonds between the soil particles. The increase in cohesion was evident due to the cementation process, but there was no clear trend over time and with an increase in the % of the binder. For fly ash samples, the substantial increase over time had recorded for KOS FA and for TENT B with cement addition, while with the addition of lime there was no further increase of cohesion after 7 days. For ash-slag mixtures, there was a scattering of cohesion results, probably due to the inhomogeneity of the samples and due to the method used for strength determination. In the direct shear test, the orientation of the failure plane was predetermined as being near the middle of the sample height. Better results might be obtained in the triaxial device where the orientation of the failure plane is governed by the soil structure.

For fly ash samples KOS FA with higher % of binders (5% cement and 6.8% lime), no further testing was carried out after 28 days, because there was a significant increase in tested parameters for a smaller binder amount. Some test results for ash-slag mixtures were omitted because the obtained values were too high.

5.2.4. Compressibility Parameters

Compression of compacted fly ash or ash-slag mixture in wide embankments can be considered as one dimensional [37]. Thus, constrained moduli from oedometer (one-dimensional consolidation) tests were obtained and results are shown in Figure 10. According to Kim et al. [37] and Carrier [70] relevant constrained moduli should be calculated for vertical stresses ranging from zero to 200 kPa, a range of stress levels typically expected in highway embankments.

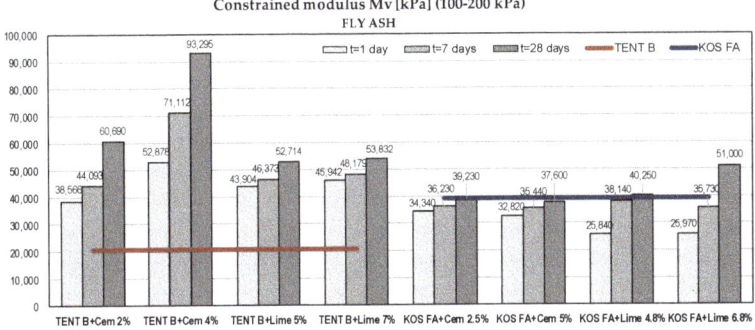

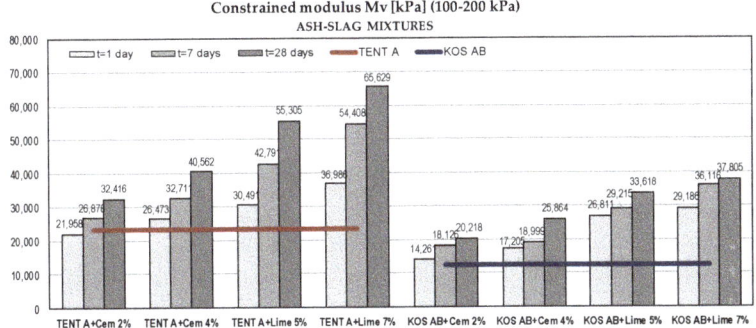

Figure 10. Compressibility parameters of waste materials with binders.

Ash-slag samples without binders exhibited slightly greater compressibility than fly ash samples. With the addition of binders constrained moduli of fly ash TENT B and ash-slag mixtures increased with time and with the percent of the binder. For fly ash samples KOS FA the stabilization effects were negligible.

Additionally, a comparison was made with sand compressibility given in [70]. Figure 11 shows typical moduli values for sand compacted at relative compaction (RC) of 85% and 99% and moduli (after 28 days) obtained from research as a function of vertical effective stresses. Constrained moduli are shown for the midpoint of the stress interval for which they are calculated. For the same compaction levels, fly ash samples TENT B with binders and ash-slag mixtures TENT A with lime are significantly less compressible than sand, while most of the other values of moduli lie near the upper limit of the sand moduli range.

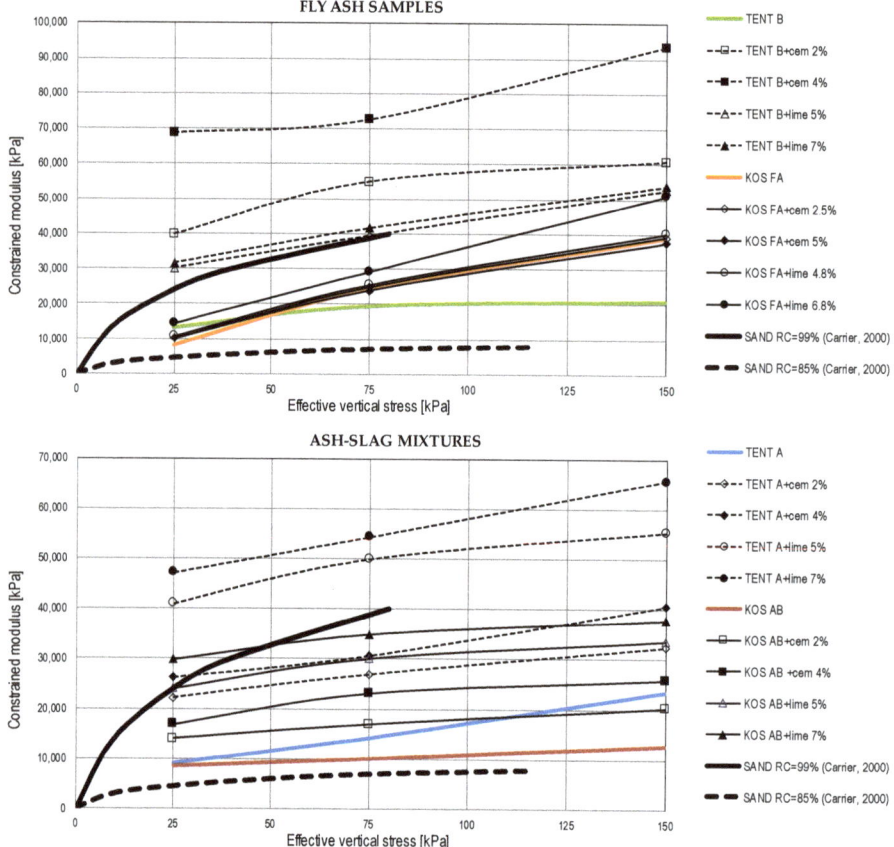

Figure 11. Constrained moduli of waste materials and sands.

5.2.5. California Bearing Ratio (CBR)

CBR values of tested materials vary within wide limits, from 7% for ash-slag mixture TENT A to 57% for fly ash KOS FA. With the use of binders, for all tested samples, there is a clear trend of CBR increase over time (Figure 12). Except for ash-slag mixture TENT A, there is no need for stabilization of tested materials for road subgrade construction purposes [40].

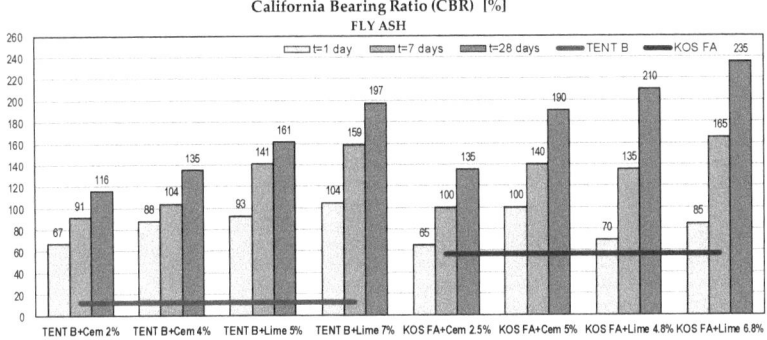

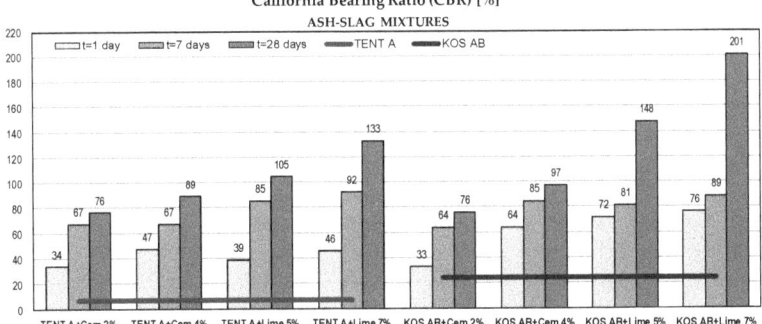

Figure 12. The California bearing ratio of waste materials with binders.

6. Conclusions

Thermal power plants have multiple negative effects on the environment: They pollute air with harmful gases and fly ash; landfills of ash and slag occupy large areas of mainly agriculture land; deposited ash can potentially pollute land and water due to the presence of trace elements and radionuclides. The amount of deposited ash and slag can be significantly reduced by use in the construction industry.

In order to assess the applicability of fly ash and ash-slag mixtures for subsoil stabilization and embankments construction, the laboratory tests on different mixtures of soil, ash, slag and binders were performed. The soil stabilization efficiency of a non-self-cementing class F fly ash without a binder was tested, as well as the effects of adding a binder as the cementation agent. The characteristics of fly ashes and ash-slag mixtures as the construction material were also investigated.

Considering the results of CH clay stabilization, fly ashes from power plants Kolubara and Kostolac could be successfully used as an additive that improves all mechanical characteristics of the soil required for the subsoil. Due to improved mechanical parameters, the stabilized soil has better bearing capacity and low compressibility. The increase in CBR values and elimination of swell potential make tested soil usable for road construction. With the addition of binders, all tested engineering properties were significantly improved. The addition of lime yields more significant stabilization results compared to cement.

Regarding the embankments and road subgrade design purposes, fly ash and ash-slag mixtures from Serbian thermal power plants have comparable mechanical properties with sands. The use of binder contributes to the substantial increase of shear strength parameters, compressibility modulus and stiffness of tested materials. Achieved high shear strength of waste materials ensures higher bearing capacity and slope stability. Low compressibility also makes waste materials suitable for embankment construction. The use of ashes and ash-slag mixtures as construction material provides multiple benefits: Reduced amount of ash on landfills, preservation of natural resources, lower price of embankment construction, lower energy consumption and CO_2 emission.

Author Contributions: Conceptualization, Data Curation, Funding Acquisition, Supervision, M.V.; Investigation, Validation, Writing (Review & Editing), M.V., M.M., V.P. and S.J.; Methodology, M.V., M.M. and V.P.; Visualization S.J.; Writing (Original Draft), M.V., M.M. and S.J.

Funding: This research was funded by Public Company "Elektroprivreda Srbije" (Electric Power Industry of Serbia) via studies "Use of fly ash of thermal power plants for soil stabilization, self-compacting and rolling (RCC) concrete with emphasis on durability of cement mortar and fine grained concrete" (2014) and "The use of fly ash and slag from "Elektroprivreda Srbije" thermal power plants in railways" (2015). This research is partially sponsored by Serbian Government, Ministry of Education, Science and Technological Development via Project TR 36046: Research on the effects of vibration on people and objects for the sustainable development of cities.

Conflicts of Interest: The authors declare no conflict of interest. The funders had no role in the design of the study; in the collection, analyses, or interpretation of data; in the writing of the manuscript, or in the decision to publish the results.

References

1. Electricity Production from Oil, Gas and Coal Sources. Available online: https://data.worldbank.org/indicator/EG.ELC.FOSL.ZS (accessed on 1 July 2019).
2. European Coal Combustion Products Association. Available online: http://www.ecoba.com/ecobaccpprod.html (accessed on 1 July 2019).
3. Cmiljanić, S.; Jotić, S.; Tošović, S. Prethodni rezultati istraživačko razvojnog programa-primena elektrofilterskog pepela u putogradnji. In Proceedings of the Simpozijum o Istraž. i Primeni Savremenih Dostignuća u Našem Građevinarstvu u Oblasti Materijala i Konstrukcija, Divčibare, Serbia; 2008; pp. 3–13. (In Serbian).
4. Cmiljanić, S.; Vujanić, V.; Rosić, B.; Vuksanović, B.; Tošović, S.; Jotić, S. Physical-mechanical properties of fly-ash originating from thermo-electric power plants of Serbia. In Proceedings of the 14th Danube-European Conference on Geotechnical Engineering: From Research to Design in European Practice, Bratislava, Slovakia, 2–4 June 2010.
5. Cmiljanić, S. *Study: Use of Fly Ash and Slag Produced in the Thermal Power Plants "Nikola Tesla A/B" and "Kostolac A/B" for the Needs of Road Construction*; Highway Institute, Contractor: Public Enterprise "Roads of Serbia": Belgrade, Serbia, 2008. (In Serbian)
6. Šušić, N. *Study: Application and Placement of Ash Produced at "Elektroprivreda Srbije" Power Plants*; Institute for testing materials, Contractor: Public Company "Elektroprivreda Srbije": Belgrade, Serbia, 2011. (In Serbian)
7. Despotović, J. *Study: The Use of Fly Ash of Thermal Power Plants for Soil Stabilization, Self-Compacting and Rolling (RCC) Concrete with Emphasis on Durability of Cement Mortar and Fine Grained Concrete*; University of Belgrade—Faculty of Civil Engineering, Contractor: Public Company "Elektroprivreda Srbije": Belgrade, Serbia, 2014. (In Serbian)
8. Vukićević, M. *Study: The Use of Fly Ash and Slag from "Elektroprivreda Srbije" Thermal Power Plants in Railways*; Institute for testing materials/University of Belgrade—Faculty of Civil Engineering, Contractor: Public Company "Elektroprivreda Srbije": Belgrade, Serbia, 2015. (In Serbian)
9. *Uredba o Tehničkim i Drugim Zahtevima Za Pepeo, Kao Građevinski Materijal Namenjen Za Upotrebu u Izgradnji, Rekonstrukciji, Sanaciji I Održavanju Infrastrukturnih Objekata Javne Namene*; Official Gazette of RS, 56/2015: Belgrade, Serbia, 2015. (In Serbian)
10. *Pravilnik o Granicama Radioaktivne Kontaminacije Životne Sredine i o Načinu Sprovođenja Dekontaminacije*; Official Gazette of FRY: Belgrade, Serbia, 1999. (In Serbian)
11. Vilotijević, M.; Vukićević, M.; Lazarević, L.; Popović, Z. Sustainable railway infrastructure and specific environmental issues in the Republic of Serbia. *Teh. Vjesn.* **2018**, *25*, 516–532. [CrossRef]
12. Izquierdo, M.; Querol, X. Leaching behaviour of elements from coal combustion fly ash: An overview. *Int. J. Coal Geol.* **2012**, *94*, 54–66. [CrossRef]
13. *ASTM C618-15: Standard Specification for Coal Fly Ash and Raw or Calcined Natural Pozzolan for Use in Concrete*; ASTM International: West Conshohocken, PA, USA, 2015.

14. Ferguson, G.; Leverson, S.M. *Soil and Pavement Base Stabilization with Self-Cementing Coal Fly Ash*; American Coal Ash Association International: Alexandria, VA, USA, 1999.
15. *SRPS EN 14227-4: Hydraulically Bound Mixtures—Specifications—Part 4: Fly Ash for Hydraulically Bound Mixtures*; Institute for Standardization of Serbia: Belgrade, Serbia, 2014.
16. Ferguson, G. Use of self-cementing fly ashes as a soil stabilization agent. *ASCE Geotech. Spec. Publ.* **1993**, *36*, 1–14.
17. Çokça, E. Use of class C fly ashes for the stabilization of an expansive soil. *J. Geotech. Geoenviron.* **2001**, *127*, 568–573. [CrossRef]
18. Acosta, H.A.; Edil, T.B.; Benson, C.H. Soil Stabilization and Drying using Fly Ash. In *Geo Engineering Report No. 3*; Geo Engineering Program, University of Wisconsin-Madison: Madison, WI, USA, 2003.
19. Mackiewicz, S.M.; Ferguson, E.G. *Stabilization of Soil with Self-Cementing Coal Ashes*; 2005 World of Coal Ash (WOCA): Lexington, KY, USA, 2005; pp. 1–7.
20. Parsons, R.L. *Subgrade Improvement through Fly Ash Stabilization*; Miscellaneous Report; Kansas University Transportation Center, University of Kansas: Lawrence, KS, USA, 2002.
21. Parsons, R.L.; Kneebone, E. Field performance of fly ash stabilized subgrades. *Proc. ICE Ground Improv.* **2005**, *9*, 33–38. [CrossRef]
22. White, D.J.; Harrington, D.; Thomas, Z. Fly ash soil stabilization for non-uniform subgrade soils, Volume I: Engineering properties and construction guidelines. In *Report No. IHRB Project TR-461, FHWA Project 4*; Center for Transportation Research and Education, Iowa State University: Raleigh, NC, USA, 2005.
23. Senol, A.; Edil, T.B.; Bin-Shafique, M.S.; Acosta, H.A.; Benson, C.H. Soft subgrades' stabilization by using various fly ashes. *Resour. Conserv. Recycl.* **2006**, *46*, 365–376. [CrossRef]
24. American Coal Ash Association. Fly ash facts for highway engineers. In *Technical Report No. FHWA-IF-03-019*; FHWA: Washington, DC, USA, 2003.
25. Pandian, N.S.; Krishna, K.C.; Leelavathamma, B. Effect of fly ash on the CBR behaviour of soils. In Proceedings of the Indian Geotechnical Conference, Allahabad, India, 20–22 December 2002; Volume 1, pp. 183–186.
26. Ramadas, T.L.; Kumar, N.D.; Yesuratnam, G. A study on strength and swelling characteristics of three expansive soils treated with fly ash. In Proceedings of the International Symposium on Ground Improvement (IS-GI Brussels 2012)—Recent research, Advances & Execution Aspects of Ground Improvement Works, Brussels, Belgium, 30 May–1 June 2012; pp. 459–466.
27. Kolay, P.K.; Sii, H.Y.; Taib, S.N.L. Tropical peat soil stabilization using class F pond ash from coal fired power plant. *IJCEE* **2011**, *3*, 79–83. [CrossRef]
28. Kumar, S.P. Cementitious compounds formation using pozzolans and their effect on stabilization of soils of varying engineering properties. In Proceedings of the International Conference on Environment Science and Engineering IPCBEE, Singapore, 26–28 February 2011; Volume 8, pp. 212–215.
29. Vukićević, M.; Maraš-Dragojević, S.; Jocković, S.; Marjanović, M.; Pujević, V. Research results of fine-grained soil stabilization using fly ash from Serbian electric power plants. In Proceedings of the 18th International Conference on Soil Mechanics and Geotechnical Engineering "Challenges and Innovations in Geotechnics", Paris, France, 2–6 September 2013; pp. 3267–3270.
30. Nalbantoglu, Z.; Gucbilmez, E. Utilization of an industrial waste in calcareous expansive clay stabilization. *Geotech. Test. J.* **2002**, *25*, 78–84. [CrossRef]
31. Edil, T.B.; Acosta, H.A.; Benson, C.H. Stabilizing soft fine-grained soils with fly ash. *J. Mater. Civ. Eng.* **2006**, *18*, 283–294. [CrossRef]
32. Zia, N.; Fox, P.J. Engineering properties of loess-fly ash mixtures for roadbase construction. *Transp. Res. Rec.* **2000**, *1714*, 49–56. [CrossRef]
33. Meij, R.; Berg, J. Coal fly ash management in Europe trends, regulations and health & safety aspects. In Proceedings of the International Ash Utilization Symposium, Lexington, KY, USA, 20–24 October 2001.
34. Murthy, A.V.S.R.; Guru, U.K.; Havanagi, V.G. Construction of road embankments using fly ash. In *Fly Ash Disposal and Deposition: Beyond 2000 AD*; Narosa Publishing House: New Delhi, India, 2000.
35. McKerall, W.C.; Ledbetter, W.B.; Teague, D.J. *Analysis of Fly Ashes Produced in Texas*; Texas Transportation Institute, Research Report No. 240-1; Texas A&M University: College Station, TX, USA, 1982.
36. DiGioia, A.M.; McLaren, R.J.; Burns, D.L.; Miller, D.E. *Fly Ash Design Manual for Road and Site Application*; Vol. 1: Dry or conditioned placement. Manual Prepared for EPRI, CS-4419, Research, Project 2422-2, Interim Report; Electric Power Research Institute: Palo Alto, CA, USA, 1986.

37. Kim, B.; Prezzi, M.; Salgado, R. Geotechnical properties of fly and bottom ash mixtures for use in highway embankments. *J. Geotech. Geoenviron.* **2005**, *131*, 914–924. [CrossRef]
38. Sridharan, A.; Pandian, N.S.; Rao, P.S. Shear strength characteristics of some Indian fly ashes. *Proc. Inst. Civ. Eng. Ground Improv.* **1998**, *2*, 141–146. [CrossRef]
39. Chesner, W.H.; Collins, R.J.; MacKay, M.H.; Emery, J. *User Guidelines for Waste and By-Product Materials in Pavement Construction*; No. FHWA-RD-97-148, Guideline Manual, Rept No. 480017; Recycled Materials Resource Center: Madison, WI, USA, 1998.
40. Yoder, E.J.; Witczak, M.W. *Principles of Pavement Design*, 2nd ed.; John Wiley and Sons: Hoboken, NJ, USA, 1975.
41. Kaniraj, S.R.; Gayathri, V. Permeability and consolidation characteristics of compacted fly ash. *J. Energy Eng.* **2004**, *130*, 18–43. [CrossRef]
42. Kaniraj, S.R.; Havanagi, V. Geotechnical characteristics of fly ash-soil mixtures. *Geotech. Eng.* **1999**, *30*, 129–147.
43. McLaren, R.J.; DiGioia, A.M. The typical engineering properties of fly ash. In *Proceedings of the Conference on Geotechnical Practice for Waste Disposal, Ann Arbor, MI, USA, 15–17 June 1987*; Woods, R.D., Ed.; ASCE: New York, NY, USA, 1987; pp. 683–697.
44. Kim, B.; Prezzi, M. Evaluation of the mechanical properties of class-F fly ash. *Waste Manag.* **2008**, *28*, 49–59. [CrossRef] [PubMed]
45. Moreno-Maroto, J.M.; Alonso-Azcárate, J. Plastic limit and other consistency parameters by a bending method and interpretation of plasticity classification in soils. *Geotech. Test. J.* **2017**, *40*, 467–482. [CrossRef]
46. Moreno-Maroto, J.M.; Alonso-Azcárate, J. What is clay? A new definition of "clay" based on plasticity and its impact on the most widespread soil classification systems. *Appl Clay Sci.* **2018**, *161*, 57–63. [CrossRef]
47. Barnes, G.E. An apparatus for the plastic limit and workability of soils. *Proc. Inst. Civ. Eng. Geotech. Eng.* **2009**, *162*, 175–185. [CrossRef]
48. Seed, H.B.; Woodward, R.J.; Lundgren, R. Prediction of swelling potential for compacted clays. *J. Soil Mech. Found Div.* **1962**, *88*, 53–88.
49. SuperRoads—Innovative Technology in Road Construction. Available online: http://superroads.rs/technology.html (accessed on January 2015).
50. Institute for Standardization of Serbia. *SRPS U.B1.014: Testing of Soils—Determination of Density*; Institute for Standardization of Serbia: Belgrade, Serbia, 1988.
51. Institute for Standardization of Serbia. *SRPS U.B1.018: Testing of Soils—Determination of Particle Size Distribution*; Institute for Standardization of Serbia: Belgrade, Serbia, 2005.
52. Institute for Standardization of Serbia. *SRPS U.B1.020: Testing of Soils—Determination of Atterberg Limits*; Institute for Standardization of Serbia: Belgrade, Serbia, 1980.
53. Institute for Standardization of Serbia. *SRPS U.B1.038: Testing of Soil—Determination of the Relation Moisture Content—Density of Soil*; Institute for Standardization of Serbia: Belgrade, Serbia, 1997.
54. Institute for Standardization of Serbia. *SRPS.U.B1.029: Testing of Soils—Triaxial Compression Test*; Institute for Standardization of Serbia: Belgrade, Serbia, 1996.
55. Institute for Standardization of Serbia. *SRPS.U.B1.028: Testing of Soils—Direct Shear Test*; Institute for Standardization of Serbia: Belgrade, Serbia, 1996.
56. Institute for Standardization of Serbia. *SRPS.U.B1.032: Testing of Soils—Determination of Soil Compressibility*; Institute for Standardization of Serbia: Belgrade, Serbia, 1970.
57. Institute for Standardization of Serbia. *SRPS EN 13286-47: Unbound and Hydraulically Bound Mixtures—Part 47: Test Method for the Determination of California Bearing Ratio, Immediate Bearing Index and Linear Swelling*; Institute for Standardization of Serbia: Belgrade, Serbia, 2012.
58. Institute for Standardization of Serbia. *SRPS U.B1.050: Testing of Soils—Resistance Cement Stabilized Soils to Freezing*; Institute for Standardization of Serbia: Belgrade, Serbia, 1970.
59. Terrel, R.L.; Epps, J.A.; Barenberg, E.J.; Mitchell, J.K.; Thompson, M.R. *Soil Stabilization in Pavement Structures, A Users Manual*; Mixture Design Considerations FHWA-IP-80-2; Federal Highway Administration, Department of Transportation: Washington, DC, USA, 1979; Volume 2.
60. American Coal Ash Association. *Flexible Pavement Manual*; American Coal Ash Association: Washington, DC, USA, 1991.

61. Sharma, N.K.; Swain, S.K.; Umesh, C.S. Stabilization of a clayey soil with fly ash and lime: A micro level investigation. *Geotech. Geol. Eng.* **2012**, *30*, 1197–1205. [CrossRef]
62. STO 69646750-001-2011: *Soil and Asphalt-Granules-Concrete Mixtures Reinforced with Soil Stabilizer Polybond^{TM} for Use during Automobile Road, Railroad and Airfield Constructions*; SuperRoadRus: Moscow, Russia, 2011.
63. Terzaghi, K.; Peck, R.B.; Mesri, G. *Soil Mechanics in Engineering Practice*, 3rd ed.; John Wiley and Sons: New York, NY, USA, 1967; pp. 241–290.
64. Bowles, J.E. *Foundation Analysis and Design*, 5th ed.; McGraw-Hill: New York, NY, USA, 1996.
65. Davis, E.H. The California bearing ratio method for the design of flexible roads and runways. *Géotechnique* **1949**, *1*, 249–263. [CrossRef]
66. Briaud, J.L. *Geotechnical Engineering: Unsaturated and Saturated Soils*, 1st ed.; John Wiley and Sons: New York, NY, USA, 2013; p. 122.
67. Jones, D.E., Jr.; Holtz, W.G. Expansive soils-the hidden disaster. *Civ. Eng.* **1973**, *43*, 49–51.
68. Nalbantoğlu, Z. Effectiveness of class C fly ash as an expansive soil stabilizer. *Constr. Build. Mater.* **2004**, *18*, 377–381. [CrossRef]
69. U. S. Navy. *Design Manual—Soil Mechanics, Foundations, and Earth Structures*; NAVFAC DM 7; U. S. Navy: Alexandria, VA, USA, 1986.
70. Carrier, W.D. Compressibility of a compacted sand. *J. Geotech. Geoenviron.* **2000**, *126*, 273–275. [CrossRef]

© 2019 by the authors. Licensee MDPI, Basel, Switzerland. This article is an open access article distributed under the terms and conditions of the Creative Commons Attribution (CC BY) license (http://creativecommons.org/licenses/by/4.0/).

Article

Effect of Cement Type on the Mechanical Behavior and Permeability of Concrete Subjected to High Temperatures

Izabela Hager [1,*], Tomasz Tracz [1], Marta Choińska [2] and Katarzyna Mróz [1]

[1] Cracow University of Technology, Faculty of Civil Engineering, Chair of Building Materials Engineering, 24 Warszawska St., 31-155 Cracow, Poland
[2] Nantes University—IUT Institut Universitaire de Technologie, Research Institute in Civil and Mechanical Engineering GeM—UMR CNRS Unité Mixte de Recherche du Centre National de la Recherche Scientifique 6183, 44600 Saint-Nazaire, France
* Correspondence: ihager@pk.edu.pl; Tel.: +48-12-628-23-67

Received: 5 August 2019; Accepted: 16 September 2019; Published: 18 September 2019

Abstract: The paper presents experimental investigations concerning the influence of the cement type (CEMI 42.5 R Portland cement and CEMIII/A 42.5 N slag cement—with 53% granulated blast furnace slag) on the mechanical and transport properties of heated concretes. The evolution of properties due to high temperature exposure occurring during a fire was investigated. High temperature exposure produces changes in the transport and mechanical properties of concrete, but the effect of cement type has not been widely studied in the literature. In this paper, concretes were made with two cement types: CEMI and CEMIII, using basalt (B) and riverbed aggregates (RB). The compressive and tensile strength, as well as the static modulus of elasticity and Cembureau permeability, were tested after high temperature exposure to 200, 400, 600, 800, and 1000 °C. The evaluation of damage to the concrete and crack development due to high temperature effects was performed on the basis of the change in the static modulus of elasticity. The test results clearly demonstrated that permeability increases with damage, and it follows an exponential type formula for both types of cement.

Keywords: high temperature; damage; permeability; CEMI and CEMIII; mechanical properties

1. Introduction

Cements with granulated blast furnace slag are widely employed due to their lower carbon footprint, as a strategy for sustainable development in the field of construction. The use of ground granulated blast furnace slag (GGBFS), which presents an amorphous structure and shows pozzolan characteristics, in concrete as an additive has a positive effect on the properties of fresh and hardening concrete [1,2]. The use of GGBFS provides the important advantage of helping to avoid thermal cracks in concrete due to the low hydration process [2]. In fact, as previous results have shown, the hydration of GGBFS is slower than that of ordinary CEMI cement. Concrete with ground granulated blast furnace slag has a later setting time and a lower stiffness [3].

When equal amounts of cement and water binder (w/b) are used, concretes with slag content have a lower compressive strength at early ages and higher compressive strength at late ages than Portland cement [2]. Furthermore, with a specific compressive strength, slag concrete has a better mechanical performance in terms of tension than concrete made with Portland cement [2]. However, the study by Shumuye et al. [4] showed that the compressive strength of the concrete decreased as the slag content increased.

Environmental conditions and the temperature exposure during curing has a strong effect on concrete mechanical properties [3,5]. When the material is subjected to heating at higher temperatures up to 1000 °C, like during a fire, thermal damage occurs due to dehydration of the cement paste and the thermal mismatch of strains between the shrinking cement paste and expanding aggregates, which induces cracking [6,7]. Moreover, during the phase of cooling down to the ambient temperature, stresses induced by inversed thermal gradients result in the development of cracks within the cement paste that will affect permeability, and also may compromise the durability of the material after a fire [8,9]. The changes of concrete's mechanical properties at high temperatureshave been widely investigated [6–12], helping us to better understand the behavior of concrete structuresin a fire situation and to determine parameters influencing its behavior. The evolution of concrete mechanical properties in fire depends on the concrete composition: presence of mineral additions [4,13], w/c water cement ratio [14,15], the nature and type of aggregates [10–12]. Moreover, the concrete heating conditions: heating rate and maximum temperature of exposure play a major role in concrete strength evolution, as well as the testing procedure: hot tested concrete or tested after temperature exposure and cooling down to the ambient temperature [7,15]. Nevertheless, for material mechanical properties testing, a slow heating rate is recommended in order to ensure limitation of the thermal gradient inside the specimen. In the literature investigations the heating rates of 0.1–10 °C/min are employed. Nevertheless, the heating rates recommended by RILEM International Union of Laboratories and Experts in Construction Materials, Systems and Structures [16] depend on the specimen diameter and are from 0.5 to 2.0 °C/min for accidental conditions (fires).

The existing knowledge regarding the behavior of high performance concrete in a fire was recently reviewed by the RILEM Technical Committee HPB-227 [8], however, there are still no clear reports as to whether the properties of concretes subjected to high temperatures change in a similar or a very different way, depending on the cement type used.

According to Shumuye et al. [4], the addition of GGBFS seems to improve the resistance of concrete to fire conditions. It was highlighted that, when the exposure to fire temperature increased from 200 to 400 °C, the compressive strength increased for concrete with slag (70% ordinary Portland cement OPC and 30% slag cement, as well as 50/50 proportions). For the group of concretes with 30% OPC and 70% slag cement, the opposite behavior was observed. The concrete mix containing GGBFS usually has a lower thermal expansion coefficient than Portland cement. The 15% and 30% replacement of CEMI by GGBFS gives coefficients of thermal expansion of 22.7×10^{-6}/°C and 17.2×10^{-6}/°C, respectively, which is 99.2% and 75.5% of the value obtained for Portland cement paste [4]. However, a recent study by Asamoto et al. [17] highlighted that the reduction in the elastic modulus and increase in permeability of the concrete with GGBFS subjected to 65°Cwere larger than those of concrete without slag. Indeed, astonishingly, this can be attributed to a larger thermal expansion coefficient and larger cement paste shrinkage with the slag, leading to the formation of microcracks around the aggregate.

Moreover, it can be concluded that the addition of aluminosilicate minerals like fly ash, ground granulated blast furnace slag (GGBFS), and silica fume (SF) can affect concrete behavior at high temperatures in a way that may produce spalling of heated concrete in material that is denser, and thus less permeable [18–20]. Lower permeability leads to moisture clog occurrence and increase of vapor pore pressures inside the heated concrete [20]. The moisture clog effect was explained and linked with the permeability decrease observed in temperature from 100 to 200 °C but this effect is observed when the permeability is tested at hot stage and not after cooling down when the residual values of permeability are determined, like in present study. An important finding on gas pore pressure development were provided by works of Kalifa et al. [19,20] and linked with the permeability.

Cases of fires that took place in engineering facilities (Gotthard tunnel, Chunnel tunnel, or Mont Blanc tunnel, for example) have caused numerous fatalities, but also significant financial losses. During these fires an important loss of concrete in tunnel linings was observed. The load-bearing capacity of the structural elements was reduced due to the explosive spalling. The spalling may take different forms, from small concrete pieces chipping, known as the popcorn effect, to explosive behavior when larger pieces of concrete are separated from the concrete element with great energy [21–24]. In all cases, concrete fire spalling leads to the exposure of steel reinforcement, which is sensitive to high temperatures [24,25]. So far, it has been confirmed that the type and composition of concrete, including the aggregate type, water cement ratio, pozzolanic mineral material, and moisture content of concrete, affect its behavior in fire conditions [8,13,19]. Research aimed at understanding the causes of the spalling phenomenon, as well as determination of material parameters affecting its intensity, has been carried out by experiments [21,22,24] and numerical analysis [21,25,26]. Thus, concrete spalling is one of the most interesting and complex phenomena occurring in concrete exposed to fire conditions. The RILEM Technical Committee 256-SPF: Spalling of concrete due to fire: Testing and modelling has been established, and is mainly dedicated to studying this specific behavior.

During heating, the permeability usually progressively increases [27–29], exceptwhenthe permeability of concrete may decrease [30] due to the moisture clog effect. In this situation the water vapor pressure increases in the material's pore network, which may lead to spalling behavior. It is believed that the interaction of high temperature, an increase in water vapor pressure in the material pores, and the internal stress state is responsible for the occurrence of concrete spalling [19–26]. It seems that the key parameter governing the occurrence of spalling is its permeability. In denser and less permeable concretes the risk of spalling is higher. Researchers have shown that in fire conditions, concretes that are modified with the addition of mineral additives like silica fume and calcareous filler are prone to spalling behavior. As the spalling behavior of concrete is mainly governed by its permeability, researchers have been testing the influence of GGBFS addition on concrete permeability. Recently, Karahan [27] showed an increase of concrete transport properties after exposure to temperatures of 400 °C, accompanied by compressive strength reduction. Moreover, the conclusion of the authors indicated an optimum GGBFS/cement blend from the point of view of material behavior in a fire of 50–70% slag content as the cement replacement.

Hence, the results available do not reflect all the relevant aspects of this topic, and additional investigation is required. The literature results cannot be compared to each other due to the fact that the mixes differ. A research programme was therefore proposed which would allow for a clear comparison of the influence of cement type on the mechanical and physical properties of concrete at high temperatures. For this we performed various tests on identical concrete mixes, for which the only changing factor was the cement. Therefore, the main goal of this work is to present the comparison of the changes in mechanical and physical properties of concretes made with two different cement types; CEMI and CEMIII. For all four concretes, the composition of cement paste, as well as the volume of cement paste and mortar, remained the same. Thus, the study reflected solely the cement type effect of Portland cement versus slag cement on the mechanical performances and permeability of concretes made with two types of aggregates: crushed basalt (B) and riverbed gravel (RB). For all the concretes tested, the amount of all components (cement paste and mortar volume) and aggregate type and nature, as well as the particle size distribution, was identical, apart from the type of cement.

This research investigates the mechanical performances and permeability of concretes made with different cements, to compare their reference mass transport capacities, strength, and stiffness after high temperature exposure. The reference values of permeability enable one to assess their potential for spalling in fire conditions, as denser and less permeable materials are prone to this behavior. Furthermore, the evolution of permeability with heating temperature was investigated, as well as the compressive strength and splitting tensile strength. Moreover, the stress strain curves were determined, and the modulus of elasticity was determined. All residual mechanical performances (f_{cT}, f_{tT}, E_T) were evaluated after heating to temperature T (°C), which corresponds to the post-fire performance of concrete in situations where the assessment of material properties is required. In this specific situation, the residual permeability of concrete is also an issue because it governs all aspects of durability, and there may be a need for assessment when a decision must be made on the further use of concrete elements after a fire.

2. Materials, Specimen Preparation, Curing, and Heating

The concretes investigated in this research were manufactured with the following components: Portland cement CEMI 42.5R and CEMIII/A 42.5 N containing 53% GGBFS, quartz sand 0/2mm, and one of two types of coarse aggregate: (B)basalt or (RB) riverbed gravel.

Cements from Lafarge (Małogoszcz, Poland) were used for both the CEMI 42.5 R Portland cement and CEMIII/A 42.5 N slag cement. The chemical characteristics of these cements are provided in Table 1, the physical characteristics in Table 2, and the mechanical characteristics in Table 3.

Table 1. The chemical characteristics of CEM I and CEM III cements(oxide analysis, % by mass).

Component	CEMI 42.5 R	CEMIII/A 42.5 N
SiO_2	18.6	30.0
Al_2O_3	5.3	6.2
Fe_2O_3	2.9	1.7
CaO	62.7	50.3
MgO	1.50	4.98
SO_3	3.22	2.41
Na_2O	0.19	0.37
K_2O	0.96	0.70
$eqNa_2O$	0.82	0.83
Cl^-	0.060	0.016
Portland clinker content	96	45
GGBFS	0	53
Gypsum	4	2

Table 2. Physical characteristics of CEMI and CEMIII cements.

Parameter	CEMI 42.5 R	CEMIII/A 42.5 N
Specific area (Blaine method), m^2/kg	340	465
True density, g/cm^3	3.09	2.97
Setting time, minutes		
-initial	199	221
-final	270	266

Table 3. Mechanical characteristics of CEM I and CEM III cements.

Parameters	CEMI 42.5 R	CEMIII/A 42.5 N
Compressive strength, MPa		
-after 2 days	29.3	13.7
-after 28 days	55.1	50.7

Two types of aggregates were used in this research programme: gravel from Dunajec River (Dwudniaki, Poland) and crushed basalt.

In Table 4, the concrete mixes are presented. The cement paste volume was 300 dm^3/m^3 and the mortar volume was 550 dm^3/m^3. The concretes are denominated as B CEMI, B CEMIII, RB CEMI, and RB CEMIII. Plasticizer (BASF BV 18 (Myślenice, Poland) and superplasticizer (BASF Glenium SKY 591 (Myślenice, Poland) were used and the water-cement ratio (w/c) of the concretes was equal to 0.3.

Table 4. Mix composition of the test concretes.

Component	Unit	B CEMI	B CEMIII	RB CEMI	RB CEMIII
CEM I 42.5 R	kg/m³	482		482	
CEM III/A 42.5 N	kg/m³		482		482
Water	dm³/m³		145		
w/c ratio	–		0.30		
Riverbed quartz sand 0–2 mm		662	662	663	663
RB, riverbed 2–8 mm		–	–	610	610
RB, riverbed 8–16 mm	kg/m³	–	–	558	558
B, basalt 2–8 mm		709	709	–	–
B, basalt 8–16 mm		648	648	–	–
Plasticizer BASF BV 18		0.90	0.90	0.90	0.90
Superplasticizer BASF Glenium SKY 591	% mc	2.20	2.35	2.20	2.35
Cement paste content			300		
Mortar content	dm³/m³		550		
Coarse aggregate content			450		
Slump (consistency)	mm		120–150		
Air content in concrete mix	% vol.		1.7–2.0		

All concrete cubic and cylindrical specimens were cast in plastic molds and stored for 24 h. After preliminary 24 h curing, the molds were covered with plastic lids for 7 days to prevent water evaporation. Samples were stored in laboratory conditions at T = 20 ± 5 °C and relative humidity HR = 50% ± 5%. Cylindrical specimens dedicated to permeability measurements were cut into discs with a diameter of 150 mm and thickness of 50 mm at the age of 28 days. At 90 days, all specimens for mechanical performance testing and permeability were heated in an electric furnace to T = 200, 400, 600, 800, and 1000 °C. As recommended by RILEM [16], a heating rate of 0.5 °C/min was applied. A slow heating rate is applied for concrete mechanical behavior testing at high temperatures in order to ensure limitation of the thermal gradient inside the specimen. When the target temperature was reached it was maintained for three consecutive hours in order to obtain a homogenous temperature in the whole cross section of the specimen. Afterwards, all specimens were cooled down inside of the furnace chamber.

3. Testing Procedures

3.1. Concrete Permeability

The permeability test used nitrogen as a gas media and the Cembureau method was applied [31]. The testing set-up used is presented in detail in Figure 1.

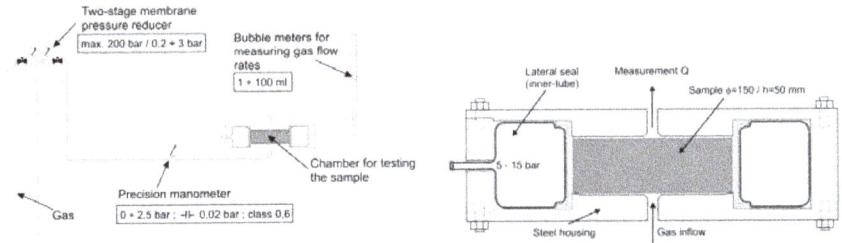

Figure 1. Permeability Cembureau testing set-up of the sample chamber.

In Equation (1), the permeability (k) was determined:

$$k = \frac{2QP_a \eta L}{A(P^2 - P_a^2)} (m^2), \qquad (1)$$

where:

Q: the measured gas flow intensity Q = V/t (m³/s);
V: gas volume (m³)
t: time (s)
P_a: atmospheric pressure (1 bar = 10^5 Pa);
P: absolute pressure (Pa);
A: cross section area of the specimen (m²);
η: nitrogen viscosity; η = 17.15 (Pa·s);
L: thickness of the specimen (m).

The initial reference permeability of the concrete in the samples at 90 days old was determined on the specimens that were not pre-dried, in order to represent the non-dried condition in the real structure. Subsequently, the samples were heated to a temperature ranging from 200 to 1000 °C, and after cooling the permeability was measured. Each measurement value represented the mean value from three samples.

3.2. Mechanical Tests

The cubic specimens with side a = 150 mm were used for compressive strength determination, with a diameter (d) of 100 mm and height (h) of 200 mm for the cylindrical samples for the splitting tensile strength tests. Three samples were used to test unheated concrete and two were used to test heated concrete. The modulus of elasticity was determined from the stress–strain (σ-ε) using one cylindrical sample (d = 100 mm; h = 200 mm). All E values were expressed in GPa and calculated from σ-ε curves as the stress to strain and strain ratio in the range of 10% to 40% of the ultimate stresses. For all properties six temperature levels were studied: T = 20, 200, 400, 600, 800, and 1000 °C. The compressive strength test procedures applied were presented in EN 12390-3 [32], and the splitting Brazilian tests were done according to EN 12390-6 [33].

4. Test Results and Discussion

4.1. Initial Properties

For B CEMI, B CEMIII, RB CEMI, and RB CEMIII concretes, the initial physical properties of bulk density $\rho_{o20°C}$ and permeability k, and the mechanical properties of compressive strength $f_{c20°C}$ tensile strength $f_{t20°C}$ and modulus of elasticity $E_{20°C}$ were determined after 90 days. The initial measurements, obtained for non-heated concrete properties, are presented in Table 5 and marked with the symbol 20 °C.

Table 5. Initial properties and parameters of the test concretes.

Property	Unit	B CEMI	B CEMIII	RB CEMI	RB CEMIII
		B Basalt Coarse Aggregate		RB Riverbed Coarse Aggregate	
Bulk density $\rho_{o20°C}$	kg/m³	2558.8	2533.2	2300.7	2315.6
Compressive strength $f_{c20°C}$	MPa	84.9	96,2	77.0	87.4
Splitting tensile strength $f_{t20°C}$	MPa	6.2	6.9	6.0	5.6
Modulus of elasticity $E_{20°C}$	GPa	44.4	48.9	30.6	29.7
Permeability $k_{20°C}$	m²	0.70×10^{-17}	0.52×10^{-17}	1.20×10^{-17}	1.00×10^{-17}

4.2. Evolution of Bulk Density with Temperature

The progressive increase of the temperature resulted in free water evaporation and progressive dehydration of the material. The C-S-H, as well as portlandite and calcium carbonate decomposition, were progressive in higher temperatures. As a result, weight loss was observed and the progressive density changes were recorded. The bulk density of B CEMI, B CEMIII, RB CEMI, and RB CEMIII concretes decreased as a function of the temperature. The mean values of bulk density are presented in Figure 2.

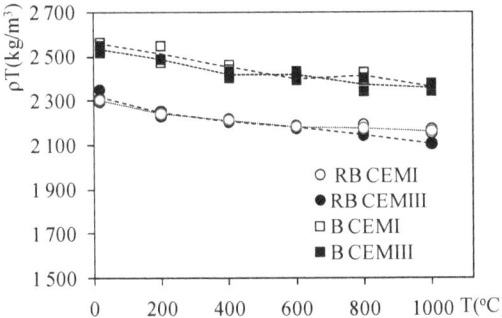

Figure 2. Bulk density of riverbed aggregates (RB) and basalt (B) concretes made with CEMI and CEMIII concretes; mean value of three samples.

In Figure 2 the bulk densities of the test concretes are presented. The values are mainly related to the type of aggregate: basalt or riverbed. The density of basalt CEMI concrete was 2558.8 kg/m^3 and the B CEMIII 2533.2 kg/m^3. The RB CEMI and RB CEMIII concrete were 2300.7 and 2315.6 kg/m^3, respectively. Apart from the initial values of density observed in the non-heated pristine concrete, the changes of the density with the temperature were quite similar for both cement types.

4.3. Evolution of Compressive Strength and Splitting Tensile Strength with Temperature Exposure

Figure 3 depicts the average and individual values of compressive strength. From the figure it can be concluded that the compressive strength of unheated concrete was higher for both CEMIII concretes made with basalt and riverbed aggregates. This tendency is maintained at 200 °C. When the temperature is higher than 400 °C, there are few differences in strength between B CEMI and B CEMIII, as well as between RB CEMI and RB CEMIII concretes. They all presented almost the same strength of 60 MPa.

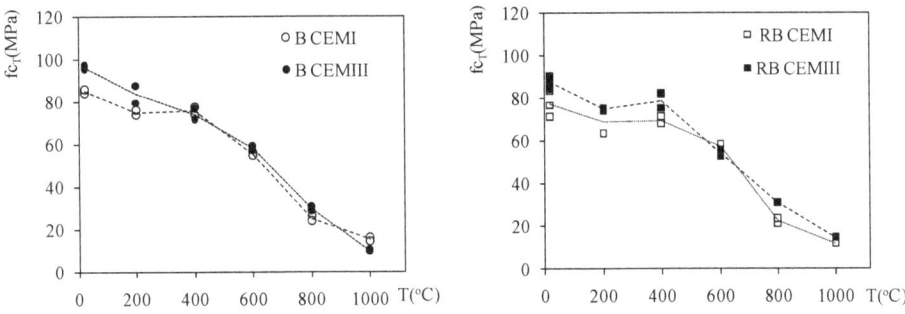

Figure 3. The compressive strength evolution CEMI and CEMIII concretes on basalt and riverbed aggregate.

In Figure 4, the average and individual values of f_{tT} are presented. Heating resulted in a progressive reduction in strength, nevertheless, the differences between CEMI and CEMIII concretes over a whole range of temperatures may be considered insignificant, in the scope of measurement error, or the scatter of results for this mechanical property.

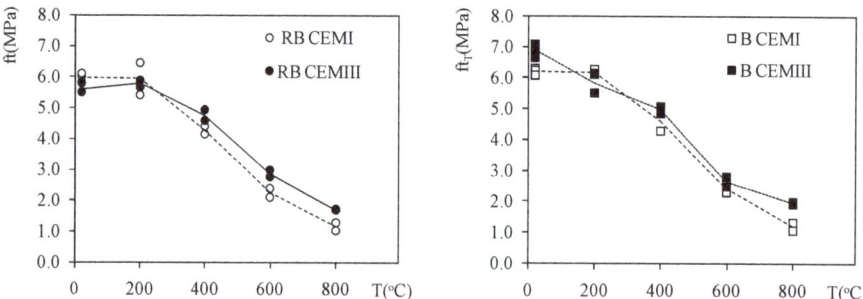

Figure 4. The changes in the splitting tensile strength of heated CEMI and CEMIII concretes on basalt and riverbed aggregate.

As has already been shown in previous research, an important aspect in the high temperature behavior of concrete is the thermal stability of aggregates at high temperatures. This can be evaluated by thermo-gravimetric and differential thermal analysis, which indicate the physical or chemical transformation of aggregates. As has already been reported [10], basalt is thermally stable up to 1000 °C; above this temperature melting is observed at 1050 °C and expansion and gas release both occur.

4.4. Relationship between Stress and Strain, and the Modulus of Elasticity Evaluation

The stress–strain relationships for the tested concretes are presented in Figure 5. Along with the temperature increase, a change of concrete stiffness was observed, as represented by the slope of the stress–strain curve. For the specimens heated to 600 °C and above, the stress–strain curve presents nonlinear behavior in compression due to the presence of cracks, which are closing partially when a compressive load is applied during the test. The similar stress–strain behavior of concrete in compression was observed for hot tested and tested after cooling down [8,15], an important cracking of samples was observed, especially for concretes with siliceous aggregates, heated without loading. The cracking of unloaded concrete was confirmed by the thermal strain evolution observation during heating [6].

The static modulus of elasticity values (E_T) of heated B CEMI and B CEMIII, as well as RB CEMI and RB CEMIII, are shown in Figure 6. The pristine non-heated concretes' modulus of elasticity ($E_{20°C}$) were 44.4 and 48.9 GPa, respectively, for B CEMI and B CEMIII. For riverbed aggregate RB CEMI and RB CEMIII they were 30.6 and 29.7 GPa. These results show clearly that for concretes with the same volume of cement paste, the modulus of elasticity is related to the nature of the aggregate and is strongly related to concrete density. Higher values of E_T were observed for both CEMIII concretes with RB and B aggregates.

A quasi linear decrease in the E_T value over the whole range of heating temperatures was observed. The slope of E_T decrease is most pronounced in the range of temperatures between 400 and 1000 °C (Figure 5). This sharp decrease of stiffness was attributed to crack development due to a mismatch of the strains between the cement paste and aggregates that is observed in this range of temperatures, and an increase in thermal strains resulting from cracking [6,7].

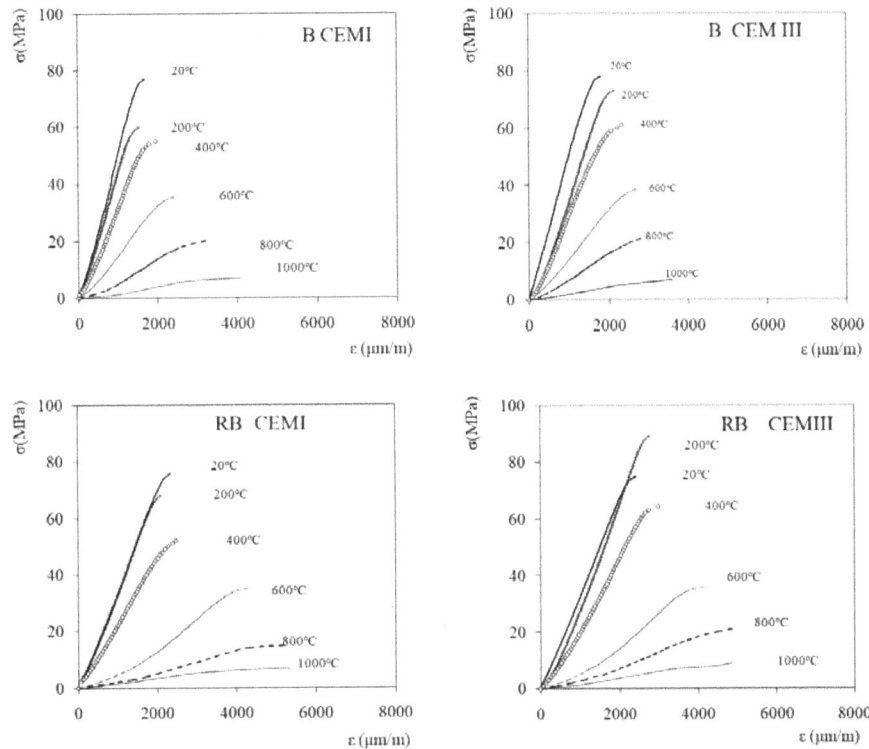

Figure 5. curves of heated concretes.

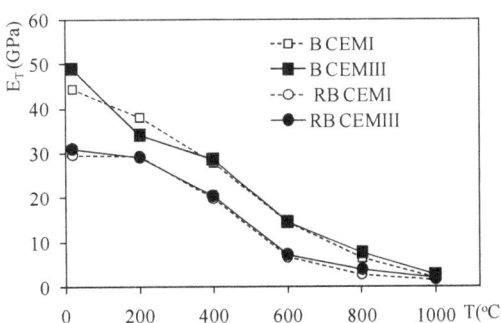

Figure 6. Modulus of elasticity change with the temperature of CEMI and CEMIII concretes.

From Figure 6 it can be concluded that the relative change of the modulus of elasticity is quasi identical for the concretes tested, and does not depend on cement type. The differences between the modulus of elasticity values of RB CEMI and RB CEMIII are not significant except for differences occurring at 20 °C.

4.5. Heated Concrete Permeability Evolution

For RB CEMI and RB CEMIII, the initial reference permeability, measured on non-heated concrete after exposure to 20 °C, reached values of 1.20×10^{-17} m^2 and 1.00×10^{-17} m^2, respectively. For B CEMI and B CEMIII this permeability was 0.70×10^{-17} m^2 and 0.52×10^{-17} m^2. With the increase of heating temperature residual permeability was increased. For the specimens heated to 1000 °C the permeability could not be measured due to crack development, and the gas flows could not be stabilized, so the permeability could not be measured with the Cembureau set-up. The results of the permeability measurements are presented in Figure 7. For B CEMIII and RB CEMIII concretes generally, lower values of permeability were observed. For the riverbed aggregate concrete RB CEMIII, permeability measured after exposure to high temperatures at 200, 400, 600, and 800 °C was systematically slightly lower than for RB CEMI. Basalt aggregate-based concretes provide lower permeability than riverbed ones. Nevertheless, these differences could not be considered significant. For all the concretes heated up to 1000 °C, the permeability could not be measured with the Cembureau method due to the significant damage to the concrete and crack development.

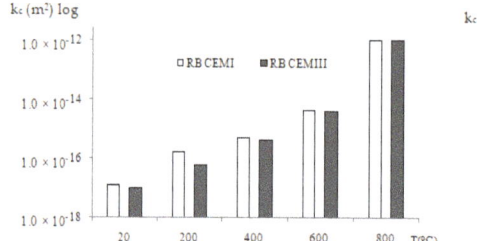 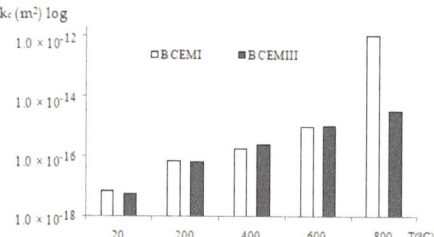

Figure 7. Effect of heating on the permeability of the test materials: RB CEMI and RB CEMIII, B CEMI and CEMIII. The reference permeability at 20 °C and permeability after heating to 200, 400, 600, and 800 °C.

4.6. Permeability vs. High Temperature Damage Factor

Previous studies [34,35] have indicated that the degradation of concrete at high temperatures, arising from a coupled hygro-thermal, chemical (dehydration) and mechanical interaction, can be modelled by means of the isotropic damage theory of Mazars [36]. Following Gawin et al. [9], the total damage D may be described by a multiplicative format of mechanical and thermo-chemical damage components, as shown in Equation (2):

$$D = 1 - \frac{E(T)}{E_0(T_0)} = 1 - \frac{E(T)}{E_0(T)} \frac{E_0(T)}{E_0(T_0)} = 1 - (1-d) \times (1-V), \quad (2)$$

where V corresponds to the thermo-chemical damage and d to the mechanical damage. The term $(1-d)$ corresponds to $\frac{E(T)}{E_0(T)}$, and $(1-V)$ to $\frac{E_0(T)}{E_0(T_0)}$. In the above equation $E_0(T_0)$ is the initial value of the static modulus of elasticity, $E_0(T)$ is the modulus for mechanically undamaged material expressed in a function of heating temperature, and $E(T)$ represents the static modulus of elasticity of mechanically damaged heated concrete.

Following this approach, in Figure 8 the effect of temperature on the damage parameter for heated concretes is presented. The damage factor was calculated on the basis of the change in the modulus of elasticity with temperature (see Figure 6), leading to Equation (3), and this evaluates the deterioration of the stiffness of the heated concrete samples by comparing them with the parameters found in non-heated concrete:

$$D_E = 1 - E_T/E_{20\,°C}, \quad (3)$$

where $E_{20°C}$ is the static modulus of elasticity tested at 20 °C and E_T is the value obtained for heated concrete.

The damage factor follows a comparable increasing change for all tested materials and almost reaches the value of 0.9, which means that 90% of the concrete has deteriorated. However, at 400 °C the damage value becomes much higher for the basalt aggregate concretes in comparison with the riverbed aggregate ones. Overall, the damage values for the CEMIII concretes appear to be slightly lower than for the CEMI concretes, especially for the basalt-based materials.

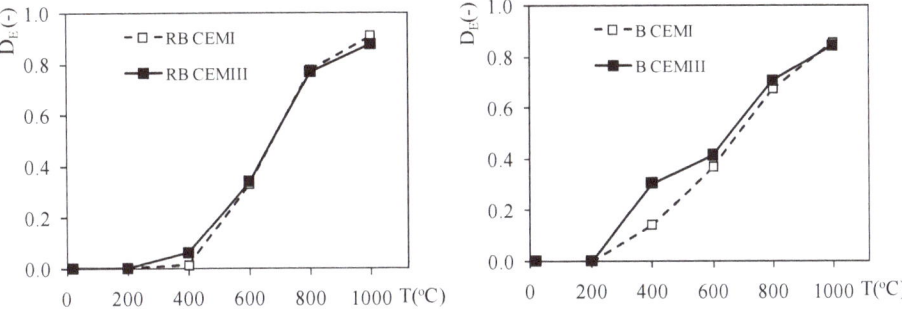

Figure 8. Damage factor (D_E) as a function of temperature.

These changes may be qualitatively compared to the change of total damage with temperature of a high performance concrete [9]. However, the damage values obtained and cited in this study are higher (damage of 0.8 at 600 °C). The reason for this difference may be due to the heating conditions, and notably the heating rate, which was four times higher in the study by Gawin et al. [9] than in our procedure, and which may provide stronger thermal gradients and therefore greater degradation.

It has already been noted that the changes to the inner micro-structure and permeability of the concrete may be characterized using this mechanistic approach, using damage evaluation to describe the high temperature degradation and/or micro-cracking effects [9,26,37,38]. The results of such a correlation are presented in Figure 9 for all the test materials. One may observe that all the data follow a single master law, independent of cement type or aggregate type. The results follow an exponential relationship, except for the permeability values obtained at 800 °C (Equation (4)):

$$\frac{k}{k_0} = \exp[C_{DE} \cdot D_E]. \qquad (4)$$

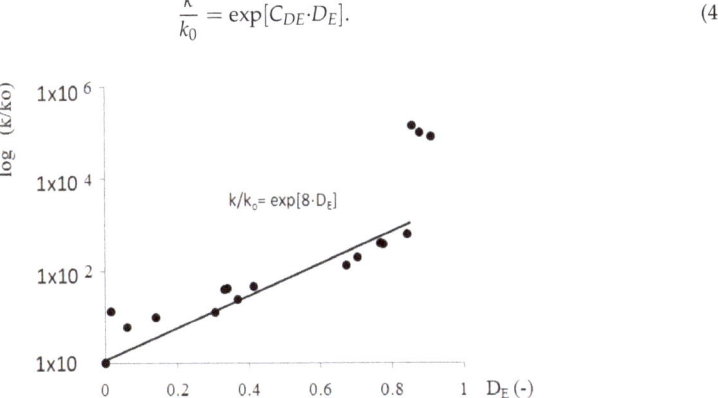

Figure 9. Juxtaposition of permeability (log scale) and damage to unheated concretes (20 °C) and concretes heated to 200, 400, 600, and 800 °C.

In Equation (4) k is the permeability of the heated material, k_0 the initial reference permeability, D_E the damage factor, and C_{DE} is the material dependent parameter, here equal to 8, which confirms the value obtained for another high performance concreteat elevated temperatures, but based only on the CEMI cement [9]. The C_{DE} value being equal to 8was obtainedfrom the regression curve with the coefficient of determination R^2 of 0.86 value. Therefore, the proposed regression curve is limited in the range of temperature from 20 to 600 °C. Three points that do not follow the trend in Figure 9 correspond to permeability values obtained at 800 °C. At this temperature, important cracking occurs following the already mentioned nonlinear mechanical behavior.

5. Conclusions

This paper intended to present the study of the influence of the cement types CEMI and slag cement CEMIII, in which the GGBFS amount reaches 54%, on the mechanical and physical performances of heated concretes with riverbed (RB) and basalt (B) aggregates. Four concretes with the same volume of cement paste and mortar were investigated. The only parameter differentiating RB and B concretes was the cement type. Analysis of the experimental data obtained concerned the mechanical tests, stiffness, and the permeability test results of the four concretes subjected to high temperature exposure (up to 1000 °C). The following main conclusions were drawn:

1. Type of cement influences compressive strength and permeability of 90 day concrete. Concretes with CEMIII presented lower permeability and higher compressive strength for both basalt and riverbed aggregate concretes;
2. High temperature exposure strongly influences the mechanical and physical properties of concretes, and the damage to concrete increases with exposure temperature. A temperature increase leads to the reduction of strength and modulus of elasticity. The splitting tensile strength decrease is more pronounced than the compressive strength evolution.
3. Minor differences between the mechanical properties of heated CEMI and CEMIII concretes were observed. The bulk density values, as well as the mechanical properties f_{cT}, f_{tT} and E_T, were very close or the differences were within the range of measurement error or the scatter of results of the properties tested;
4. The nature of the aggregate has a dominant influence on the material physical density and mechanical properties of the tested concretes. The compressive and tensile strengths depend on the aggregate nature for temperature up to 400 °C; above this temperature level, similar values of strength are observed;
5. The decrease in the mechanical properties is the result of progressive cement paste damage due to dehydration and chemical changes in the cement paste. Moreover, crack development due to the thermal mishmash of aggregate and cement paste results in nonlinear behavior of heated concretes;
6. The course of changes of the relative value in the elastic modulus for all the concretes investigated was very similar, except for the temperature of 400 °C. The riverbed aggregate concretes RB CEMI and RB CEMIII hada lower damage parameter than that observed for basalt aggregate concretes (B CEMI and B CEMIII) at this temperature. For 200, 600, 800, and 1000 °C, the damage levels were similar;
7. Important changes of up to six orders of magnitude were observed in permeability values following heating. However, the differences between the concretes could not be considered as significant. Indeed, CEMIII concretes presented slightly lower values of permeability in comparison with the CEMI ones in whole range of temperatures. On the other hand, basalt aggregate-based concretes have slightly lower permeability than riverbed ones. Concretes with CEMI: riverbed 1.2×10^{-17} vs. basalt 0.7×10^{-17}; concretes with CEMIII: riverbed 0.99×10^{-17} vs. basalt CEMIII 0.53×10^{-17}. That difference was be explained by lower permeability of basalt aggregate itself. This relation was also observed for the temperatures of 200, 400, and 600 °C;
8. Analysis of the results allowed the formulation of the constitutive exponential law, presenting the relationship between the permeability of concrete and damage, which is valid up to 600 °C.
9. It can be considered that heating induces damage, which may be represented by changes in the initial modulus of elasticity, that depends to a small degree on the type of cement. In this range of damage, the effects of aggregate type are also non-significant.

Author Contributions: I.H. conception, T.T. and K.M. data curation, M.C. formal analysis, I.H. founding acquisition, T.T., M.C. and K.M. investigation, T.T. and M.C. methodology, I.H. project administration, I.H. supervision, I.H. and M.C. validation. I.H. and M.C. writing—original draft, I.H. M.C., T.T. and K.M. writing—review and editing.

Funding: The study was supported. This research was funded in part from the EMMAT project E-mobility and sustainable materials and technologies PPI/APM/2018/1/00027 financed by the Polish National Agency for Academic Exchange (NAWA) and Polish National Research Centre Grant "Multi-parameter assessment of cement concretes exposed to fire temperature" N N506045 040.

Conflicts of Interest: The authors declare no conflict of interest.

References

1. Pal, S.C.; Mukherjee, A.; Pathak, S.R. Investigation of hydraulic activity of ground granulated blast furnace slag in concrete. *Cem. Concr. Res.* **2003**, *33*, 1481–1486. [CrossRef]
2. Topçu, I.B. High-volume ground granulated blast furnace slag (GGBFS) concrete. In *Eco-Efficient Concrete*; Pacheco-Torgal, F., Jalali, S., Eds.; Elsevier: Cambridge, UK, 2013; pp. 218–240.
3. Lura, P.; van Breugel, K.; Maruyama, I. Effect of curing temperature and type of cement on early-age shrinkage of high-performance concrete. *Cem. Concr. Res.* **2001**, *31*, 1867–1872. [CrossRef]
4. Shumuye, E.D.; Zhao, J.; Wang, Z. Effect of fire exposure on physico-mechanical and microstructural properties of concrete containing high volume slag cement. *Constr. Build. Mater.* **2019**, *213*, 447–458. [CrossRef]
5. Farzampour, A. Compressive Behavior of Concrete under Environmental Effects. In *Compressive Strength of Concrete*; IntechOpen: London, UK, 2019; pp. 1–12.
6. Mindeguia, J.-C.; Hager, I.; Pimienta, P.; Borderie, C.L.; Carré, H. Parametrical study of transient thermal strain of high performance concrete. *Cem. Concr. Res.* **2013**, *48*, 40–52. [CrossRef]
7. Hager, I. Behaviour of cement concrete at high temperature. *Bull. Pol. Acad. Sci. Tech. Sci.* **2013**, *61*, 145–154. [CrossRef]
8. Pimienta, P.; Jansson McNamee, R.; Mindeguia, J.-C. (Eds.) *Physical Properties and Behaviour of High-Performance Concrete at High Temperature. State-of-the-Art Repport of the RILEM Technical Committee HPB-227*; Springer: Cham, Switzerland, 2019; pp. 1–130.
9. Gawin, D.; Alonso, C.; Andrade, C.; Majorana, C.E.; Pesavento, F. Effect of damage on permeability and hygro-thermal behaviour of HPCs at elevated temperatures: Part 1. Experimental results. *Comput. Concr.* **2005**, *2*, 189–202. [CrossRef]
10. Hager, I.; Tracz, T.; Śliwiński, J.; Krzemień, K. The influence of aggregate type on the physical and mechanical properties of high-performance concrete subjected to high temperature. *Fire Mater.* **2016**, *40*, 668–682. [CrossRef]
11. Xing, Z.; Beaucour, A.-L.; Hebert, R.; Noumowe, A.; Ledesert, B. Influence of the nature of aggregates on the behaviour of concrete subjected to elevated temperature. *Cem. Concr. Res.* **2011**, *41*, 392–402. [CrossRef]
12. Mindeguia, J.-C.; Pimienta, P.; Carré, H.; Borderie, C.L. On the influence of aggregate nature on concrete behaviour at high temperature. *Eur. J. Environ. Civ. Eng.* **2012**, *16*, 236–253. [CrossRef]
13. Ju, Y.; Tian, K.; Liu, H.; Reinhardt, H.-W.; Wang, L. Experimental investigation of the effect of silica fume on the thermal spalling of reactive powder concrete. *Constr. Build. Mater.* **2017**, *155*, 571–583. [CrossRef]
14. Al-Jabri, K.S.; Waris, M.B.; Al-Saidy, A.H. Effect of aggregate and water to cement ratio on concrete properties at elevated temperature. *Fire Mater.* **2016**, *40*, 913–925. [CrossRef]
15. Hager, I.; Pimienta, P. Mechanical properties of HPC at high temperatures. In *Fire Design of Concrete Structures: What Now? What Next?* Gambarova, P.G., Felicetti, R., Meda, A., Riva, P.Eds.; Starrylink: Brescia, Italy, 2004; pp. 95–100.
16. RILEM TC: Test methods for mechanical properties of concrete at high temperatures, Recommendations, Part 3: Compressive strength for service and accident conditions. *Mater. Struct.* **1995**, *28*, 410–414. [CrossRef]
17. Asamoto, S.; Yuguchi, R.; Kurashige, I.; Chun, P.J. Effect of high temperature at early age on interfacial transition zone and material properties of concrete. In Proceedings of the RILEM Proc. 121 618 SynerCrete18 Int. Conf. Interdiscip. Approaches Cem.-Based Mater. Struct. Concr, Funghal, Portugal, 24–26 October 2018; pp. 683–688.

18. Śliwiński, J.; Tracz, T.; Zdeb, T. Influence of selected parameters of cement concrete composition on its gas permeability. In *Recent Advances in Civil Engineering: Building Materials and Building Physics*; Śliwiński, J., Politech, K., Eds.; Elsevier: Cracow, Poland, 2015; Volume 479, pp. 97–119.
19. Kalifa, P.; Menneteau, F.D.; Quenard, D. Spalling and pore pressure in HPC at high temperature. *Cem. Conc. Res.* **2000**, *30*, 1915–1927. [CrossRef]
20. Kalifa, P.; Chéné, G.; Gallé, C. High-temperature behaviour of HPC with polypropylene fibers: From spalling to microstructure. *Cem. Conc. Res.* **2001**, *31*, 1487–1499. [CrossRef]
21. Mindeguia, J.C.; Pimienta, P.; Noumowe, A.; Kanema, M. Temperature, pore pressure and mass variation of concrete subjected to high temperature—Experimental and numerical discussion on spalling risk. *Cem. Conc. Res.* **2012**, *40*, 477–487. [CrossRef]
22. Lo Monte, F.; Felicetti, R.; Meda, A.; Bortolussi, A. Assessment of concrete sensitivity to fire spalling: A multi-scale experimental approach. *Constr. Build. Mater.* **2019**, *212*, 476–485. [CrossRef]
23. Mróz, K.; Hager, I. Causes and mechanism of concrete spalling under high temperature caused by fire. *Cem. Wap. Bet.* **2017**, *6*, 445–456.
24. Zheng, W.Z.; Hou, X.M.; Shi, D.S.; Xu, M. Experimental study on concrete spalling in prestressed slabs subjected to fire. *Fir. Saf. J.* **2010**, *45*, 283–297. [CrossRef]
25. Zhang, Y.; Zeiml, M.; Pichler, C.H.; Lackner, R. Model-based risk assessment of concrete spalling in tunnel linings under fire loading. *Eng. Struct.* **2014**, *77*, 207–215. [CrossRef]
26. Gawin, D.; Pesavento, F.; Guerrero Castells, A. On reliable predicting risk and nature of thermal spalling in heated concrete. *Arch. Civ. Mech. Eng.* **2018**, *18*, 1219–1227. [CrossRef]
27. Karahan, O. Transport properties of high volume fly ash or slag concrete exposed to high temperature. *Constr. Build. Mater.* **2017**, *152*, 898–906. [CrossRef]
28. Choinska, M.; Khelidj, A.; Chatzigeorgiou, G.; Pijaudier-Cabot, G. Effects and interactions of temperature and stress-level related damage on permeability of concrete. *Cem. Concr. Res.* **2007**, *37*, 79–88. [CrossRef]
29. Hager, I.; Tracz, T. The impact of the amount and length of fibrillated polypropylene fibers on the properties of HPC exposed to high temperature. *Arch. Civ. Eng.* **2010**, *1*, 57–68. [CrossRef]
30. Schneider, U.; Herbst, H.J. *Permeabilität und Porosität von Betonbeihohen Temperaturen*; Deutscher Ausschuss für Stahlbeton: Berlin, Germany, 1989; Volume 403, pp. 1–140.
31. Kollek, J.J. The determination of the permeability of concrete to oxygen by the Cembureau method—RILEM recommendation. *Mater. Struct.* **1989**, *22*, 225–230. [CrossRef]
32. *EN 12390-3 Testing hardened concrete—Part 3: Compressive Strength of Test Specimens*; British Standards Institution (BSI): London, UK, 2019; pp. 1–24.
33. *EN 12390-6 Testing hardened concrete—Part 6: Tensile Splitting Strength of Test Specimens*; British Standards Institution (BSI): London, UK, 2009; pp. 1–14.
34. Gawin, D.; Majorana, C.; Pesavento, F.; Schrefler, B. A fully coupled multiphase FE model of hygro-thermo-mechanical behaviour of concrete at high temperature. In *Computational Mechanics, New Trends and Applications, Proceedings of the 4th World Congress on Computational Mechanics, Buenos Aires, Argentina, 29 June–2 July 1998*; Onate, E., Idelsohn, S.R., Eds.; CIMNE: Barcelona, Spain, 1998; pp. 1–19.
35. de Borst; et al. On modelling of thermo-mechanical concrete for the finite element analysis of structures submitted to elevated temperatures. In *Fract. Mech. Concr. Struct*; de Borst; et al. Swets&Zeitlinger: Lisse, Nedherlands, 2001; Volume 1, pp. 271–278.
36. Mazars, J. Application de la Mecanique de L'endommagement au Comportement non Lineaire et la Rupture du Beton de Structure. Ph.D. Thesis, Universite de Paris, Paris, France, 1984.
37. Gawin, D.; Pesavento, F.; Schrefler, B.A. Simulation of damage-permeability coupling in hygro-thermo-mechanical analysis of concrete at high temperature. *Commun. Numer. Methods Eng.* **2002**, *18*, 113–119. [CrossRef]
38. Pijaudier-Cabot, G.; Dufour, F.; Choinska, M. Damage and Permeability in Quasi-brittle Materials: From Diffuse to Localized Properties. In *Multiscale Modeling of Heterogenous Materials: From Microstructure to Macro-scale Properties*; Wiley-ISTE: London, UK, 2010; pp. 277–292.

© 2019 by the authors. Licensee MDPI, Basel, Switzerland. This article is an open access article distributed under the terms and conditions of the Creative Commons Attribution (CC BY) license (http://creativecommons.org/licenses/by/4.0/).

Article

The Influence of Fluidized Bed Combustion Fly Ash on the Phase Composition and Microstructure of Cement Paste

Michał A. Glinicki, Daria Jóźwiak-Niedźwiedzka *and Mariusz Dąbrowski

Institute of Fundamental Technological Research, Polish Academy of Sciences, Pawińskiego 5b, 02-106 Warsaw, Poland
* Correspondence: djozwiak@ippt.pan.pl; Tel.: +48-22-8261281 (ext. 310)

Received: 2 July 2019; Accepted: 2 September 2019; Published: 3 September 2019

Abstract: Fly ashes from coal combustion in circulating fluidized bed boilers in three power plants were tested as a potential additive to cement binder in concrete. The phase composition and microstructure of cement pastes containing fluidized bed fly ash was studied. The fractions of cement substitution with fluidized bed fly ash were 20% and 30% by weight. X-ray diffraction (XRD) tests and thermal analyses (derivative thermogravimetry (DTG), differential thermal analysis (DTA) and thermogravimetry (TG)) were performed on ash specimens and on hardened cement paste specimens matured in water for up to 400 days. Quantitative evaluation of the phase composition as a function of fluidized bed fly ash content revealed significant changes in portlandite content and only moderate changes in the content of ettringite.

Keywords: clean coal combustion; fluidized bed fly ash; microstructure; phase composition; portlandite; unburned carbon

1. Introduction

Constantly evolving technologies of coal combustion, with the aim of improving energy efficiency, include among others the more efficient fluidized bed combustion (FBC) systems. A well designed boiler can combust coal with relatively high efficiency and an acceptable level of gas emissions. FBC boilers operate at lower temperatures (800–900 °C) compared to conventional pulverized coal combustion (1400–1700 °C) [1]. The diversity of work conditions of increasingly used fluidized bed boilers makes it possible to use them to burn high-sulfur fuels [2]. The fuel flexibility is an important characteristic of FBC boilers [3]. The best example is circulating fluidized bed combustion (CFBC), which provides low emissions of SO_2 and NO_x to the atmosphere [3–5] while maintaining the efficiency of coal combustion.

A systematic increase of coal combustion by-products from fluidized boilers has been observed in numerous countries. One of the ways of managing CFBC by-products is to use them as an additive to cement and concrete. Studies on the possibility of the use of CFBC fly ash as a cement or concrete additive have been carried out for several years [1,6–12], but still there are no normative regulations regarding the use of this kind of fly ash as a cement additive.

Some researchers have performed investigations on the material characteristics and mechanical properties of CFBC fly ash in cement-based composites. It has been shown that CFBC fly ash is characterized by high variability of chemical and phase composition [13]. The dominant CFBC particles were coarse and had an irregular flaky shape characterized by a broad particle size range [12] which resulted in high demand of water and admixture [14]. The high content of anhydrite in CFBC resulted in excessive expansion in cement matrix composites due to abundant ettringite formations [15]. Moreover, most of the CFBC fly ash had less pozzolanic activity than the conventional siliceous fly ash [7] and some of the CFBC fly ashes do not meet the standard requirements to be classified as either Class C or Class F fly ash [11]. The main reason for limiting the use of CFBC fly ash in blended cements is a high content of sulfates (mostly anhydrite), and free calcium oxide, whose presence contribute to creation of secondary ettringite—negative from point of view of the durability of cement matrix composites [16]. The overview of the effect of reactive mineral additives on the microstructure and phase composition of the cement hydration products was presented in [17]. Rajczyk et al. [18] used the DTA method to follow the hydration process of cement containing fluidized bed combustion fly ash. Based on endothermic peaks attributed to the dehydration of phases formed on hydration, the conditions leading to the formation of so-called delayed ettringite were found. Gazdiča et al. [1], based on the progress of hydration of blended cements, proved that ettringite formed from anhydrite contained in fluidized bed ash was one of the hydration products built during the formation of the cement stone structure, thus ettringite was not the cause of negative volume changes.

However, despite these observations, it has also been shown that application of CFBC fly ash simultaneously with siliceous fly ash or slag as a binder component makes it possible to obtain a durable cement-based composite. Nguyen et al. [19] used CFBC fly ash as a sulfate activator which significantly improved the mechanical properties of the modified high-volume fly ash cement pastes at early ages. The accelerated hydration of C_3S and more precipitated ettringite formation of hardened pastes confirmed the important role of CFBC fly ash to enhance the mechanical properties of the cement pastes. Hlaváček et al. [20] used CFBC fly ash in a ternary binder (plus siliceous fly ash and calcium hydroxide) and they obtained a compressive strength of paste equal to 32 MPa after 28 days of curing. The ternary hydraulic binder containing CFBC fly ash was also analyzed by Škvára et al. [21]. They stated that a ternary binder possessed strength values comparable to those of Portland cement. Lin et al. [22] investigated the properties of controlled low strength material with circulating fluidized bed combustion ash and recycled aggregates. They showed that CFBC hydrated ash resulted in a higher compressive strength when compared with desulfurized slag, but had a lower compressive strength than coal bottom ash. Chi [23] characterized the mortars with CFBC fly ash and ground granulated blast-furnace slag and he found that this kind of fly ash had the potential to partially replace the cementing materials, but the proportion of CFBC fly ash was recommended to be limited to a maximum of 20% due to the decreasing of compressive strength and the increasing of initial setting time of mortars. Dung et al. [24] analyzed the application of raw circulating fluidized bed combustion fly ash and slag as eco-binders for a novel concrete without ordinary Portland cement. They found that this new binder behaves similar to ordinary Portland cement. The pastes showed good workability, proper setting times, and sufficient compressive and tensile strength, which were improved with the decrease of water/binder (w/b) ratio and with the increase of hydration time. Chen et al. [25] tested expansion properties of cement paste with circulating fluidized bed fly ash. They stated that the curing method is of great importance. The expansion development in its early days was dominated by the ettringite, and its quantity and morphology were seriously affected by the inadequateness of water. The application of CFBC fly ash improved concrete durability. Jóźwiak-Niedźwiedzka [26] analyzed the influence of CFBC fly ash on the chloride and scaling resistance of concrete. She showed that the cement replacement by 15% and 30% CFBC fly ash provided higher chloride resistance, but only replacement by 15% CFBC fly ash can be used to achieve concrete scaling durability. Kubissa et al. [27] showed that use of 25 wt % fluidized fly ash as cement replacement improved concrete water permeability and sorptivity, as well as chloride resistance. On the other hand, Czarnecki et al. [28] found that the depth of concrete carbonation increased with increasing of the content of CFBC fly ash but at the same time they obtained similar results for ordinary siliceous fly ash.

Juenger et al. [29] have recently presented a review paper on emerging supplementary cementitious materials (SCM). They concluded that with the increase in demand for conventional SCM and the simultaneous reduction in supply, there is a great need to find new sources of materials that provide comparable or better properties to highly used fly ash and slag sources. So, the research into new sources of SCMs (like fly ash from bed fluidized coal combustion that does not meet current specifications) and their impact on cement matrix properties has been increasing in recent years, and has great likelihood to further increase as the demand for these materials grows.

Despite many publications on the possible application of the CFBC fly ash in concrete technology, the problem of its proper use remains unresolved. An increasingly technical rigorous regime during the combustion of coal in fluidized bed boilers has contributed to improving the stability of the composition and reducing the amount of undesirable ingredients. Research on the new kinds of fly ash from circulating fluidized bed combustion has considered their specific characteristics of phase composition and paste microstructure. This paper aims to show some relationships between the differences in mineral composition and origin of burning coal. The phase composition as well as microstructure of cement pastes with various content of CFBC fly ash have been studied. A prolonged time of curing has been applied to determined how the replacement of Portland cement by CFBC fly ash would change the microstructure and phase composition of paste stored in high-humidity conditions.

2. Experimental Program

2.1. Materials and Methods

The fly ash used in the research came from the circulating fluidized bed combustion process in three Polish power stations: Katowice (K), Siersza (S) and Turów (T). Ashes were formed from the combustion of hard coal (K, S) and lignite (T). CFBC fly ash chemical analysis was carried out in accordance with European standards and is presented in Table 1.

Table 1. Chemical composition of circulating fluidized bed combustion (CFBC) fly ash determined by X-Ray Fluorescence Spectroscopy XRF method (wt %).

Component	CFBC Fly Ash		
	Hard Coal Burning		Lignite Burning
	S	K	T
LOI *	5.71	3.40	2.73
SiO_2	38.80	47.18	36.47
CaO	9.80	5.84	15.95
CaO_{free} **	not tested	3.40	4.75
Fe_2O_3	9.59	6.80	4.40
Al_2O_3	23.26	25.62	28.4
MgO	2.28	0.15	1.65
Na_2O	2.01	1.18	1.64
K_2O	2.22	2.36	0.62
SO_3	5.82	3.62	3.8
TiO_2	not tested	1.08	3.84
Cl-	0.40	0.10	0.03

* Loss on ignition at 1000 °C; ** Glycol method according to PN-EN 451-1.

Comparing the chemical composition to the requirements of PN EN 450-1, an excess of SO_3 caused by anhydrite in the fly ash is shown. Ordinary Portland cement 32.5 R from "Małogoszcz" cement plant was used. The composition and physical properties are given in Table 2. Binders of the above ingredients were prepared by replacing 20% and 30% of the cement mass by fluidized fly ash K, S or T. For example, T20 corresponds to 20% fly ash from Turów (lignite burning), and K30 corresponds to 30% fly ash from Katowice (hard coal burning). The detailed composition of pastes is presented in Table 3. The difference in density between fly ash and cement was neglected. The cement paste with a constant water to binder ratio 0.5 was formed using prisms with dimensions 40 × 40 × 160 mm. Storage of the molds took place for 48 hours in a chamber with a humidity of 95%; after this period of time the specimens were kept in water and matured for 400 days at 20 ± 2 °C.

Table 2. Chemical composition and physical properties of cement, determined by XRF method (wt %).

Component	Cement CEM I 32.5 R
LOI	3.40
SiO_2	21.40
CaO	65.06
CaO_{free} **	1.00
Fe_2O_3	2.53
Al_2O_3	4.80
MgO	1.37
Na_2O_{eq}	0.76
SO_3	2.60
Cl-	0.07
Specific surface, cm^2/g	3210
Density, g/cm^3	3.09

** Glycol method according to PN-EN 451-1.

Table 3. Composition of pastes prepared with CEM I and CFBC (g).

Designation of Specimens	Cement	CFBC	Water
CEM I	420	0	210
S20	336	84	210
T20	336	84	210
K20	336	84	210
S30	294	126	210
T30	294	126	210
K30	294	126	210

2.2. Test Methods

The CFBC fly ash was analyzed at macro and micro levels. For the analysis of mineral composition X-ray (XRD) (TUR-M62/VEB TuR Dresden, Germany) and thermal analysis (DTA, TG and DTG) (SDTQ 600, TA Instruments, Artisan Technology Group, Champaign, USA) were used. Tests were carried out with powder specimens which were also divided by fraction: above and below 0.045 mm, separated by measuring the fineness of ash in accordance with PN-EN 451-2. The content of free lime CaO_{free} was measured according to standard glycol method, PN-EN 451-1. After the chemical and physical characterization of CFBC fly ash, the samples proceeded to the second part of the research aimed to determine the phase composition of cement paste. The cement–fly ash binder was investigated by thermal analysis and X-ray diffraction, as well as microscopic observation after 28, 200 and 400 days of maturity.

Macroscopic analysis was performed with the naked eye. In order to identify differences in behaviour and colour of specimens, they were tested by reaction with an aqueous solution of HCl (1:3) and phenolphthalein (1% in ethanol), respectively. Determination of color specimens was carried out by applying a sheet of colours recommended by PN-EN 12407.

The flexural strength testing was carried out after 28 days of maturity, on three 40 × 40 × 160 mm beams for each paste composition according to the three-point bending test presented in the Standard PN-EN 191-1, [30]. The test was carried out using the Lloyd Instruments testing machine, type EZ50 (Lloyd Instruments LTD, Hampshire, UK). Compressive strength of analyzed pastes was determined on halves of beams broken during bending test (after 28 days of curing). The test was performed using appropriate inserts, enabling measurements in accordance with PN-EN 196-1.

The microstructure of the powder and paste specimens was analyzed using scanning electron microscope (SEM) LEO 1530 (Carl Zeiss Microscopy GmbH, Jena, Germany). The surface of specimen observation in SEM was not less than 1.0 cm^2 over the range of magnifications from 200× to 50,000×. The specimens were coated with a layer of carbon with a thickness of about 10 nm using the device

BalTec SCD 005, CEA 035 adapter (BAL-TEC GmbH, Schalksmühle, Germany). Thermal analysis was performed at the following setup: heat rate 10 °C/min in platinum crucibles for the standard sample of Al_2O_3 in the air atmosphere. X-ray diffraction analysis was performed using diffractometer TUR-M62 (TUR-M62/VEB TuR Dresden, Germany), with the following conditions: CuK_α radiation, monochromator filter, 40 kVA voltage and 20 mA current of the X-ray lamp, BDS-7 meter, 0.05 s step, 5 s time constant.

3. Test Results and Discussion

Results of macroscopic analysis are presented in Table 4. The most important observations made in a part of this study was the colour test variation from brownish-grey (5YR4/1) for CFBC fly ash from Siersza to light olive-grey (5Y6) for CFBC fly ash from Turów. It is probably related to the amount of unburned coal in the analyzed specimen and the type of burned material: Siersza (hard coal) and Turów (lignite). The darker colour corresponds to the higher coal content, which is also visible on the basis of LOI results in the analyzed fly ashes.

Table 4. Result of macroscopic analysis of the CFBC ashes: S, T and K.

	Attribute	CFBC Fly Ash		
		S	T	K
	>0.045 mm, Yield (%)	22.0	36.1	19.5
Color	Whole lot	brownish-grey 5YR 4/1	light olive-grey 5Y 6	olive-grey 5Y4/1
	Fraction > 0.045 mm	light brownish-grey 5YR 6/1	light brownish-grey 5YR 6/1	brownish-grey 5
	Fraction < 0.045 mm	brownish-grey 5YR 4/1	not tested	not tested

Reaction with HCl, for each of the fly ashes passed with the same intensity of the characteristic light "effervescence". So it should therefore be expected these specimens had similar content of CaO_{free}. Higher fineness (36.1%) was exhibited by CFBC fly ash (T); it was several age points higher than the other tested fly ashes (K and S).

Observations in the scanning electron microscope showed typical grains corresponding to fly ash from fluidized bed combustion boilers. In Figure 1 a high surface development and irregular shape of the fly ash particles is visible. The largest observed particle size reached 120 μm for T fly ash and 80 μm for K and S fly ash. Also, relics of coal (often like char, which was verified by EDS analysis) are shown, which are disadvantageous from the standpoint of application of these fly ashes in cement (Figure 2). The non-hydroxylated clay minerals and combustion sorbent calcium carbonate have been found.

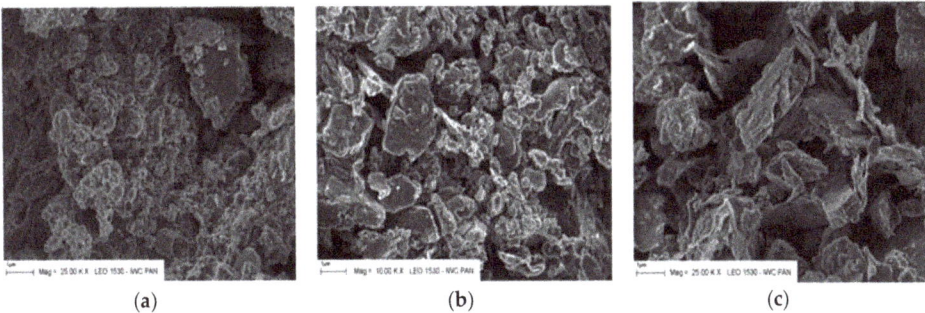

Figure 1. Microstructure of CFBC ashes from different power stations: (**a**) T, lignite burning; (**b**) S, hard coal burning; and (**c**) K, hard coal burning. Magnification 25,000×, scale bar = 1 μm.

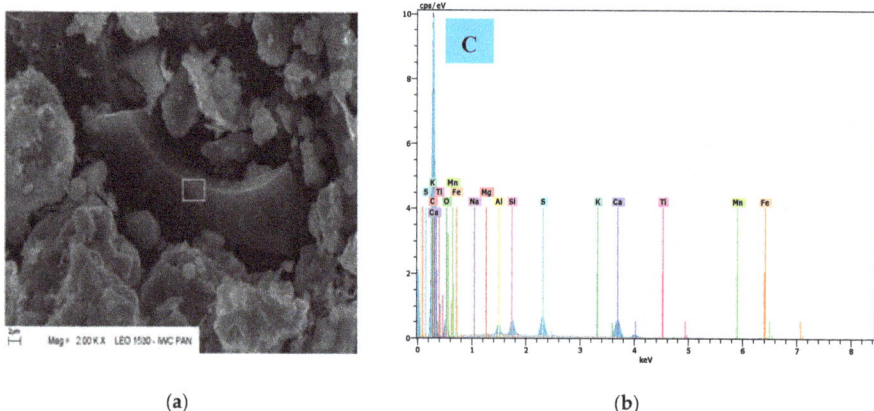

(a) (b)

Figure 2. Microstructure of CFBC fly ash grain with visible unburned coal particle (**a**) and analysis in microarea (**b**); magnification 20,000×, scale bar = 2 μm.

The results of flexural strength and compressive strength of tested pastes are shown in Table 5. The results of flexural strength for all pastes with CFBC were in the range of 5.2–6.4 MPa and differences were statistically negligible. The reference paste revealed a lower value of flexural strength (4.6 MPa). A much higher (about 50%) 28-day compressive strength of pastes made with the 20% replacement of cement by CFBC fly ash from hard coal burning (K and S) was observed compared to the CEM I reference paste. Specimens with CFBC fly ash from lignite burning (T) achieved slightly lower values of compressive strength compared to specimens with fly ash from hard coal burning. The increase in CFBC fly ash content influenced the decrease of compressive strength of pastes. This tendency is much more visible with increase of CFBC addition to 30%. The obtained results are consistent with the literature data [31,32]. Šiler et al. [31] showed that the replacement of cement by 10, 20 and 40 wt % of fluidized bed combustion fly ash influenced on the increase of compressive strength compared to ordinary cement paste in the early age of hydration (up to 28 days). Hanisková et al. [32] analyzed the influence of fly ash from fluidized bed combustion on mechanical properties of pastes. They showed that the highest values for 28-day compressive strength were achieved by pastes with 30–40% replacement of cement by CFBC fly ash, twice as much as the reference paste without fly ash.

Table 5. The results of compressive and flexural strength of pastes stored in water in 20 ± 2 °C after 28 days of curing (MPa).

Designation of Specimens	Flexural Strength	Comressive Strength
CEM I	4.6 ± 0.3	33.8 ± 0.6
S20	5.8 ± 0.2	47.8 ± 1.8
T20	5.2 ± 0.3	39.8 ± 2.2
K20	6.4 ± 0.4	53.3 ± 2.3
S30	5.9 ± 0.3	46.0 ± 1.4
T30	5.9 ± 0.7	35.9 ± 1.0
K30	5.9 ± 0.6	53.1 ± 2.0

The phase composition was evaluated using the thermal analysis method. The loss on ignition and weight loss of analyzed specimens were associated with the relics of clay minerals (temperature up to 350 °C) as well as the oxidation of unburned coal residue, and they were used to determine the content of $Ca(OH)_2$ and $CaCO_3$. The total loss on ignition was identified by heating fly ash to a temperature of 1000 °C. The results of thermal analysis are presented in Table 6. The obtained measurements show that a total loss on ignition depends not only on the loss of the unburned coal relics, but also the distribution of carbonates, portlandite and disposal of residual clay mineral water. Therefore, the LOI determined according to PN-EN 450-1 does not clearly reflect a precise content of unburned coal. This phenomena is the most visible for CFBC fly ash from Turów power plant (lignite burning), where the presence of the LOI (at 1000 °C) is mainly due to the calcium carbonate content (4.80%), and the relics of unburned coal accounted for only a fraction of a percent in the tested ash (0.8%).

Table 6. Selected properties of CFBC fly ash separated for the fractions below and above 0.045 mm.

Specimen	LOI *, wt %	Mass Loss at <350 °C, wt %	Relics of Coal, wt %	Portlandite, wt %	Calcium Carbonate, wt %
S					
Fraction > 0.045 mm	1.7	0.45	0.8	not detected	not detected
Fraction < 0.045 mm	5.8	0.74	4.4	not detected	not detected
T					
Fraction > 0.045 mm	3.7	0.79	0.8	not detected	4.80
Fraction < 0.045 mm	5.9	1.05	not detected	1.2	10.35
K					
Fraction > 0.045 mm	3.4	0.89	1.8	not detected	1.60
Fraction < 0.045 mm	5.7	0.87	4.4	not detected	1.05

* Loss on ignition at 1000 °C.

The X-ray analysis performed on whole fraction of fly ash revealed the presence of crystalline phases such as anhydrite II ($CaSO_4$), portlandite ($Ca(OH)_2$), quartz (SiO_2) and small quantities of clay minerals and calcium (CaO) and magnesium (MgO) oxide. A clear difference between the CFBC fly ash from the hard coal (K and S) and lignite (T) combustion is visible. There is much more portlandite ($Ca(OH)_2$) and calcite ($CaCO_3$) in the fly ash from the combustion of lignite (fly ash T) compared to hard coal combustion. Diffraction patterns of the analyzed specimens are presented in Figure 3.

For the further study two CFBC fly ashes were selected: K and T as representatives of the waste materials from the fluidized bed combustion of hard coal and lignite, respectively, with a low content of SO_3. The analysis of the phase composition of cement pastes with 20% and 30% replacement of the cement mass by CFBC fly ash after 28, 200 and 400 days of maturation in water was conducted.

Based on the results of thermal analysis of cement paste, the chemical-bound water content in hydration products (HI) was determined.

The volume of HI is considered as the loss of water at temperatures up to 400 °C, which is associated with hydrated calcium silicates (C–S–H) and calcium aluminosulfate. Additionally, the presence of calcium carbonate (CC) was detected in the hardened cement paste, and it was considered as the mass loss at 800–1000 °C. XRD analysis of the above hydration products revealed that small amounts of ettringite and calcium sulfates mainly in the form of gypsum were present. Diffraction patterns of the cement paste specimens are shown in Figure 4. After the qualitative analysis in order to present the quantitative changes occurring during the hydration process, the change index (WZ) was introduced, which is used as standard procedure [33]. The above method was described in [12].

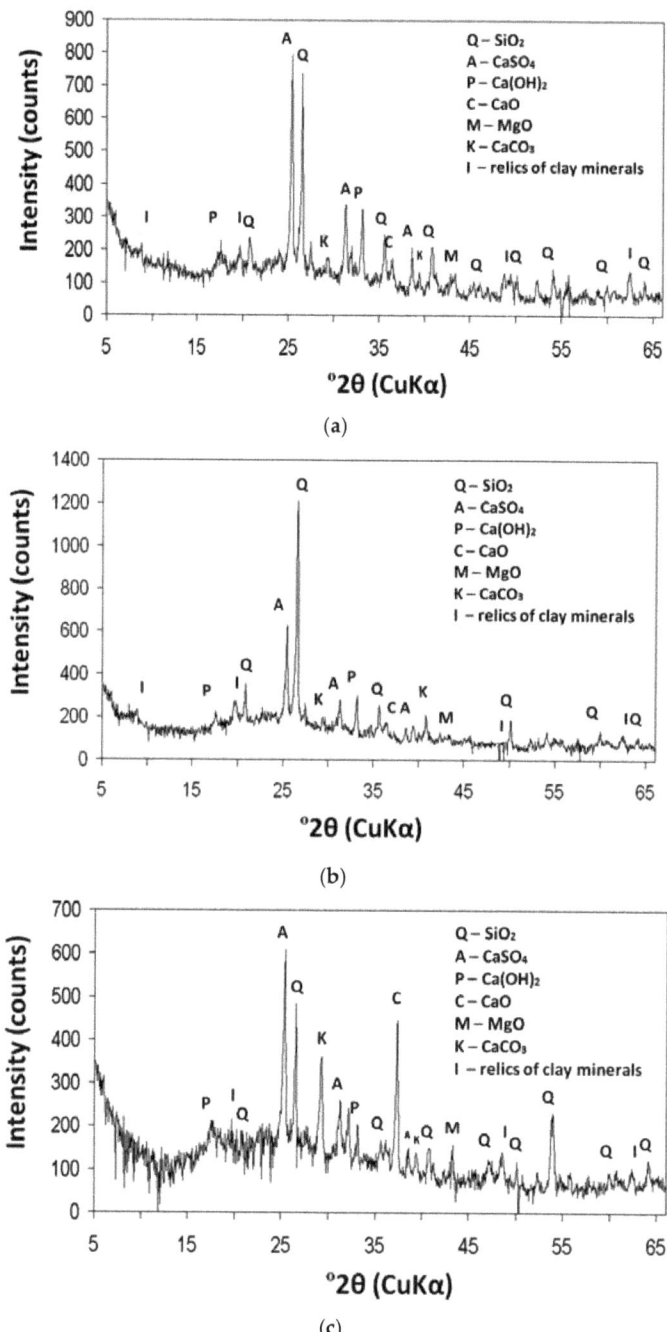

Figure 3. X-ray (XRD) patterns of CFBC fly ash from (a) S (hard coal); (b) K (hard coal); and (c) T (lignite).

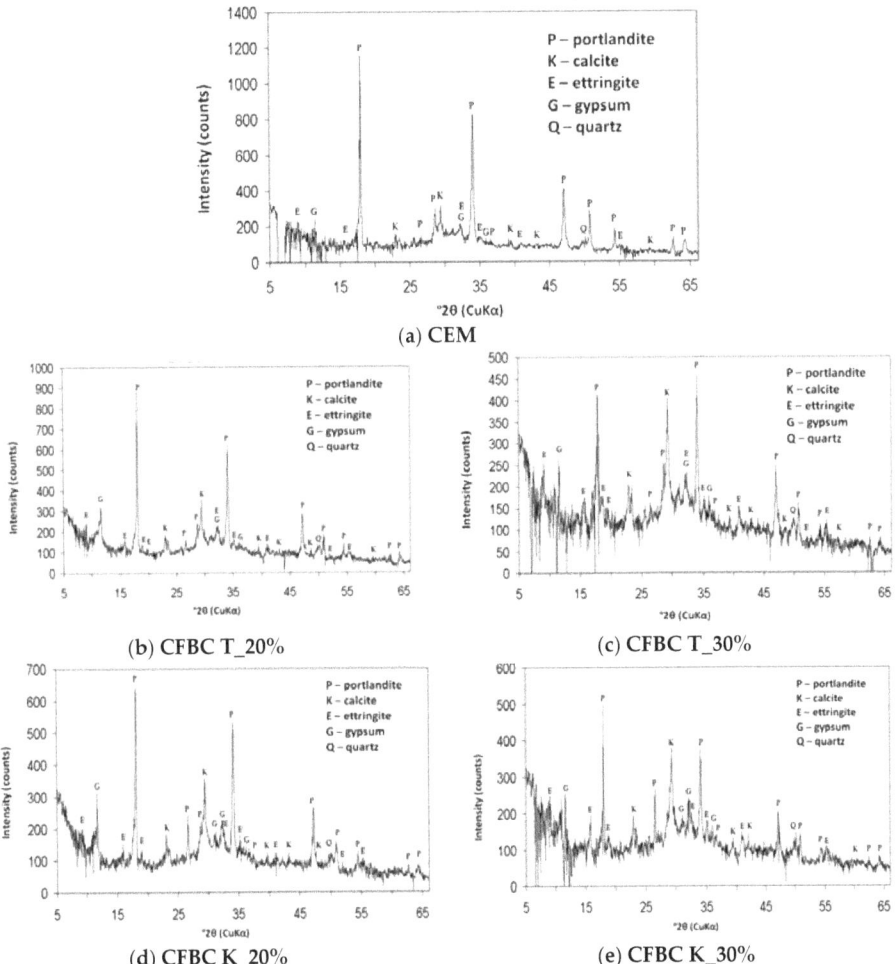

Figure 4. XRD patterns of cement pastes matured in water for 400 days. (**a**) Without any addition; (**b**) with 20% CFBC fly ash T; (**c**) with 30% CFBC fly ash T; (**d**) with 20% CFBC fly ash K; (**e**) with 30% CFBC fly ash K.

The XRD pattern revealed portlandite as a major phase in the reference cement paste specimen analyzed after 400 days curing in water. The products of the cement hydration in the forms of portlandite, gypsum and ettringite, and products of the carbonation of the hydrated cement in the form of calcite were found in the specimens made with addition of the CFBC fly ash. The content of calcite was increased with increasing fly ash content.

Value of change index WZ was defined as the ratio of selected characteristic of the tested cement paste with CFBC fly ash to the same characteristic of the reference cement paste, expressed as a percentage. As a reference, a cement paste without addition of CFBC fly ash matured in the same period of time as specimens with addition of fly ash was chosen. The results of the comparison of phase changes in the tested cement pastes are presented in Table 7. Due to the increased sulfate and calcium ion content in the analyzed fly ashes, the most relevant in the examination of cement paste are ettringite content and amount of calcium carbonate introduced. It was revealed that the increase of ettringite and calcium carbonate content in the hardened cement paste was related to the increase in CFBC fly ash content.

Table 7. XRD estimated composition of hardened cement paste after 28, 200 and 400 days of curing (CFBC T—paste with fly ash from lignite burning. CFBC K—paste with fly ash from hard coal burning. 20/30–percent content of fly ash addition).

Age, Days	Composition Parameter	Reference Paste CEM *	WZ Index for Hardened Pastes Containing CFBC Fly Ash in Relation to the Reference Paste (%)			
			T20	T30	K20	K30
28	Hydration products HI, %	17.8 = 100%	116.3	133.1	119.7	130.9
	Portlandite CH, %	16.0 = 100%	66.9	41.3	41.3	33.1
	Calcium carbonate CC, %	10.7 = 100%	100.0	91.6	112.1	103.7
	Ettringite $\sum$ I, a.u. **	326 = 100%	127.9	131.3	121.2	124.2
200	Hydration products HI, %	23.0 = 100%	111.7	104.3	110.0	104.3
	Portlandite CH, %	15.2 = 100%	53.9	34.9	53.9	27.0
	Calcium carbonate CC, %	13.6 = 100%	72.0	111.8	78.7	122.0
	Ettringite $\sum$ I, a.u. **	459 = 100%	97.8	104.1	102.8	112.9
400	Hydration products HI, %	18.0 = 100%	118.0	123.9	112.7	131.1
	Portlandite CH, %	16.4 = 100%	65.2	39.6	37.8	17.1
	Calcium carbonate CC, %	10.2 = 100%	89.2	119.6	147.1	133.3
	Ettringite $\sum$ I, a.u. **	380 = 100%	105.0	115.5	97.1	112.1

* The ordinary cement paste without fly ash addition after 28, 200 and 400 days is equal to 100%; ** Absolute intensity ettringite planes of symmetry (d = 9.73 i d = 5.61) in conventional units.

It was noted that the specimens without the addition of CFBC fly ash (i.e., reference specimens) had less than 15% and 13% ettringite content after 400 days of maturing than specimens with 30% fly ash, T30 and K30. respectively. It was surprising that the largest difference in ettringite content between reference specimens and those with addition of CFBC fly ash was visible for the first 28 days of the maturation period. Ettringite decline in subsequent periods is probably associated with the progress of hydration and transformation of ettringite in monosulfate or connected the SO_4^{2-} ions to the rising C–S–H phase with a low C/S ratio. Hydrated calcium silicates formed by the pozzolanic reaction decrease the content of the portlandite in the cement paste. This is consistent with the results regarding siliceous fly ash. It is known that the addition of the siliceous fly ash to Portland cement generally reduces the amount of portlandite and this is often accompanied with an increase in the amount of C–S–H with reduced Ca/Si ratio and AFm phases due to a higher content of Al_2O_3 in fly ash. The AFm phase of Portland cement refers to a family of hydrated calcium aluminates based on the hydrocalumite-like structure of $4CaO \cdot Al_2O_3 \cdot 13-19H_2O$, [34]. Also, the content of ettringite varies depending on the reactivity of the siliceous fly ash used. Studies to characterize the microstructure of concrete modified with addition of calcareous fly ash were performed by Glinicki et al. [35]. They showed that the addition of calcareous fly ash reduced the content of portlandite in the matrix by 45–74%. Results of tests conducted by Tishmack et al. [36] showed that the products of cement hydration incorporating calcareous fly ash included lower amounts of ettringite and higher content of AFm phases, including mainly monosulfates.

Microphotographs of the reference specimens and cement paste with addition of CFBC fly ash after 28, 200 and 400 days of maturation are presented in Figures 5 and 6. It is visible as a fine-grained and fine-porous microstructure in micro-areas occupied by C–S–H, verified by EDS analysis. The C–S–H phase created a spongy mass of the conformation of small grains, generally forming single fibrils with lengths less than 0.1 μm. Ettringite needles and micro-tubes with lengths of up to 2 μm occurred in air-voids.

The addition of the CFBC fly ash, both from hard coal and lignite burning, caused an increase in the content of ettringite, assessed on the basis of the X-ray diffraction analysis. After 28 days of curing the ettringite content in the specimens with 30% T fly ash was 31% higher and with 30% K fly ash, 24% higher than in reference specimens. The content of portlandite decreased with increasing CFBC fly ash content. In specimens with 30% ash from hard coal K burning, an increase of the content of calcium carbonate over time is clearly visible. Similar observations regarding the microstructure of paste with CFBC fly ash from hard coal burning were made by Lee and Kim [37], who investigated the hydration reactivity of the CFBC fly ash. They concluded that the microstructure of pastes with CFBC fly ash (the mixing ratio of CFBC fly ash to water was set at 1.0) after 1 day of hydration consisted mainly of fibrous ettringite and various sizes of hexagonal-plate portlandite. After 91 days, the CFBC fly ash was hydrated to a considerable degree, as the reaction ratio of the anhydrous gypsum was more than 80%. The microstructure of CFBC fly ash pastes contained portlandite, ettringite, gypsum and C–S–H [37].

The phase composition of the hardened cement paste was not affected by prolonged exposure in water and temperatures of 20 ± 2 °C. Irrespective of the content of CFBC fly ash in concrete, the microstructure of the presented cement pastes is similar to the reference specimen. It can be assumed that the efficiency of mineral additives, like fluidized bed combustion fly ash in cement paste, can be similar to other non-standard fly ash (e.g., calcareous fly ash). It was found in [38] that the addition of calcareous fly ash resulted in an improvement of concrete durability. A beneficial reduction of chloride migration coefficient was observed, while the effect on the water and air permeability was similar to its effect on the compressive strength of concrete.

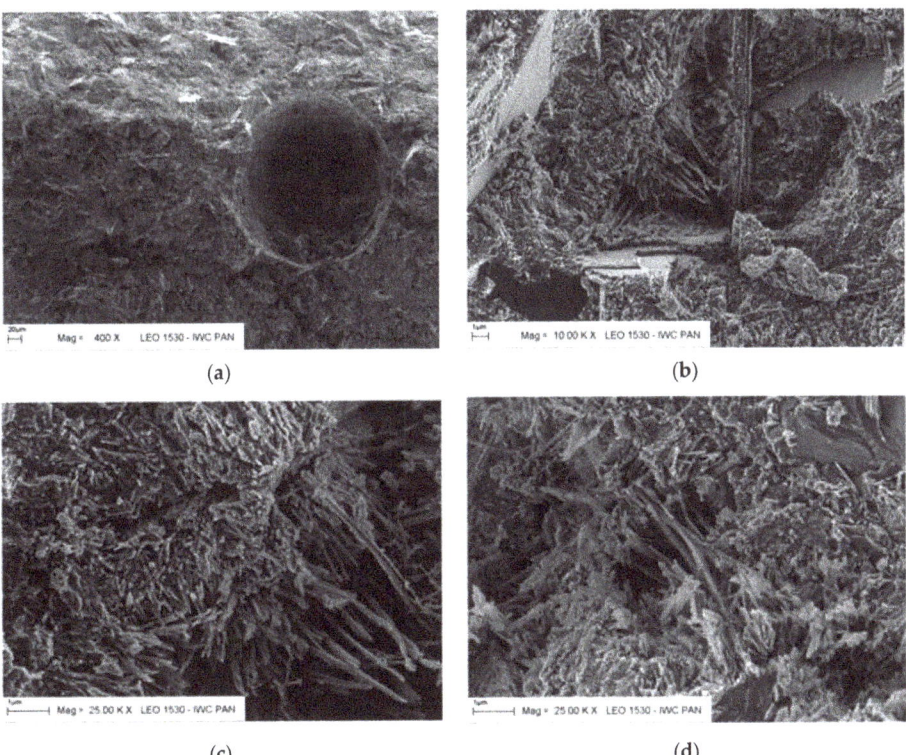

Figure 5. Microphotographs of hardened cement paste microstructure without CFBC fly ash addition, matured in water for 400 days. (**a**) Empty air void without crystalline hydration products; (**b**) C–S–H, ettringite, relics of clinker; (**c**) C–S–H and cluster of elongated ettringite needles; (**d**) C–S–H, ettringite, relics of clinker.

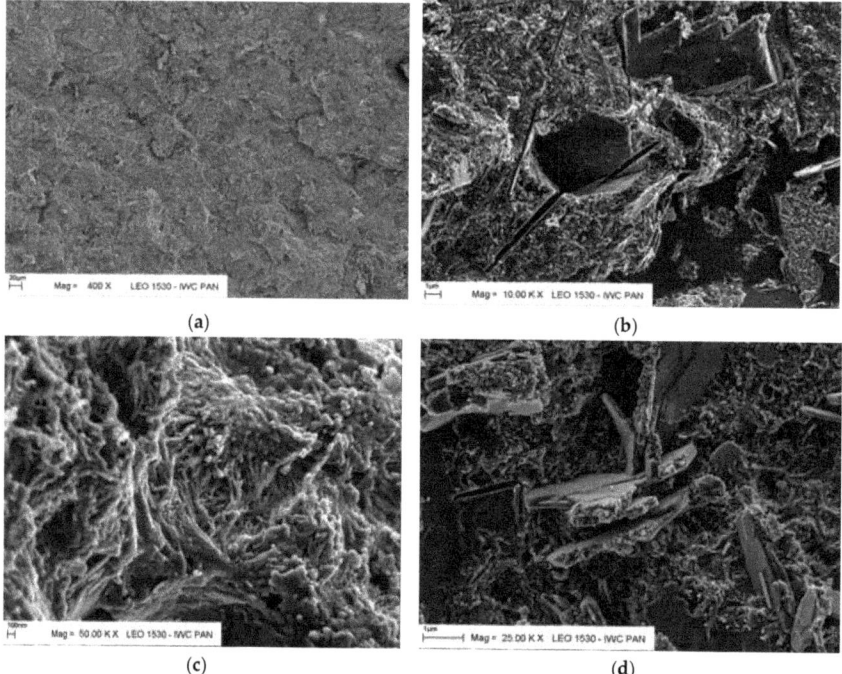

Figure 6. Microphotographs of hardened cement paste with CFBC fly ash (K30), matured in water for 400 days. (**a**) Cement paste; (**b**) C–S–H and relics of clinker, microcracks; (**c**) carbonated hydration products; (**d**) C–S–H and relics of clinker.

4. Conclusions

The performed investigation revealed the following conclusions:

- Macroscopic analysis revealed differences in colour of fluidized bed combustion fly ashes, which was assumed to be correlated to carbon content. This method can be applied for preliminary evaluation of CFBC fly ash suitability as concrete additive;
- The major components in the investigated CFBC fly ash consisted of the following (in descending order of content): SiO_2, Al_2O_3, CaO, Fe_2O_3 and SO_3.
- The largest difference between analyzed fly ash was visible in CaO content, which was the result of the type of fuel. The content of the CaO in CFBC fly ash from lignite burning was two to three times higher than in fly ash from hard coal burning.
- A proper determination of the unburned carbon content in fluidized bed fly ash required separation of $CaCO_3$, portlandite and non-hydrated clay minerals content from the loss on ignition data.
- The addition of CFBC fly ash for replacement of cement by 20% or 30% by weight did not induce significant changes in qualitative phase composition of hardened cement paste cured in water up to 400 days in regard to curing for 28 days.
- The addition of the CFBC fly ash resulted in increasing content of C–S–H gel and crystalline ettringite, which was indicated by the increase of water bound in hydration products and the decrease of portlandite content.
- The crystalline ettringite content in hardened cement paste containing 30% CFBC fly ash from lignite burning was higher by about 20% in comparison to cement paste without ash at 28 days of curing.

Author Contributions: Conceptualization, M.A.G. and D.J.-N.; methodology, M.A.G., D.J.-N. and M.D.; analysis, D.J.-N. and M.D.; investigation, M.A.G., D.J.-N. and M.D.; writing—original draft preparation, M.A.G., D.J.-N.; writing—review and editing, D.J.-N. and M.D.

Funding: This research received no external funding.

Conflicts of Interest: The authors declare no conflict of interest.

References

1. Gazdiča, D.; Fridrichováa, M.; Kulíseka, K.; Vehovská, L. The potential use of the FBC ash for the preparation of blended cements. *Procedia Eng.* **2017**, *180*, 1298–1305. [CrossRef]
2. Anthony, E.J.; Granatstein, D.L. Sulfation phenomena in fluidized bed combustion systems. *Prog. Energy Combust. Sci.* **2001**, *27*, 215–236. [CrossRef]
3. Nowak, W. Clean coal fluidized–bed technology in Poland. *Appl. Energy* **2003**, *74*, 405–413. [CrossRef]
4. Zhu, Q. *Developments in Circulating Fluidized Bed Combustion*; IEA Clean Coal Centre: London, UK, 2013; ISBN 978-92-9029-539-6. 60p.
5. Cai, R.; Ke, X.; Lyu, J.; Yang, H.; Zhang, M.; Yue, G.; Ling, W. Progress of circulating fluidized bed combustion technology in China: A review. *Clean Energy* **2017**, *1*, 36–49. [CrossRef]
6. Strigáč, J.; Števulová, N.; Mikušinec, J.; Sobolev, K. The fungistatic properties and potential application of by-product fly ash from fluidized bed combustion. *Constr. Build. Mater.* **2018**, *159*, 351–360. [CrossRef]
7. Anthony, E.J. Fluidized bed combustion of alternative solid fuels, status, successes and problems of the technology. *Prog. Energy Combust. Sci.* **1995**, *21*, 239–268. [CrossRef]
8. Brandstetr, J.; Havlica, J.; Odler, I. Properties and use of solid residue from fluidized bed coal combustion. In *Waste Materials Used in Concrete Manufacturing*; Chandra, S., Ed.; Noyes Publications Press: Westwood, NJ, USA, 1997; pp. 1–47.
9. Conn, R.E.; Sellakumar, K.; Bland, A.E. Utilization of CFB fly ash for construction applications. In Proceedings of the 15th International Conference on Fluidized Bed Combustion, Savannah, Georgia, 16–19 May 1999.
10. Roszczynialski, W.; Nocuń-Wczelik, W.; Gawlicki, M. Fly ash from fluidized bed coal combustion as complex cement addition. In Proceedings of the Three-Day CANMET/ACI International Symposium on Sustainable Development and Concrete Technology, San Francisco, CA, USA, 16–19 September 2001; pp. 415–430.
11. Marks, M.; Jóźwiak-Niedźwiedzka, D.; Glinicki, M.A.; Olek, J.; Marks, M. Assessment of scaling durability of concrete with CFBC ash by automatic classification rules. *J. Mater. Civ. Eng.* **2012**, *24*, 860–867.
12. Glinicki, M.A.; Zieliński, M. The influence of CFBC fly ash addition on phase composition of air-entrained concrete. *Bull. Pol. Acad. Sci. Tech. Sci.* **2008**, *56*, 45–52.
13. Giergiczny, Z. Fly ash and slag. *Cem. Concr. Res.* **2019**, *124*, 15. [CrossRef]
14. Fu, X.; Li, Q.; Zhai, J.; Sheng, G.; Li, F. The physical–chemical characterization of mechanically—Treated CFBC fly ash. *Cem. Concr. Compos.* **2008**, *30*, 220–226. [CrossRef]
15. Sheng, G.; Zhai, J.; Li, Q.; Li, F. Utilization of fly ash doming from a CFBC boiler co-firing coal and petroleum coke in Portland cement. *Fuel* **2007**, *86*, 2615–2631. [CrossRef]
16. Rajczyk, K.; Giergiczny, E. Research on the possibility of application of the waste products from energy. In Proceedings of the 3rd International Scientific Conference Energia i Środowisko w Technologiach Materiałów Budowlanych, Opole, Poland, 27–29 September 2004. (In Polish).
17. Lothenbach, B.; Scrivener, K.; Hooton, R.D. Supplementary cementitious materials. *Cem. Concr. Res.* **2011**, *41*, 1244–1256. [CrossRef]
18. Rajczyk, K.; Giergiczny, E.; Glinicki, M.A. Use of DTA in the investigations of fly ashes from fluidized bed boilers. *J. Therm. Anal. Calorim.* **2004**, *77*, 165–170. [CrossRef]
19. Nguyen, H.A.; Chang, T.P.; Shih, J.Y.; Chen, C.T.; Nguyen, T.D. Influence of circulating fluidized bed combustion (CFBC) fly ash on properties of modified high volume low calcium fly ash (HVFA) cement paste. *Constr. Build. Mater.* **2015**, *91*, 208–215. [CrossRef]
20. Hlaváček, P.; Šulc, R.; Šmilauer, V.; Rößler, C.; Snop, R. Ternary binder made of CFBC fly ash, conventional fly ash, and calcium hydroxide: Phase and strength evolution. *Cem. Concr. Compos.* **2018**, *90*, 100–107. [CrossRef]
21. Škvára, F.; Šulc, R.; Snop, R.; Peterová, A.; Šídlová, M. Hydraulic clinkerless binder on the fluid sulfocalcic fly ash basis. *Cem. Concr. Compos.* **2018**, *93*, 118–126. [CrossRef]

22. Lin, W.T.; Weng, T.L.; Cheng, A.; Chao, S.J.; Hsu, H.M. Properties of controlled low strength material with circulating fluidized bed combustion ash and recycled aggregates. *Materials* **2018**, *11*, 715. [CrossRef]
23. Chi, M. Synthesis and characterization of mortars with circulating fluidized bed combustion fly ash and ground granulated blast-furnace slag. *Constr. Build. Mater.* **2016**, *123*, 565–573. [CrossRef]
24. Tien Dung, N.; Chang, T.P.; Chen, C.H.T. Circulating fluidized bed combustion fly-ash-activated slag concrete as novel construction material. *ACI Mater. J.* **2014**, *111*, 105.
25. Chen, X.; Gao, J.; Yan, Y.; Liu, Y. Investigation of expansion properties of cement paste with circulating fluidized bed fly ash. *Constr. Build. Mater.* **2017**, *157*, 1154–1162. [CrossRef]
26. Jóźwiak-Niedźwiedzka, D. Effect of fluidized bed combustion fly ash on the chloride resistance and scaling resistance of concrete. In *Concrete in Aggressive Aqueous Environments—Performance, Testing, and Modeling*; Alexander, M.G., Bertron, A., Eds.; RILEM Publications SARL: Toulouse, France, 2009; pp. 556–563.
27. Kubissa, W.; Pacewska, B.; Wilińska, I. Comparative investigations of some properties related to durability of cement concretes containing different fly ashes. *Adv. Mater. Res.* **2014**, *1054*, 154–161. [CrossRef]
28. Czarnecki, L.; Woyciechowski, P.; Adamczewski, G. Risk of concrete carbonation with mineral industrial by-products. *KSCE J. Civ. Eng.* **2018**, *22*, 755–764. [CrossRef]
29. Juenger, M.C.G.; Ruben Snellings, R.; Bernal, S.A. Supplementary cementitious materials: New sources, characterization, and performance insights. *Cem. Concr. Res.* **2019**, *122*, 257–273. [CrossRef]
30. *Methods of Testing Cement. Determination of Strength*; PN-EN 196-1:2016-07; The Polish Committee for Standardization: Warsaw, Poland, 2016; 34p.
31. Šiler, P.; Bayer, P.; Sehnal, T.; Kolářová, I.; Opravil, T.; Šoukal, F. Effects of high-temperature fly ash and fluidized bed combustion ash on the hydration of Portland cement. *Constr. Build. Mater.* **2015**, *78*, 181–188. [CrossRef]
32. Hanisková, D.; Bartoníčková, E.; Koplík, J.; Opravil, T. The ash from fluidized bed combustion as a donor of sulfates to the Portland clinker. *Procedia Eng.* **2016**, *151*, 394–401. [CrossRef]
33. Krzywobłocka-Laurów, R.; Siemaszko-Lotkowska, D. Determination of the phase composition of common cements. *Instr. Build. Res. Inst.* **2006**, *419*.
34. Matschei, T.; Lothenbach, B.; Glasser, F.P. The AFm phase in Portland cement. *Cem. Concr. Res.* **2007**, *37*, 118–130. [CrossRef]
35. Glinicki, M.A.; Krzywobłocka-Laurów, R.; Ranachowski, Z.; Dąbrowski, M.; Wołowicz, J. Microstructure analysis of concrete modified with addition of calcareous fly ash. *Roads Bridges Drog. i Mosty* **2013**, *12*, 173–189.
36. Tishmack, J.K.; Olek, J.; Diamond, S. Characterization of high-calcium fly ashes and their potential influence on ettringite formation in cementitious systems. *Cem. Concr. Aggreg.* **1999**, *21*, 82–92.
37. Lee, S.H.; Kim, G.S. Self-cementitious hydration of circulating fluidized bed combustion fly ash. *J. Korean Ceram. Soc.* **2017**, *54*, 128–136. [CrossRef]
38. Gibas, K.; Glinicki, M.A.; Nowowiejski, G. Evaluation of impermeability of concrete containing calcareous fly ash in respect to environmental media. *Roads Bridges Drog. i Mosty* **2013**, *12*, 159–171.

© 2019 by the authors. Licensee MDPI, Basel, Switzerland. This article is an open access article distributed under the terms and conditions of the Creative Commons Attribution (CC BY) license (http://creativecommons.org/licenses/by/4.0/).

Article

The Effect of Wood Ash as a Partial Cement Replacement Material for Making Wood-Cement Panels

Viet-Anh Vu [1], Alain Cloutier [1,*], Benoit Bissonnette [2], Pierre Blanchet [1] and Josée Duchesne [3]

1. Department of Wood and Forest Sciences, Laval University, Quebec, QC G1V 0A6, Canada
2. Department of Civil Engineering, Laval University, Quebec, QC G1V 0A6, Canada
3. Department of Geology and Geological Engineering, Laval University, Quebec, QC G1V 0A6, Canada
* Correspondence: alain.cloutier@sbf.ulaval.ca; Tel.: +1-418-656-5851

Received: 9 August 2019; Accepted: 23 August 2019; Published: 28 August 2019

Abstract: The aim of this study was to consider the use of biomass wood ash as a partial replacement for cement material in wood-cement particleboards. Wood-cement-ash particleboards (WCAP) were made with 10%, 20%, 30%, 40%, and 50% of wood ash as a partial replacement for cement with wood particles and tested for bending strength, stiffness, water absorption, and thermal properties. Test results indicate that water demand increases as the ash content increases, and the mechanical properties decrease slightly with an increase of the ash content until 30% of replacement. On the other hand, the heat capacity increases with the wood ash content. The WCAP can contribute to reducing the heat loss rate of building walls given their relatively low thermal conductivity compared to gypsum boards. The replacement of cement to the extent of approximately 30% by weight was found to give the optimum results.

Keywords: biomass; wood ash; fibrocement; strength; mortar

1. Introduction

Fiber cement panels have been on the market for a long time. Originally, asbestos was used as the reinforcing material, but due to the health hazards involved, it was replaced by cellulose in the 1980s. Nowadays, these panels are used as exterior siding, roof shingles, and tiles for exterior applications. Wood-cement particleboard has several benefits, since it is resistant to termites, does not rot, is impact resistant, and has fireproof properties. However, studies carried out on the compatibility of wood with cement [1–3] show that not all species are equally suited for the manufacturing of wood-cement particleboard. Softwood species actually show the greatest potential for this type of application. The results of Tittelein et al. [4] show that it is possible to make low-density (specific gravity of about 0.7) wood-cement particleboards with better bending properties than gypsum boards and a screw-withdrawal resistance that is 1.7 times higher. Moreover, these panels can be cut with a knife in a similar manner as gypsum boards. Therefore, the panel installation process is essentially the same. Thanks to its high porosity, the thermal conductivity of wood-cement particleboards is about three times lower than that of gypsum boards.

Environmental concerns and economic pressure are amongst the driving forces of today's industrial development. Therefore, several research projects are being conducted worldwide on the use of waste materials to reduce threats to the environment and to streamline present waste disposal and recycling methods by making them more affordable [5].

Manufacturing of ordinary Portland cement (OPC) ranks third in the world among the producers of anthropogenic CO_2, after transport and power generation. The emission of CO_2 by the cement industry represents 5%–7% of the total worldwide CO_2 emissions from fuel combustion and industrial activities [6]. The use of additives and substitutes to OPC has been so far one of the most successful solutions to decrease CO_2 emission generated by cement production.

Wood ash (WA) is produced by the combustion of wood in domestic wood stoves or in industrial power generation plants. At the end of the 80s, an estimated quantity of 45,000 tons of wood ash was produced annually in the Province of Québec, Canada by the pulp and paper industry [7]. In 2006, more than 300,000 tons of wood ash were produced per year, two-thirds coming from pulp and paper plants and the remaining from cogeneration plants, sawmills, and other wood-related industries. WA chemical characteristics differ with species of wood, but it mainly contains lime and silica [8]. Ash production is likely to expand further with the increasing interest for bioenergy.

In 2007, 150,000 tons of residual ash were used as fertilizers in Quebec [9]. Most of the residual ash (54%) was used in agriculture. The rest was used for the revegetation of degraded sites, soil mix manufacturing, composting, and other uses. Half of the wood ash resource produced annually is still landfilled [9]. When favorable conditions are met, the wood ash may have some pozzolanic potential that can be taken advantage of in Portland cement-based systems.

Several studies have investigated the suitability of wood ash as a supplementary cementing material in the production of ordinary and self-compacting concretes. Subramaniam [10] reported an optimum dosage of wood ash of 15% in the replacement of cement (by weight) for the production of concrete having a sufficiently high compressive strength for the casting of blocks. Abdulladi [11] found an optimum replacement rate of 20% and showed that the water requirement increases as the wood ash content increases. Chowdhury et al. [12] characterized the mechanical strength (compression, tensile, and flexural) of concrete incorporating wood ash. The presence of essential pozzolanic compound (as required by the ASTM C618-15 standard), the content in small size particles, and the large surface area of the particles qualify the wood ash investigated in their study as a pozzolanic material.

The aim of the present study was to evaluate the physical, thermal, and mechanical properties of wood-cement particleboards prepared using wood ash as a supplementary cementing material.

2. Materials and Methods

2.1. Materials

The main binder used was an ordinary CSA (Canadian Standards Association) type 10 (GU, General Use) Portland cement.

The wood ash selected for investigation was supplied from the thermal energy production plant of "La Cité Verte", a residential development in Quebec, QC, Canada.

The wood-cement mixtures were prepared with air-dried wood chips obtained from white spruce (*Picea glauca* (Moench), Voss, Norway) trees harvested at the Petawawa Research Forest in Mattawa (ON), Canada. The wood chips were refined in a Pallmann PSKM8-400 ring refiner (Ludwig Pallmann K.G, Zweibrücken, Germany). The particles supplied were screened, and those ranging between 1 and 3 mm in size were retained.

2.2. Wood-Cement Mixtures

The wood-cement particleboard mixtures were all prepared with a wood-to-binder ratio of 0.35 by weight, where the binder phase is the sum of cement and wood ash. A total of six mixtures were investigated, the variables being essentially the fraction of cement replaced by wood ash. Assessing mixtures with different percentages of wood ash was intended to determine the maximum amount of wood ash that could be used without significantly affecting the properties of the material in comparison with those of the reference wood-cement mixture. The corresponding mixtures are referred to as P0, P1, P2, P3, P4, and P5, respectively. The control mixture (P0) was prepared with cement and wood particles only, while mixtures P1, P2, P3, P4, and P5 were prepared by incorporating wood ash as a partial replacement of cement at a rate of 10%, 20%, 30%, 40%, and 50%, respectively.

The mixing sequence was observed to have a critical influence upon the material rheology, with slight changes altering the fresh mixture behavior significantly. The mixing sequence retained after the preliminary tests is presented in Table 1.

Table 1. Mixing sequence.

Step	Mixer Rotor Speed (rpm)	Cumulative Time (s)
1. Addition of cement and wood ash	140	0
2. Addition of water	140	60
3. Addition of wood particles	140	120
4. Change of speed	285	180
5. End of mixing	0	270

Directly after mixing, the workability of each mixture was determined using the slump test in accordance with the ASTM C143/C143M-15a standard [13].

2.3. Preparation of Test Specimens

After mixing in a mortar mixer (HOBART A-120, Hobart Canada Inc, Don Mills, ON, Canada), each wood-cement-ash-water mixture was cast into a 450 × 330 × 15 mm wooden mold. After pouring the mixture, the mold was closed with a lid held in place by C-clamps. This set-up allowed to pour material up to a thickness of 15 mm. The wet mixture was poured into the mold, the surface was levelled off with a wood screed and the lid was finally secured in place. From the pressure of the lid, the panel thickness was reduced to 14 mm. The hardened panels were stripped from the mold at the age of 3 days and then stored in a conditioning chamber at 23 °C and 60% relative humidity. The various test specimens were cut from the panels (3 panels per mixture) on the day of testing.

2.4. Test Methods

The panels were cured and tested to determine their mechanical performance after 3, 7, and 28 days of curing following the ASTM D 1037-12 standard [14]. The bending modulus of rupture (MOR) and the modulus of elasticity (MOE) were determined at the same ages by MTS QTest-5 Universal Test Frames (MTS systems corporation, Eden Prairie, MN, USA) featuring The Elite Modular Control System. Screw-withdrawal resistance, water absorption, and thickness swelling were also tested following the ASTM D 1037-12 standard [14]. The thermal properties of the wood-cement particleboards were measured by FOX 314 Heat Flow Meter (TA instruments-LaserComp Inc, Wakefield, MA, USA) following the ASTM C518 [15] standard. The board was placed between two plates at a regulated temperature and a flux meter was glued on each side of it so that temperature and heat flux could be measured at the board surface, which can be submitted to temperature variations. Heat capacity and thermal conductivity can be calculated from these four parameters (two temperatures and two heat fluxes). The solubility of the WA was evaluated by the mass loss measured on 15 g of WA placed in 100 mL of distilled water and stirred for one hour at 23 °C. The residue is then filtered under vacuum and rinsed with distilled water. The residue of WA is placed in an oven overnight then the loss of mass is measured. The soluble proportion corresponds to the average mass loss of tree samples. Finally, solid samples were observed under a JEOL JSM-840A Scanning Electron Microscope (JEOL USA Inc, Peabody, MA, USA) (SEM) equipped with an energy dispersive X-ray analysis system (EDS). For SEM observations, the specimens were mounted intact on double-sided adhesive tape and coated with a thin alloy of Au-Pd. Operating conditions were set at 15 kV.

3. Results

3.1. Material Characterization

3.1.1. Wood Particles

Wood particle size distribution was evaluated using five sieve sizes: 1.19, 1.4, 1.7, 2.38, 2.8, and 3 mm. According to the results shown in Figure 1, 100% of the particles were less than 3 mm in size, and particles with a diameter of 1.7 mm make for the highest mass fraction (57%).

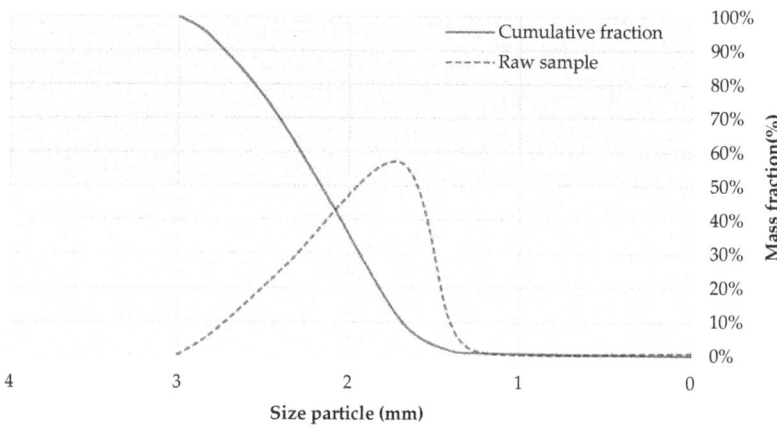

Figure 1. Particle size distribution of wood particles.

3.1.2. Wood Ash

Particle Size and Shape Analysis

Shape analysis by scanning electron microscopy observations revealed that the ash particles were irregular in shape and spherical (Figure 2b). Wood ash is suitable for use as a filler/partial replacement of cement in high-performance concrete due to an enhanced "ball bearing" effect given from the spherical shape of WA. The "ball bearing" effect of wood ash creates a lubricating effect when concrete is in its plastic state. According to the results shown in Figure 3, the D10, D50, and D90 values of the WA were 2.5, 18.5, and 114.1 µm, respectively. Wood ash contains an amount of ultrafine particles of 18% (particle diameter ϕ < 5 µm).

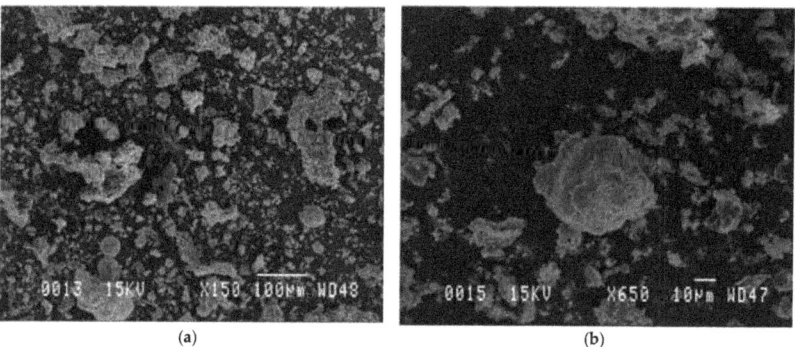

Figure 2. Low magnification (**a**) and high magnification (**b**) scanning electron microscopy of wood ash (WA).

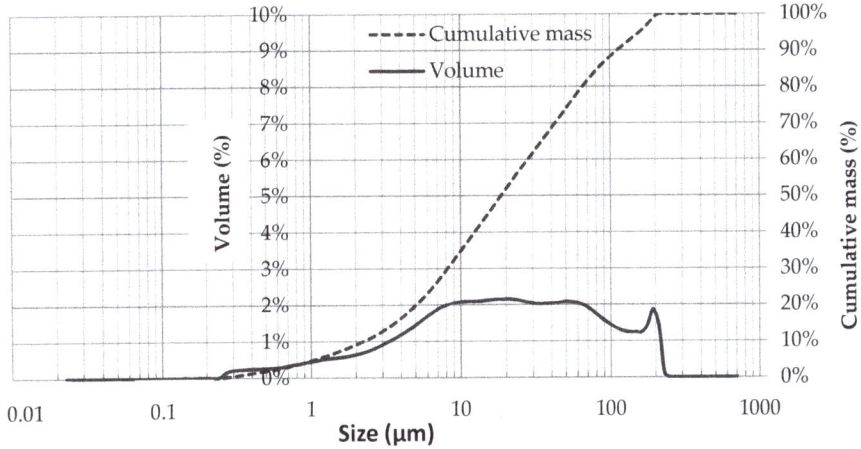

Figure 3. Particle size analysis of WA.

Chemical Composition

The results of the chemical analysis carried out on the investigated wood ash are shown in Table 2. The combined content in iron oxide (Fe_2O_3 = 1.22%), aluminum oxide (Al_2O_3 = 2.25%), and silicium dioxide (SiO_2 = 7.80%) is found to be 11.27%, which is considerably less than the minimum amount required to qualify a material as a pozzolan, established at 70% [16].

Table 2. Physical and chemical properties of wood ash.

Properties	Value	Chemical Composition (%)	
Conventional parameters		SiO_2	7.80
Organic material (mg/kg)	<10	Al_2O_3	2.25
pH	13	Fe_2O_3	1.22
		MgO	7.47
Physical properties		CaO	46.70
Density (kg/m^3)	2970	Na_2O	0.86
Specific surface (m^2/kg)	261	K_2O	9.61
		TiO_2	0.11
		MnO	4.51
		P_2O_5	2.34
		Cr_2O_3	<0.01
		V_2O_5	<0.01
		ZrO_2	<0.02
		ZnO	0.04
		Loss on ignition	14.20

The recorded loss on ignition at 950 °C was 14.2%, which exceeds the 12% maximum requirement for pozzolans [16]. This means that the ash contains a significant amount of unburnt carbon, which reduces its pozzolanic activity. The alkali content (%Na_2O + 0.658 × %K_2O) was found to be 7.18%, a value higher than the maximum alkali content of 1.5% required for pozzolana. The specific gravity of wood ash was found to be 2.97, which is far less than the Portland cement density (3.15). WA contains more than 99% (by weight) of inorganic material and yields a pore solution with a high pH.

Solubility Test

Table 3 shows the percentage of the wood ash dissolved in water during the solubility test. The solubility of WA is estimated to be 7% including lime and alkali hydroxides that are readily soluble in water in laboratory conditions. This soluble component plays an important role in the hydration reaction.

Table 3. Solubility test of wood ash in water.

	Wood Ash (g)	Mass Loss (g)	Material Dissolved (%)
1	14.10	0.90	6.30
2	15.00	1.20	8.00
3	14.30	0.90	6.30
	Average		6.90

3.2. Change in Density

The weight of all panels was recorded at the beginning and at the end of the curing period (3 days in the mold) to determine changes in the panel-specific gravity. It decreased by about 5% during that period, owing to the fact that the mold being used was not perfectly impervious. Some water was probably absorbed by the mold itself, as it was made of plywood.

The panel mass reached a plateau about 6 days after removal from the mold, meaning that most of the free water in the cement paste had evaporated in the conditioning chamber at 23 °C and 60% RH by then.

3.3. Workability

Table 4 shows the results obtained for the consistency test. The results reveal that the water demand increased with the wood ash content. The wood ash introduced into the cement increased the carbon content, thereby increasing the amount of water required to achieve satisfactory workability.

Table 4. Consistency test results.

Mass Ratio	P0	P1	P2	P3	P4	P5
Wood ash/Cement	0.00	0.10	0.20	0.30	0.40	0.50
Wood/Binder	0.35	0.35	0.35	0.35	0.35	0.35
Water/Binder	1.00	1.04	1.08	1.12	1.16	1.20

3.4. Bending Properties of the Raw Wood-Cement Particleboard

As described previously, the panels were tested in bending at 3, 7, and 28 days after manufacturing. Each test was performed on three specimens and the mean value is presented in Table 5.

Table 5. Average bending strength test results of wood-cement-ash particleboards (WCAP). Mean values with the same superscript are not significantly different for $p = 0.05$; standard deviation is given in parentheses.

		P0	P1	P2	P3	P4	P5
3 days	MOR (MPa)	$0.92_{(0.16)}$	$0.85_{(0.04)}$	$0.75_{(0.02)}$	$0.68_{(0.07)}$	$0.53_{(0.04)}$	$0.35_{(0.08)}$
	MOE (GPa)	$1.04_{(0.21)}$	$0.90_{(0.21)}$	$0.84_{(0.24)}$	$0.75_{(0.08)}$	$0.58_{(0.07)}$	$0.54_{(0.08)}$
7 days	MOR (MPa)	$1.35_{(0.21)}$	$1.28_{(0.24)}$	$1.22_{(0.17)}$	$1.15_{(0.17)}$	$0.74_{(0.05)}$	$0.43_{(0.05)}$
	MOE (GPa)	$1.12_{(0.14)}$	$1.12_{(0.15)}$	$1.05_{(0.13)}$	$1.01_{(0.18)}$	$0.87_{(0.03)}$	$0.70_{(0.08)}$
28 days	MOR (MPa)	$1.36^{(x)}_{(0.32)}$	$1.30^{(x)}_{(0.33)}$	$1.24^{(x)}_{(0.21)}$	$1.20^{(x)}_{(0.16)}$	$0.78^{(y)}_{(0.25)}$	$0.47^{(z)}_{(0.21)}$
	MOE (GPa)	$1.40^{(a)}_{(0.17)}$	$1.39^{(a)}_{(0.12)}$	$1.07^{(b)}_{(0.07)}$	$1.12^{(b)}_{(0.12)}$	$0.82^{(c)}_{(0.14)}$	$0.50^{(d)}_{(0.24)}$

Table 5 and Figure 4 show the bending behavior of the WCAP at different curing times. It shows that the bending strength and stiffness values of the sample panels increase with the curing time. They changed little after 7 days of curing, as generally observed for Portland cement-based materials. The statistical analysis results showed that there is a significant difference among samples in terms of bending strength and stiffness at all stages of curing (3 days of curing: $p < 0.001$, 7 days of curing: $p < 0.001$, 28 days of curing: $p < 0.05$). The bending strength and stiffness of P4 and P5 were significantly lower than for the other panels at all curing stages. Optimum bending strength observed in these tests was obtained at 30% wood ash replacement (P3) after 28 days of moist curing.

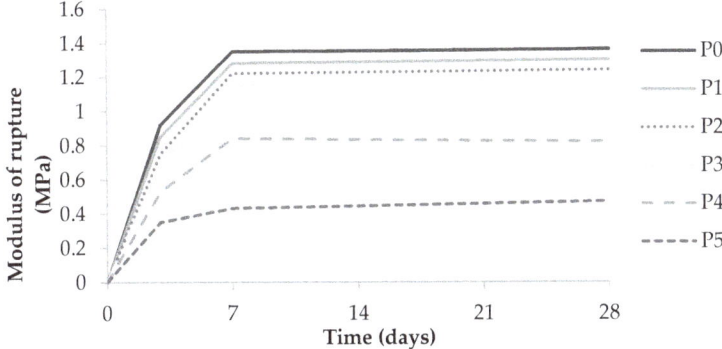

Figure 4. Evolution of the modulus of rupture in bending of WCAP as a function of the moist curing duration.

3.5. Screw-Withdrawal Resistance

Figure 5 shows the screw-withdrawal resistance of WCAP as a function of the WA content. It shows that the screw-withdrawal resistance decreases as the WA replacement rate increases. The results of the statistical analysis show that the screw withdrawal-resistance is slightly affected up to a replacement rate of 30% in wood ash. However, beyond that value, it decreases rapidly.

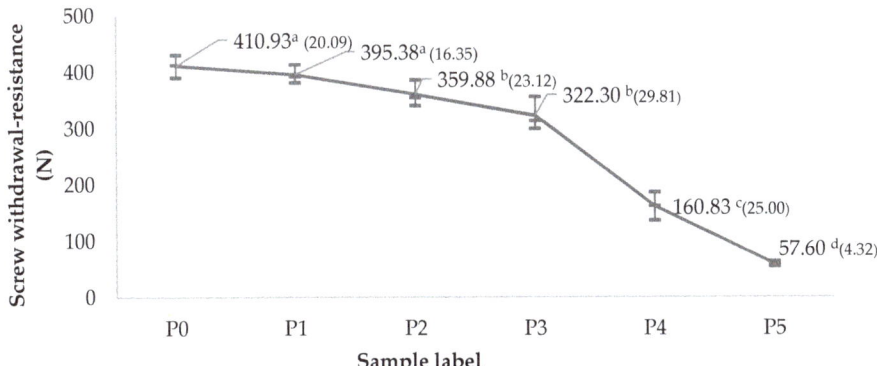

Figure 5. Effect of wood ash replacement rate on the screw-withdrawal resistance of the WCAP (mean values with the same superscript are not significantly different for $p = 0.05$; standard deviation is given in parentheses).

3.6. Water Absorption

The water absorption test results are shown in Figure 6. The value of water absorption increases with the percentage of WA replacement and time of immersion in water. Table 6 shows that the thickness swelling of WCAP in water is small (<2%). According to the results, the water absorption of all boards incorporating wood ash is higher than that of the control sample after 28 days of curing.

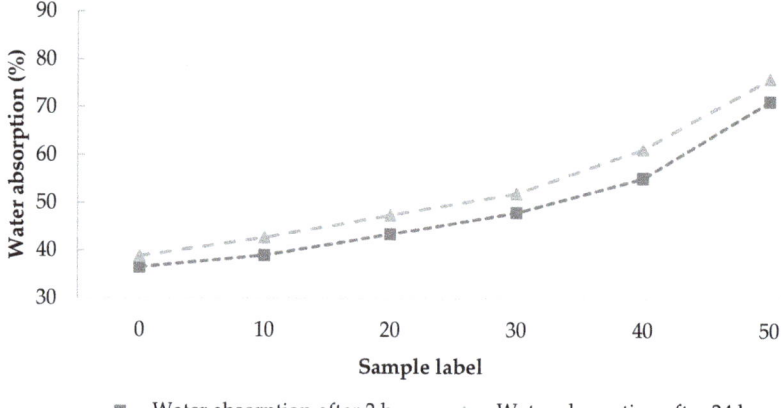

Figure 6. Water absorption and thickness swelling of WCAP recorded as a function of the WA content.

Table 6. Average water absorption and swelling of WCAP as a function of the WA content.

		P0	P1	P2	P3	P4	P5
Water absorption (%)	2 h	36.5	39.0	43.4	48.0	60.3	76.9
	24 h	38.8	42.7	47.5	52.0	61.6	76.1
Thickness swelling (%)	2 h	0.4	0.8	0.5	0.9	0.9	0.7
	24 h	2.0	0.9	0.7	1.6	1.6	0.8

3.7. Thermal Properties

Table 7 shows the WCAP heat capacity and thermal conductivity test results. It is interesting to note that the heat capacity increases with the wood ash content. It can contribute to reducing the heat loss rate of building walls given its relatively low thermal conductivity when used as interior partition. Wood ash level P3 yields a heat capacity 7% higher than that of the control panel. Conversely, the thermal conductivity does not change importantly between 0% and 30% wood ash replacement levels.

Table 7. Average thermal properties and density of WCAP as a function of the WA content.

	P0	P1	P2	P3	P4	P5
Specific gravity	0.63	0.61	0.59	0.57	0.43	0.39
Thermal conductivity (W/m·K)	0.13	0.12	0.12	0.11	0.08	0.07
Heat capacity (J/g·K)	1304	1334	1368	1390	1424	1470

3.8. Microstructure of Mortars

According to the results shown in Figure 7, there are no clear differences in microstructure between the two samples. They both exhibit a low porosity and pore sizes smaller than 10 μm. The occurrence of spherical particles that have the shape of WA can be observed in Figure 7b as shown by the white arrows.

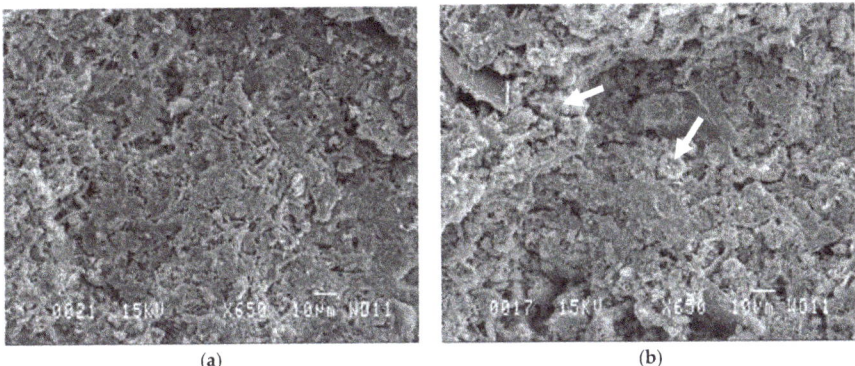

Figure 7. Scanning electron microscopy images of cement control (**a**) and cement + 30% WA (**b**).

4. Discussion

Although the investigated wood ash does not qualify as a pozzolan, it can be used in replacement of cement up to significant amounts without affecting the physical and mechanical properties of the wood-cement particleboards significantly. In previous studies, maximum wood ash proportions in the order of 15%–20% were reported [10,11]. Compared to the control sample (P0), WCAP prepared with 30% of wood ash in replacement of cement (P3) showed moderate mechanical properties reductions of 10% for bending MOR and 21% for screw-withdrawal resistance. The pH value increases with the hydration of the cement. A high alkaline solution promotes the reactivity of the silica present in the WA, which enhances the pozzolanic activity at the initial stage. Increased pH levels favor the formation of hydrous silica. This compound reacts with Ca^{2+} ions and produces insoluble compounds, which are secondary cementitious products [10]. Moreover, WA can act as a filler in the mixtures.

The density of the samples is found to decrease as the WA replacement rate increases, due to the slightly lower density of the ash and, most importantly, the increased amount of water (Tables 4 and 7). As a result of the larger volume of capillary pores, the mechanical and physical properties including density decline. Indeed, water absorption increased significantly from 30% of WA in replacement. It can be explained by the lower amount of cement particles with increasing wood ash contents. Therefore, the hydration reaction was reduced, and the water evaporated quickly in a porous medium with high porosity due to the presence of the wood fibers.

A fraction of the ash of about 7% dissolves in water and contributes to the hydration process. The large surface area associated to the ash particles could also be a factor, as it acts to some degree as nucleation sites for cement hydration. Indeed, based upon SEM examination, no significant difference in the microstructure of a mixture of neat cement and a mixture containing 30% of WA in replacement was found, both exhibiting a dense and uniform microstructure.

The increase in the heat capacity of WCAP after replacement of cement by wood ash has shown that it has the potential to reduce the heat losses of building walls, given the improved insulation it provides. Indeed, WCAP has a low thermal conductivity, about three times lower than that of gypsum boards (0.32 W/m·K) [4]. This low thermal conductivity is mainly due to the higher WCAP porosity compared to that of gypsum because the thermal conductivity of empty voids is very low (about 0.025 W/m·K).

5. Conclusions

This project studied the physical, thermal, and mechanical properties of wood-cement particleboards incorporating wood ash. Wood ash was found to have an excellent potential for use as partial replacement to Portland cement. Based on the results generated in this study, the optimum replacement rate is about 30% by weight. At this replacement level, the engineering properties of

WPCA were moderately reduced (bending MOR by 12%; bending MOE by 20%; screw-withdrawal resistance by 21%) compared to a neat wood-cement control sample. Beyond 30% in replacement, the mechanical and physical properties start to decrease at a significantly higher rate (bending MOR by 43%, bending MOE by 41%, and screw-withdrawal resistance by 60% at a 40% replacement rate). The use of wood ash improves the heat capacity of the WCAP by 11% compared to a neat wood-cement control sample.

The work reported herein is quite promising in view of producing eco-friendly wood cement panels with improved characteristics compared to those of standard gypsum boards. Future work should include the fire-resistance and acoustic properties measurement of this material. The formulation and the processing phases could also be further improved. Notably, the use of a paper surface layer should be studied to enhance the mechanical properties of the panel.

Author Contributions: Conceptualization, V.-A.V., A.C., B.B. and P.B.; Data curation, V.-A.V.; Formal analysis, V.-A.V.; Funding acquisition, P.B.; Investigation, V.-A.V., A.C., B.B., P.B. and J.D.; Methodology, V.-A.V., A.C., B.B., P.B. and J.D.; Project administration, A.C. and P.B.; Supervision, A.C. and B.B.; Validation, A.C., B.B., P.B. and J.D.; Writing—original draft, V.-A.V.; Writing—review & editing, A.C., B.B., P.B. and J.D.

Funding: This work is part of the research program of the Natural Sciences and Engineering Research Council of Canada (NSERC) Industrial Research Chair on Eco-Construction in Wood (CIRCERB) through programs IRC (IRCPJ 461745-12) and CRD (RDCPJ 445200-12).

Acknowledgments: The authors are also grateful to the industrial partners of the NSERC Industrial Chair on Eco-Responsible Wood Construction (CIRCERB) and the SSQ insurance company for providing wood ash from "La Cité Verte".

Conflicts of Interest: The authors declare no conflict of interest.

References

1. Sauvat, N.; Sell, R.; Mougel, E.; Zoulalian, A. A study of ordinary Portland cement hydration with wood by isothermal calorimetry. *Holzforschung* **1999**, *53*, 104–108. [CrossRef]
2. Sha, W.; O'Neill, E.; Guo, Z. Differential scanning calorimetry study of ordinary Portland cement. *Cem. Concr. Res.* **1999**, *29*, 1487–1489. [CrossRef]
3. Maurice, D.; Cloutier, A.; Bernard, R. Wood-cement compatibility of some Eastern Canadian woods by isothermal calorimetry. *For. Prod. J.* **2004**, *10*, 49.
4. Tittelein, P.; Cloutier, A.; Bissonnette, B. Design of a low-density wood–cement particleboard for interior wall finish. *Cem. Concr. Compos.* **2012**, *34*, 218–222. [CrossRef]
5. Rajamma, R.; Senff, L.; Ribeiro, M.J.; Labrincha, J.A.; Ball, R.J.; Allen, G.C.; Ferreira, V.M. Biomass fly ash effect on fresh and hardened state propeties of cement bases material. *Compos. Part B Eng.* **2015**, *77*, 1–9. [CrossRef]
6. Barcelo, L.; Kline, J.; Walenta, G.; Gartner, E.M. Cement and carbon emissions. *Mater. Struct.* **2013**, *47*, 1055–1065. [CrossRef]
7. AIFQ. *Pourquoi Gaspiller nos Déchets*; Association des Industries Forestières du Québec ltée: Québec, QC, Canada, 1990.
8. Swaptik, C.; Mishra, M.; Om, S. The incorporation of wood waste ash as a partial cement replacement material for making structural grade concrete: An overview. *Ain Shams Eng. J.* **2015**, *6*, 429–437.
9. Hébert, M.; Busset, G.; Groeneveld, E. *Bilan 2007 De La Valorisation Des Matières Résiduelles Fertilisantes*; Government of Quebec: Quebec, QC, Canada, 2008.
10. Subramaniam, P.; Subasinghe, K.; Fonseka, W.R.K. Wood ash as an effective raw material for concrete blocks. *Int. J. Res. Eng. Technol.* **2015**, *4*, 228–233.
11. Abdullahi, M. Characteristics of wood ash/OPC concrete. *Leonardo Electron. J. Pract. Technol.* **2006**, *8*, 9–16.
12. Chowdhury, S.; Maniar, A.; Suganya, O. Strength development in concrete with wood ash blended cement and use of soft computing models to predict strength parameters. *J. Adv. Res.* **2015**, *6*, 907–913. [CrossRef] [PubMed]
13. ASTMC143-15. *Standard Test Method for Slump of Hydraulic-Cement Concrete, American Standard Test of Materials*; ASTM International: West Conshohocken, PA, USA, 2015.

14. ASTMD1037-12. *Standard Test Method for Evaluating the Properties of Wood-Base Fiber and Particle Panel Materials, American Standard Test of Materials*; ASTM International: West Conshohocken, PA, USA, 2012.
15. ASTMC518-17. *Standard Test Method for Steady-State Thermal Transmission Properties by Means of the Heat Flow Meter Apparatus, American Standard Test of Materials*; ASTM International: West Conshohocken, PA, USA, 2017.
16. ASTMC618-15. *Standard Specification for Coal Fly Ash and Raw or Calcined Natural Pozzolan for Use in Concrete American Standard Test on Materials*; American Standard Test of Materials; ASTM International: West Conshohocken, PA, USA, 2015.

© 2019 by the authors. Licensee MDPI, Basel, Switzerland. This article is an open access article distributed under the terms and conditions of the Creative Commons Attribution (CC BY) license (http://creativecommons.org/licenses/by/4.0/).

Article

Prediction of Carbonation Progress in Concrete Containing Calcareous Fly Ash Co-Binder

Piotr Woyciechowski [1], Paweł Woliński [2] and Grzegorz Adamczewski [1,*]

1. Department of Building Materials Engineering, Warsaw University of Technology, 00-637 Warszawa, Poland
2. Faculty of Applied Sciences, Collegium Mazovia Innovative School, 08-110 Siedlce, Poland
* Correspondence: p.woyciechowski@il.pw.edu.pl; Tel.: +48-602-444-978

Received: 18 July 2019; Accepted: 19 August 2019; Published: 21 August 2019

Abstract: According to the European Standards (EN 450-1, EN 206), it is not permissible to use calcareous fly ash as an additive to concrete. However, other standards (for example, the American and Canadian ones) allow the use of high-calcium fly ash (type C) in concrete. As a result of brown coal combustion, a large amount of this type of fly ash is produced, and considerations on their use in concrete are in progress. Research into the influence of high-calcium fly ash on concrete durability is fundamental for dealing with that issue. The aim of the present research was to develop a new model of carbonation over time, also including calcareous fly ash content in the binder. The self-terminating model of carbonation is new, and not developed by other authors. In the current research, the former simplest model (a function of w/c ratio and time) is expanded with the calcareous fly ash to cement ratio. The basis is a statistically planned experiment with a large scope of two material variables (w/c ratio and fly ash to cement ratio). The main measured property is the carbonation depth after exposure to 4% of CO_2 concentration (according to CEN/TS 12390-12). The model of carbonation obtained from this experiment is an output of the paper. Also, the idea of developing similar models for concrete families as a tool for designing concrete cover thickness for reinforced elements is described in the paper.

Keywords: concrete; durability; carbonation modeling; calcareous fly ash

1. Introduction

A common cause of damage to concrete structures with steel reinforcement is the corrosion of steel resulting from insufficient protection. The protective abilities of concrete decrease due to the effect of physical and chemical factors over time. An important factor is carbonation, which decreases the concrete's pH value. The approach of using the carbonation model elaborated in the research (on the basis of the collected data) and statistic curve-fitting for the results obtained for the tested concrete, enables us to design the concrete cover thickness for individual cases on the basis of the actual protective abilities of the concrete used.

The development of universal carbonation models for various types of concrete can lead to the creation of useful tools for designing durable structures in XC (carbonation threat) classes of exposure according to Eurocode EC2.

The course of carbonation in typical atmospheric conditions depends primarily on the material characteristics of the concrete, including the w/c ratio and the qualitative as well as quantitative composition of the binder. Most of the test results confirm that the use of fly ash in appropriate proportions not only impairs the protective properties of concrete against reinforcing steel [1,2], but also increases the concrete tightness [3]. Especially when chlorides from seawater or deicing salts, etc. cause a threat, the introduction of ash to cement or concrete can be very beneficial. However, in a heavily contaminated environment with high concentrations of carbon dioxide and chlorides, fly ash used as a partial substitution for the cement can accelerate the corrosion process of the reinforcement. For these reasons, the use of fly ash in concrete technology should be approached with great caution and preceded by a series of tests confirming the correctness of the solution [1].

The resistance to carbonation of concretes containing fly ash depends on its microstructure. Fly ash partially binds $Ca(OH)_2$ due to the pozzolanic reaction. It leads to a decrease in the content of $Ca(OH)_2$ in concrete. Due to the carbonation depth being higher, the same amount of CO_2 could carbonate a larger volume of concrete, as Neville states [4] on the basis of Bier's research [5]. Bier's study proves that the carbonation rate is higher when the quantity of $Ca(OH)_2$ in the cement paste is lower. In such a way, the fly ash presence in concrete may accelerate the rate of carbonation and increase the carbonated zone in concrete. However, the reverse effect connected with the formation of a denser, hardened paste with fly ash could also be observed. This leads to a reduction in the diffusivity and rate of carbonation limitation. Therefore, it could be concluded that the influence of fly ash on carbonation includes two contradictory effects:

- accelerating—connected with the shortage of $Ca(OH)_2$ used for the pozzolanic reaction, which effects a deeper diffusion of carbon dioxide due to the concentration gradient;
- inhibitory—connected with the denser microstructure of the paste thanks to the physical effect of fine grains of ash and the chemical effect of the products of the pozzolanic reaction.

The subject of the research presented in this article is concretes made with fly ash, classified as calcareous according to PN-EN 197-1 [6] and according to the USA [7] and Canadian [8] standards, due to the high content of reactive lime, i.e., 15% by mass. The terms "calcareous fly ash" and "high-calcium fly ash" are used interchangeably in the literature [9–11], while from a formal point of view this nomenclature is regulated internationally by various standards, in which different criteria for ash differentiation are applied (Table 1).

Table 1. Requirements for high-calcium (calcareous) fly ash according to different standards.

Requirements		ASTM 618	CAN/CSA A3000			PN EN 450-1		PN EN 197-1	
		Class C	Class CI	Class CH		Case 1	Case 2	Case 1	Case 2
total CaO content, %		≥10.0	8 ÷ 20	≥20.0		-	-	-	-
reactive CaO content, %		-	-	-		≤10	≤10	10 ÷ 15	>15
free CaO content, %		-	-	-		≤1.5	>1.5	-	-
volume stability, mm		-	-	-		-	≤10	≤10	≤10
reactive SiO_2 content, %		-	-	-		-	-	≥25	-
$SiO_2 + Al_2O_3 + Fe_2O_3$ content, %		≥ 50.0	-	-		≥70.0	≥70.0	-	-
SO_3 content, %		≤5.0	≤5.0	≤5.0		≤3.0	≤3.0	≤3.0	-
Activity index, %	7 days	≥75.0	≥68.0	≥68.0		-	-	-	-
	28 days	-	-	-		≥75.0	≥75.0	-	-
	90 days	-	-	-		≥85.0	≥85.0	-	-
Water demand, %		≤105.0	-	-		≤95	≤95	-	-

There are different requirements, according to PN-EN 197-1 [6], for fly ash used as a main component of ordinary cements and ash used as a type II additive for concrete according to PN-EN 206 [12]. A reactive CaO content above 10% by mass formally excludes their use as a substitute for cements in concrete in EU countries, in the light of the requirements of PN-EN 450-1 [13]. Standard regulations in the USA or Canada [7,8], however, do not limit the use of high-CaO fly ash. In EU countries, the use of fly ash as a component of cement is not limited by the content of CaO. According to PN-EN 197-1, two types of fly ash are distinguished in this respect, siliceous and calcareous, and the limit of CaO is 10%.

In this article the term "calcareous fly ash" is used, defining it as fly ash for the production of cement in accordance with PN-EN 197-1, but using it as a component of concrete, despite not meeting the requirements of PN-EN 450.

It is worth noting that the results of published research are difficult to compare directly due to the different types of ash used, their content in relation to the cement mass, different concrete formulas and curing conditions, and different methods of conducting the experiments. There is a lot of research testing the influence of siliceous fly ash on concrete properties, among them carbonation: for example, the studies published by Kurda at al. (2019) [14], Ghorbani at al. (2019) [15], Carevic (2019) [16], Hussain at al. (2017) [17], Cai-feng Lu at al. (2018) [18], Branch at al. (2018) [19], Ying Chen at al. (2018) [20], and many others. However, during the last few years there have been only a few publications concerning the carbonation of concrete containing high-calcium fly ash as an additive [21–26].

The results of the published studies do not agree [21,22] as to which effect (accelerating or inhibitory) is dominant in the case of the carbonation of concrete with calcareous ash [27]. It depends on the interaction of compounds and external agents; however, one of the most important factors is the curing regime. Proper curing of concrete is crucial for pozzolanic reactions and advantageous for achieving the microstructure densifying effect. It was found that fly ash concrete that was not cured in the first days after concreting could very rapidly carbonate—even a 20 mm depth of carbonation was observed after a one-year exposure in the urban atmosphere [27].

First of all, the method of introducing ash into the concrete mix is important, i.e., whether the additive is introduced as a substitute of part of the cement or as an increase in the amount of binder. In the first case, the effect of the $Ca(OH)_2$ deficit strongly influences the carbonation progress; in the second case the role of the densifying effect is most important [28].

There are only a few publications on the research into carbonation of concretes with calcareous ash. The high content of free lime in the ash suggests that the availability of $Ca(OH)_2$ for the carbonation reaction will be high at the beginning of the process. It is worth noting that the free lime in the ash is highly reactive [23,24,29,30]. We would expect, then, a high initial rate of carbonation. On the other hand, the intensity of the pozzolanic reaction will reduce the calcium hydroxide amount, while at the same time filling pores with reaction products. Thus, a reduction of CO_2 diffusion capacity occurs, which inhibits the rate of carbonation. The research results presented in [7,31] indicate that the replacement of a part of the cement with calcareous fly ash in an amount of 10–20% does not significantly affect the course of electrochemical processes on the surface of reinforcing steel, without deteriorating the protective properties of concrete against reinforcement. At the same time, there are also published studies indicating an increase in the dynamics of the progress of carbonation of concrete with calcareous fly ash, especially when it has a high content in relation to cement [26,30]. The research work carried out by Wolinski et al. (2018) [30] also showed that the calcareous ash used as a substitute for a part of the aggregate (corresponding to 20-50% of the cement mass) allowed them to obtain concrete with a very low carbonation depth (less than 10 mm after 70 days in 4% CO_2 concentration). These studies indicated that, with a constant cement content and constant w/c ratio, there is an optimal ash content, leading to the lowest susceptibility to carbonation. The dynamics of carbonation depth development change when the ash content increases, in such a way that intensive progress is noticed after the 56th day in 4% of CO_2 if the ash content is high.

Research on the development of universal models of carbonation and different material and technological variables has been conducted by many authors [3,32–44]. When searching for a model of carbonation, one difficult issue is describing the intensity of the carbon dioxide flow in concrete. The first Fick's law, which is used to describe diffusion, assumes the constant microstructural characteristics of concrete over the time. In this way, the following carbonation model could be developed, in the form of a power function of carbonation depth over time:

$$x = \sqrt{\frac{2D\varphi_{ext}}{a}} \times \sqrt{t} \qquad (1)$$

where: x—depth of carbonation; D—diffusion coefficient; φ_{ext}—external concentration of CO_2; t—time of carbonation; a—coefficient determining the amount of CO_2 bound in the way of carbonation by unit volume of concrete in kg/m^3, calculated according to the CEB Bulletin 238 [45] as: $a =$

0.75·C·[CaO]·αH·(MCO₂/MCaO) (C—content of cement in concrete, kg/m³; [CaO]—CaO content in the cement composition; αH—degree of hydration of cement; MCO₂, MCaO—molar masses).

In practice, the most widely used model is simplified. It relates to an average constant relative humidity RH and carbon dioxide concentration in the environment and can be expressed in the form:

$$x = A \times t^{1/2} + B \qquad (2)$$

where A is a constant depending on the diffusion coefficient, the ability of concrete to bind CO_2 and CO_2 concentration in the air, and B is an empirical factor accounting for the initiation period of carbonation. This model is used by most researchers [3,31–44], but it treats the carbonation as a process unlimited in concrete space and unlimited in time. The abovementioned models based on the first Fick's law assume that concrete in which diffusion takes place will not change its microstructure over time. The assumption of a constant diffusion flux in Equation (1) in such a circumstances is not valid. This significant simplification of the description of the carbonation process does not take into account additional factors that lead to changes in diffusivity. The effect of the saturation of the pores with carbonation products makes it impossible to describe the process using Fick's law. Carbonation decreases porosity, in particular capillary porosity, and reduces the permeability of the concrete. This self-terminating nature of the phenomenon was mentioned by Bakker [46], Hergenröder [47], Nilsson [48], and Fagerlund [49]. The idea was further developed by Czarnecki, Woyciechowski at al. [50–55]. According to their findings, concrete carbonation can be described with a hyperbolic function of its depth over time (reciprocal square root of time), which has an asymptotic value parallel to the time axis. This asymptote is the limit of carbonation depth. The traditional and hyperbolic models of carbonation are shown in Figure 1.

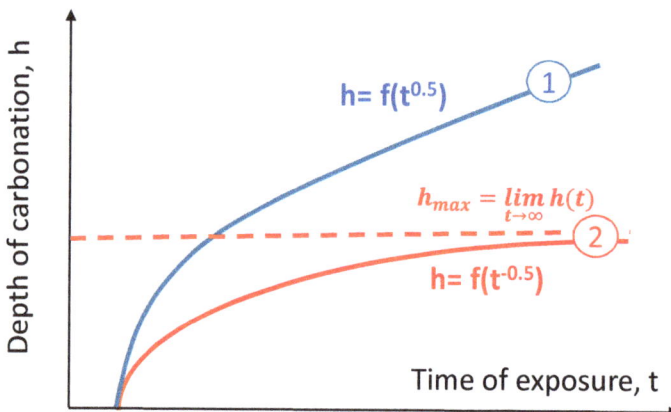

Figure 1. Traditional" power (1) and hyperbolic (2) models of carbonation phenomena.

The hyperbolic carbonation model is expressed in the general formula:

$$h = f(t^{-0.5}) \qquad (3)$$

All results published in the literature [50–55] are well fitted to the hyperbolic model expressed in Equation (3), enriched with w/c ratio and early curing time factors:

$$h = a(w/c) + b(cp) + c(t^{-0.5}) \qquad (4)$$

where: h—depth of carbonation, mm, w/c—water-cement ratio, cp—early curing with water period, days, t—time of exposure, days, a, b, c—coefficients describing relevance of influence of w/c ratio, early curing and exposure time on depth of carbonation. It was stated that parameters (a, b, c) mainly depend on the binder properties, the presence of mineral additives, and, especially, on the CO_2 concentration. This type of model was elaborated for many types of concrete, particularly with Portland, slag, and siliceous fly ash cement. SEM analyses published in the literature [51,52,55] show a different density of concrete in carbonated and non-carbonated zones for all the tested binder compositions. The hyperbolic model allows us to calculate the maximum depth of carbonation (the limits of the hyperbolic model) and compare it with the reinforcement cover thickness in the analyzed element. This allows us to predict the risk of corrosion due to the carbonation and to calculate the time when the carbonation front will reach the reinforcement surface. This moment could be interpreted as the time of corrosion initiation.

The aim of the research presented in this article was to develop a model of carbonation of concrete with calcareous ash as a function of two basic variables in the composition of concrete: the water/cement ratio and the ratio of mass content of fly ash to cement. The implementation of this objective required the assumption of a preliminary general form of the model and then conducting an experiment in accordance with the principles of statistical planning, enabling the determination of a model with high conformity to the test results.

A model for predicting the long-term depth of carbonation in natural conditions on the basis of short-term tests in accelerated conditions was developed in the paper for concrete containing calcareous fly ash. The article also presents a simplified scheme of actions, enabling the use of a developed model for determining the optimal thickness of the reinforcement cover in exposure to carbonation.

2. Materials and Methods

We studied concretes with CEM I 32,5 R cement, river sand 0/2 mm (fineness modulus: 2.96), gravel aggregate 2/16 mm and fly ash, classified as calcareous according to PN-EN 197-1 [6], due to the high content of reactive lime: higher than 15% by mass. The fly ash used in the research was obtained from the combustion of brown coal in power plants. The grain size distribution was typical for this type of fly ash, according to analyses published in [24] and [56].

Basic physical (Table 2, Figure 2) and chemical (Table 3) properties allow us to categorize the fly ash as a component of cement according to PN-EN 197-1, for common cements.

Table 2. Basic properties of calcareous fly ash used in research.

Property	Density, ρ g/cm^3	Min Grain Size, D_{min} μm	Max Grain Size, D_{max} μm	Specific Area, cm^2/cm^3
Result	2.27	0.2	174.6	8677

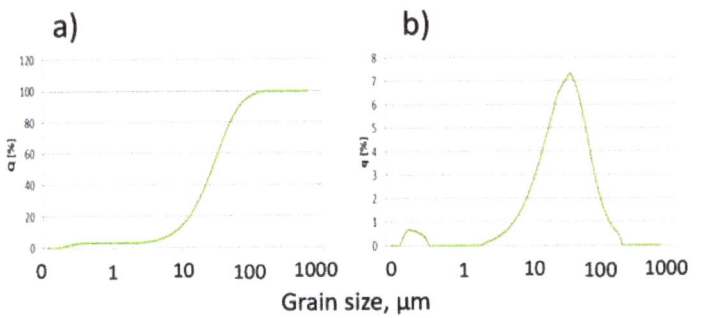

Figure 2. Calcareous fly ash grain size distribution (laser granulometry): (a): cumulative curve (b): population curve.

Table 3. Characteristics of the chemical composition of calcareous fly ash.

Lp.	Composition Characteristic	Mass Content, %
1.	SiO_2	47.06
2.	Al_2O_3	18.40
3.	Reactive CaO	15.20
4.	SO_3	1.94
5.	CO_2	0.58
6.	Loss of ignition	3.60
7.	Other compounds	15.70

The study of the activity index (Table 4) according to PN-EN 450-1 [11] indicates that it is an active material. The literature [7,23,24] indicates that calcareous ashes are materials with both pozzolanic and latent hydraulic properties, and the test result is the combined effect of both properties.

Table 4. Activity index (according to PN-EN 450-1) for calcareous fly ash, f_{c28}, f_{c90}—compressive strength after 28 and 90 days; WA_{28}, WA_{90}—calculated as a compressive strength of reference mortar and fly ash mortar ratio in %.

Lp.	Type of Material	f_{c28}, MPa	WA_{28}, %	f_{c90}, MPa	WA_{90}, %
1	Reference mortar	41.4	83.0	49.5	85.9
2	Mortar with 25% of fly ash as cement substitution	49.9		57.6	

As variables in the basic research program, w/c ratio (from 0.35 to 0.55) and ash to cement mass ratio (from 0.20 to 0.50) were assumed. The consistency was constant (12 ± 2 cm of slump) and adjusted with the help of a superplasticizer. The experimental program was developed on the basis of a two-factor, partial, quasi-uniform plan with a double repetition of the experiment at the central point (Table 5). The values x1 and x2 are the coded values of variables characteristic of the abovementioned experimental plan that lead to the best statistically representative answer with minimal testing [57]. The compositions of concrete mixtures, determined empirically to achieve the assumed consistency, are presented in Table 6.

Table 5. Range of variables used in experiment.

Composition Symbol	Coded Variables Values		Real Variables Values	
	x1	x2	x1' (w/c)	x2' (p/c)
1	−1	−1	0.379	0.24
2	1	1	0.521	0.46
3	−1.414	0	0.35	0.35
4	1.414	0	0.55	0.35
5	0	−1.414	0.45	0.20
6	0	1.414	0.45	0.50
7	0	0	0.45	0.35
8	−1	1	0.379	0.46
9	1	−1	0.521	0.24
10	0	0	0.45	0.35

Table 6. Concrete mix compositions (constant value of cement and fly ash to aggregate mass coefficient = 0.24; constant consistence 12 ± 2 cm of slump), w/c—water/cement mass ratio, p/c—fly ash/cement mass ratio.

Composition Symbol	Components Content, kg/m^3				w/c	p/c	Superplasticizer, % of Cement Mass
	Cement	Water	Aggregate	Fly Ash			
1	361.2	137	1872	88.1	0.379	0.244	2.5
2	299.6	156	1817	136.6	0.521	0.456	1
3	338.6	118.5	1904	118,5	0.35	0.35	2.2
4	317.3	174.5	1784	111.0	0.55	0.35	0.1
5	363.3	163.5	1816	72.6	0.45	0.2	0.4
6	297.8	134	1861	148.8	0.45	0.5	1.8
7	327.3	147.3	1841	114.5	0.45	0.35	1.44
8	312.4	118.5	1895	142.5	0.379	0.456	2.55
9	343.8	179	1781	83.5	0.521	0.244	0.48
10	327.3	147.3	1841	114.5	0.45	0.35	1.44

Concrete specimens for all tests were demolded after one day and then cured for 27 days in water, under laboratory conditions.

The progress of carbonation depth under accelerated conditions as well as the compressive strength after 28, 56, and 90 days and the tensile splitting strength after 28 days (i.e., at the start of accelerated carbonation exposure) were investigated. The compressive strength was determined on cubic specimens of 150 mm per side according to PN-EN 12390-3 [58], with the mean values determined from no fewer than five samples. The tensile strength for splitting was tested on cubic specimens of 150 mm per side according to PN-EN 12390-6 [59]. The depth of carbonation was tested according to the draft CEN TC 12390-12 [60]: CO_2 concentration = 4%, t = 20 ± 2, RH = 50–60%, front measurement with phenolphthalein indicator, after exposure time of 14, 28, 56, 70, and 90 days. Beams 100 × 100 × 500 mm were used for the tests, with the two opposite long side surfaces of the specimen exposed to carbon dioxide. The remaining surfaces were covered with a paraffin coating just before placing the samples in the carbonation chamber. Three samples were tested for each composition.

Scanning electron microscopy (SEM, TM3000, Hitachi, Japan) was used to characterize the microstructure of the chosen compositions of carbonized and noncarbonized two-year-old concrete specimens.

3. Results

The strength characteristics of the tested concrete (Table 7) indicate that concretes with the addition of calcareous ash show an increase in strength between 28 and 90 days at 10–20% depending on the w/c ratio and the content of ash in the binder. The presence of ash in the binder at a constant w/c ratio caused a slight increase in the compressive and tensile strength. These results are in line with the literature [9,23,24,61].

The main goal of the research was to analyze the progress of carbonation of concrete over time and to attempt to mathematically describe this phenomenon in a practically useful way. The results of the depth measurement of carbonation with the phenolphthalein test after subsequent exposure periods in the carbonation chamber are summarized in Table 8. The given values were calculated as arithmetic averages of 10 measurements on each of the three specimens of the series (five on each of the two lateral surfaces of a single sample) [62].

Table 7. Compressive and tensile strength results.

Mix Composition Symbol and Characteristics	Average Compressive Strength (MPa) Standard Deviation (MPa) Concrete Age (days)			Average Tensile Strength (MPa) Standard Deviation (MPa)
	28	56	90	28
1 w/c = 0.379; p/c = 0.24	65.6	66.1	68.4	4.50
	0.67	2.20	0.80	0.07
2 w/c = 0.521; p/c = 0.46	54.4	57.6	58.7	3.40
	1.45	0.72	1.30	0.14
3 w/c = 0.35; p/c = 0.35	68.0	68.4	73.3	4.40
	0.65	2.92	1.76	0.04
4 w/c = 0.55; p/c = 0.35	43.2	45.5	46.6	2.90
	1.73	0.62	2.47	0.19
5 w/c = 0.45; p/c = 0.2	62.3	65.3	65.8	4.60
	0.86	1.11	0.46	0.06
6 w/c = 0.45; p/c = 0.5	61.4	67.9	73.4	3.80
	1.02	1.30	0.96	0.05
7 w/c = 0.45; p/c = 0.35	66.4	68.6	70.5	4.50
	2.25	1.32	2.26	0.10
8 w/c = 0.379; p/c = 0.46	72.0	73.0	78.8	5.20
	1.37	0.99	1.53	0.14
9 w/c = 0.521; p/c = 0.24	44.2	47.8	48.2	3.30
	1.01	0.54	1.98	0.28

Table 8. Average values of measured depths of carbonation.

Exposure Time, Days	Mix Composition Symbol								
	1	2	3	4	5	6	7	8	9
	Average Depth of Carbonation (mm)								
14	1.47	0.86	0.23	0.99	1.04	1.06	1.32	0.25	0.85
28	4.68	2.03	0.54	3.44	2.20	2.89	1.57	1.17	2.32
56	5.05	8.48	0.63	7.82	6.31	5.16	3.58	1.76	6.81
70	5.70	8.58	0.60	10.42	7.80	8.89	7.19	2.02	9.51
90	6.22	8.29	2.45	11.63	9.25	9.30	7.25	2.09	10.44

It has been observed that, in the case of a low cement ratio, the presence of unreacted calcareous fly ash particles was more clearly observed (Figure 3 versus Figures 4 and 5 versus Figure 6). After carbonation in concrete with a low w/c ratio (Figure 5), there are still many fly ash grains, which are not involved in the carbonation process. In the case of w/c = 0.52 after carbonation, fewer fly ash particles and portlandite crystals were visible due to the generally higher homogeneity of the microstructure. These observations are correlated with the higher positive effect of fly ash addition on the resistance of carbonation with a higher w/c ratio. It can be concluded that an optimal content of fly ash from the point of view of carbonation resistance exists and is dependent on the w/c ratio of concrete. This observation was discussed by Wolinski et al. (2015) [63].

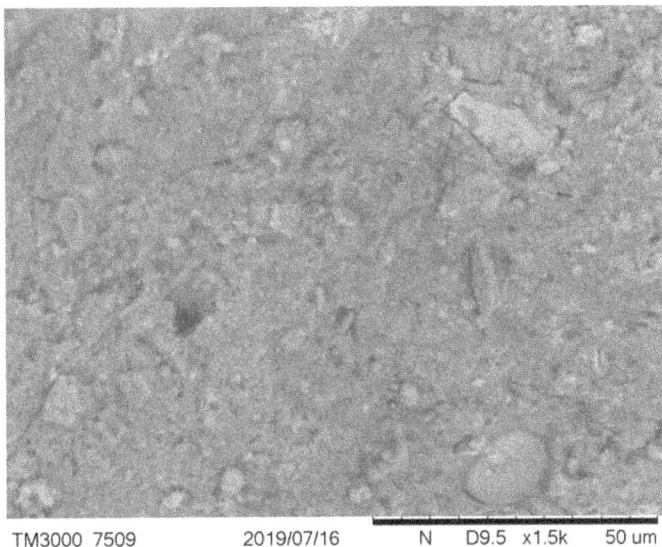

Figure 3. Noncarbonated concrete microstructure—mix symbol: 8, w/c = 0.38, p/c = 0.46.

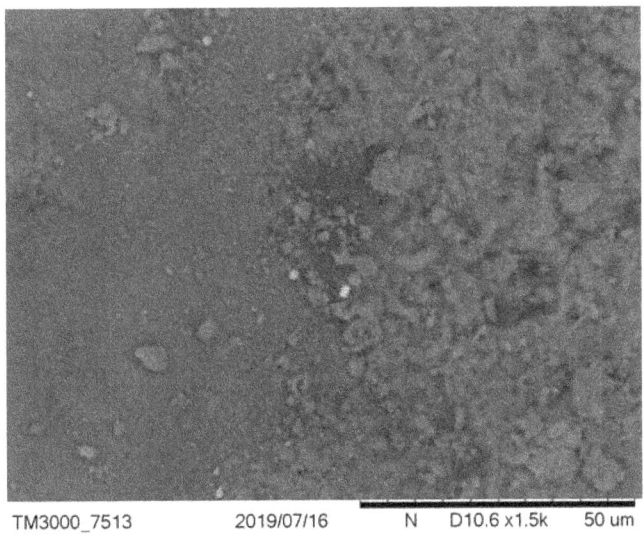

Figure 4. Noncarbonated concrete microstructure—mix symbol: 2, w/c = 0.52, p/c = 0.46.

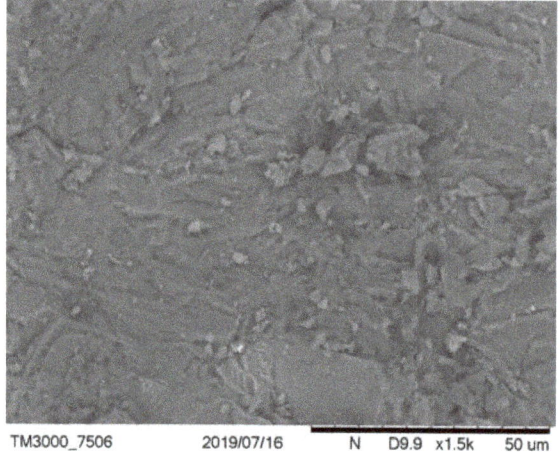

Figure 5. Carbonated concrete microstructure—mix symbol: 8, w/c = 0.38, p/c = 0.46.

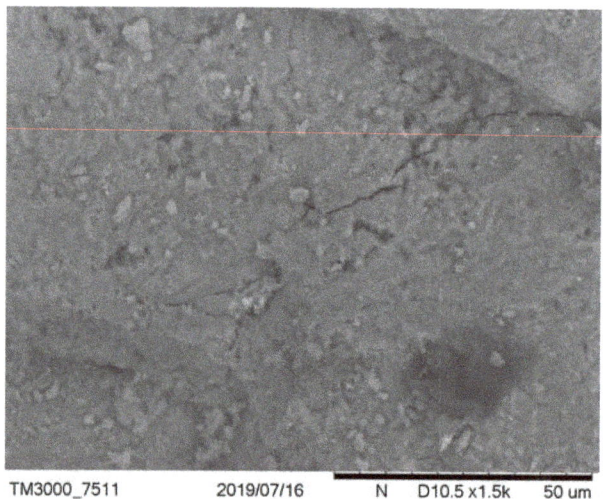

Figure 6. Carbonated concrete microstructure—mix symbol: 2 w/c = 0.52, p/c = 0.46.

4. Discussion

The results of carbonation depth were used to formulate a model expressing the depth of carbonation h_t (after a predetermined time t under assumed exposure conditions) as a function of the water-cement ratio (w/c) and the mass proportion of ash and cement in the binder (p/c), as follows:

$$h_t(w/c; p/c) = a + b(w/c) + c(p/c) + d(w/c)^2 + e(w/c)(p/c) + f(p/c)^2 \tag{5}$$

Response surfaces, including the total range of w/c ratio values from 0.35 to 0.55 and p/c values from 0.2 to 0.50 assumed in the tests, are presented in Figures 7–9, corresponding to selected exposure times t, i.e., 56, 70, and 90 days, with the determination coefficient in all cases significantly exceeding the value of 0.9.

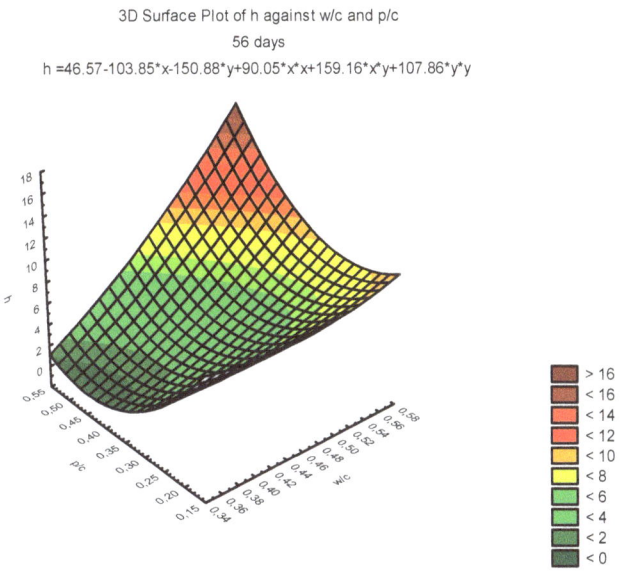

Figure 7. Depth of carbonation as a function of w/c and p/c, after exposure time $t = 56$ days, $R^2 = 0.93$.

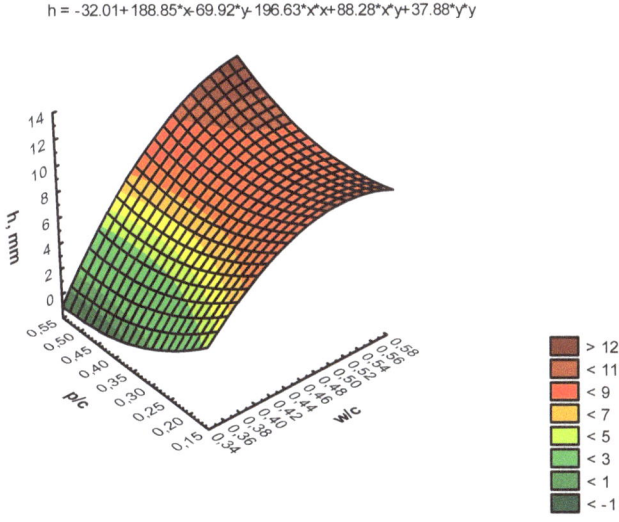

Figure 8. Depth of carbonation as a function of w/c and p/c, after exposure time $t = 70$ days, $R^2 = 0.92$.

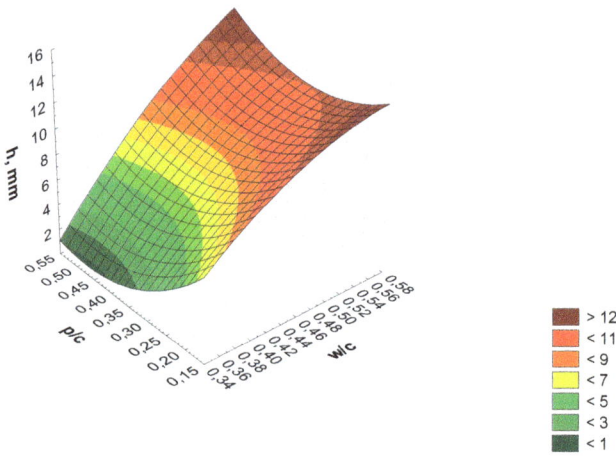

Figure 9. Depth of carbonation as a function of w/c and p/c, after exposure time $t = 90$ days, $R^2 = 0.95$.

For a complete set of results, a general concrete carbonation model with fly ash was developed, taking into account both material variables (w/c ratio and p/c ratio) and the time of carbonation under specific exposure conditions. The general form of the model was assumed to take into account the finite nature of the carbonation process over time, consistent with the considerations presented in the introduction to this article. Finally, the following general form of the model was adopted:

$$h(w/c; p/c; t) = a + b_1 \times (w/c) + b_2 \times (p/c) + b_3/sqrt(t) \tag{6}$$

where: h—depth of carbonation, mm; w/c—water-cement ratio; p/c—powder to cement ratio; a, b_1, b_2, b_3—material and technological coefficients, t—time of exposure, days.

The model does not take into account the essential technological factor, which is the time of early curing of concrete, because all tests were carried out with reference to samples hardening for up to 28 days in water. The possible practical use of models of the proposed type requires taking into account this factor in the form of an additional expression or limiting the validity of the equation to a specific regime of concrete curing. The detailed form of the proposed model determined by the curve fitting method, based on the experimental results, is as follows:

$$h(w/c; p/c; t) = 1.07 + 25.28 \times (w/c) - 3.53 \times (p/c) - 41.07/sqrt(t) \tag{7}$$

and is characterized by the determination coefficient 0.85. This indicates the good fit of the adopted model to the results of laboratory tests obtained under accelerated carbonation conditions in 4% CO_2. The presented detailed model can be applied only to concrete with the range of constant and changed material and technological variables used in the research, but the method of its determination is an example that can be repeated in the case of various material and technological assumptions.

For further verification of model fitting to the experimental results, values calculated from the model and obtained from the experiments over 90 days of exposure in 4% of CO_2 concentration were compared, as in Table 9. The accuracy of the prediction is ± 15%, which is an acceptable value from an engineering point of view.

Table 9. Comparison of experimental and calculated values of depth of carbonation.

Mix Symbol	Average Depth of Carbonation after 90 Days in 4% of CO_2		
	Calculated from Model (7), mm	Measured during Experiment, mm	Relative Difference, % $((c - b)/b) \times 100\%$
(a)	(b)	(c)	(d)
1	5.7	6.2	8
2	8.3	8.3	0
3	2.8	2.5	−14
4	9.9	11.6	15
5	7.9	9.3	15
6	8.0	9.3	14
7 and 10 (average)	7.9	8.8	10
8	2.4	2.1	−15
9	9.1	10.4	13

Predicting the course of carbonation in a real structure using the developed model requires establishing an equation expressing the dependence of the progress of carbonation on the CO_2 concentration in the research environment and in the exploitation environment. The dependence of carbonation depth on CO_2 concentration is described in many publications [37,40,64], with a form similar to the following:

$$X_1 = X_2 \times \frac{c_1 \times t_1}{c_2 \times t_2} \qquad (8)$$

where: X_1—depth of carbonation after exploitation in c_1 concentration after t_1 time of exploitation; X_2—depth of carbonation in accelerated conditions, t_2—time of exposure in accelerated conditions; c_2—CO_2 concentration in accelerated conditions; t_1—expected service life of construction; c_1—CO_2 concentration in exploitation conditions. Please check that intended meaning has been retained

Thus, having the developed detailed model in the general form (Equation (6)), i.e., a form that assumes the finite character of the carbonation process, we can calculate the limit of functions at t reaching to infinity, i.e., the ordinate of the model asymptote denotes the maximum possible depth of carbonation in the test conditions. In the analyzed case it means under accelerated conditions (CO_2 concentrations equal to c_2 = 4%). Assuming this value as X_2 in Equation (8), and as t_2 (the maximum test time under accelerated carbonation conditions), i.e., 90 days, one can determine the predicted carbonation depth after time t_1 (e.g., assumed service life of 50 or 100 years) at a given concentration of carbon dioxide c_1 (e.g., atmospheric concentration—approximately 0.04%). In Table 10 a simulation of the carbonation depth using the above dependence after 50 and 100 years of service life in the atmosphere with a concentration of 0.04% carbon dioxide is shown.

Table 10. Calculated depths of carbonation (mm) after 50 and 100 years in a natural atmosphere.

w/c	p/c				
	0.2	0.24	0.35	0.46	0.5
	50 years				
0.35	19	20	18	17	16
0.379	20	20	19	18	18
0.45	24	24	23	22	22
0.521	27	27	26	26	25
0.55	29	29	28	27	27
	100 years				
0.35	37	39	35	34	33
0.379	40	39	38	37	36
0.45	48	47	45	44	43
0.521	54	54	53	51	50
0.55	58	57	56	54	53

The values presented in Table 10 can be considered as the starting point for determining the required reinforcement cover in a construction made of concrete with given w/c and p/c ratios. These values should be increased by a safety factor, taking into account random factors and those related to the curing of concrete. This is an alternative approach to the standard procedure for determining the thickness of the reinforcement cover indicated by the EC2 standard [65], which is favorable due to the optimal method of determining the minimum safe thickness due to durability, taking into account the concrete material specification. The proposed procedure for determining the required cover is presented in the form of the diagram (Figure 10) for the concrete family. As a concrete family, a set of technologically similar concretes was defined, characterized by a constant type of cement, type of ash, chemical admixture, initial care method, and the following variables: any type of aggregate (except lightweight), any type of water (in accordance with PN-EN 1008 [66]), the specified range of variability of w/c and p/c ratios, and dosage of the same admixture in the amount necessary to the required consistency.

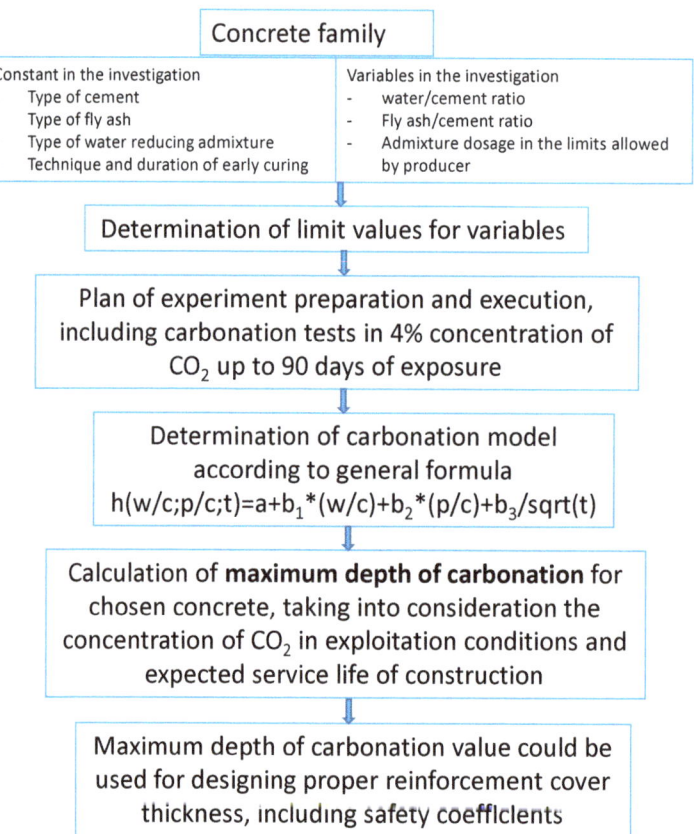

Figure 10. Proposed procedure for determining the thickness of the reinforcement cover with use of the developed model of carbonation.

5. Conclusions

As a result of the research presented in this article, the following research objectives have been met:

1. It was found that an increase in the content of calcium fly ash used as a partial cement substitute (a fly ash to cement ratio between 0.2 and 0.5) in concrete at a constant w/c ratio caused a slight increase in the compressive and tensile strength;
2. A carbonation range model was developed as a function of the w/c and calcareous fly ash to cement ratio in the form of $h_{ti} = f(w/c, p/c)$, after different exposure times t_i, i.e., after 56, 70, and 90 days of exposure;
3. A model of the progress of carbonation in the form of $h = f(1/sqrt(t))$ has been formulated and verified in relation to concretes with calcareous fly ash of various w/c and p/c ratios;
4. A general model for concrete carbonation with calcareous fly ash $h = f(w/c; p/c; 1/(sqrt(t)))$ has been developed, taking into account all the variables used in the research;

It was shown that the developed general model of carbonation for concrete with a wide range of water/cement and ash to cement ratios and a constant set of components as well as fixed technological procedures could be a useful tool for the experimental design of optimal cover thickness.

Author Contributions: Conceptualization, P.W. (Piotr Woyciechowski); Methodology, P.W. (Paweł Woliński); Formal Analysis, P.W. (Piotr Woyciechowski) and G.A.; Investigation, P.W. (Paweł Woliński); Writing—Original Draft Preparation, P.W. (Piotr Woyciechowski); Writing—Review and Editing, P.W. (Piotr Woyciechowski); Visualization, G.A. and P.W. (Paweł Woliński)

Funding: This research was funded by the Department of Building Materials Engineering of the Faculty of Civil Engineering at Warsaw University of Technology.

Conflicts of Interest: The authors declare no conflict of interest.

References

1. Jiang, L.; Liu, Z.; Ye, Y. Durability of concrete incorporating large volumes of low-quality fly ash. *Cem. Concr. Res.* **2004**, *34*, 1467–1469. [CrossRef]
2. Saraswathy, V.; Muralidharan, S.; Thangavel, K.; Srinivasun, S. Influence of activated fly ash on corrosion-resistance and strength of concrete. *Cem. Concr. Compos.* **2003**, *25*, 673–680. [CrossRef]
3. Hossain, K.M.A.; Lachemi, M. Development of model for the prediction of carbonation in pozzolanic concrete. In Proceedings of the Third International Conference on Construction Materials: Performance, Innovations and Structural Implications, University of British Columbia, Vancouver, BC, Canada, 2005.
4. Neville, A.M. *Properties of Concrete*, 5th ed.; Prentice Hall: London, UK, 2012.
5. Bier, T.A. Influence of type of cement and curing on carbonation progress and pore structure of hydrated cement paste. In Proceedings of the Materials Research Society Symposium, Erlangen, Germany, 1987; pp. 123–134.
6. PN-EN 197-1:2012 Cement. Composition, Specifications and Conformity Criteria for Common Cements. Available online: https://infostore.saiglobal.com/en-au/Standards/PN-EN-197-1-201241848_SAIG_PKN_PKN_2216693/ (accessed on 20 August 2019).
7. ASTM International. *ASTM C618-15, Standard Specification for Coal Fly Ash and Raw or Calcined Natural Pozzolan for Use in Concrete*; ASTM C618-15; ASTM International: West Conshohocken, PA, USA, 2015.
8. Cloud Security Alliance. *CAN/CSA–A3000-13, Cementitious Materials Compendium*; CAN/CSA–A3000-13; Cloud Security Alliance: Toronto, ON, Canada, 2013.
9. Fan, W.J.; Wang, X.Y.; Park, K.B. Evaluation of the chemical and mechanical properties of hardening high-calcium fly ash blended concrete. *Materials* **2015**, *8*, 5933–5952. [CrossRef] [PubMed]
10. Nowoświat, A.; Gołaszewski, J. Influence of the variability of calcareous fly ash properties on rheological properties of fresh mortar with its addition. *Materials* **2019**, *12*, 1942. [CrossRef] [PubMed]
11. Jóźwiak-Niedźwiedzka, D. Influence of blended cements on the concrete resistance to carbonation. In *Brittle Matrix Composites 10*; Brandt, A.M., Olek, J., Glinicki, M.A., Leung, C.K.Y., Eds.; Woodhead Publishing: Cambridge, UK, 2012; pp. 125–134.
12. Polish Committee of Standardization (PKN). *PN-EN 206:2014 Concrete. Specification, Performance, Production and Conformity*; Polish Committee of Standardization (PKN): Warszawa, Poland, 2014.

13. Polish Committee of Standardization (PKN). *PN-EN 450-1:2012 Fly ash for concrete-Part 1: Definition, Specifications and Conformity Criteria*; PN-EN 450-1:2012; Polish Committee of Standardization (PKN): Warszawa, Poland, 2012.
14. Kurda, R.; de Brito, J.; Silvestre, J.D. Carbonation of concrete made with high amount of fly ash and recycled concrete aggregates for utilization of CO_2. *J. CO_2 Util.* **2019**, *29*, 12–19. [CrossRef]
15. Ghorbani, S.; Sharifi, S.; Ghorbani, S.; Tam, V.W.Y.; De Brito, J.; Kurda, R. Effect of crushed concrete waste's maximum size as partial replacement of natural coarse aggregate on the mechanical and durability properties of concrete. *Resour. Conserv. Recycl.* **2019**, *149*, 664–673. [CrossRef]
16. Carević, V.; Ignjatović, I.; Dragaš, J. Model for practical carbonation depth prediction for high volume fly ash concrete and recycled aggregate concrete. *Constr. Build. Mater.* **2019**, *213*, 194–208. [CrossRef]
17. Hussain, S.; Dipendu, B.; Singh, S.B. Comparative study of accelerated carbonation of plain cement and fly-ash concrete. *J. Build. Eng.* **2017**, *10*, 26–31. [CrossRef]
18. Cai-feng, L.; Wei, W.; Qing-tao, L.; Ming, H.; Yuan, X. Effects of micro-environmental climate on the carbonation depth and the pH value in fly ash concrete. *J. Clean. Prod.* **2018**, *181*, 309–317.
19. Branch, J.L.; Epps, R.; Kosson, D.S. The impact of carbonation on bulk and ITZ porosity in microconcrete materials with fly ash replacement. *Cem. Concr. Res.* **2018**, *103*, 170–178. [CrossRef]
20. Ying, C.; Peng, L.; Zhiwu, Y. Effects of Environmental Factors on Concrete Carbonation Depth and Compressive Strength. *Materials* **2018**, *11*, 2167. [CrossRef]
21. Bogas, J.A.; Real, S.; Ferrer, B. Biphasic carbonation behaviour of structural lightweight aggregate concrete produced with different types of binder. *Cem. Concr. Compos.* **2016**, *71*, 110–121. [CrossRef]
22. Dąbrowski, M.; Glinicki, M.A.; Gibas, K.; Jóźwiak-Niedźwiedzka, D. Effects of calcareous fly ash in blended cements on chloride ions migration and strength of air entrained concrete. *Constr. Build. Mater.* **2016**, *126*, 1044–1053. [CrossRef]
23. Glinicki, M.A.; Jóźwiak-Niedźwiedzka, D.; Gibas, K.; Dąbrowski, M. Influence of blended cements with calcareous fly ash on chloride ion migration and carbonation resistance of concrete for durable structures. *Materials* **2016**, *9*, 18. [CrossRef] [PubMed]
24. Giergiczny, Z. The hydraulic activity of high calcium fly ash. *J. Therm. Anal. Calorim.* **2006**, *83*, 227–232. [CrossRef]
25. Jóźwiak-Niedzwiedzka, D.; Gibas, K.; Sobczak, M. Carbonation of concretes containing calcareous fly ashes. *Roads Bridges* **2013**, *12*, 131–144.
26. Papadakis, V.G. Effect of fly ash on Portland cement systems part II high calcium fly ash. *Cem. Concr. Res.* **2002**, *30*, 1647–1654. [CrossRef]
27. Woyciechowski, P. *Model of Concrete Carbonation*; Series: Scientific works. Buildings; Warsaw University of Technology: Warsaw, Poland, 2013.
28. Papadakis, V.G. Effect of supplementary cementing materials on concrete resistance against carbonation and chloride ingress. *Cem. Concr. Res.* **2000**, *30*, 291–299. [CrossRef]
29. Giergiczny, Z. *Fly Ash as a Component of Cement and Concrete*; Edition of Silesian Technical University: Gliwice, Poland, 2013.
30. Woliński, P.; Woyciechowski, P.P.; Jaworska, B.; Adamczewski, G.; Tokarski, D.; Grudniewski, T.; Chodyka, M.; Nitychoruk, A. The influence of the mineral additives on the carbonation of cement composites. *MATEC Web Conf.* **2018**, *196*, 04062. [CrossRef]
31. Bary, B.; Sellier, A. Coupled moisture-carbon dioxidecalcium transfer model for carbonation of concrete. *Cem. Concr. Res.* **2004**, *34*, 1859–1872. [CrossRef]
32. Burkan Isgor, O.; Ghani Razaqpur, A. Finite elements modeling of coupled heat transfer, moisture transport and carbonation processes in concrete structures. *Cem. Concr. Compos.* **2004**, *26*, 57–73. [CrossRef]
33. Ishida, T.; Maekawa, K.; Soltani, M. Theoretically identified strong coupling of carbonation rate and thermodynamic moisture states in micropores of concrete. *J. Adv. Concr. Technol.* **2004**, *2*, 213–222. [CrossRef]
34. Maekawa, K.; Ishida, T. Modeling of structural performances under coupled environmental and weather action. *Mater. Struct.* **2002**, *35*, 591–602. [CrossRef]
35. Loo, Y.H.; Chin, M.S.; Tam, C.T.; Ong, K.C.G. A Carbonation prediction model for accelerated carbonation testing of concrete. *Mag. Concr. Res.* **1994**, *46*, 191–200. [CrossRef]
36. Masuda, Y.; Tanano, H. Mathematical model on process of carbonation of concrete. *Concr. Res. Technol.* **1991**, *2*, 99–107.

37. Ming-Te, L.; Wen-Jun, Q.; Chih-Hsin, L. Mathematical modeling and prediction method of concrete carbonation and its applications. *J. Mar. Sci. Technol.* **2002**, *10*, 128–135.
38. Monteiro, I.; Branco, F.A.; de Brito, J.; Neves, R. Statistical analysis of the carbonation coefficient in open air concrete structures. *Constr. Build. Mater.* **2012**, *29*, 263–269. [CrossRef]
39. Steffens, A.; Dinkler, D.; Ahrens, A. Modeling carbonation for corrosion risk prediction of concrete structures. *Cem. Concr. Res.* **2002**, *32*, 935–941. [CrossRef]
40. Papadakis, V.G.; Vayenas, C.G.; Fardis, M.N. Fundamental modeling and engineering investigation of concrete carbonation. *ACI Mater. J.* **1991**, *88*, 363–373.
41. Muntean, A. On the interplay between fast reaction and slow diffusion in the concrete carbonation process: A matched-asymptotics approach. *Meccanica* **2009**, *44*, 35–46. [CrossRef]
42. Fu, C.; Ye, H.; Jin, X.; Jin, N.; Gong, L. A reaction-diffusion modeling of carbonation process in self-compacting concrete. *Comput. Concr. Int. J.* **2015**, *15*, 847–864. [CrossRef]
43. Medeiros-Junior, R.A.; Lima, M.G.; Yazigi, R.; Medeiros, M.H.F. Carbonation depth in 57 years old concrete structures. *Steel Compos. Struct. Int. J.* **2015**, *19*, 953–966. [CrossRef]
44. Ekolu, S.O. A review on effects of curing, sheltering, and CO_2 concentration upon natural carbonation of concrete. *Constr. Build. Mater.* **2016**, *127*, 306–320. [CrossRef]
45. Schiesl, P. *New Approach to Durability Design. An Example for Carbonation induced Corrosion*; CEB Bulletin 238; Comitée Euro-International du Béton CEB 238: Lausanne Switzerland, 1997.
46. Bakker, R.F.M. Initiation period. In *Corrosion of Steel in Concrete: Report of the Technical Committee 60-CSC RILEM*; Schiessl, P., Ed.; Chapman and Hall: London, UK, 1988; pp. 22–54.
47. Hergenröder, M. Zur statistichen Instandhaltungsplanung für bestehende Betonbauwerke bei Karbonatisierung des Betons und möglicher der Bewerhung. Doctoral Thesis, Technische Universität, München, Germany, 1992.
48. Nilsson, L.-O. Interaction between microclimate and concrete–A perquisite for deterioration. *Constr. Build. Mater.* **1997**, *10*, 301–308. [CrossRef]
49. Fagerlund, G. *Durability of Concrete Structures*; Arkady: Warszawa, Poland, 1997.
50. Czarnecki, L.; Woyciechowski, P. Concrete carbonation as a limited process and its relevance to concrete cover thickness. *ACI. Mater. J.* **2012**, *109*, 275–282.
51. Czarnecki, L.; Woyciechowski, P. Prediction of the reinforced concrete structure durability under the risk of carbonation and chloride aggression. *Bull. Pol. Acad. Sci. Tech. Sci.* **2013**, *61*, 173–181. [CrossRef]
52. Czarnecki, L.; Woyciechowski, P. Modelling of concrete carbonation; is it a process unlimited in time and restricted in space? *Bull. Pol. Acad. Sci.: Tech. Sci.* **2015**, *63*, 43–54. [CrossRef]
53. Czarnecki, L.; Woyciechowski, P.; Adamczewski, G. Risk of concrete carbonation with mineral industrial by-products. *KSCE J. Civ. Eng.* **2018**, *22*, 755–764. [CrossRef]
54. Woyciechowski, P.P.; Sokolowska, J.J. Self-terminated carbonation model as a useful support for durable concrete structure designing. *Struct. Eng. Mech.* **2017**, *63*, 55–64.
55. Czarnecki, L.; Woyciechowski, P. Effect of fluidal fly ash on concrete carbonation. In *Zastosowanie popiołów lotnych z kotłów fluidalnych w betonach konstrukcyjnych*; Brandt, A.M., Ed.; Polish Academy of Science: Warszawa, Poland, 2010; pp. 209–252.
56. Itskos, G.; Itskos, S.; Koukouzas, N. Size fraction characterization of highly-calcareous fly ash. *Fuel Process. Technol.* **2010**, *91*, 1558–1563. [CrossRef]
57. Box, G.E.P.; Hunter, W.G.; Hunter, J.S. *Statistics for Experimenters: Design, Innovation, and Discovery*, 2nd ed.; Wiley: Hoboken, NJ, USA, 2005.
58. Polish Committee of Standardization (PKN). *PN-EN 12390-3:2009 Testing Hardened Concrete. Compressive Strength of Test Specimens*; PN-EN 12390-3; Polish Committee of Standardization (PKN): Warszawa, Poland, 2009.
59. Polish Committee of Standardization (PKN). *PN-EN 12390-6:2011 Testing Hardened Concrete. Tensile Splitting Strength*; PN-EN 12390-6:2011; Polish Committee of Standardization (PKN): Warszawa, Poland, 2011.
60. Polish Committee of Standardization (PKN). *CEN/TS 12390-12 Testing hardened concrete—Part 12: Determination of the Potential Carbonation Resistance of Concrete: Accelerated Carbonation Method*; CEN/TS 12390-12; Polish Committee of Standardization (PKN): Warszawa, Poland, 2012.
61. Ponikiewski, T.; Gołaszewski, J. The effect of high-calcium fly ash on selected properties of self-compacting concrete. *Arch. Civ. Mech. Eng.* **2014**, *14*, 455–465. [CrossRef]

62. Wolinski, P. Influence of Calcareous Fly Ash on Concrete Carbonation. Doctoral Thesis, Warsaw University of Technology, Warsaw, Poland, 2018.
63. Woliński, P.; Woyciechowski, P.; Adamczewski, G. Effect of calacreous fly ash on the carbonation progress in concrete. *Mater. Bud.* **2015**, *12*, 24–25.
64. Wieczorek, G. *Reinforcement Corrosion Initiated by Chlorides or Carbonation of Concrete Cover*; Dolnośląskie Wydawnictwo Edukacyjne: Wrocław, Poland, 2002.
65. Polish Committee of Standardization (PKN). *PN-EN 1992-1-1 Eurocode 2: Design of Concrete Structures*; PN-EN 1992-1-1; Polish Committee of Standardization (PKN): Warszawa, Poland, 2008.
66. Polish Committee of Standardization (PKN). *PN-EN 1008:2004 Mixing Water for Concrete*; PN-EN 1008; Polish Committee of Standardization (PKN): Warszawa, Poland, 2004.

© 2019 by the authors. Licensee MDPI, Basel, Switzerland. This article is an open access article distributed under the terms and conditions of the Creative Commons Attribution (CC BY) license (http://creativecommons.org/licenses/by/4.0/).

Article

Influence of Crystalline Admixtures on the Short-Term Behaviour of Mortars Exposed to Sulphuric Acid

Victoria Eugenia García-Vera [1], Antonio José Tenza-Abril [2,*], José Miguel Saval [2] and Marcos Lanzón [1]

1 Departamento de Arquitectura y Tecnología de la Edificación, Universidad Politécnica de Cartagena, 30203 Murcia, Spain; victoria.eugenia@upct.es (V.E.G.-V.); marcos.lanzon@upct.es (M.L.)
2 Department of Civil Engineering, University of Alicante, 03690 Alicante, Spain; jm.saval@ua.es
* Correspondence: ajt.abril@ua.es; Tel.: +34-96-5903-400 (ext. 2729)

Received: 7 December 2018; Accepted: 22 December 2018; Published: 27 December 2018

Abstract: Using durable materials is a sustainable solution for extending the lifetime of constructions. The use of crystalline admixtures makes cementitious materials more durable. They plug pores, capillary tracts and microcracks, blocking the entrance of water due to the formation of crystals that prevent the penetration of liquids. The literature has covered the performance of these admixtures on concrete, but studies on mortars are still scarce. The aim of this study is to investigate the effect of an aggressive environment (sulphuric acid solution—3 wt%) on mortars produced with different percentages of a crystalline admixture (1%, 1.5% and 2% by weight of cement content). Physical and mechanical properties were studied after immersing the mortars in a H_2SO_4 solution for 90 days. It was found that, after a 90-day sulphuric acid exposure, mortars with the crystalline admixture showed greater compressive strength than the control mortar, besides exhibiting lower mass loss. However, the crystalline admixture did not produce any significant effect on the capillary water absorption coefficient. In a nonaggressive environment, and in the short term, the crystalline admixture did not have a significant effect on the compressive strength, the capillary water absorption coefficient or the ultrasonic pulse velocity.

Keywords: crystalline admixture; chemical exposure; sulphuric acid attack; durability; Xypex

1. Introduction

Nowadays, cement-based materials are exposed to acids because of environmental pollution, contact with some soils, ground waters, industrial waste-waters or sewers. These conditions can be found in some foundations, sanitation networks, treatment plants, agricultural and farm facilities, etc. [1–3]. Materials that are designed to work in aggressive chemical conditions must be durable materials, that is, materials with the capacity to withstand, for its service life, the physical and chemical conditions to which they are exposed [4]. Architects and engineers are vested with the responsibility of designing and building constructions by optimising the existing but limited resources in nature. Working with durable materials contributes to sustainability since they help to maintain the existing resources and avoid causing harm to the environment.

Concrete is susceptible to acid attack because of its alkaline nature. Sulphuric acid is particularly corrosive due to the sulphate ion in addition to the dissolution caused by the hydrogen ion causing the deterioration of cement-based materials. The reaction of the sulphuric acid (H_2SO_4) with the chemical compounds of cement hydration (calcium silicate hydrate, portlandite and calcium sulphoaluminate hydrate) generates gypsum, amorphous hydrous silica and aluminum sulphate. [3]. These chemical reactions, located in the regions close to the surface, provoke a profound degradation of the hydrated cement paste in terms of physical and mechanical aspects, such as compressive strength loss, mass variation, cracking, softening and decohesion. This process leads to spalling and exposure of the interior of the concrete structure [5,6].

To improve acid attack resistance in cement-based materials, the matrix must be dense, with low permeability, and it is advisable to use in the production a sulphate-resistant cement, high alumina cement, etc. [7–9]. Some authors state that effective strategies for manufacturing cementitious materials that are durable when exposed to acid attacks involve the following: (i) controlling the materials used during the manufacturing process; (ii) producing them with a low water/cement ratio; and (iii) guaranteeing suitable curing conditions [10,11].

A very popular strategy used to produce durable cement-based material is adding admixtures and additives [12–15]. Some studies utilise nanomaterials to reduce the porosity and water absorption of cement-based materials [16,17], which leads to enhanced durability and improved performance when these materials are exposed to acid attacks. Examples of such nanotechnologies are nano silica, nano alumina, titanium oxide, carbon nanotubes and polycarboxylates [18–22]. Water-repellent materials are also used to increase durability in acid attacks, since they reduce the affinity of capillary pore surfaces to moisture [23,24]. Examples of hydrophobic compounds used as admixtures are powdered stearates, oleates and products based on silanes and silicones [25–27]. When adequate recycling materials are available, another strategy to enhance the concrete durability is the use of aggregates derived from sanitary ceramics. Such concretes are characterised by low water absorption, higher mechanical strength and higher water-resistance compared to conventional concrete. Moreover, they have more resistance to chemically aggressive environments compared to conventional concretes [28–31].

Crystalline admixtures can also help increase the durability of cement-based materials, especially when cementitious materials are in acid environments. According to various manufacturers of products for crystalline waterproofings for concrete—Xypex, Kryton, Penetron, etc.—there are three ways of applying the waterproofing systems: (i) a coating applied as a cementitious slurry to the surface of existing concrete structures; (ii) an admixture added to the concrete at the time of batching; and (iii) a dry-shake to the fresh concrete surface. In all cases, the protection system is a blend of Portland cement, fine treated silica sand and undisclosed chemicals that are the intellectual property of the manufacturer. The catalytic reaction of the chemicals in the admixture occurs as long as there is moisture in the cement-based materials. These chemicals react with calcium hydroxide and other products resulting from cement hydration. The reaction generates nonsoluble crystalline formations that fill the pores and capillary tracks of cement-based materials. The crystalline structure becomes a permanent and integral part of the concrete matrix. Thus, the concrete becomes permanently sealed against the penetration of water or liquids from any direction [32].

The crystallisation process of these admixtures is not immediate, and the complete waterproofing effect is not achieved until approximately 12 days after the concrete creation [33], as long as there is enough moisture for chemical reactions to occur. Moreover, it is especially important to have a moist curing environment to obtain better performance [34]. There are several studies that focused on testing the waterproofing effect of these admixtures [33,35], as well as the impact of an aggressive medium on concretes made with crystalline-based waterproofing products [36,37].

However, it is still a challenge to find high durability mortars manufactured with these types of admixtures in aggressive environments. In this study, a crystalline protection product (applied as an admixture) was used to make mortars with different percentages of admixture (1 wt%, 1.5 wt% and 2 wt%). Their behaviour when they were exposed to an aggressive environment (sulphuric acid solution—3 wt%) was evaluated. Physical (mass loss, ultrasonic pulse velocity and capillary water absorption coefficient) and mechanical (compressive strength) properties were studied after the mortars were immersed in an H_2SO_4 solution for 28 and 90 days. Moreover, the textural alterations due to the acid attack were evaluated using microscopy.

2. Materials and Methods

2.1. Materials

The cement used for the manufacture of the mortars was a high early strength type belonging to the European cement class CEM I 52.5 R [38]. Limestone sand was used as a fine aggregate ($D_i/d_i = 0/4$). The particle-size distribution was calculated according to the procedure of UNE-EN 933-1 standard [39]. The distribution was obtained for three samples of sand and the mean values were calculated (Figure 1a). Xypex Admix C-1000 NF (Xypex Chemical Corporation, Richmond, BC, Canada) was added in three different concentrations (1 wt%, 1.5 wt% and 2 wt% of the cement weight) depending on the kind of mortar. The specific surface areas of the cement and crystalline admixture (Xypex Admix) used were 1.48 and 0.642 g/m^2, respectively (determined by the Brunauer–Emmett–Teller—BET—method), and their particle-size distributions are plotted in Figure 1b. In total, four types of mortars were made: (i) control mortar (A mortar); (ii) mortar with 1 wt% of Xypex (B Mortar); (iii) mortar with 1.5 wt% of Xypex (C Mortar); and (iv) mortar with 2% of Xypex (D Mortar). The dosage used for all mortars was 1 part cement, 0.65 parts water and 3 parts limestone sand.

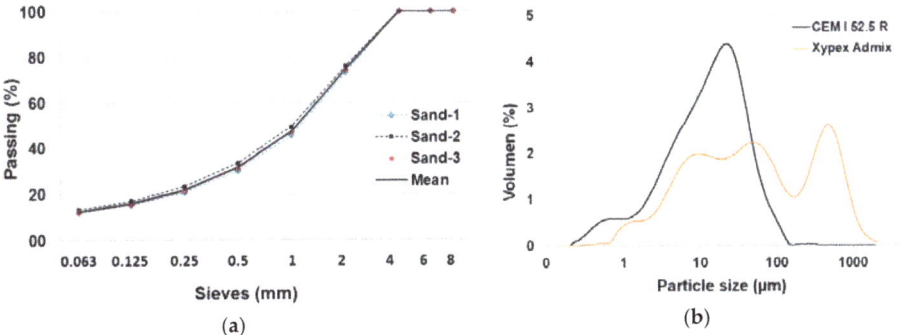

Figure 1. (a) Particle-size distribution of the limestone sand. (b) Particle-size distribution (done by laser granulometry) of the cement (CEM I 52.5 R) and the admixture (Xypex Admix).

Chemical–Physical Characterisation

The crystallographic phases present in the composition of the cement and the admixture (Xypex Admix C-1000 NF) were analysed with X-ray diffraction (XRD) using a Bruker D8-Advance diffractometer (Bruker Española S.A., Madrid, Spain) with mirror Goebel, and a generator of X-ray KRISTALLOFLEX K 760-80F (power: 3000W, voltage: 20–60 KV and current: 5–80 mA) equipped with a X-ray tube with copper anode. The spectra were registered with angles from 4° to 60° at 0.05° stepping intervals in Θ-Θ mode, and the X-ray tube was utilised at 40 kV and 40 mA. The diffraction patterns were evaluated with HighScore software (Malvern Panalytical, Madrid, Spain) and powder diffraction database PDF4+. Moreover, the chemical compositions of the binder and the admixture were analysed with X-ray fluorescence (XRF) using an X-ray sequential spectrometer PHILIPS MAGIX PRO (Philips Ibérica, Madrid, Spain) equipped with a rhodium X-ray tube and beryllium window. The spectrometer was controlled with the software package SuperQ (Malvern Panalytical, Madrid, Spain) that also stored the measurements and results.

The XRF results (Table 1) show that both materials were mainly composed of CaO (cement: 55.4%; Xypex: 59.8%) and SiO$_2$ (cement: 14.9%; Xypex: 8.1%), containing moderate proportions of Al$_2$O$_3$, SO$_3$ and Fe$_2$O$_3$ (percentages between 1.9% and 4.6%). Only the cement contained moderate proportions of MgO (3.3%). The XRD spectra (Figure 2) revealed that both materials had common crystalline compounds, such as alite (C$_3$S), belite (C$_2$S), tetracalcium aluminoferrite (C$_4$AF) and calcite (CaCO$_3$). Gypsum was only detected in the cement, and portlandite and quartz were only detected in the Xypex admixture. The peak of portlandite in Xypex Admix may indicate that this admixture was partially constituted by hydrated cement.

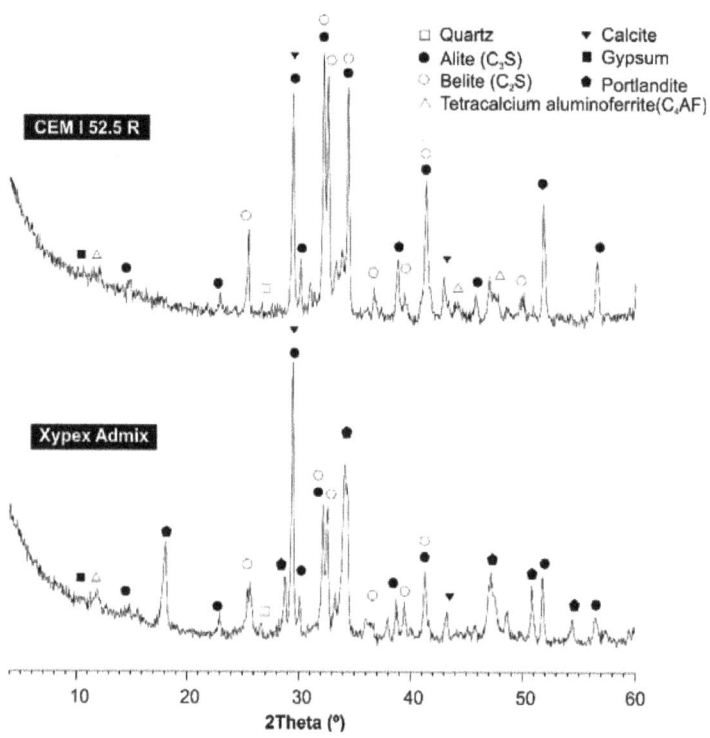

Figure 2. X-ray diffraction (XRD) spectra of cement type CEM I 52.5 R and Xypex Admix.

Table 1. X-ray fluorescence (XRF) characterisation of CEM I 52.5 R and Xypex Admix.

Oxides	CEM I 52.5 R (%)	Xypex Admix (%)
Na_2O	0.23	1.20
MgO	3.29	0.82
Al_2O_3	3.36	1.98
SiO_2	14.89	8.10
P_2O_5	0.14	0.06
SO_3	4.62	2.09
Cl	0.11	0.03
K_2O	1.06	0.44
CaO	55.36	59.77
TiO_2	0.25	0.17
MnO	0.04	0.06
Fe_2O_3	3.06	2.08
SrO	0.12	0.06
Other elements	<0.3	<0.20

2.2. Methods

2.2.1. Manufacturing and Curing Process of the Mortars and Acid Attack Simulation

The different tests were performed on normalised mortar specimens—4 × 4 × 16 cm. The Xypex powder admixture was previously dry-mixed with the cement for 1 min with a laboratory mortar mixer.

After that, the mortars were manufactured following the procedure of the standard UNE-EN 196-1 [40]. A total of 48 specimens were manufactured, 12 per each type of mortar (A, B, C and D). The specimens were cured in a humidity chamber for 28 days at 95% ± 2% of relative humidity and 20 ± 2 °C to complete the hardening process. After 28 days of curing, 24 of the specimens (representing half of the total of 48, that is, 6 specimens for each type of mortar) were exposed to a sulphuric acid solution for 90 days, and the other 24 were immersed in water for the same amount of time for reference purposes.

There are no European standards to test the chemical resistance of cement-based materials. However, Sokolowska et al. [41] studied the tests on cement-based materials according to ASTM (American Society for Testing and Materials) standards and they concluded that there is a lack of clear criteria for the evaluation of research results via ASTM methods. Due to the absence of standardised tests, in this study, the sulphuric acid attack was performed by immersing the specimens into an H_2SO_4 solution (3% w/w) in hermetically closed containers. This procedure has been used in previous studies to analyse the effect of acidic environments on cement-based materials [23,24]. A high concentration of sulphuric acid was chosen in order to accelerate their effects on the mortars and obtain the same degradation in less exposure time [12]. The performance of mortars against the acid attack was evaluated taking into account common parameters used in the literature for this purpose, such as mass variation and mechanical properties decrease [41]. The ultrasonic pulse velocity and the capillary water absorption were also studied [24]. The volume of solution was approximately four times the volume of the samples, as suggested by the ASTM C 1012-04 standard [42]. The H_2SO_4 solution was replaced weekly for a new solution (3% w/w) so that the concentration of sulphuric acid had minimal variation. After removing the specimens from the acid solution, they were brushed under a flow of water in order to remove the superficial layer of adhered material. After that, the specimens were introduced into a new solution. Once the acid exposure was finished, the specimens were dried at 105 ± 2 °C for 24 h, and then they were kept for 1 h at laboratory conditions before continuing with the tests.

2.2.2. Scanning Electron Microscopy (SEM) Examination

The microstructural changes on the mortars, due to the action of the crystalline admixture and the effect of the sulphuric acid attack, were examined with scanning electron microscopy (SEM, Hitachi High-Technologies Canada, Inc., Toronto, ON, Canada) (Hitachi S3000N). Before the examination, small fragments from the mortar surfaces were removed, then softly dried at 60 °C for 24 h and finally metallised with Au–Pd (30 nm) in order to improve the image quality. The images were taken with the following conditions: secondary electrons mode, ultrahigh vacuum, 15 kV of accelerating voltage and variable working distance.

2.2.3. Physical and Mechanical Properties of the Mortars after the Sulphuric Acid Attack

The impact of a sulphuric acid attack on cementitious material can be studied by evaluating its mass loss over the acid exposure time [43]. In this work, the mass loss was studied in seven stages of the acid simulation. To do this, the samples were weighed at the following intervals: (i) after completing their curing (i.e. at 28 days after their manufacture and 0 days of acid attack ($t_{28(0)}$)); (ii) at 7 days of acid exposure ($t_{35(7)}$); (iii) at 14 days ($t_{42(14)}$); (iv) at 21 days ($t_{49(21)}$); (v) at 28 days ($t_{56(28)}$); (vi) at 56 days ($t_{84(56)}$); and (vii) at 90 days ($t_{118(90)}$) of acid exposure. The percentages of mass loss of the mortars were calculated taking into account the initial weights.

Compressive strength is a characteristic of the cement-based materials commonly used in the literature for analysing their performance against a chemical attack [12,43,44]. The reference mortars (that were kept in a nonaggressive environment) and the mortars exposed to the sulphuric acid attack were tested at $t_{56(28)}$ and $t_{118(90)}$. The compressive strength tests were performed according to the UNE-EN 196-1 standard [40]. The conventional mortar testing machine used had a load cell of 20 T capacity and was operated at a speed of 2.4 kN/s until failure.

Ultrasonic pulse velocity is a parameter that can be correlated with the elasticity modulus and therefore provides information about the stiffness of the material [45]. The ultrasonic pulse velocity test was performed according to the UNE-EN 12504-4 standard [46]. A total of four determinations were made per sample and the mean value was adopted. The test consisted of measuring the propagation time of the ultrasonic waves when crossing the longest dimension of the specimen (160 mm). Contact transducers emitting ultrasonic pulses at 54 kHz were coupled to the end sides of the specimens using a coupling agent. The wave speed was obtained from the propagation time and the length of the sample.

The impact of the sulphuric acid attack on the mortars was also evaluated studying the capillary water absorption of the specimens. The tests were conducted following the UNE-EN 1015-18 standard [47] for all the mortars (nonattacked and attacked) at $t_{56(28)}$ and $t_{118(90)}$. According to the standard, the water absorption coefficient is the slope of the line that joins the points corresponding to 10 and 90 min in the curve, representing the mass variation of water absorbed per unit area as a function of the square root of time; that is, the coefficient was computed using the formula

$$C = \frac{M_2 - M_1}{A\left(t_2^{0.5} - t_1^{0.5}\right)} \qquad (1)$$

where:

C is the capillary water absorption coefficient, $kg/(m^2 \cdot min^{0.5})$;
M_1 is the specimen mass after the immersion for 10 min, kg;
M_2 is the specimen mass after the immersion for 90 min, kg;
A is the surface of the specimen face immersed in the water, m^2;
t_2 = 90 min;
t_1 = 10 min.

3. Results and Discussion

3.1. Compressive Strength

The results of the compressive strength tests showed that for a nonaggressive environment (Figure 3a) at t_{56} (56 days from the manufacture of the mortars), the mortars with the highest compressive strength were the C (54.3 MPa) and D (53.7 MPa) types compared with the compressive strength of the A (42.2 MPa) and B (40.8 MPa) mortars. Besides the obvious advantages in the use of the admixtures for mortars, there were drawbacks to take into account. One of those was the lowered final compressive strength compared with nonmodified concretes and mortars using some types of admixtures [48]. In the interval from t_{56} to t_{118}, the only mortar that increased its compressive strength was the reference mortar (A mortar, without crystalline admixture), from 42.2 to 42.8 MPa, whereas in that same period, the three mortars with the crystalline admixture decreased their compressive strength. Usually, admixtures with an accelerated setting effect can reduce the strength of the concrete at later ages. In the high-strength mortars, a mechanism of deterioration of the hardened cement paste phase at the microscopic scale seems to lead to reduced strength at long ages [49]. The C and D mortars clearly had more compressive strength than the reference mortar at t_{56} (B mortar had 28.7% more compressive strength than the A and C mortars (27.7%)), although at t_{118}, this difference was smaller (C mortar: 2%, D mortar: 2.3%) (Figure 3c). It was observed that the C and D mortars exhibited similar compressive strength to the reference mortar at t_{118}, which was in accordance with previous research [33], where the crystalline admixture does not significantly affect the compressive strength of the concretes studied. However, the compressive strength of the B mortar was lower than that of the reference mortar, both at t_{56} and t_{118}.

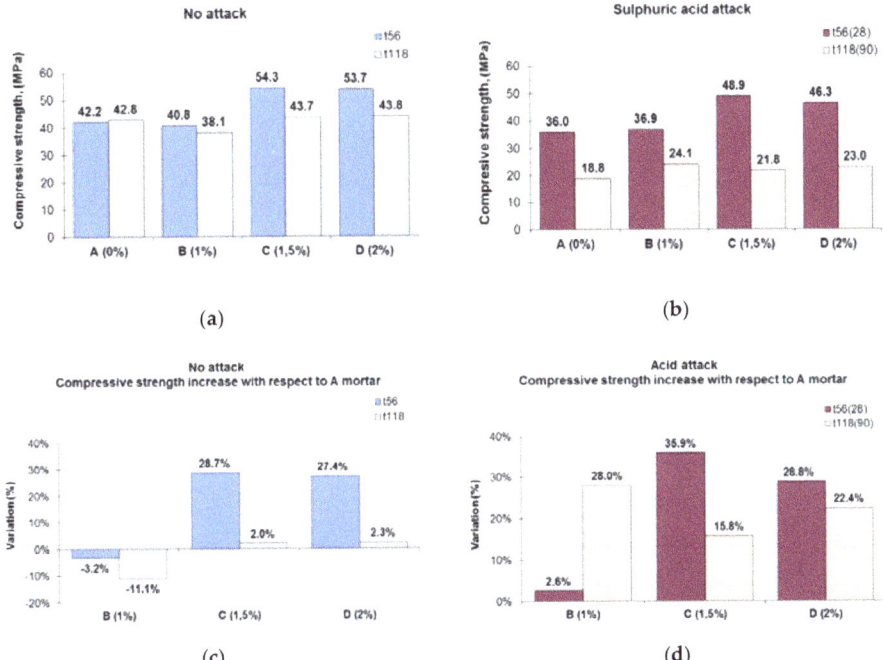

Figure 3. (a) Compressive strength in a nonaggressive environment at t_{56} and t_{118}. (b) Compressive strength in an aggressive environment (sulphuric acid exposure) at $t_{56(28)}$ and $t_{118(90)}$. (c) Compressive strength differences of the mortars with the crystalline admixture with respect to the reference mortar in a nonaggressive environment. (d) Compressive strength differences of the mortars with the crystalline admixture with respect to the reference mortar exposed to sulphuric acid.

The compressive strength of the mortars exposed to 28 days of acid attack (56 days after manufacture) followed a pattern similar to that of mortars without acid attack (Figure 3b). The mortars with the highest compressive strength were types C (48.9 MPa) and D (46.3 MPa), and the lowest were types A (36.0 MPa) and B (36.9 MPa). The compressive strength at $t_{56(28)}$ decreased in the same way for all mortar types (due to the acid attack), given as a reference the compressive strength without acid attack, for each mortar with the same curing time (t_{56}). The reduction in compressive strength was a direct effect of the acid attack due to the microcracking caused by the formation of expansive compounds [50,51]. As expected, the compressive strength of mortar type B slightly exceeded the compressive strength of mortar A. When increasing the exposure time of the attack to 90 days ($t_{118(90)}$), the compressive strength of all mortars decreased due to the effects of the attack. In this case, mortars with the admixture (B, C and D mortars) clearly had higher compressive strength compared with the reference mortar (B mortar had 28% more strength than mortars A, C (15.8%) and D (28.8%)) (Figure 3d).

On the other hand, a behavioural change in the compressive strength of the mortars was observed for 90 days of acid exposure. B mortar had the highest compressive strength, whereas for a 28-day exposure, the one with the highest compressive strength was C mortar.

3.2. Mass Loss Due to the Sulphuric Acid Attack

The results of the mass loss test due to the sulphuric acid exposure (Figure 4a) showed that an acid attack caused a mass loss in mortars so that, as the acid attack continued longer, the mass loss of the mortars was higher. The compressive strength loss described above and the increment in mass loss when the exposure time increased were consistent with the results obtained in previous studies [43]. Moreover, a linear correlation (R^2 = 0.7724) between the decrease in compressive strength and the mass loss was found (Figure 4b). Progression of the acid attack front caused an increase in porosity and permeability, leading to mass and strength loss [52]. Nevertheless, the mortars with the crystalline admixture (B, C and D mortars) behaved clearly better than the mortars without this admixture (A mortar) since they presented lower mass loss (B mortar: 10.5%; C mortar: 9.5 %; D mortar: 10.1%) than the mortar without the admixture (A mortar: 15%).

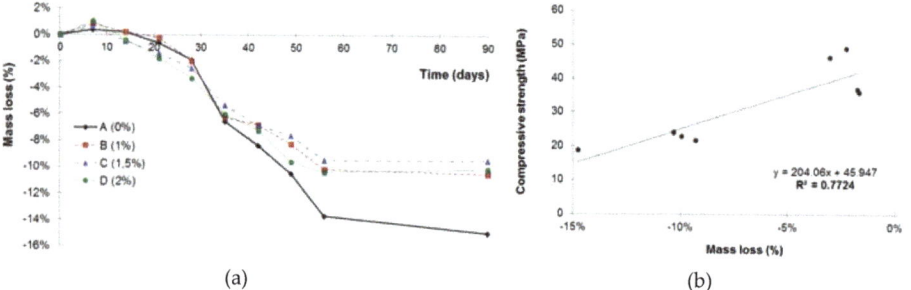

Figure 4. (a) Mass loss due to the sulphuric acid attack. (b) Correlation between compressive strength (MPa) and mass loss (%) after the acid attack.

As a result of the sulphuric acid attack, calcium sulphate (gypsum) was formed by the reaction of the acid with the calcium hydroxide (chemical reaction 2) and calcium silicate hydrate (chemical reaction 3) that were present in the hydrated Portland cement [3,5] and limestone sand (chemical reaction 4) [45,53]. The gypsum coating could also be observed with the naked eye (Figure 5b). The formation of gypsum after a sulphuric acid attack has been confirmed in the literature with XRD studies [52,54]. The chemical reactions produced by the acid attack resulted in a profound degradation of the hydrated cement paste, associated with a loss of compressive strength. When the concrete surface in addition to the acid attack was exposed to flowing water, the products of the degradation were carried away to a significant degree, causing a mass loss. Generally, an attack by free sulphuric acid is more severe than any with a neutral sulphate solution [3]. As mentioned in Section 2.2.1, the specimens were brushed and cleaned weekly to remove the gypsum formed on the specimen surface. The mass loss shown in Figure 4a was mainly associated with the amount of gypsum removed from the mortar surfaces during the brushing process. However, as the literature states [54], at the beginning of an acid attack, there is an increase in mass, which can be explained by the generation of gypsum in the pores and the cement–aggregate interface. This gypsum was difficult to remove even though the surface of the samples was brushed; therefore, at the first stage of the attack, a mass increase was found. To corroborate the existence of gypsum on the surface of the attacked mortars, an XRD analysis was performed. The samples were taken from fragments obtained from the surface of the specimens (4 × 4 × 16 cm) that had been broken in the compressive tests. The XRD analysis showed the presence of bassanite ($CaSO_4 \cdot 1/2H_2O$), which was the result of the thermal decomposition of gypsum, and it happened at 110 °C (chemical reaction 5) [55]. Therefore, it was consistent to find bassanite (hemihydrate phase of gypsum) after having exposed the specimens at 105 ± 2 °C continuously for 24 h. This result confirmed the previous existence of gypsum.

$$Ca(OH)_2 + H_2SO_4 \rightarrow CaSO_4 \cdot 2H_2O \tag{2}$$

$$xCaO \cdot SiO_2 \cdot aq + xH_2SO_4 + xH_2O \rightarrow xCaSO_4 \cdot 2H_2O + SiO_2 \cdot aq \tag{3}$$

$$CaCO_3 + H_2SO_4 + 2H_2O \rightarrow CaSO_4 \cdot 2H_2O + CO_2 + H_2O \tag{4}$$

$$CaSO_4 \cdot 2H_2O \rightarrow CaSO_4 \cdot 1/2H_2O + 3/2H_2O \tag{5}$$

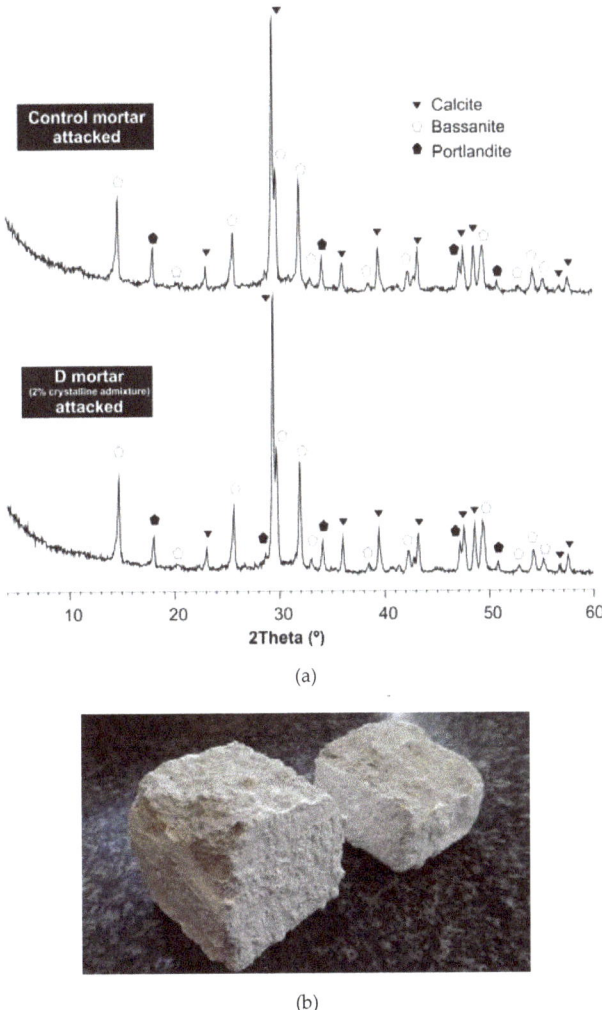

Figure 5. (**a**) X-ray diffraction spectra of control mortar and D mortar (with crystalline admixture) after 90 days of sulphuric acid attack; (**b**) Specimens exposed to the acid attack after compressive strength testing at $t_{118(90)}$, where the massive formation of gypsum in the surface of the specimen can be seen.

3.3. Ultrasonic Pulse Velocity

Figure 6 shows the results of the ultrasonic pulse velocity test. This test consisted of obtaining the velocity (m/s) that the ultrasonic waves needed to cross the longest length of the specimen. The test was performed for mortars maintained in nonaggressive conditions (Figure 6a) and the ultrasonic velocity obtained was similar for all mortars in both t_{56} (from 4248 to 4340 m/s) and t_{118} (from 4198 to 4364 m/s). For mortars exposed to acid attack (Figure 6b), the ultrasonic velocity decreased for all mortars with respect the velocity obtained for a nonaggressive environment for both $t_{56(28)}$ (from 4091 to 4206 m/s) and $t_{118(90)}$ (from 3525 to 3794 m/s). To obtain a measure of the effect of the acid attack on the propagation velocity, the difference between ultrasonic pulse velocity at t_{118} for the mortars with and without attack was calculated. The measurements were taken after 90 days of acid attack, which is an action time long enough for the effects to manifest. It was found that the mortars with the admixture had less velocity decrease (B mortar: 13.0%, C mortar: 14.3% and D mortar: 11.9%) than the reference mortar (16.0%). In addition, the longer the acid exposure time, the lower the ultrasonic pulse velocity. The results demonstrate that the ultrasonic technique reflects the deterioration caused by sulphuric acid attack, providing information about the internal condition and quality of the mortars [56,57]. This is due to the large amount of cracks in specimens [58]. Indeed, in this work, two relationships between the involved variables were found: (i) correlation between the decrease in ultrasonic pulse velocity and the mass loss, and (ii) correlation between the ultrasonic pulse velocity and the compressive strength (Figure 6c,d). Recent studies have demonstrated [59,60] that the variation of the ultrasonic pulse velocity indicates a variation in Young's modulus. Lower speed means lower Young's modulus and greater deterioration of mortars.

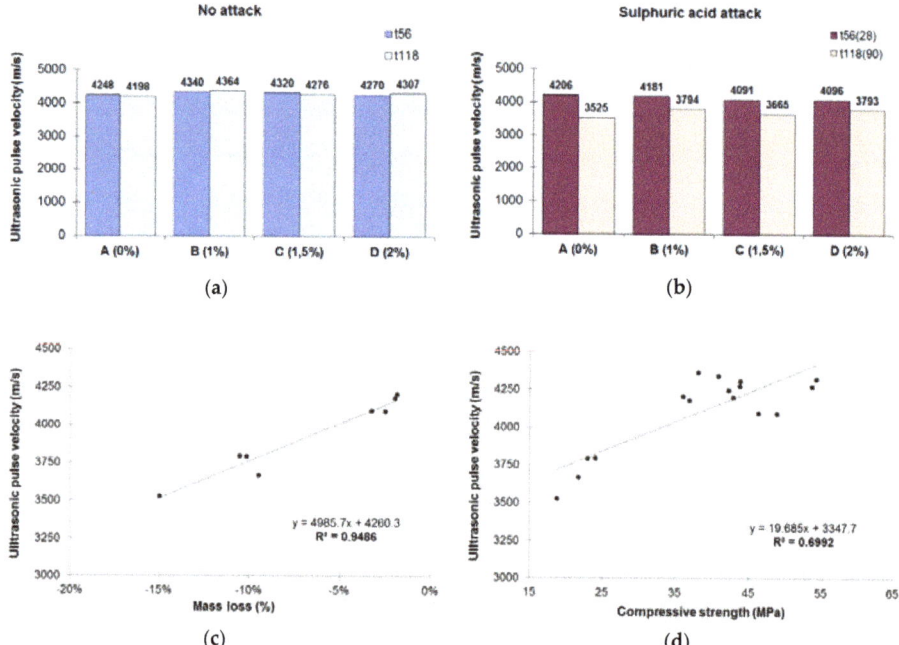

Figure 6. (a) Ultrasonic pulse velocity of mortars kept in a nonaggressive environment at t_{56} and t_{118}. (b) Ultrasonic pulse velocity of mortars exposed to a sulphuric acid solution for 28 and 90 days. (c) Correlation between ultrasonic pulse velocity (m/s) and mass loss (%) after the acid attack. (d) Correlation between ultrasonic pulse velocity (m/s) and compressive strength (MPa).

3.4. Capillary Water Absorption Coefficient

The results of the capillary water absorption test are shown in Figure 7. For nonaggressive conditions and early ages (56 days from its manufacture), the absorption coefficient of the mortars with the crystalline admixture was slightly lower (B mortar: 0.20; C mortar: 0.19; D mortar: 0.20) than the absorption coefficient of the reference mortar (A mortar: 0.21). At t_{118}, the absorption coefficients for all the mortars reduced but without reaching the values obtained for a water-repellent mortar [25]. These results agree with other studies which found that concretes treated with crystalline materials had slightly lower depth of pressure water penetration than the reference concrete in the short term [36,61] and negligible effects in terms of water vapor permeability [33]. In a sulphuric acid exposure, all mortars slightly reduced their capillary coefficients when compared to the values obtained for those same times in a nonaggressive environment. This reduction could be due to the formation of a gypsum coating on the surface so that the gypsum crystals obstructed the capillary net, reducing the absorption coefficient [24].

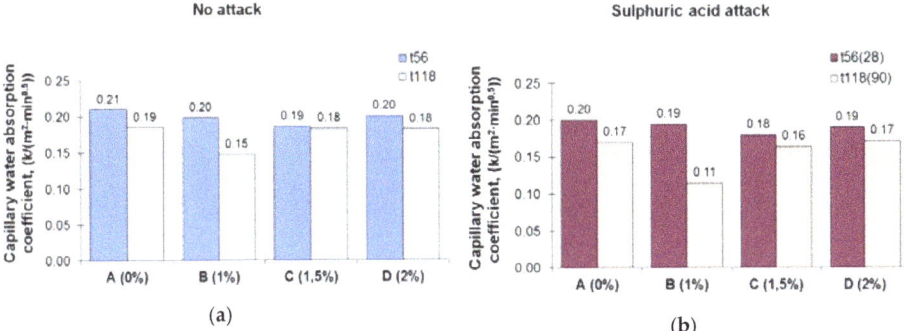

Figure 7. (a) Capillary water absorption coefficients of mortars kept in a nonaggressive environment at t_{56} and t_{118}. (b) Capillary water absorption coefficients of mortars exposed to a sulphuric acid solution for 28 and 90 days.

3.5. Scanning Electron Microscopy

In the SEM images, crystalline structures were observed in the pores of the nonattacked mortars containing the crystalline admixture (Figure 8a,b). These structures had very similar shapes (needlelike crystal) to those that the admixture manufacturer assures are formed when the crystalline chemical reacts with the calcium hydroxide and other by-products of cement hydration. These crystalline formations are insoluble in water and fill and plug pores, capillaries and microcracks of the cementitious materials [62]. In addition, in the mortar manufactured with the Xypex admixture, the typical hexagonal shapes of the portlandite were observed (arrows in Figure 8d). Figure 8c,d shows images of the mortar with the crystalline admixture after being exposed to the sulphuric acid solution for 56 days (8 weeks). It is also possible to observe crystalline formations in the pore but with lower density than for the case of the nonattacked C mortar. The decrease in crystal density may be due to the acid attack that could be reducing the crystalline formations. In the SEM images corresponding to the nonattacked control mortar (Figure 9a,b), the needlelike crystalline shapes described above did not appear. After the acid attack, crystalline forms appeared in the control mortar (Figure 9c,d), which were different from those observed in the mortar with the admixture, and they could correspond to bassanite (hemihydrate phase of gypsum). The bassanite was the result of the effect of the sulphuric acid on the products of cement hydration and was detected in the diffractogram (Figure 5a).

Figure 8. Cont.

(c)   (d)

Figure 8. Scanning electron microscopy (SEM) images performed at 84 days (12 weeks) from the mortar manufacture. Left (**a**,**c**): images at 500× magnification. Right (**b**,**d**): zoom at 1500× magnification on the left image square. Top (**a**,**b**): mortar type C (with 1.5% of crystalline admixture) kept in a nonaggressive environment. Bottom (**c**,**d**): mortar type C exposed to a sulphuric acid attack for 56 days (8 weeks).

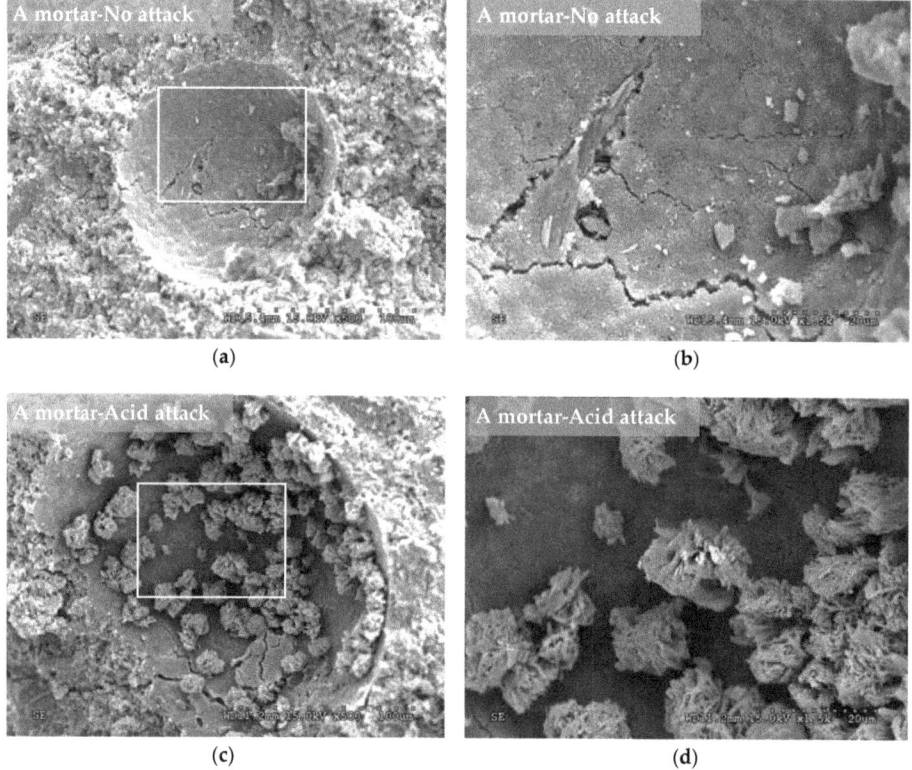

Figure 9. SEM images performed at 84 days (12 weeks) from the mortar manufacture. Left (**a**,**c**): images at 500× magnification. Right (**b**,**d**): zoom at 1500× magnification on the left image square. Top (**a**,**b**): mortar type A (0% of crystalline admixture) kept in a nonaggressive environment. Bottom (**c**,**d**): mortar type A exposed to a sulphuric acid attack for 56 days (8 weeks).

4. Conclusions

This study analysed the effects of a sulphuric acid exposure on the physical and mechanical properties of mortars manufactured with a crystalline admixture in the short term (up to 90 days). The acid attack had significant consequences on the microstructure and the physical and mechanical properties of mortars. According to the results obtained in this study, the following can be concluded:

1. In a nonaggressive environment, the use of crystalline admixtures did not produce significant effects on the compressive strength of mortars. This is an expected result, since this type of admixture is not designed to increase the compressive strength of cement-based materials but their durability. After an acid attack, both the control mortar and the treated mortars experienced a decrease in their compressive strength. The longer the sulphuric acid exposure, the greater the compressive strength loss. However, mortars with the crystalline admixture showed greater compressive strength than the control mortar after 90 days of sulphuric acid attack. The compressive strength decrease, compared to the control mortar, was reduced between 16% and 28% depending on the amount of the admixture concentration.
2. Likewise, after a sulphuric acid attack, all the mortars (control and treated mortars) presented mass loss, being greater at longer exposure time. However, the treated mortars had lower mass loss after 90 days of sulphuric acid exposure. Their mass loss percentages were 9.5%–10.5% for the treated mortars and 15% for the control mortar.
3. In a nonaggressive environment, the crystalline admixture did not produce significant effects on the capillary water absorption coefficient of mortars, since it only produced a slight reduction of their absorption coefficients. Therefore, it should not be concluded that it is a hydrophobic material. After sulphuric acid exposure (90 days), all mortars reduced their absorption coefficients. In this case, the use of the crystalline admixture did not have a positive effect on the mortars as expected, since no significant differences were found between the control mortar and the treated ones. The reduction of the water absorption can be due to the formation of a gypsum coating that protects the surface of mortars. The coating of gypsum alters the superficial colour of mortars and can be appreciated with the naked eye.
4. The exposure to sulphuric acid had the effect of decreasing the ultrasonic pulse velocity, which implies a reduction of Young's modulus of mortars and a probable loss of mechanical qualities. The decrease in velocity was greater when the exposure time increased. After 90 days of acid exposure, the mortars with the crystalline admixture showed a smaller decrease in the ultrasonic pulse velocity (between 11.9% and 14.3%) than the untreated mortar (16%) and thus a lower decrease of Young's modulus.
5. SEM images corroborated the formation of insoluble crystalline products in the treated mortars. These formations had needlelike shapes and were located into the mortars' pores.

This study has verified that the incorporation of crystalline admixtures improves the durability of mortars exposed to aggressive environments (sulphuric acid attack). Compared to normal mortars, crystalline admixtures added to mortars increase compressive strength and reduce mass loss. Consequently, a lower reduction of ultrasonic velocity is observed. Furthermore, incorporating crystalline admixtures in the mixing of the cementitious materials does not have the drawbacks of superficial treatments, such as the difficulty of adhesion with the substrate, the uneven thickness layer or the cost of applying the coating. Although, in the short term, the use of durable materials implies a higher initial cost without apparent benefit, in the long term, the decrease of maintenance and rehabilitation actions as well as the increase in the construction's lifecycle is more cost efficient and, importantly, reduces the negative environmental impact.

Author Contributions: Conceptualization, V.E.G.-V. and M.L.; Data curation, V.E.G.-V. and A.J.T.-A.; Formal analysis, V.E.G.-V. and M.L.; Funding acquisition, A.J.T.-A. and J.M.S.; Investigation, V.E.G.-V., A.J.T.-A. and J.M.S.; Methodology, V.E.G.-V. and A.J.T.-A.; Project administration, J.M.S.; Resources, A.J.T.-A. and J.M.S.; Supervision,

A.J.T.-A., J.M.S. and M.L.; Validation, M.L.; Writin—original draft, V.E.G.-V.; Writing—review & editing, V.E.G.-V., A.J.T.-A., J.M.S. and M.L.

Funding: This research was funded by the University of Alicante (GRE13-03) and (VIGROB-256).

Acknowledgments: The authors would like to thank Imrepol S.L. and Xypex España for providing the admixture used in this study. The authors also would like to thank Lidia Adeva Gil and Wanda Cecilia Candia Santos for their technical support.

Conflicts of Interest: The authors declare no conflict of interest.

References

1. Alexander, M.G.; Bertron, A.; de Belie, N. *Performance of Cement-Based Materials in Aggressive Aqueous Environments*; Springer: Berlin/Heidelberg, Germany, 2013; ISBN 9789400754126.
2. Ortega Álvarez, J.; Esteban Pérez, M.; Rodríguez Escribano, R.; Pastor Navarro, J.; Sánchez Martín, I. Microstructural Effects of Sulphate Attack in Sustainable Grouts for Micropiles. *Materials* **2016**, *9*, 905. [CrossRef] [PubMed]
3. Skalny, J.; Marchand, J.; Odler, I. *Sulfate Attack on Concrete*; Spon Press: London, UK; New York, NY, USA, 2002; ISBN 0-419-24550-2.
4. Ministerio de Fomento. *EHE-08-Code on Structural Concrete. Articles and Annexes*; Ministerio de Fomento: Madrid, Spain, 2008.
5. Hadigheh, S.A.; Gravina, R.J.; Smith, S.T. Effect of acid attack on FRP-to-concrete bonded interfaces. *Constr. Build. Mater.* **2017**, *152*, 285–303. [CrossRef]
6. Wang, Z.H.; Zhu, Z.M.; Sun, X.; Wang, X.M. Deterioration of fracture toughness of concrete under acid rain environment. *Eng. Fail. Anal.* **2017**, *77*, 76–84. [CrossRef]
7. Irassar, E.F.; Di Maio, A.; Batic, O.R. Sulfate attack on concrete with mineral admixtures. *Cem. Concr. Res.* **1996**, *26*, 113–123. [CrossRef]
8. González, M.A.; Irassar, E.F. Ettringite formation in low C3A Portland cement exposed to sodium sulfate solution. *Cem. Concr. Res.* **1997**, *27*, 1061–1071. [CrossRef]
9. Irassar, E.F. Sulfate attack on cementitious materials containing limestone filler—A review. *Cem. Concr. Res.* **2009**, *39*, 241–254. [CrossRef]
10. Monteiro, P.J.; Kurtis, K.E. Time to failure for concrete exposed to severe sulfate attack. *Cem. Concr. Res.* **2003**, *33*, 987–993. [CrossRef]
11. Ortega, J.M.; Sánchez, I.; Climent, M.A. Durability related transport properties of OPC and slag cement mortars hardened under different environmental conditions. *Constr. Build. Mater.* **2012**, *27*, 176–183. [CrossRef]
12. Ortega, J.M.; Esteban, M.D.; Williams, M.; Sánchez, I.; Climent, M.A. Short-term performance of sustainable silica fume mortars exposed to sulfate attack. *Sustainability* **2018**, *10*, 2517. [CrossRef]
13. Ponikiewski, T.; Golaszewski, J. The effect of high-calcium fly ash on selected properties of self-compacting concrete. *Arch. Civ. Mech. Eng.* **2014**, *14*, 455–465. [CrossRef]
14. Glinicki, M.A.; Jozwiak-Niedzwiedzka, D.; Gibas, K.; Dabrowski, M. Influence of Blended Cements with Calcareous Fly Ash on Chloride Ion Migration and Carbonation Resistance of Concrete for Durable Structures. *Materials* **2016**, *9*, 18. [CrossRef] [PubMed]
15. Estokova, A.; Kovalcikova, M.; Luptakova, A.; Prascakova, M. Testing Silica Fume-Based Concrete Composites under Chemical and Microbiological Sulfate Attacks. *Materials* **2016**, *9*, 324. [CrossRef]
16. Norhasri, M.S.M.; Hamidah, M.S.; Fadzil, A.M. Applications of using nano material in concrete: A review. *Constr. Build. Mater.* **2017**, *133*, 91–97. [CrossRef]
17. Aïtcin, P.-C. Cements of yesterday and today. *Cem. Concr. Res.* **2000**, *30*, 1349–1359. [CrossRef]
18. Yu, R.; Spiesz, P.; Brouwers, H.J.H. Effect of nano-silica on the hydration and microstructure development of Ultra-High Performance Concrete (UHPC) with a low binder amount. *Constr. Build. Mater.* **2014**, *65*, 140–150. [CrossRef]
19. Adak, D.; Sarkar, M.; Mandal, S. Effect of nano-silica on strength and durability of fly ash based geopolymer mortar. *Constr. Build. Mater.* **2014**, *70*, 453–459. [CrossRef]

20. Massa, M.A.; Covarrubias, C.; Bittner, M.; Fuentevilla, I.A.; Capetillo, P.; Von Marttens, A.; Carvajal, J.C. Synthesis of new antibacterial composite coating for titanium based on highly ordered nanoporous silica and silver nanoparticles. *Mater. Sci. Eng. C* **2014**, *45*, 146–153. [CrossRef]
21. Morsy, M.S.; Alsayed, S.H.; Aqel, M. Hybrid effect of carbon nanotube and nano-clay on physico-mechanical properties of cement mortar. *Constr. Build. Mater.* **2011**, *25*, 145–149. [CrossRef]
22. Navarro-Blasco, I.; Pérez-Nicolás, M.; Fernández, J.M.; Duran, A.; Sirera, R.; Alvarez, J.I. Assessment of the interaction of polycarboxylate superplasticizers in hydrated lime pastes modified with nanosilica or metakaolin as pozzolanic reactives. *Constr. Build. Mater.* **2014**, *73*, 1–12. [CrossRef]
23. Soroushian, P.; Chowdhury, H.; Ghebrab, T. Evaluation of Water-Repelling Additives for Use in Concrete-Based Sanitary Sewer Infrastructure. *J. Infrastruct. Syst.* **2009**, *15*, 106–110. [CrossRef]
24. García-Vera, V.E.; Tenza-Abril, A.J.; Lanzón, M.; Saval, J.M. Exposing Sustainable Mortars with Nanosilica, Zinc Stearate, and Ethyl Silicate Coating to Sulfuric Acid Attack. *Sustainability* **2018**, *10*, 3769. [CrossRef]
25. Lanzón, M. Evaluation of capillary water absorption in rendering mortars made with powdered waterproofing additives. *Const. Build. Mater.* **2009**, *23*, 2391. [CrossRef]
26. Li, W.; Wittmann, F.; Jiang, R.; Zhao, T.; Wolfseher, R. Metal Soaps for the Production of Integral Water Repellent Concre. In Proceedings of the Hydrophobe VI: Water Repellent Treatment of Building Materials, Rome, Italy, 12–13 May 2011; pp. 145–154.
27. Nunes, C.; Slížková, Z. Freezing and thawing resistance of aerial lime mortar with metakaolin and a traditional water-repellent admixture. *Constr. Build. Mater.* **2016**, *114*, 896–905. [CrossRef]
28. Ogrodnik, P.; Zegardło, B.; Szeląg, M.; Ogrodnik, P.; Zegardło, B.; Szeląg, M. The Use of Heat-Resistant Concrete Made with Ceramic Sanitary Ware Waste for a Thermal Energy Storage. *Appl. Sci.* **2017**, *7*, 1303. [CrossRef]
29. Zegardło, B.; Szeląg, M.; Ogrodnik, P. Concrete resistant to spalling made with recycled aggregate from sanitary ceramic wastes—The effect of moisture and porosity on destructive processes occurring in fire conditions. *Constr. Build. Mater.* **2018**, *173*, 58–68. [CrossRef]
30. Debieb, F.; Kenai, S. The use of coarse and fine crushed bricks as aggregate in concrete. *Constr. Build. Mater.* **2008**, *22*, 886–893. [CrossRef]
31. López, V.; Llamas, B.; Juan, A.; Morán, J.M.; Guerra, I. Eco-efficient Concretes: Impact of the Use of White Ceramic Powder on the Mechanical Properties of Concrete. *Biosyst. Eng.* **2007**, *96*, 559–564. [CrossRef]
32. Xypex Chemical Corporation. *ADMIX C-1000 NF*; Xypex Chemical Corporation: Houston, TX, USA, 2005.
33. Pazderka, J.; Hájková, E. Crystalline admixtures and their effect on selected properites of concrete. *Acta Polytech.* **2016**, *56*, 291. [CrossRef]
34. Weng, T.-L.; Cheng, A. Influence of curing environment on concrete with crystalline admixture. *Monatshefte für Chemie* **2014**, *145*, 195–200. [CrossRef]
35. Wang, K.L.; Hu, T.Z.; Xu, S.J. Influence of Permeated Crystalline Waterproof Materials on Impermeability of Concrete. *Adv. Mater. Res.* **2012**, *446–449*, 954–960. [CrossRef]
36. Bohus, S.; Drochytka, R. Cement Based Material with Crystal-Growth Ability under Long Term Aggressive Medium Impact. *Appl. Mech. Mater.* **2012**, *166–169*, 1773–1778. [CrossRef]
37. Dao, V.T.N.; Dux, P.F.; Morris, P.H.; Carse, A.H. Performance of Permeability-Reducing Admixtures in Marine Concrete Structures. *ACI Mater. J.* **2010**, *107*, 291–296.
38. AENOR. *UNE-EN 197-1:2011. Cement-Part 1: Composition, Specifications and Conformity Criteria for Common Cements*; AENOR: Madrid, Spain, 2011.
39. AENOR. *UNE-EN 933-1:2012. Tests for Geometrical Properties of Aggregates-Part 1: Determination of Particle Size Distribution-Sieveing Method*; AENOR: Madrid, Spain, 2012.
40. AENOR. *UNE-EN 196-1:2005. Methods of Testing Cement-Part 1: Determination of Strength*; AENOR: Madrid, Spain, 2005.
41. Sokołowska, J.J.; Woyciechowski, P.; Adamczewski, G. Influence of Acidic Environments on Cement and Polymer-Cement Concretes Degradation. *Adv. Mater. Res.* **2013**, *687*, 144–149. [CrossRef]
42. ASTM. *Standard Test Method for Length Change of Hydraulic-Cement Mortars Exposed to a Sulfate Solution*; ASTM C 1012-04; ASTM: West Conshohocken, PA, USA, 2004.
43. Deb, P.S.; Sarker, P.K.; Barbhuiya, S. Sorptivity and acid resistance of ambient-cured geopolymer mortars containing nano-silica. *Cem. Concr. Compos.* **2016**, *72*, 235–245. [CrossRef]

44. Bonakdar, A.; Mobasher, B. Multi-parameter study of external sulfate attack in blended cement materials. *Constr. Build. Mater.* **2010**, *24*, 61–70. [CrossRef]
45. García-Vera, V.E.; Lanzón, M. Physical-chemical study, characterisation and use of image analysis to assess the durability of earthen plasters exposed to rain water and acid rain. *Constr. Build. Mater.* **2018**, *187*. [CrossRef]
46. AENOR. *UNE-EN 12504-4:2006. Testing Concrete-Part 4: Determination of Ultrasonic Pulse Velocity*; AENOR: Madrid, Spain, 2006.
47. AENOR. *UNE-EN 1015-18:2003. Methods of Test for Mortar for Masonry-Part 18: Determination of Water Absorption Coefficient Due to Capillary Action of Hardened Mortar*; AENOR: Madrid, Spain, 2003.
48. Pizoń, J. Long-term Compressive Strength of Mortars Modified with Hardening Accelerating Admixtures. *Procedia Eng.* **2017**, *195*, 205–211. [CrossRef]
49. Igarashi, S.; Kawamura, M. Reduction in strength in high strength mortars at long ages. In Proceedings of the Third International Conference on Fracture Mechanics of Concrete and Concrete Structures, Gifu, Japan, 12–16 October 1998; pp. 243–252.
50. Huang, Q.; Wang, C.; Yang, C.; Zhou, L.; Yin, J. Accelerated sulfate attack on mortars using electrical pulse. *Constr. Build. Mater.* **2015**, *95*, 875–881. [CrossRef]
51. Ma, H.; Li, Z. Microstructures and mechanical properties of polymer modified mortars under distinct mechanisms. *Constr. Build. Mater.* **2013**, *47*, 579–587. [CrossRef]
52. Kwasny, J.; Aiken, T.A.; Soutsos, M.N.; McIntosh, J.A.; Cleland, D.J. Sulfate and acid resistance of lithomarge-based geopolymer mortars. *Constr. Build. Mater.* **2018**, *166*, 537–553. [CrossRef]
53. Lanzon, M.; Garcia-Ruiz, P.A. Deterioration and damage evaluation of rendering mortars exposed to sulphuric acid. *Mater. Struct.* **2010**, *43*, 417–427. [CrossRef]
54. Monteny, J.; Vincke, E.; Beeldens, A.; De Belie, N.; Taerwe, L.; Van Gemert, D.; Verstraete, W. Chemical, microbiological, and in situ test methods for biogenic sulfuric acid corrosion of concrete. *Cem. Concr. Res.* **2000**, *30*, 623–634. [CrossRef]
55. Borrachero, M.V.; Payá, J.; Bonilla, M.; Monzó, J. The use of thermogravimetric analysis technique for the characterization of construction materials. *J. Therm. Anal. Calorim.* **2008**, *91*, 503–509. [CrossRef]
56. Yaman, I.; Inci, G.; Yesiller, N.; Aktan, H. Ultrasonic pulse velocity in concrete using direct and indirect transmission. *ACI Mater. J.* **2001**, *98*, 450–457.
57. Toutanji, H. Ultrasonic wave velocity signal interpretation of simulated concrete bridge decks. *Mater. Struct.* **2000**, *33*, 207–215. [CrossRef]
58. Chen, F.; Gao, J.; Qi, B.; Shen, D. Deterioration mechanism of plain and blended cement mortars partially exposed to sulfate attack. *Constr. Build. Mater.* **2017**, *154*, 849–856. [CrossRef]
59. Zhou, Y.; Li, M.; Sui, L.; Xing, F. Effect of sulfate attack on the stress–strain relationship of FRP-confined concrete. *Constr. Build. Mater.* **2016**, *110*, 235–250. [CrossRef]
60. Atahan, H.N.; Arslan, K.M. Improved durability of cement mortars exposed to external sulfate attack: The role of nano & micro additives. *Sustain. Cities Soc.* **2016**, *22*, 40–48.
61. Al-Kheetan, M.J.; Rahman, M.M.; Chamberlain, D.A. Development of hydrophobic concrete by adding dual-crystalline admixture at mixing stage. *Struct. Concr.* **2018**. [CrossRef]
62. Xypex Chemical Corporation. *Special Print. Concrete Technology. Crystalline Technology for Enhancing the Performance of Precast Concrete in Marine and Sewer Structures*; Xypex Chemical Corporation: Houston, TX, USA, 2014.

© 2018 by the authors. Licensee MDPI, Basel, Switzerland. This article is an open access article distributed under the terms and conditions of the Creative Commons Attribution (CC BY) license (http://creativecommons.org/licenses/by/4.0/).

MDPI
St. Alban-Anlage 66
4052 Basel
Switzerland
Tel. +41 61 683 77 34
Fax +41 61 302 89 18
www.mdpi.com

Materials Editorial Office
E-mail: materials@mdpi.com
www.mdpi.com/journal/materials

www.ingramcontent.com/pod-product-compliance
Lightning Source LLC
LaVergne TN
LVHW070247100526
838202LV00015B/2188